International Nuclear Science, Technology and Engineering Conference (iNuSTEC2017)

IOP Conference Series: Materials Science and Engineering
Volume 298

Selangor, Malaysia
25 - 27 September 2017

Editors:

Abdul Aziz Mohamed
Faridah Mohmad Idris
Hassan Mohamed
Nasri A. Hamid
Nahrul Khair Alang Md Rashid

ISBN: 978-1-5108-5837-4
ISSN: 1757-8981

Printed from e-media with permission by:

Curran Associates, Inc.
57 Morehouse Lane
Red Hook, NY 12571

Printed by Curran Associates, Inc. (2018)

For permission requests, please contact the Institute of Physics
at the address below.

Institute of Physics
Dirac House, Temple Back
Bristol BS1 6BE UK

Phone: 44 1 17 929 7481
Fax: 44 1 17 920 0979

techtracking@iop.org

Additional copies of this publication are available from:

Curran Associates, Inc.
57 Morehouse Lane
Red Hook, NY 12571 USA
Phone: 845-758-0400
Fax: 845-758-2633
Email: curran@proceedings.com
Web: www.proceedings.com

Table of contents

Volume 298

International Nuclear Science, Technology and Engineering Conference 2017 (iNuSTEC2017)

25–27 September 2017, Selangor, Malaysia

Accepted papers received: 9 January 2018
Published online: 12 February 2018

Preface

Papers

Science: Environment

Science: Sustainability

Engineering: Energy

IOP Conf. Series: Materials Science and Engineering **298** (2018) 011001 doi:10.1088/1757-899X/298/1/011001

IOP Conference Series: Materials Science and Engineering

International Nuclear Science, Technology and Engineering Conference 2017 (iNuSTEC2017)

"Crafting Nuclear Knowledge for Global Well-Being"

Universiti Tenaga Nasional
Kajang, Selangor, Malaysia
25 – 27 September 2017

Editors

Abdul Aziz Mohamed
Universiti Tenaga Nasional Malaysia (UNITEN)
Faridah Mohamad Idris
Malaysian Nuclear Agency (Nuclear Malaysia)
Hassan Mohamed
Universiti Tenaga Nasional Malaysia (UNITEN)
Nasri A. Hamid
Universiti Tenaga Nasional Malaysia (UNITEN)
Nahrul Khair Alang Md Rashid
Universiti Teknologi Malaysia (UTM)

All papers have been peer-reviewed

Supporting/Sponsoring Organisations

Ministry of Education Malaysia; Ministry of Science Technology and Innovation Malaysia; Malaysian Nuclear Agency; Hitachi-GE Nuclear Energy, Ltd; ASME Malaysia Section; Malaysian Welding and Joining Society; Malaysia Nuclear Power Corporation; Reactor Interest Group.

IOP Conf. Series: Materials Science and Engineering **298** (2018) 011001 doi:10.1088/1757-899X/298/1/011001

CONTENTS

IOP Conf. Series: Materials Science and Engineering **298** (2018) 011001 doi:10.1088/1757-899X/298/1/011001

PREFACE

All praises to God the Al-Mighty for the strengths and His blessing in making Nuclear Science, Technology and Engineering Conference 2017 (iNuSTEC2017) a success.

The history in organizing Nuclear Science, Technology and Engineering Conference started back in 2010. During that year, it was called NuSTEC2010 and was organized only at national and regional levels. From that year onwards, we succeeded in organizing our second (iNuSTEC2011), third (iNuSTEC2012), fourth (international iNUSTEC2013), fifth (NuSTEC2014), sixth (international iNuSTEC2015) and (international iNuSTEC2016) conferences. Based on these achievements, we have decided to reconvene the conference sustainability by organizing regional/international iNuSTEC on annual basis.

The iNuSTEC2017 was held from 25 - 27 September 2017 at Universiti Tenaga Nasional (UNITEN) at Putrajaya campus in Kajang, Selangor. The conference continued the tradition of continuous effort by Malaysian Nuclear society to bring together the nuclear communities in an annual event to discuss and to discourse on progress in nuclear researches and related fields. The conference particularly encouraged the interaction of research students, research institutions and academics with the more established community. iNuSTEC2017 also played an essential role in introducing and fostering new research talents and in advancing the future profile of nuclear science, technology and engineering regionally and globally.

iNuSTEC2017 comprised of the conference, WIN-MNS seminar and Nuclear Youth competition. UNITEN had also organized a post-conference workshop on neutronic simulation on the 28th of September 2017. At the national forefront of science and energy development, we were honoured to be able to highlight the usage of science and engineering in the nuclear fields and to promote the use of nuclear science and technology for peaceful use through these conference, workshop and side-meetings.

We would like to acknowledge the Ministry of Science, Technology and Innovation Malaysia (MOSTI), Ministry of Education Malaysia (MOE), Malaysian Nuclear Agency (Nuclear Malaysia), International Atomic Energy Agency (IAEA), Hitachi-GE Nuclear Energy Ltd., Japan Atomic Energy Agency (JAEA), Tokyo Institute Technology, Asia Oceana Neutron Science Association (AONSA), ASME Section Malaysia, Malaysian Welding and Joining society, NGOs, universities and companies for their moral and financial support. Special thanks to iNuSTEC2017 organizing committees for their effort and commitment during this conference. We look forward to meeting all of you in 2018 – iNuSTEC2018 @ UTM.

Assoc Prof Dr Nasri A Hamid (UNITEN)
Dr. Faridah Mohamad Idris (MNS)
Chairs of iNuSTEC2017,
Kajang, Selangor,
Malaysia.

IOP Conf. Series: Materials Science and Engineering **298** (2018) 011001 doi:10.1088/1757-899X/298/1/011001

MESSAGE FROM :
THE PRESIDENT, MALAYSIAN NUCLEAR SOCIETY (MNS)

On behalf of the Malaysian Nuclear Society (MNS), I would like to express our thanks to all participants of the International Nuclear Science, Technology and Engineering Conference 2017 (iNuSTEC2017). MNS is honoured to receive supports from distinguished international and national experts in the field of nuclear (neutron) science and engineering as International Advisory and National Advisory Panels for iNuSTEC2017. We are indeed pleased to have Organizing Committees of iNuSTEC2017 and the Secretariats working with along to ensure the success of iNuSTEC2017. At the forefront of nuclear sicence and engineering MNS collaborates with universities and related NGOs to bring the success of iNuSTEC2017. MNS members shall strive to make it better and stronger in future.

In conjunction with this conference, MNS is also honoured to organise 2nd Nuclear Youth Competition 2017 (NYC2017) to help bring the young nuclear generation into the main stream of nuclear development and issues. In addition, this is the second time that Women in Nuclear Malaysia (WiN Malaysia), as Chapter of MNS and Women in Nuclear (WiN) Global, is organizing Women in Nuclear Malaysia 2017 Annual Conference to highlight the importance of women factor in the development of nuclear sector in Malaysia. A post conference workshop on neutronic simulation had been organised to give essential knowledge for Malaysian nuclear researchers in using computers as a tool of a simulation method.

This conference brings together experienced and young researchers to discuss and debate latest development in all areas of nuclear science, engineering and technology applications for friendly purposes at international level. This would spur and catalyse further research and development leading to the sustainable and improvement of the quality of life without compromising the quality of the environment in Malaysia and in the world.

We would like to express our gratitude to UNITEN for hosting iNuSTEC2017, WiN Conference 2017 and NYC2017 and to all the personnel involved for their unwavering efforts to make this event a reality. I wish to also record my appreciations and thanks to all sponsors for their generous support. May all these efforts bring benefits to all of us and I congratulate everyone involved in making this conference succesful event.

"Crafting Nuclear Knowledge for Global Well-Being "

Thank you.

Assoc Prof Dr Abdul Aziz Mohamed
President, MNS.

IOP Conf. Series: Materials Science and Engineering **298** (2018) 011001 doi:10.1088/1757-899X/298/1/011001

PROCEEDING EDITORIAL COMMITTEE

(PAPERS)

Science:

Engineering:

IOP Conf. Series: Materials Science and Engineering **298** (2018) 011001 doi:10.1088/1757-899X/298/1/011001

iNuSTEC2017 ORGANISING COMMITTEE

Patrons:
Prof. Dato' Ir. Dr. Kamal Nasharuddin Mustapha,
Vice Chancellor Universiti Tenaga Nasional (UNITEN).

Advisors:
Prof. Dr. Izham Bin Zainal Abidin,
Dean, College of Engineering, Universiti Tenaga Nasional
Prof. Dr. Nahrul Khair Alang Md Rashid (Universiti Teknologi Malaysia),
Advisor, Malaysian Nuclear Society.
Assoc. Prof. Abdul Aziz Mohamed (UNITEN);
Head, Nuclear Engineering and Energy Group.

Chairman:
Assoc. Prof. Dr. Nasri A Hamid (UNITEN)
Head, Nuclear Engineering and Energy Group

Co-Chairman:
Dr Faridah Mohamad Idris (MNS/ Nuklear Malaysia).

Deputy Chairman:
Dr Mohd Syukri Yahya (UNITEN).

Secretary:
Anas Muhamad Pauzi (UNITEN).
Azraf Azman (MNS).

Treasurer:
Dr Julie Andrianny (MNS Treasurer).
Finance Department (UNITEN).

Committee Members:

UNITEN
Assoc. Prof. Dr. Nasri Bin A. Hamid
AP Dr Siti Fadzlili Abdullah
Dr Hassan Mohamed
Dr Juniza Md Saad
Rogemah Ramli
Madihah Mujaini
Hj Zainudin Yahya
Marketing and Corporate Communication (MCC) Department

MALAYSIA NUCLEAR SOCIETY
Dr Muhamad Rawi Mohd Zin
Nazaratul Ashifa Bt. Abdullah Salim
Mohd Rizal B. Mamat @ Ibrahim
Anwar B. Abdul Rahman
Dr. Hafizal Yazid
Azraf Bin Azman
Shalina Bt. Sheik Muhamad
Rafhayudi Jamro

INTERNATIONAL ADVISORY COMMITTEE
Prof. Emiretus Dr. Noramly Muslim, Universiti Tenaga Nasional, Malaysia/ Western Australia University , Australia.
Prof Misaki Saito, Tokyo Institute of Technology, Japan.
Prof Il Hwang Soon, Seoul National University, Republic of Korea.
Director General, Malaysian Nuclear Agency, Malaysia.
Mr. Hamrah bin Mohd Ali, Director General, Atomic Energy Licensing Board (AELB) of Malaysia.
Dr Mohd Zamzam bin Jaafar, CEO, Malaysia Nuclear Power Corporation (MNPC).
Prof. Dr. Oprea Cristiana, Joint Institute for Nuclear Research, Rusia.
Prof. Dr. Thamir A. AL-Kazimi, Al-Karakh University of Science, Bagdad Iraq.
Prof. Dr. Ir. Sunarno, Gadjah Mada University, Indonesia.
Dr Hadj Hamma Tadjine, IAV GmbH, Germany.
Assoc. Prof. Dr. Burak Dikici, Yuzuncu Yil University, Turkey.

NATIONAL ADVISORY COMMITTEE
Prof. Dr. Bachok M.Taib, Universiti Sains Islam Malaysia.
Prof. Dr. Hj. Khudzir Hj Ismail, Universiti Teknologi MARA.
Prof. Emeritus Dato' Wira Ir. Dr. Md Noor bin Salleh, International Islamic Unviversity Malaysia.
Prof Dr Shahrim Ahmad, Universiti Kebangsaan Malaysia.
Prof Dr. Zanariah Abdullah, Universiti Malaya.
Assoc Prof Eng Dr Rahmat Mohsin, Universiti Teknologi Malaysia.
Prof Dr Wan Ahmad Tajuddin Wan Abdullah, Universiti Malaya.
Prof Madya Dr Izham bin Datuk Zainal Abidin, UNITEN.
Dr Mokhtar Awang (Universiti Teknologi PETRONAS /ASME Malaysia Section).
Assoc Prof Dr Samsudin Bani, President of Malaysian Welding and Joining Society (MWJS).

SCIENTIFIC AND PUBLICATION
Dr. Siti Fazlili bt Abdullah (MNS/ UNITEN)
Dr. Julia Abdul Karim (MNS/Nucklear Malaysia)
Dr Hafizal Yazid (MNS/Nucklear Malaysia)
Dr. Chantara Thevy Ratnam (MNS/Nucklear Malaysia)
Dr Hishamuddin Hussain (MNS/Nucklear Malaysia)
Assoc. Prof. Dr. Ahmad Taufek bin Abdul Rahman (UiTM)

IOP Conf. Series: Materials Science and Engineering **298** (2018) 011001 doi:10.1088/1757-899X/298/1/011001

Dr. Norhasliza Yusof (MNS/Universiiti Malaya)
Prof. Dr. Hasan Abu Kassim (MNS/Universiiti Malaya)
Assoc. Prof. Dr. Zaidi Embong (MNS/ UTHM)
Dr Aznan Fazli Ismail (MNS/ UKM)
Assoc. Prof. Dr. Irman Abdul Rahman (MNS/ UKM)
Dr. Norsyahidah Mohd. Hidzir (MNS/ UKM)
Dr. Suhairul Hashim (UTM)

WIN MALAYSIA COMMITTEE

Chairperson: Dr. Noor Hasnah Mohamed Khairullah
Secretary: Dr. Siti A'iasah Hashim

Technical Committee:
Dr. Hasni Hasan (Chair)
Dr. Noor Hasnah Mohamed Khairullah
Dr. Siti A'iasah Hashim
Dr. Chantara Thevy Ratnam
Dr. Zainah Adam
Rokiah Mohd Sabri

Session Chairperson:
Suzilawati Muhd Sarowi
Salmah Moosa

Treasurer:
Dr. Susan Maria Sipaun

IOP Conf. Series: Materials Science and Engineering **298** (2018) 011001 doi:10.1088/1757-899X/298/1/011001

ACKNOWLEDGEMENT

iNuSTEC2017 was organized by

Malaysian Nuclear Society (MNS)
and
Universiti Tenaga Nasional (UNITEN)

In Collaboration with
Malaysian Nuclear Agency (Nuclear Malaysia)
Universiti Tenaga Nasional (UNITEN)
Universiti Teknologi Malaysia (UTM)
Universiti Islam Antarabangsa Malaysia (IUM)
Universiti Kebangsaan Malaysia (UKM)
Universiti Malaya (UM)
Universiti Tun Hussien Onn Malaysia (UTHM)

MNS COUNCIL MEMBERS 2017/2019

Advisor / Past President	Dr. Nahrul Khair Bin Alang Md Rashid (UTM)
President	Dr. Abdul Aziz Bin Mohamed (UNITEN)
Vice President	Dr. Faizal Bin K.P. Kunchi Mohammed (UKM)
Secretary	Dr. Faridah Bt Mohamad Idris (NUCLEAR MALAYSIA)
Deputy Secretary	Dr. Irman B Abdul Rahman (UKM)
Treasurer	Dr Julie Andrianny Bt. Murshidi (NUCLEAR MALAYSIA)
Deputy Treasurer	Dr Nasri B A Hamid (UNITEN)

Members
Azraf Bin Azman (NUCLEAR MALAYSIA)
Shalina Bt. Sheik Muhamad (NUCLEAR MALAYSIA)
Dr Abu Hassan Husin (UiTM)
Dr Julia Bt. Abdul Karim (NUCLEAR MALAYSIA)
Dr. Muhammad Rawi Bin Mohamed Zin (NUCLEAR MALAYSIA)
Dr Mohd Syukri Yahya (UNITEN)

Auditors
Dr. Mohd Yuhyi Mohd Tadza (UMP)
Safwan Shalbi (UTM)

MNS Chapters

RadTech (Radiation Tech)
XApp (X-ray App)
NuEnergy (Nuclear Energy)
Neutron Photon (Neutron- Photon)
Student Chapter (UKM, UiTM, UNITEN)
Nuclear Engineering Student Society (NESS)
Women in Nuclear Malaysia (WiN Malaysia)
Institute of Nuclear Engineers (INE)

Visit us at http://www.nuklearmalaysia.org

Monozukuri — Technology to protect energy and tomorrow
Trees bathe in the warm sunshine while soaking up water.
They feed on the precious nutrients, put out buds, extend trunks, spread branches,
and put out leaves as they grow slowly but steadily over the time.
The flourishing tree produces clean air, clear water and rich harvest soil,
furthermore provides many blessing to humankind.
The various trees breathing firmly and perpetuating life
provide hope for the earth and humankind.
We at Hitachi-GE Nuclear Energy provide wide-ranging support for the nuclear power
industry to open up a new horizon for a promising tomorrow of the earth,
using highly reliable monozukuri in the integrated Hitachi Group.
Nuclear energy produces no carbon dioxide and has a low environmental load
while providing a stable supply of electricity to support a comfortable life.
Our aim is an abundant, secure, and promising future
based on the theme of a human society that coexists with nature.
From nuclear energy to new reliable energy
Hitachi-GE Nuclear Energy, Ltd.
http://www.hitachi-hgne.co.jp/nuclear/index.html

Peer review statement

All papers published in this volume of *IOP Conference Series: Materials Science and Engineering* have been peer reviewed through processes administered by the proceedings Editors. Reviews were conducted by expert referees to the professional and scientific standards expected of a proceedings journal published by IOP Publishing.

IOP Conf. Series: Materials Science and Engineering **298** (2018) 012001 doi:10.1088/1757-899X/298/1/012001

An assessment of absorbed dose and radiation hazard index from soil around repository facility at Bukit Kledang, Perak, Malaysia

M I Abdul Adziz [1] and K S Khoo [1]

[1] School of Applied Physics, Faculty of Science and Technology, Universiti Kebangsaan Malaysia (UKM), 43600, Bangi, Selangor, Malaysia.

khoo@ukm.edu.my

Abstract. The process of natural decay of radionuclides that emit gamma rays can infect humans and other living things. In this study, soil samples were taken at various locations which have been identified around the Long Term Storage Facility (LTSF) in Bukit Kledang, Perak. In addition, the respective dose rates in the sampling sites were measured at 5cm and 1m above the ground using a survey meter with Geiger Muller (GM) detector. Soil samples were taken using a hand Auger and then brought back to the laboratory for sample pre-preparation process. The measuring of radioactivity concentration in soil samples were carried out using gamma spectrometer counting system equipped with HPGe detector. The obtained results show, the radioactivity concentration ranged from 11.98 – 29.93 Bq/kg for Radium-226 (^{226}Ra), 20.97 – 41.45 Bq/kg for Thorium-232 (^{232}Th) and 5.73 – 59.41 Bq/kg for Potassium-40 (^{40}K), with mean values of 20.83 ± 5.88 Bq/kg, 32.87 ± 5.88 Bq/kg and 21.50 ± 2.79 Bq/kg, respectively. To assess the radiological hazards of natural radioactivity, radium equivalent activity (R_{aeq}), the rate of absorption dose (D), the annual effective dose and external hazard index (H_{ex}) was calculated and compared to the world average values.

1. Introduction

The human environment in this world is exposed to radiation generated from various sources including cosmic rays; natural radionuclides in water, soil and plants; man-made radioactivity from nuclear tests; and applications in medicine. Gamma radiation from natural radionuclides and cosmic radiation provides external exposure while gamma radiation derived from inhalation and intake through drinks and food provides internal exposure to humans. In 1996, the IAEA estimated that 80% of the dose received was from natural radionuclides while the remaining 20% was from cosmic radiation and nuclear processing [1]. The naturally occurring radioactive element in rock, soil and water exists in different concentrations [2]. Natural radionuclide concentrations are affected by various activities such as mining; extraction and purification of uranium minerals; phosphate and phosphate fertilizer production; charcoal burning; and oil and gas production [3].

Most of the radionuclides present in the environment are primordial radionuclides, ^{238}U, ^{232}Th, ^{40}K and their decay products [4]. This natural radionuclide contributes to external exposure risk due to the release of radon (gas) radiation and its decaying product that releases alpha radiation. Natural environmental radioactivity and the associated external exposure due to gamma radiation depend primarily on the geological and geographical conditions, and appear at different levels in the soils of each region in the world [5, 6]. Therefore the assessment of gamma radiation dose from natural sources is of particular importance as natural radiation is the largest contributor to the external dose of the world population. Knowledge of specific activity or concentration and radionuclide distribution, particularly in the soil, is an interesting thing because it provides useful information in monitoring environmental radioactivity.

The aims of this study are to determine the level of natural radionuclide concentration (^{232}Th, ^{40}K and ^{226}Ra) in the soil around the area near the repository facility and to determine the radiation hazard index in the vicinity of the repository facility by calculating radium equivalent activity (Ra_{eq}), the rate of absorption dose (D), the annual effective dose and external hazard index (H_{ex}) [7].

2. Experimental Procedure

2.1. Sampling Area

A total of 12 soil samples were collected consisting 10 samples from around the facility and 2 samples from selected residential area near the repository facility in Bukit Kledang, Perak. Originally, this facility was the Long Term Storage Facility (LTSF), which was upgraded once the decontamination and disposal process (D&D) was completed. This area is the radioactive waste storage site for the Asian Rare Earth Factory (ARE). Thorium waste and contaminated material that may contain some minor amounts of thorium hydroxide were disposed in this facility. The sampling sites are shown in figure 1 and figure 2.

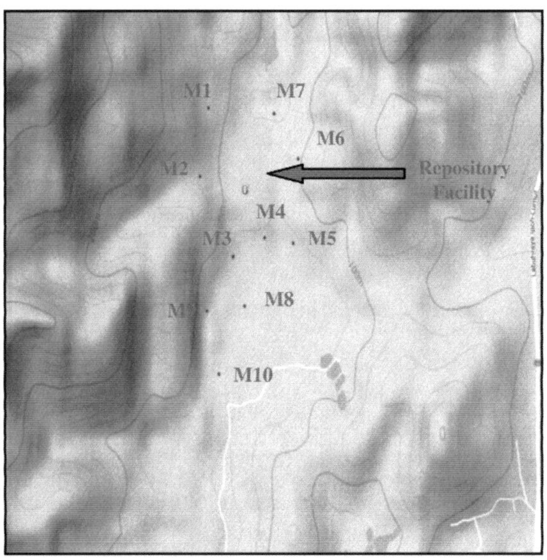

Figure 1. Location of sampling station around the repository facility.

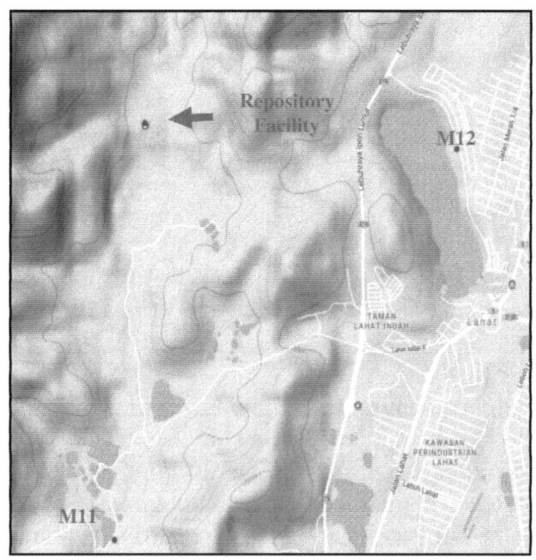

Figure 2. Location of sampling station near the residential area.

* Source: Google Map 2017.

2.2 Samples Collection and Preparation

Soil samples were collected from different locations around the repository facility. Dose monitoring was carried out using survey meter with Geiger Muller (GM) detector (μSv/h) and Sodium Iodide (NaI) detector (kcps converted to mR/h and converted to μSv/h, using conversion factor unit = 0.009335) to detect alpha, beta and gamma. In this study, the measurement had been done according to a specified distance of 5 cm and 1 m above the ground. All the samples were collected at a depth of 0-10 cm from the soil-surface using hand Auger, then were carried in dried cleaned polyethylene bags with sample codes and transferred to the laboratory. After removing the stones and vegetation, all soil samples were dried in the oven at 85^0C, sieved (500 micron), placed in 250g marinelli container and left for four weeks to reach radioactive equilibrium.

2.3. Samples Analysis

Three Marinelli beakers were prepared for measuring the average natural radioactivity. The preparation of soil samples was made according to the procedures proposed by Johar & Embong [8]. A same amount of CRM IARMA-001 (IARMA Limited, United Kingdom) prepared in the same manner was used as standard reference material. The measurements were made using a gamma ray HPGe counting system calibrated using ^{57}Co, ^{137}Cs, ^{133}Ba, ^{85}Sr, ^{54}Mn, ^{88}Y and ^{65}Zn for two hours. The spectra were analyzed using the GENIE 2000 program. The counting time interval was 43200 seconds. The activities of ^{40}K were determined from its 1460 keV γ-line, while the specific activity of ^{226}Ra will be determined based on the average concentration of the ^{214}Pb at 352 keV and ^{214}Bi at 609 keV. As for ^{232}Th, the concentrations will be determined from the average concentrations of ^{208}Tl at 583 keV and ^{228}Ac at 911.21 keV [9]. The background spectrum was recorded immediately after or before the sample counting. The recommended values of ^{226}Ra, ^{232}Th and ^{40}K concentrations contained in the CRM IARMA-001 were 44.1, 40.0 and 440 Bq/kg, respectively, at confidence intervals of 40.7 – 47.5, 33.8 – 46.2 and 410 – 470 Bq/kg, respectively. The measurements were carried out using facilities in Malaysian Nuclear Agency, Bangi.

2.4. Activity Concentration Calculation

The activity concentration of individual nuclides in the soil samples were calculated using the following equation 1 [10]:

$$A = \frac{N_{net}}{I_\gamma \varepsilon m t} \tag{1}$$

where,

A : calculated gamma activity (Bq/kg),

N_{net}: net peak area count subtract background of the sample (counts),

ε : absolute efficiency of the detector,

I_γ : emission probability of a specific energy photo peak,

t: sample count time (sec),

m : sample weight/volume.

This calculation was done by assuming that all correction factors are equal to unity. The distribution of the activity mass concentrations due to the radionuclides ^{226}Ra, ^{232}Th and ^{40}K is not uniform throughout the soils. The incompatibility with radiation exposure has been defined in terms of radium equivalent activity (Ra_{eq}) used to represent the specific activity value of the three radionuclides by only single quantity. It is calculated using the following equation 2 [11][12]:

$$Ra_{eq} = C_{Ra} + 1.43C_{Th} + 0.07C_K \tag{2}$$

where,

Ra_{eq} : radium equivalent activity in Bq/kg,

C_{Th} : ^{232}Th activity concentration in Bq/kg,

C_{Ra} : ^{226}Ra activity concentration in Bq/kg,

C_K : ^{40}K activity concentration in Bq/kg.

While defining Ra_{eq} activity, it is assumed that 10 Bq/kg of ^{226}Ra, 7Bq/kg of ^{232}Th and 130 Bq/kg of ^{40}K produce equal gamma ray dose [13]. According to [12], the maximum value of this parameter should not exceed 370 Bq/kg in order to limit the annual exposure 1.5 mGy.

From previous studies [14,15], the external gamma absorbed dose rate (D) in the air at 1 m above ground level was calculated using the factor 0.042 nGy/h per Bq/kg for ^{40}K, 0.462 nGy/h per Bq/kg for ^{226}Ra and 0.604 nGy/h per Bq/kg for ^{232}Th. The calculations were performed according to the following equation 3 [7]:

$$D = 0.462C_{Ra} + 0.604C_{Th} + 0.042C_{K} \tag{3}$$

where,

D: External gamma absorbed dose rate in nGy/h,
C_{Th} : ^{232}Th activity concentration in Bq/kg,
C_{Ra} : ^{226}Ra activity concentration in Bq/kg,
C_{K} : ^{40}K activity concentration in Bq/kg.

To calculate the effective dose rate, the conversion coefficient for the dose rate absorbed to the effective dose rate and external occupancy factor should be taken into account. The absorbed dose rate found to be converted into effective dose rate using 0.7 Sv/Gy conversion factor recommended by UNSCEAR (2000) and external occupancy factor of 0.2 assuming that the man spends 20% of their time outdoors [11]. The annual effective doses are calculated using the following equation 4:

$$\text{Annual effective dose (Sv)} = D \times 8760 \times 0.7 \times 0.2 \tag{4}$$

where,

D: External gamma absorbed dose rate in nGy/h,
0.7: conversion factor in Sv/Gy,
8760: time in a year (hour),
0.2: external occupancy factor.

External hazard is an index hazard widely used in the study as it represents external exposure to humans. The value of this index must be less than unity (unity value = 1) in order to keep the radiation hazard insignificant. The maximum value of H_{ex} equal to unity corresponds to the upper limit of radium equivalent activity (370 Bq/kg) [11]. The external hazard indexs, H_{ex}, are calculated using the following equation 5 [7]:

$$H_{ex} = \frac{C_{Ra}}{370} + \frac{C_{Th}}{259} + \frac{C_{K}}{4810} \tag{5}$$

where,

C_{Ra} : ^{226}Ra activity concentration in Bq/kg,
C_{K} : ^{40}K activity concentration in Bq/kg,
C_{Th} : ^{232}Th activity concentration in Bq/kg.

3. Results And Discussions

The results of measurement for 12 soil samples collected at different locations around the repository facility are summarised in table 1. The three most common primordial radionuclides determined in the study areas were ^{226}Ra, ^{232}Th and ^{40}K. The natural activity of ^{226}Ra and ^{232}Th of radionuclides is obtained from the most abundant photopeaks activity of uranium and thorium series decay products, while the concentration of ^{40}K natural activity is a single peak decision [13].

Table 1. Specific activities of radionuclides in soil samples at different locations around repository facility.

Samples Location	GPS Coordinates		Specific activity of radionuclides (Bq/kg)		
	Latitude (DMS)	Longitude (DMS)	^{226}Ra	^{232}Th	^{40}K
M1	N 04° 33.338'	E 101° 00.790'	17.87 ± 3.12	34.96 ± 4.66	7.59 ± 0.73
M2	N 04° 33.195'	E 101° 00.755'	16.30 ± 3.07	36.41 ± 6.54	6.51 ± 0.60
M3	N 04° 32.999'	E 101° 00.838'	20.63 ± 3.81	30.14 ± 6.52	27.95 ± 5.49
M4	N 04° 33.294'	E 101° 00.554'	22.14 ± 4.00	41.43 ± 6.80	9.30 ± 0.70
M5	N 04° 33.044'	E 101° 00.965'	25.49 ± 4.29	35.48 ± 7.25	26.96 ± 5.50
M6	N 04° 33.154'	E 101° 00.580'	**29.92 ± 5.81***	38.71 ± 8.14	**59.40 ± 1.15***
M7	N 04° 33.284'	E 101° 00.961'	24.90 ± 4.56	33.91 ± 7.00	50.84 ± 8.54
M8	N 04° 32.538'	E 101° 00.525'	17.30 ± 3.15	**41.44 ± 6.69***	10.55 ± 2.29
M9	N 04° 32.848'	E 101° 00.767'	**11.98 ± 2.20***	25.10 ± 5.52	**5.73 ± 0.50***
M10	N 04° 32.442'	E 101° 00.462'	20.80 ± 3.98	**20.97 ± 4.45***	18.33 ± 4.03
M11	N 04° 31.468'	E 101° 00.598'	21.55 ± 3.97	33.70 ± 6.92	22.31 ± 4.33
M12	N 04° 32.526'	E 101° 02.144'	12.07 ± 2.10	22.12 ± 4.18	12.47 ± 2.24
Range			11.98 – 29.92	20.97 – 41.44	5.73 – 59.407
Mean			20.83 ± 3.59	32.870 ± 5.88	21.49 ± 2.79

* The bold characters represent the minimum and maximum values.

Figure 3 shows the specific activities of ^{226}Ra was ranged from 11.98 – 29.92 Bq/kg, ^{232}Th was ranged from 20.97 – 41.44 Bq/kg and ^{40}K was ranged from 5.73 – 59.40 Bq/kg, with a mean value for ^{226}Ra, ^{232}Th and ^{40}K were 20.83 ± 3.59 Bq/kg, 32.87 ± 5.88 Bq/kg and 21.49 ± 2.79 Bq/kg, respectively. The highest ^{226}Ra and ^{40}K concentrations were recorded at M6 with a value of 29.92 ± 5.81 Bq/kg and 59.40 ± 1.15 Bq/kg, respectively, while the lowest concentrations for both radionuclides were recorded at M9 with a value of 11.98 ± 2.20 Bq/kg for ^{226} Ra and 5.73 ± 0.50 Bq/kg for ^{40}K. As for ^{232}Th, the highest and the lowest concentration were recorded at M8 and M10 with a value of 41.44 ± 6.69 Bq/kg and 20.97 ± 4.45 Bq/kg, respectively.

Hussain [16] reported that the ratio of ^{232}Th to ^{226}Ra in the earth's crust is generally greater than one. This statement supports the results of this study, where the mean activity value of ^{232}Th was higher than ^{226}Ra. The specific activity concentrations show varying values and vary according to location likely to be affected by several factors such as soil type and soil pH [13], latitude of area, soil texture, existence of underground rock and organic matter.

As explained earlier, this study focuses only on the surrounding area and near repository facility area. The results gained from this study only represent the area of the study. A comparison of a mean value of the specific activity of the radionuclides in soil samples obtained from this study with other location in Peninsular Malaysia and other countries are shown in table 2. The mean activity concentrations of ^{226}Ra and ^{232}Th were comparable with worldwide mean data [7]. However, as for ^{40}K, the value was lower (21 Bq/kg) compared to worldwide mean value (474 Bq/kg). It is also found that the average value of the three radionuclides is lower than the average value recorded by previous studies [17][18][19] for certain areas in Peninsular Malaysia.

The values obtained for radium equivalent activity, gamma dose rate, annual effective dose and external hazard index are presented in table 3.

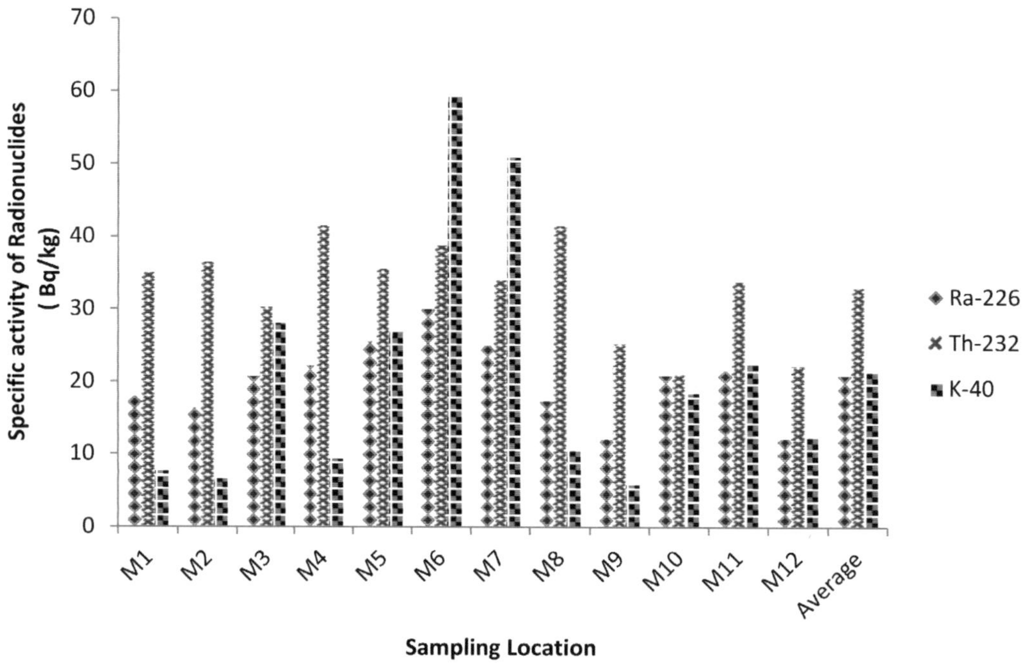

Figure 3. Specific activity concentration for ^{226}Ra, ^{232}Th and ^{40}K in soil samples

Table 2. Comparison of mean specific activities of radionuclides in soil samples at around the repository area with other location in Peninsular Malaysia and other countries [7][17][18][19].

Samples	Specific activity of radionuclides (Bq/kg)			Reference
	^{226}**Ra**	^{232}**Th**	^{40}**K**	
Repository Facility, Bukit Kledang, Perak.	**21**	**33**	21	**Present study**
Kinta District, Perak	-	246	-	Lee *et al.* [17]
Pulau Pinang	64 - 799	16 - 667	-	Almayahi *et al.* [18]
Dengkil, Selangor	-	27 - 103	-	Yasir, S. M. *et al.* [19]
Malaysia	67	82	310	UNSCEAR Report, 2000 [7]
Bangladesh	34	-	350	
Thailand	48	40	400	
Japan	33	28	310	
China	32	41	440	
India	29	64	400	
Egypt	17	18	320	
Iran	28	22	640	
Belgium	26	27	380	
Denmark	17	19	460	
Switzerland	40	25	370	
Spain	32	33	470	
Argentina	-	-	650	
United States	40	35	370	
Worldwide mean	33	36	474	

3.1. Radium Equivalent Activity (Ra_{eq})
Referring to table 3, the radium equivalent activity value for M6 (89.87 Bq/kg) is the highest among other sampling locations. Though, this value is lower than the standard set value of 370 Bq/kg. According to The Organisation of Economic Cooperation and Development (OECD) (1998) [20], the value of 370 Bq/kg is the ideal value for exposure to the public as it is equal to 1.5 mSv / year. If the value of the radium equilibrium index is high, it will cause exposure to the public to exceed the safe level of the public and may have a stochastic effect and the risk of illness that arises from radiation.

3.2. Dose Rate
The mean absorbed gamma dose in air was calculated as 30.03 nGy/h with a minimum and maximum values are 19.46 nGy/h and 39.69 nGy/h, respectively. It is found to be comparable to the world average of 57 nGy/h [7]. The main contributor to the dose rate absorbed in the air at a height of 1 m is from the ground gamma radiation generated from the naturally occurring radionuclide in the soil. Therefore, the dose rate absorbed depends on the concentrations of specific radionuclide activity. According to Al-Kaabi [21], if the concentration of radionuclide in the sample area is high, absorbed dose will also be high.

3.3. Annual Effective Dose
The calculated values of annual effective dose ranged from 0.02 to 0.05 mSv, with a mean value of 0.04 mSv. The highest value is come from M6 (48.68 μSv). Even though, the value of the area is high compared to other areas, but it is lower than the world average of 0.48 mSv [7]. This confirms that all of these research areas are at a safe level for the public.

3.4. External Hazard Index
The calculated values of external hazard index obtained in this study ranged from 0.12 to 0.24 with a mean value of 0.18. The highest value is come from M6 (0.24). The external hazard index represents external exposure to humans. Therefore, when the primordial radionuclide concentrations such as ^{226}Ra, ^{232}Th and ^{40}K are high in the area the external hazard index will also be high. However, since the values from this study are lower than unity, we can make a conclusion that the radiation hazard is low at all sampling location.

Table 3. Radium equivalent activity, dose rate, annual effective dose and external hazard index at different locations around repository facility

Samples Location	Radium equivalent activity (Bq/kg)	Dose rate (nGy/h)	Annual effective dose (10^{-6} Sv)	External hazard index
M1	68.47	29.70	36.42	0.18
M2	68.89	29.80	36.55	0.19
M3	65.89	28.91	35.45	0.18
M4	82.11	35.65	43.72	0.22
M5	78.32	34.34	42.11	0.21
M6	89.87	39.69	48.68	0.24
M7	77.31	34.11	41.83	0.21
M8	77.39	33.47	41.05	0.21
M9	48.33	20.94	26.68	0.13
M10	52.21	23.04	28.26	0.14
M11	71.46	31.24	38.32	0.19
M12	44.70	19.46	23.87	0.12
Range	44.70 – 89.87	19.46 – 39.69	23.87 – 48.68	0.12 – 0.24
Mean	68.74	30.03	36.91	0.18

IOP Conf. Series: Materials Science and Engineering **298** (2018) 012001 doi:10.1088/1757-899X/298/1/012001

4. Conclusion

Soil samples collected at different locations around the repository facility area were analysed for radioactivity due to ^{226}Ra, ^{232}Th and ^{40}K isotopes. The mean activity concentration of ^{226}Ra and ^{232}Th were comparable with worldwide mean data, however, as for ^{40}K, the value was lower (21 Bq/kg) compare to worldwide mean value (474 Bq/kg). It is also found that the average value of the three radionuclides is lower than the average value recorded by previous studies for certain areas in Peninsular Malaysia. The average value of gamma dose rate obtained in this study (30.03 nGy/h) is comparable to the world average (57 nGy/h). The calculated annual effective dose with average value 0.04 mSv is lower than the worldwide average value. All values obtained for radium equivalent activity are less than 370 Bq/kg, which are acceptable for safe use OECD 1998. The calculated values of external hazard index obtained in this study range from 0.12 to 0.24 with a mean value of 0.18 (all calculated values are lower than unity), which means that the radiation hazard is insignificant for the population living in the study area. These results can be used as an additional data to represent terrestrial radioactivity baseline data for Malaysia environment. This estimation will also serve as baseline for detection of any future related activities of contamination especially around the repository facility area.

5. References

[1] EPA 2007 *Ionizing radiation fact book* (United States Environmental Protection Agency).

[2] Palomo M, Penalver A, Aguilar C and Borrull F 2010 Presence of Naturally Occurring Radioactive Materials in Sludge Samples from Several Spanish Water Treatment Plants *J. of Hazardous Materials* **181** 716–721.

[3] Sethy N K, Jha V N, Sutar A K, Rath P, Sahoo S K, Ravi P M and Tripathi R M 2013 Assessment of Naturally Occurring Radioactive Materials in the Surface Soil of Uranium Mining Area Of Jharkhand, India *J. of Geochemical Exploration,* **57** 431-437.

[4] Tzortzis M, Svoukis E. and Tsetos H 2004 A Comprehensive Study of Natural Gamma Radioactivity Levels and Associated Dose Rates from Surface Soils in Cyprus. *Radiation Protection Dosimetry*, **109 (3)** 217–224.

[5] Iqbal M, Tufail M and Mirza S M 2000 Measurement of Natural Radioactivity in Marble Found in Pakistan Using a NaI (Tl) Gamma-Ray Spectrometer. Technical Note *J. of Environmental Radioactivity* **51(2)** 255–265.

[6] Anagnostakis M J, Hinis E P, Simopoulos S E and Angelopoulos M G 1996 Natural Radioactivity Mapping of Greek Surface Soils *Environmental International* **22 (1)** 3–8.

[7] UNSCEAR 2000 Radiation Sources and effects of ionizing radiation New York: United Nations Scientific Committee on the Effect of Atomic Radiation.

[8] Johar S M and Embong Z 2014 Terrestrial gamma radiation dose and its relationship with soil pH level in Seri Gading industrial area, Batu Pahat district, Malaysia, 20th World Congress of Soil Science, ICC Jeju, Jeju, Korea, 8-13 June, 2014
URL: http://www.dbpia.co.kr/Article/3475869.

[9] Hemby D M and Tynybekov A K 2012 Uranium, Thorium and Potassium in soils along the shore of the lake Issyk-Kyol in the Kyrghyz Republic *Environ. Monit. Assess* **73**, 101–108.

[10] Dovlete C and Povinec P P 2002 Quantification of Uncertainty in Gamma Spectrometric Analysis of Environmental samples. 2nd Regional Training Course on QA/QC of Nuclear Analytical Techniques. 12-16 August 2002. Kuala Lumpur.

[11] Beretka J and Mathew P J 1985 Natural Radioactivity of Australian Building Materials, Industrial Wastes and By-Products. *Health Physics*, **48(1)** 87–95.

[12] Tufail M, Akhtar N and Waqas M 2006 Measurement of terrestrial radiation for assessment of gamma dose from cultivated and barren saline soils of Faisalabad in Pakistan, *Radiation Measurements* **41** 443 – 451.

[13] Johar S M, Tajudin S A A and Embong Z 2015 The gamma dose assessment and pH correlation for various soil types at Batu Pahat and Kluang districts, Johor *AIP Conference Proceedings (iNUSTEC 2015).*
[14] Kocher D C and Sjoreen A L 1985 Dose-Rate Conversion Factors for External Exposure to Photon Emitters in Soil *Health Physics*, **48(1)** 193–205.
[15] Jacob P, Paretzke H G, Rosenbaum H and Zankl M 1986 Effective Dose Equivalents for Photon Exposures from Plane Sources on the Ground. *Radiation Protection Dosimetry*, **14(4)**, 299–310.
[16] Hussain A 1994 Determination of uranium and thorium concentration in rock samples. *J. Radioanal. Nucl. Chem.* **188 (4)**, 255–265.
[17] Lee S K, Husin W, Termizi R A and Nursama H A 2007 Natural gamma background radiation dose rate and its relationship with geological background in the Kinta district, Perak, Malaysia *Appl. Radiation and Isotopes* **54** 327–333.
[18] Almayahi B A, Tajuddin A A and Jaafar M S 2012 Radiation Hazard Indices of Soil and Water Samples In Northern Malaysian Peninsular. *International Journal of Applied Radiation and Isotopes* **70 (11)** 2652-2660.
[19] Yasir M S, Majid A A, Tap M S Q and Abidin M R Z 2006 Analisis ^{238}U, ^{232}Th, ^{226}Ra dan ^{40}K. Dalam Sampel Amang, Tanah Dan Air Di Dengkil, Selangor Menggunakan Spektometri Sinar Gama *Malaysia J. of Analytical Sciences* **10(1)** 35-40.
[20] OECD 1998 *Uranium 1997 - Resources, Production and Demand Paris* OECD
[21] Al-Kaabi M A and Al-Shimary A 2016 Study of the Radiological Doses and Hazard Indices In Soil Samples From Karbala City, Iraq *Chines Phys.* **16** 331-347.
[22] Google Maps, 2017 *Long Term Storage Facility, Bukit Kledang, Ipoh, Perak* https://www.google.com/maps/place/Bukit+Kledang/@4.548918,101.0051866,2084m/data=!3m1!1e3!4m5!3m4!1s0x31caea4aaf1a4261:0xd4d4d9325057a180!8m2!3d4.5803789!4d101.025359 [20 Jun 2016].

Acknowledgements
The authors wish to thank the Universiti Kebangsaan Malaysia (UKM) for research fund (DLP – 2014 – 011).

Monitoring underground water leakage pattern by ground penetrating radar (GPR) using 800 MHz antenna frequency

T S T Amran[1], M P Ismail[1], M R Ahmad[1], M S M Amin[1], M A Ismail[1], S Sani[1], N A Masenwat[1] and N S M Basri[2]

[1]Non-Destructive Testing - Material Structure Integrity group (NDT-MSI), Industrial Technology Division, Malaysian Nuclear Agency, 43000 Kajang, Selangor, Malaysia
[2]The Faculty of Science, Technology and Human Development, Universiti Tun Hussein Onn Malaysia (UTHM), 86400 Parit Raja, Batu Pahat, Johor, Malaysia

sarah@nm.gov.my

Abstract. Water is the most treasure natural resources, however, a huge amount of water are lost during its distribution that leads to water leakage problem. The leaks meant the waste of money and created more economic loss to treat and fix the damaged pipe. Researchers and engineers have put tremendous attempts and effort, to solve the water leakage problem especially in water leakage of buried pipeline. An advanced technology of ground penetrating radar (GPR) has been established as one of the non-destructive testing (NDT) method to detect the underground water pipe leaking. This paper focuses on the ability of GPR in water utility field especially on detection of water leaks in the underground pipeline distribution. A series of laboratory experiments were carried out using 800-MHz antenna, where the performance of GPR on detecting underground pipeline and locating water leakage was investigated and validated. A prototype to recreate water-leaking system was constructed using a 4-inch PVC pipe. Different diameter of holes, i.e. ¼ inch, ½ inch, and ¾ inch, were drilled into the pipe to simulate the water leaking. The PVC pipe was buried at the depth of 60 cm into the test bed that was filled with dry sand. 15 litres of water was injected into the PVC pipe. The water leakage patterns in term of radargram data were gathered. The effectiveness of the GPR in locating the underground water leakage was ascertained, after the results were collected and verified.

1. Introduction

Water is precious natural resource to human being and plant. Unfortunately, a very large amount of water goes wasted as the potential water leaking in most water-distribution systems (WDSs) is getting higher year by year, thus cause most urban and developing areas face acute water problems. The WDSs is one of the most critical and treasured utilities and infrastructure systems. According to the World Bank database report, more than 32 billion cubic meter of treated water and non-revenue water (NRW) suffered from physical leak in WDSs, each year around the world. Unaccounted water loss reportedly by International Water Supply Association (IWSA) is 20-30 percent for WDSs [1]. The result is common, however for older WDSs, the percentage can reach up to 50 percent. The reports are quite dreadful and staggering, since the actual losses is more than reported losses. Smart solution is implemented in order to solve the acute water leaking in WDSs, notably underground water pipeline system.

The technique included is smart sensor in detecting water leaking named ground penetrating radar (GPR) technology. Locating and discovering the element of WDSs such as water leaks are one of the applications of GPR [2]. The basic GPR technology had aided the development of non-destructive testing (NDT) methods. This can be proved by the growth of GPR production since 1970's [3] and GPR application in various fields. Approved physical theories and discoveries that are associated with the occurrence of frequency energy transmitter of GPR, leads to the improvement of WDSs.

2. Experimental works

Analyzing the effectiveness of ground penetrating radar (GPR) is the main goal of this study by recreating a physical laboratory experiment to detect water leaking from underground pipes. An experiment was carried out to stimulate leaks in a 4-inch PVC pipe. A 4-inch PVC pipe was chosen in this laboratory water leakage simulation because PVC is the frequently used in industrial underground pipeline of water distribution system [4]. The 4-inch PVC pipe with three different holes diameter of ¼ inch, ½ inch, and ¾ inch, was buried in a test bed with 6 tons of dry sand as shown in figure 1, which ½ inch and ¾ inch were sealed with silicone sealant. The dimensions of test bed were 2.5 m long, 1.5 m high, and 1.5 m wide as shown in figure 2. The PVC pipe was buried at depth of 0.60 m in dry sand as depicted in figure 3 (a) and the hole faced downward to simulate leaks. The test bed was constructed using wooden box without the usage of metals such as screws and nails [5]. The present of metals will cause interference with radar signal during GPR scanning. 15 liters of water was injected for the water leakage laboratory experiment. RAMAC/GPR 800 MHz shielded antenna was used for the GPR scanning. For the antenna polarization in the experiment, the long axis of the antenna was perpendicular along the PVC pipe as presented in figure 3 (b). Table 1 summarized the antenna setting employed throughout the study. Testing parameters are listed as following:

- Number of samples is 502
- Sampling frequency is 8800 MHz
- Time window is 57 ns
- Antenna separation is 0.14 m

Table 1: Antenna setting

Parameter	Value
Antenna	800 MHz shielded
Trigger	Distance
Trig int	0.005 m
Measuring wheel	Meas.wheel 300-800 MHz

Bnnnnn nmb hy 6

Figure 1. Six tons of dry sand used for the entire experiment.

Figure 2. Real photograph of test bed for the laboratory experiment.

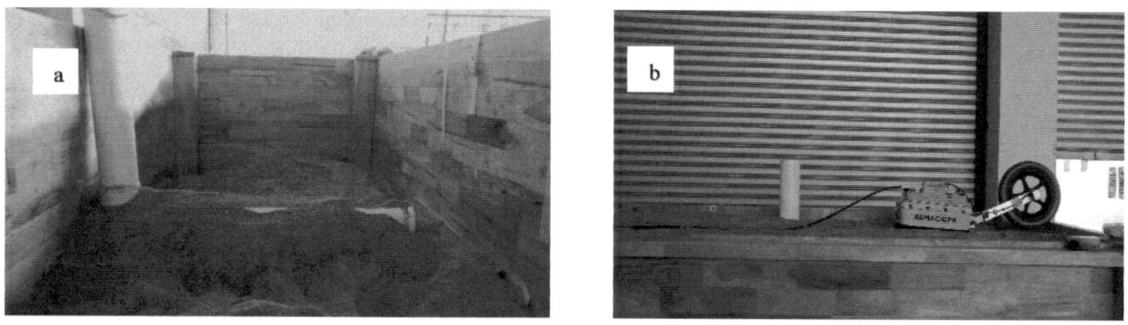

Figure 3. (a) Photograph of PVC pipe buried at the depth of 0.60 m, and the PVC pipe is in L-shape to inject water for water leaks simulation, (b) 800 MHz shielded antenna is perpendicular along the PVC pipe.

3. Result and analysis

3.1. Detecting water leakage

Several attempts of experiments were carried out to stimulate leaks in PVC pipe, in which the pipe was buried at the depth of 0.60 m in the test bed. Three different holes were drilled at different hole diameters, i.e. ¼ inch, ½ inch and ¾ inch hole diameter. 15 liter of water was injected into PVC pipe to simulate water leakage. The first section of result and discussion emphasis on comparison between time interval of water leakage at the smallest hole, i.e. ¼ inch hole diameter, the next section focus on ½ inch hole diameter for comparison between time interval of water leakage, the following section discuss on comparison at ¾ inch hole diameter, and the last section details on the comparison between ¼ inch, ½ inch and ¾ inch during water leaking.

3.1.1. Comparison between time interval of water leakage at ¼ inch hole diameter. Three different diameter of holes, ¼ inch, ½ inch and ¾ inch were chosen as variable for the water leakage simulation, to observe the variation of water pressure during water leaking. 800 MHz exhibits strong signal and provides sufficient penetration to detect the water leakage. Hyperbola pattern was exhibited as shown in figure 4 (a), (b), (c), (d), and (e), with the center is located approximately at 0.60 cm as proven in red circle. This illustrate that the position of the buried PVC pipe at approximately 0.60 m, almost 100% accurate as presented in Figure 4 (a). Generally, the hyperbolic pattern projected downward was generated to represent the buried pipe [6]. It can be seen from figure 4 (b), during water leaking at ¼ inch hole, the reflection pattern was totally clear and seems massive below the PVC reflection line, where the hyperbola intensity increased. The change in hyperbola pattern was influenced significantly by changes in dielectric medium [7]. The changes in dielectric medium were caused by the presence of water leaks. Due to highest water pressure, the water flow rate is the fastest. The hypothesis on the smaller the pipe diameter, the higher the pressure was justified. That was perhaps, the water leakage already reached at the bottom of the sand in the short period of time due to fastest water flow rate. There is no significant difference in figure 4 (c) and (d), since the water runs to the bottom at similar rate. However, the reflection of water leakage getting blurry after 24 hours as depicted in figure 4 (e). The purpose of GPR scanning after 24 hour was to analyze the water dissipation in the dry sand. The result indicates that the saturation water was already dissipated into the dry sand. The GPR scanning profiles for time interval illustrate the depth of water leakage can reach up to more than 1 m, even after 24 hours of water leaks simulation.

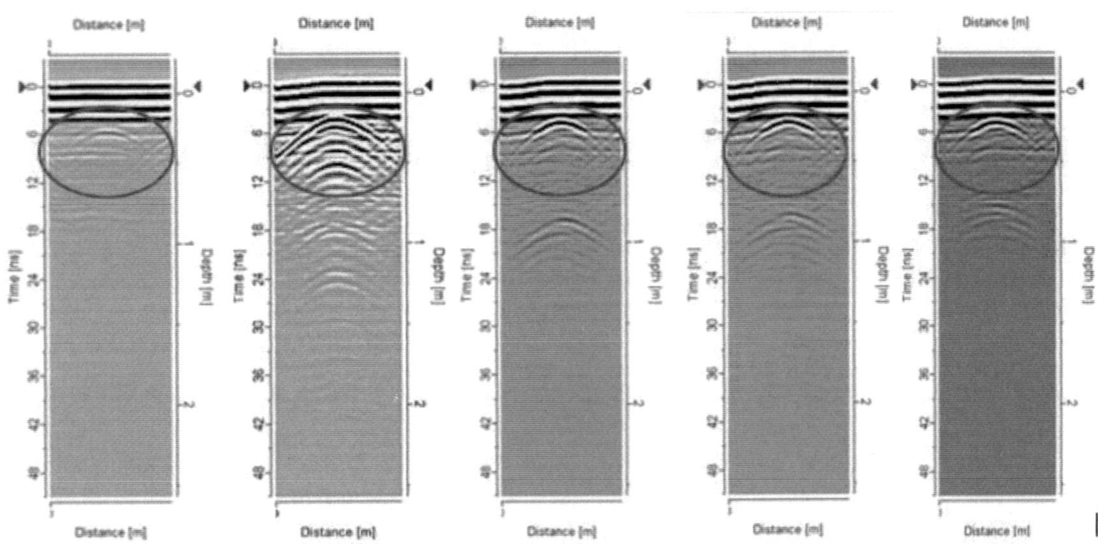

(a) (b) (c) (d) (e)

Figure 4. Radar profile of water leakage ¼ inch when (a) before water leakage, (b) during water leakage, (c) 30 minutes after water leakage, (d) 1 hour after water leakage, (e) 24 hour after water leakage.

3.1.2. Comparison between time interval of water leakage at ½ inch hole diameter. Figure 5 represents the radargram profile of water leakage simulation for hole diameter of ½ inch. The differentiating factor between the radargram profile in figure 5 (a) and other figures 5 (b), (c), (d), and (e) is the presence of water as marked with red circle. The presence of water created new hyperbolic reflection pattern in figure 5 (b). The water has substantially changed the dielectric properties of the dry sand. The pattern showed increased in intensity and the range of hyperbolic pattern started to extent after 30 minutes of water leakage as depicted in figure 5 (c). The low water pressure was caused by bigger hole diameter as compared to ¼ inch diameter. That was the reason on why the great difference in water leakage can only be seen after 30 minutes. The anomalous change then started to disappear as the phenomena demonstrates that the water leaks low started to move downward and dissolve into the dry sand as presented in figure 5 (d) and 5 (e). Figure 5 (e) illustrated that resemble to radargram in figure 5 (a), as the anomalies pattern was not notably detectable, where indicating that most of the saturation water has moved downward to the bottom of the test bed.

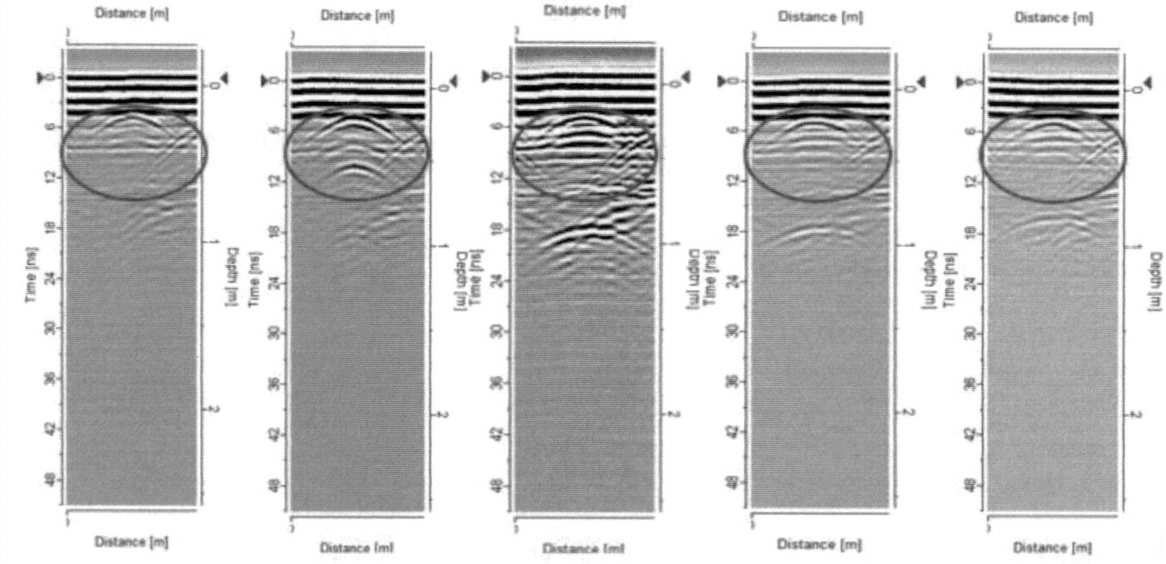

(a) (b) (c) (d) (e)

Figure 5. Radar profile of water leakage at ½ inch when (a) before water leakage, (b) during water leakage, (c) 30 minutes after water leakage, (d) 1 hour after water leakage, (e) 24 hour after water leakage.

3.1.3. Comparison between time interval of water leakage at ¾ inch hole diameter. Figure 6 illustrates the radargram images for water leakage simulation for ¾ inch hole diameter, the biggest hole for the experiment. The GPR scanning was conducted right after the injection of 15 litres of water. The radar profile during the water injection showed that more than one hyperbola were observed due to the presence of water in figure 6 (b). The reflection pattern through the GPR scanning was marked in red circle as shown in figure 6. The large amount of dielectric constant of water with the value of 80, is clearly make a huge difference between before and during of water injection. Low water velocity was caused by smallest water pressure, where only little water leaks detected through GPR scanning in figure 6 (b). The intensity of hyperbola reflection pattern increased and the water anomalous feature was sharp as illustrated in figure 6 (c), and still remain the same after 1 hour of water leakage as presented in figure 6 (d). Due to lowest water velocity, the water reflection pattern is clearly visible in figure 6 (d). The sharpness of water leaks anomalous features slightly lost after 24 hour in figure 6 (e), because the water leaks already dissipated in the dry sand and already shifted downward toward the bottom of the test bed.

IOP Conf. Series: Materials Science and Engineering **298** (2018) 012002 doi:10.1088/1757-899X/298/1/012002

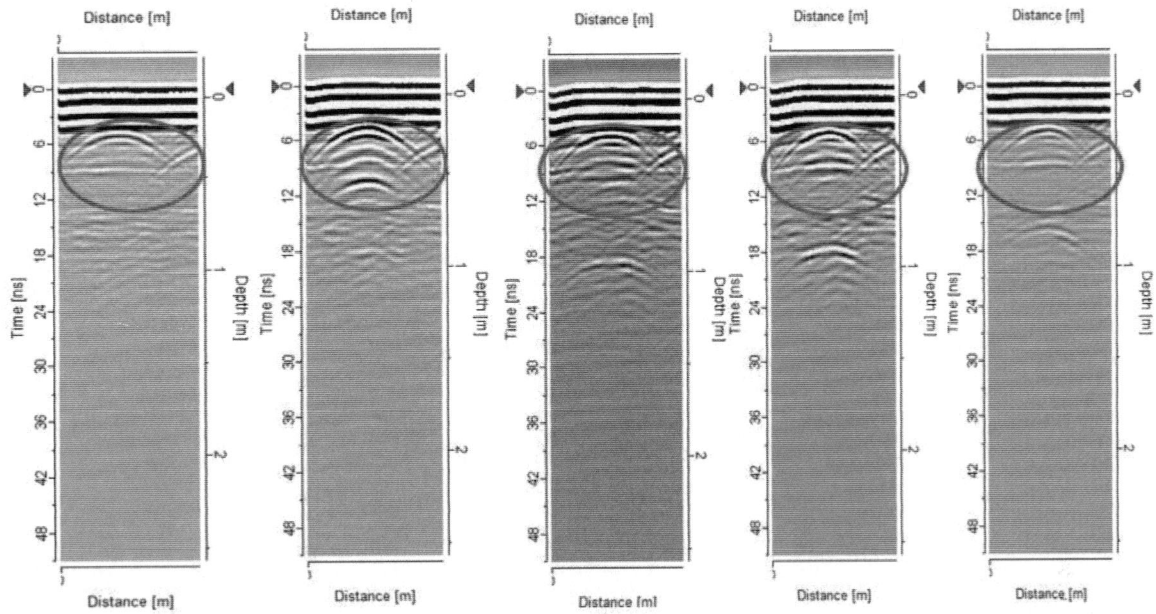

(a) (b) (c) (d) (e)

Figure 6. Radar profile of water leakage at ¾ inch when (a) before water leakage (b) during water leakage, (c) 30 minutes after water leakage, (d) 1 hour after water leakage, (e) 24 hour after water leakage.

3.1.4. Comparison between ¼ inch, ½ inch and ¾ inch during water leaking. The radargram profiles during water leakage for the three different hole-diameter were portrayed in figure 7. Significant differences of hyperbolic reflection pattern were observed during water leakage between ¼ inch hole diameter and other two holes, ½ inch hole diameter, and ¾ inch hole diameter as illustrated in red mark position. The great differences show that, at highest water pressure, the water flow rate is fastest compared to others. More hyperbolic reflection patterns were observed in figure 7 (a) compared to figure 7 (b) and figure 7 (c). One of the factors that can be weighed into the discussion was the presence of high dielectric constant of water as can be observed through the anomalies features. The dielectric constant of PVC pipe, dry sand, wet sand, and water is 3, 3-5, 30-60 and 80 respectively [8]. The intensity of hyperbolic reflection pattern is highest in figure 7 (a), as comparing to figure 7 (b) and figure 7 (c). Greatest water pressure leads to highest water velocity, hence the water flow rate is the fastest through smallest hole diameter. The phenomena then leads more hyperbola pattern can be produced during water leaking. Penetration water reaches up to 1.5 m as depicted in figure 7 (a), (b) and (c) through the GPR scanning. Hence, the argument that suggested GPR is capable on detecting water leakage was validated.

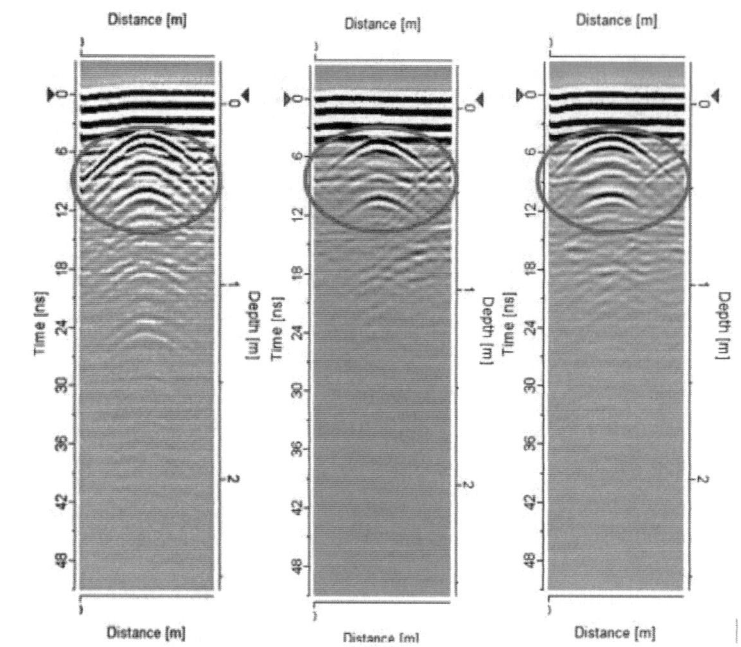

(a) (b) (c)

Figure 7. GPR radar profile of (a) ¼ inch hole diameter, (b) ½ inch hole diameter, (c) ¾ inch hole diameter.

4. Conclusion

The performance of ground penetrating radar on detecting water leaking was ascertained through the various water leakage experiments. It was revealed that the water leaks reflections was generated as similar with hyperbolic anomalies. Greatest water pressure was occurred in small hole diameter which is the ¼ inch hole diameter. The radargram pattern of underground water leakage using ¼, ½ and ¾ inch hole diameter results was clearly observed. The goal of the experiment has achieved its objective. It proves that the GPR is one of the best equipment to detect water pipe leakage without having to dig the ground and causing damages to the surrounding structures.

5. References

[1] Hunaidi O 2000 Detecting Leaks in Water Distribution Pipes *Construc. Tech. Update* **40** pp 1-6

[2] Crocco L, Soldovieri F, Millington T and Cassidy N J 2010 Bistatic Tomographic GPR Imaging for Incipient Pipeline Leakage Evaluation *Progress In Electromagnetics Research PIER* **101** pp 307-321

[3] Conyers L B 2013 *Ground Penetrating Radar for Archaeology* Rowman and Littlefield Publishers Inc.

[4] Lai W W L, Ho M L Y, Chang R K W, Sham J F C and Poon C-S 2017 Tracing and Imaging Minor Water Seepage of Concealed PVC Pipe in a Reinforced Concrete Wall by High-Frequency Ground Penetrating Radar *Construction and Building Materials* **151** pp 840-847

[5] Eyuboglu S, Mahdi H and Shukri H A 2003 Detection of Water Leaks Using Ground Penetrating Radar *Geophysics 2003 : Conf. on Appl.Geophysics* pp 8-12

[6] Yelf R J 2007 Application of Ground Penetrating Radar to Civil and Geotechnical Engineering *Electromagnetic Phenomena* **7 1** pp 102-117

[7] Hunaidi O, Chu W, Wang A, and Guan W 2000 Detecting Leaks in Plastic Water Distribution Pipes *J. AWWA* **92(2)** pp 88-94

[8] Davis J L and Annan A P 1989 Ground Penetrating Radar for High Resolution Mapping of Soil and Rock Stratigraphy *Geophysical Prospecting* **37** pp 531-551

Acknowledgements
The authors would like to acknowledge the financial support from MOSTI and Agensi Nuklear Malaysia as well as the Material and Structure Integrity Group (NDT-MSI) for the support.

IOP Conf. Series: Materials Science and Engineering **298** (2018) 012003 doi:10.1088/1757-899X/298/1/012003

Investigation of dielectric constant variations for Malaysians soil species towards its natural background dose

Khawarizmi Mohd Jafery [1,a], Zaidi Embong [1, 4,b], Yee See Khee[2,5], Samsul Haimi Dahlan[2,5], Saiful Azhar Ahmad Tajudin[3,4], Salawati Ahmad[6], Siti Kudnie Sahari[7], Omeje Maxwell[8]

[1]Faculty of Applied Science and Technology, Universiti Tun Hussein Onn Malaysia, Pagoh Campus, KM 1, Jalan Panchor, 84600 Muar, Johor, Malaysia
[2] Faculty of Electrical and Electronic Engineering, Universiti Tun Hussein Onn Malaysia
[3]Faculty of Civil and Environmental Engineering, Universiti Tun Hussein Onn Malaysia, Malaysia
[4]Research Center for Soft Soil (RECESS), Office for Research, Innovation, Commercialization and Consultancy Management (ORICC), Universiti Tun Hussein Onn Malaysia
[5]Research Centre for Applied Electromagnetic (EMCenter), Office for Research, Innovation, Commercialization and Consultancy Management (ORICC), Universiti Tun Hussein Onn Malaysia
[6]Centre for Language Studies, Universiti Tun Hussein Onn Malaysia
[7]Faculty of Engineering, Universiti Malaysia Sarawak (UNIMAS), 94300 Kota Samarahan, Sarawak, Malaysia
[8]College of Science and Technology, Department of Physics, Covenant University, Ota, Ogun State, Nigeria

[a]khawarizmijafery@hotmail.com
[b] zembong@gmail.com

Abstract. The correlation of natural background gamma radiation and real part of the complex relative permittivity (dielectric constant) for various species Malaysian soils was investigated in this research. The sampling sites were chosen randomly according to soils groups that consist of sedentary, alluvial and miscellaneous soil which covered the area of Batu Pahat, Kluang and Johor Bahru, Johor state of Malaysia. There are 11 types of Malaysian soil species that have been studied; namely Peat, Linau-Sedu, Selangor-Kangkong, Kranji, Telemong-Akob-Local Alluvium, Holyrood-Lunas, Batu Anam-Melaka- Tavy, Harimau Tampoi, Kulai-Yong Peng, Rengam-Jerangau, and Steepland soils. In-situ exposure rates of each soil species were measured by using portable gamma survey meter and ex-situ analysis of real part of relative permittivity was performed by using DAK (Dielectric Assessment Kit assist by network analyser). Results revealed that the highest and the lowest background dose rate were 94 ±26.28 μR hr^{-1} and 7 ±0.67 μR hr^{-1} contributed by Rengam Jerangau and Peat soil species respectively. Meanwhile, dielectric constant measurement, it was performed in the range of frequency between 100 MHz to 3 GHz. The measurements of each soils species dielectric constant are in the range of 1 to 3. At the lower frequencies in the range of 100 MHz to 600 MHz, it was observed that the dielectric constant for each soil species fluctuated and inconsistent. But it remained consistent in plateau form of signal at higher frequency at range above 600 MHz. From the comparison of dielectric properties of each soil at above 600 MHz of frequency, it was found that Rengam-Jerangau soil species give the highest reading and followed by Selangor-Kangkong species. The average dielectric measurement for both Selangor-Kangkong and Rengam-Jerangau soil species are 2.34 and 2.35 respectively. Meanwhile, peat soil species exhibits the lowest dielectric measurement of 1.83. It can be clearly seen that the pattern of dielectric measurement for every soil at the frequency above 600 MHz demonstrated a specific distribution which can be classified into two main regions which are higher and lower between the ranges of 1.83 to 2.35. Pearson correlation analysis between the frequency of 100 MHz and 2.6 GHz with respect to exposure rate for every soil species was r = 0.38 and r = 0.51, respectively. This indicates that there was no strong correlation between both parameter, natural background dose and soils dielectric for each soils sample. This factor could be contributed by major and minor elements contained in each soils sample species, especially Ferum, Fe and Silica, Si.

1. Introduction

Studies of natural radioactivity from the environment have become important issues around the world. The dose of radioactive received by human population is mainly contributes by radionuclides primordial and cosmic rays, thoron, radon and their progeny through food and drinks [1]. Moreover, the radiation exposure from the natural background is mainly depending on geographical features at a certain location [2,3]. In Malaysia, granite and sedimentary residual cover, most of the lands except the coastal areas of soft clay predominate in the area [4].

Naturally, the gamma radiation were emitted from radioisotopes, such as Uranium, ^{238}U and Thorium, ^{232}Th which exist at level in all grounds formations. However, many researchers have confirmed that the radiation dose is influence by the geological features and also together with the concentration of parent radionuclide, Uranium and Thorium in the soil [3,5–7]. The decay of both parent radionuclides (U and Th) in rocks or soils will lead to the emitting of gamma radiation to the environment and its concentrations in rocks and soils could lead external irradiation to human.

One of the soil properties that needs to be considered is its dielectric properties. Soil dielectric is one of the significant parameters in soil science and its influence with natural background is unclear. Many soil dielectric study had been claimed that the moisture content of vigorously influenced the soil dielectric properties due to the presence of water [8–11]. The study of these variables was conducted at microwave frequencies and many different measurement can be used for dielectric permittivity measurement [12]. Permittivity measurement consists of real part permittivity and imaginary part of permittivity. Real part relative permittivity was the ability of material to store electromagnetic energy [13] while imaginary part was represent the loss factor and attenuation of energy in the material [14]. Meanwhile, real part relative permittivity also called dielectric constant were focussed in this study. Dielectric permittivity was also related to polarization of the material to adapt the electric field when it is applied [15–17].

Even though many studies have been conducted in the soil science area, information regarding natural radiation is still lacking. Therefore, to contribute new knowledge to this area, the current study on the effect of soil dielectric toward natural background radiation is carried out. In soil science field of study, the dielectric properties cover several parameters such as resistivity, permittivity and dielectric constant [18]. In additions, most studies on the dielectric are associated with the elemental measurement of water content in the soil. As a result, the study of soil dielectric effects in the direction of natural background radiation was very significant to expose the cause other than geological formation in the study of natural radiation.

2. Material method

2.1. Soil sampling at study area

The soil sampling was conducted in several districts in Johor, Malaysia which are Batu Pahat, Kluang and Johor Bahru. Eleven species of soil have been identified and all the soil sample was measured its gamma radiation levels at the sampling site and then several kilograms of soil are used as samples to be transported to the laboratory for the next process. Each soil species sampling site was determined its background radiation reading from the portable gamma ray survey meter. Gamma radiation measurement was taken at three point for each point in order to achieve good statistics. Each sampling location was identified based its soil group as shown in table 1. The coordinate points for each sampling location were determined using a Garmin GPSMAP model 62s to record the latitude and longitude of each location.

The best exposure rate soil samples were put into container, identified with the label species and brought into laboratory for sample preparation before analysis began. Subsequent, the soil samples were dried in oven for overnight in a temperature-controlled oven at 110℃. After that, the sample were grinded into powdered and sieved through mesh size 200 µm. From that, the clay and minerals particles can be regulated to powder size. Table 1 shows the various soil species were collected according to its coordinates based on the Soil of Map [19].

Table 1. The list of several soil species with the GPS coordinate and FAO/UNESCO Legend [20][21].

Soil Names (local name)	Soil Type [22]	Soil Group	Sampling location	GPS coordinate	FAO/UNESCO Legend
Kranji	silt	Alluvial soil	Minyak Beku, Batu Pahat	1°47'50.8"N , 102°53'50.2"E	Thionic Fluvisols
Selangor-Kangkong	silty clay	Alluvial soil	Kg. Parit Selangor, Pontian	1°29'34.0"N , 103°24'07.0"E	Vertic Cambisols-eutric Gleysols
Harimau-Tampoi	sandy clay	Alluvial soil	Kg Cahaya Masai, Pasir Gudang	1°33'24.6"N 103°54'39.4"E	Ferric Acrisols
Telemong Akob	silty clay	Alluvial soil	Sungai Machap, Machap	1°52'27.0"N 103°16'40.3"E	Dystric Fluvisols-Dystric Gleysol
Gambut (Peat)	peat	Alluvial soil	Kg Parit Nipah Laut, Batu Pahat	1°49'59.4"N 103°09'48.8"E	Thionic Fluvisols
Linau Sedu	silt	Alluvial soil	Kg Parit Jelutong, Batu Pahat	1°50'46.2"N 103°04'46.9"E	Thionic Fluvisols
Holyrood Lunas	sand	Alluvial soil	Kg Baru, Sri Gading	1°50'39.9"N 103°02'28.2"E	Xanthic Ferralsol-Dystric Gleysols
Rengam-Jerangau	sandy clay	Sedentary soil	Simpang Rengam	1°51'42.2"N 103°21'93.9"E	Dystric Nitosols-Orthic Ferralsols
Kulai-Yong Peng	sand	Sedentary soil	Gelang Patah, Nusajaya	1°25'91.7"N 103°35'35.7"E	Rhodic Nitosols-Ferric Acrisols
Batu Anam-Melaka-Tavy	silty clay	Sedentary soil	Taman Perdana, Ayer Hitam	1°55'23.1"N 103°10'38.7"E	Orthic Acrisols-Plinthic Ferralsols
Steepland	sandy clay	Highland soil	Bukit Perdana, Batu Pahat	1°50'50.1"N 102°57'38.6"E	Steepland

From table 1, it can be seen that types of soil differ according to its specific location. For example, alluvial soil can be found near the riverine area or coastal areas. However, sedentary soils can be found mostly at the idled area. Alluvial and sedentary represents the group of the soil which located nearby the riverine area or and coastal area while sedentary soil formed in the midland of the Peninsular area which consists of wide range of rocks [23]. Each soil species were classified based on FAO Legend which represent the lateral cross section layer of each soil [20][24] and the soil texture was claimed by [22].

2.2. Dielectric Measurement Setup

In this current study, the real part permittivity of every soil species was measured in the range of 100 MHz to 3 GHz EM wave using Dielectric Assessment Kit (DAK). The open-ended coaxial probe technique as a base technique is used in this system and its probe was manufactured for fast and precision purposes. Furthermore, its offers a high-dielectric measurement of complex permittivity, conductivity and also loss tangent. DAK will be interfaced with vector network analyser (VNA) to measure the scattering parameter in the open-ended probe. It was assisted to DAK as a computer system that calculate the dielectric constant of samples. DAK was also a technique measurement for dielectric suitable for liquids, solids, and semi –solids [25][26].

Prior to dielectric measurement, the sample need to be process into fine powder to and passes through a 2 mm sieve to exclude present larger particle that are less reactive [27]. The dielectric probe will be put in the corresponding container to the size of the probe to prevent from the disturbance during the measurement. Prior the measurement were conducted, this equipment will be undergoing calibration procedure that involves the measurement of air (open), a coaxial cable connection (short) and deionized water measurement needed for storing the standard material reference.

3. Discussion of results

All the discussion of the results were based on exposure rate of gamma survey meter and dielectric constant measurement for every each of soil species. All data are presented in table 2 and figure 1 - 3

3.1. Exposure rate of various soil species

The background reading was taken at four to five points for each of soil species location and the highest exposure rate was taken as a sample. A sampling location was determined by using GPS based on the Reconnaissance Soil Map released by the Department of Agriculture, Malaysia [19] and recorded in table 2 below.

Table 2. A detail of soil type, GPS coordinate and survey meter reading

Soil Types	GPS	Survey Meter Reading μR hr $^{-1}$
Kranji	1°47'50.8"N, 102°53'50.2"E	15 ± 0.58
Selangor-Kangkong	1°29'34.0"N, 103°24'07.0"E	14 ± 1.45
Harimau-Tampoi	1°33'24.6"N, 103°54'39.4"E	20 ± 2.84
Telemong Akob	1°52'27.0"N, 103°16'40.3"E	16 ± 0.57
Holyrood Lunas	1°50'39.9"N, 103°02'28.2"E	16 ± 1.20
Peat	1°49'59.4"N, 103°09'48.8"E	7 ± 0.67
Linau Sedu	1°50'46.2"N, 103°04'46.9"E	16 ± 3.12
Rengam-Jerangau	1°51'42.2"N, 103°21'93.9"E	94 ± 26.28
Kulai-Yong Peng	1°25'91.7"N, 103°35'35.7"E	16 ± 2.97
Batu Anam-Melaka-Tavy	1°55'23.1"N, 103°10'38.7"E	24 ± 3.05
Steepland	1°50'50.1"N, 102°57'38.6"E	56 ± 3.33

All the sampling location were determined at Batu Pahat, Kluang and Simpang Rengam and Johor Bahru. Roughly the range of the survey meter reading for all soil sample were in range of 7 μR hr $^{-1}$ to 94 μR hr $^{-1}$. Based on the table 2, it is clearly seen that the highest readings are 94 ± 26.28 μR hr $^{-1}$ for Rengam Jerangau soil types followed by the second highest is Steepland, 56 ± 3.33 μR hr $^{-1}$ while for the lowest readings are 7 ± 0.67 μR hr $^{-1}$ which is Peat that collected in the area of oil palm plantation at Parit Nipah Laut, Batu Pahat. Significant differences are seen between the highest reading and lowest reading around 87 μR hr $^{-1}$. In addition, there are four types of soil, Telemong Akob, Holyrood Lunas, Linau Sedu and Kulai -Yong Peng exhibit the similar reading volume of 16 μR hr $^{-1}$ but in the different sampling location. Nevertheless, the parent material that exists in every each soil species also influences the radioactivity level [20]. For example, the originality of soil that come from granite will decay higher radioactivity level compared to other family of rocks and according to Badawy [28] effective dose rate of clay soil were higher compared to sandy soils due to compositions of ^{226}Ra, ^{232}Th and ^{40}K. This results agree to peat soil that the compose from partially decay plant have a lower radioactivity [20,29]. It can be conclude that the Rengam Jerangau have contain a large scale of granite compared to other soil types.

3.2. Dielectric constant measurement for each soil species
Dielectric constant measurement are present in this subtopic were based on wide range measurement frequency 100 MHz to 3 GHz, comparison between certain frequency (100 MHz, 600 MHz and 3 GHz) and correlation between exposure rates and dielectric constant for every soil species.

3.2.1. Dielectric Constant at Wide Range of Frequency from 100 MHz to 3 GHz. The dielectric constant for every soil species were performed in this research by using vector network analyser (VNA) and assisted by DAK probe and its software for signal processing. Figure 1 shows the polynomial dielectric constant graph for each soil type in the wide range frequency from 100 MHz to 3 GHz. The dielectric constant of different soils are different due to the composition of the soil. As shown in the graph, the dielectric constant was found to be linear in between 100 MHz to 600 MHz. In the present discovery, there is good agreement with [30] which discussing ground state at low frequency is the occurrence of frequency distortion due to oriental polarizations. These mechanisms were referred to the positive and negative ion were aligned to the electric field applied to the material. After entering the 600 MHz value, the results produced by all soil types are in the relaxation process. This stated that the dipole moment in the particle were completely aligned the direction of the applied electric field and its dominant frequency by the ionic conduction in that frequency range [31]. The graph depict that the results generated by this tool have resulted in two different areas at the top (2.2-2.5) and the bottom (1.6-2.1). At 3 GHz, the above group consists of Steepland, Kranji, Rengam Jerangau, Harimau Tampoi Selangor Kangkong, while the other are Peat, Linau Sedu, Telemong Akob, Holyrood Lunas and Batu Anam Melaka Tavy. In addition, the below area was produced when the value entered 600 MHz frequency readings. Finally, from 600 MHz to 3 GHz it can be clearly seen that the range of dielectric constant was stable.

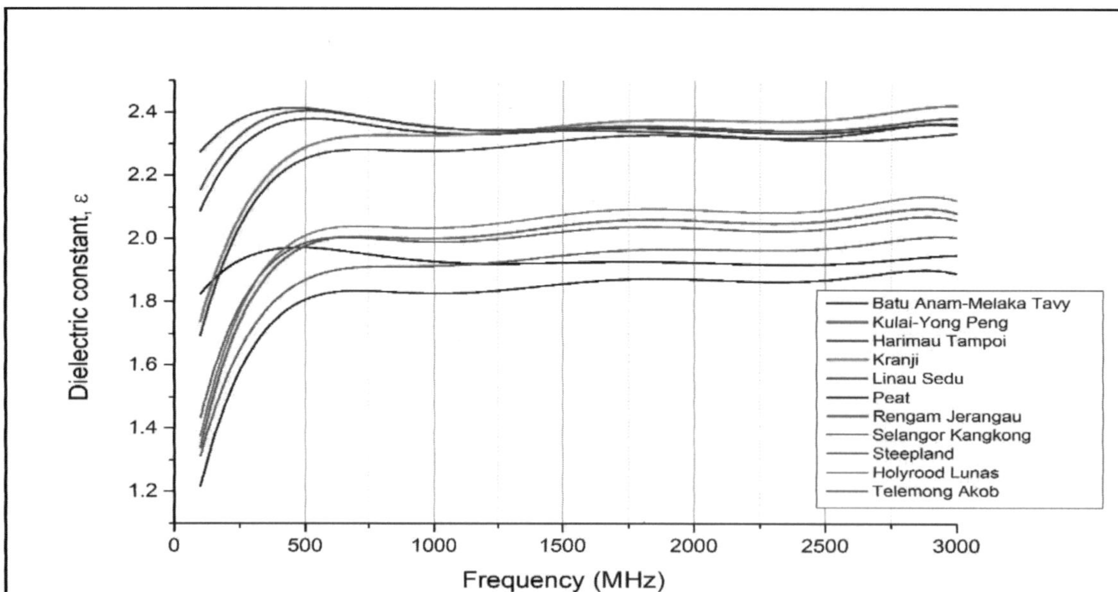

Figure 1. The wide range frequency of the dielectric constant for every soil species.

3.2.2. The comparison the value of dielectric constant at frequency 100 MHz, 600 MHz and 3 GHz. In this section, the comparison value of dielectric constant of each soil at frequency of 100 MHz, 600 MHz and 3 GHz will be discussed as follow. These three frequencies are emphasized because they show importance in this study, where the 100 MHz frequency is a low initial setting and the situation was strong in relation to polarization of particles caused by positive and negative ions in soils. While the results start to show stability at 600 MHz until 3 GHz. At 3 GHz, was the highest frequency value produced by VNA and roughly the results of all three are different. At 100 MHz, the dielectric constant of the type of Selangor Kangkong land of 2.50 was highest. The lowest was Kulai-Yong Peng, 0.99. Next, to the 600 MHz frequency, the results for each land showed that Harimau Tampoi explained that it was the highest reading value of 2.33 and the lowest was Peat 1.79. At the highest frequency value of 3 GHz, it appears an oddity occurs where the Kranji soil type dielectric constant was the highest at 2.44. The questioned was that the exposure rate for Kranji is the lowest in table 1 and it have a high value of dielectric constant at frequency 3 GHz. For Peat soil, the results for all three frequency values were the lowest. This also influence by the factor that affecting in soil dielectric constant measurement which is the soil composition in the soil beside moisture content [32]. It specifies the soil compositions in the soil contain a diversity of elements such as iron, calcium and etc. That could influence the measurement of dielectric constant [33][34]

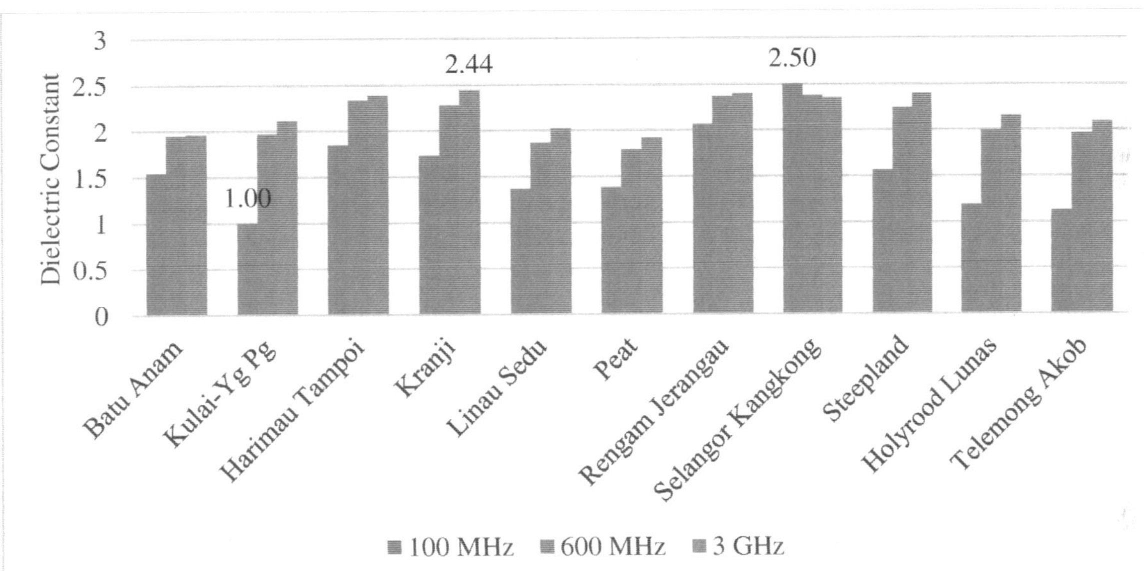

Figure 2. The dielectric constant comparison between 100 MHz, 600 MHz and 3 GHz for every soil species

3.2.3 The correlation between natural background radiation and dielectric constant of various soil species. The effects natural background radiation and dielectric constant were analysed through linear correlations and obtained results are presented in figure 3 (a-b). The linear correlation shows a statistical result of either the association between the two variables by yielding in the -1 to 1 [35]. Figure 3a, displays the correlation natural background radiation and the dielectric constant at 100 MHz are weak when observed from the correlation results, $r = 0.38$. Although for 2.6 GHz frequency (Fig. 3b), the correlation results are moderate, $r = 0.51$. From a statistical point of view, the relationship between the two variables is very weak because strong correlation must have a value greater than 0.8 [35]. This obtain that the occurrence of anomalies between radiation levels and dielectric constant of soil cannot be assume that the high level exposure rate also have a high value dielectric constant. Moreover, there were many factors that can influence the measurement of dielectric constant such as temperature ,samples, humidity and size of sample [36] .

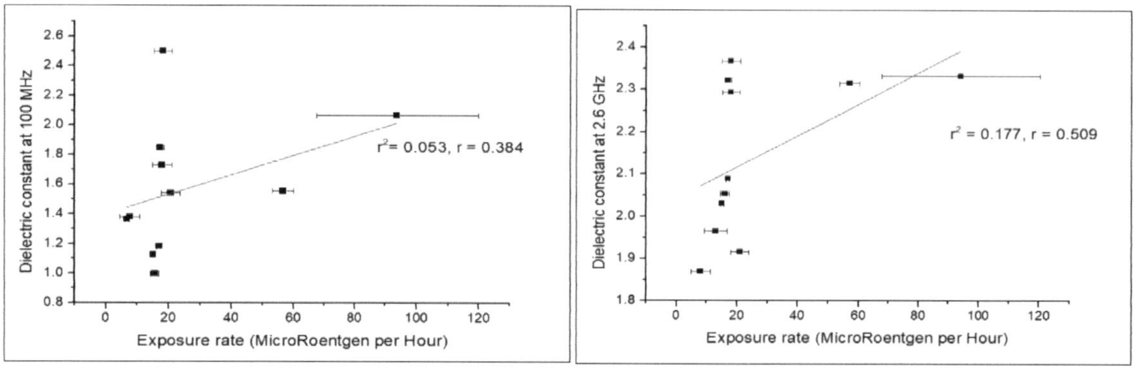

(a) (b)

Figure 3 (a-b). Correlation analysis between dielectric constant and gamma exposure rate for frequency 100 MHz and 2.6 GHz

4. Conclusion

Study on the correlation between the natural background radiation and dielectric constant for various species of soil was conducted.

- The highest dose readings recorded was soil series Rengam Jerangau, 94 ± 26.28 μR hr $^{-1}$ and lowest was Peat 7 ± 0.67 μR hr $^{-1}$.
- Good dielectric constant reading does not affect exposure dose rate readings for each soil. For example, the Rengam Jerangau soil series that has high exposure dose rate readings but the reading value of the dielectric constant is the opposite.
- The results of correlation between value dielectric constant and gamma exposure rate was not solid enough to prove it if they have correlations because of some other factor such as moisture content, soil compositions and temperature also influence the dielectric constant measurement.

5. References

[1] Ahmad N, Jaafar M S, Bakhash M and Rahim M 2015 An overview on measurements of natural radioactivity in Malaysia *J. Radiat. Res. Appl. Sci.* **8** 136–41

[2] Johar S M, Embong Z and Dalimin M N 2012 A Study of Natural Background Radiation and pH Level Distribution of Malaysian Soil Species along Batu Pahat and Kluang, Johor

[3] Ramli A T, Hussein A W M . and Lee M. 2001 Geological influence on terrestrial gamma radiation dose rate in the Malaysian State of Johore *Appl. Radiat. Isot.* **54** 327–33

[4] Salih A G 2012 Review on granitic residual soils' geotechnical properties *Electron. J. Geotech. Eng.* **17** 2645–58

[5] Saleh M A, Ramli A T, Alajerami Y, Mhareb M H A, Aliyu A S, Gabdo H T and Garba N N 2014 Assessment of radiological health implicat from ambient environment in the Muar district, Johor, Malaysia *Radiat. Phys. Chem.* **103** 243–52

[6] Sanusi M S M, Ramli A T, Basri N A, Heryanshah A, Said M N, Lee M H, Wagiran H and Saleh M A 2017 Thorium distribution in the soils of Peninsular Malaysia and its implications for Th resource estimation *Ore Geol. Rev.* **80**

[7] Saleh M A, Ramli A T, Hamzah K bin, Alajerami Y, Moharib M and Saeed I 2015 Prediction of terrestrial gamma dose rate based on geological formations and soil types in the Johor State, Malaysia *J. Environ. Radioact.* **148** 111–22

[8] Bittelli M 2011 Measuring soil water content: A review *Horttechnology* **21** 293–300

[9] Gadani D H, Vyas A D and Rana V A 2014 Dielectric properties of wet and fertilized soils at radio frequencies *Indian J. Pure Appl. Phys.* **52** 399–410

[10] Curtis J O 2001 Moisture Effects on the Dielectric Properties of Soils **39** 125–8

[11] Peplinski N R, Ulaby F T and Dobson M C 1995 Dielectric Properties of Soils in the 0.3–1.3-GHz Range *IEEE Trans. Geosci. Remote Sens.* **33** 803–7
[12] Abdelgwad A H and Said T M 2016 Measured dielectric permittivity of contaminated sandy soil at microwave frequency *J. Microwaves, Optoelectron. Electromagn. Appl.* **15**
[13] Navar Khele V V, Shaikh A A and Ramshetti R S 2009 Dielectric properties of black soil with organic and inorganic matters at microwave frequency *Indian J. Radio Sp. Phys.* **38** 112–5
[14] Tzortzis Tsertos, H. M 2004 Determination of thorium, uranium and potassium concentrations in surface soils in Cyprus *J. Environ. Radioact.* **77** 325–38
[15] Dospatliev L K, Ivanov I T, Paarvanova B K, Katrandzhiev N T and Popova R T 2014 Determining the relationship between the dielectric properties and the basic physical and chemical parameters of the air-dry soil *Int. J. Sci. Res. Publ.* **4** 1–7
[16] Neelakanta P S 1995 *Handbook of Electromagnetic Materials: Monolithic and Composite Version and Their Applications* (New York: CRC Press)
[17] Sternberg B K and Levitskaya T M 2001 Electrical parameters of soils in the frequency range from 1 kHz to 1 GHz, using lumped-circuit methods *Radio Sci.* **36** 709–19
[18] Sternberg B K and Levitskaya T M 2001 Electrical parameters of soils in the frequency range from 1 kHz to 1 GHz, using lumped-circuit methods *Radio Sci.* **36** 709–19
[19] Division of Agriculture. Ministry of Agriculture and Fisheries 1968 *Reconnaissance Soil Map of Peninsular Malaysia. Sheet 1. Series L 40A.*
[20] Saleh M A, Ramli A T, Alajerami Y and Aliyu A B S 2013 Assessment of Natural Radioactivity Levels and Associated Dose Rates in Soil Samples From Northern *J. Ovonic Res.* **9** 1–6
[21] Johar S M, Embong Z and Tajudin S A A 2016 The gamma dose assessment and pH correlation for various soil types at Batu Pahat and Kluang districts, Johor, Malaysia *AIP Conf. Proc.* **1704**
[22] Pushrajah E and Amin L L 1977 *Soils under Hevea in Peninsular Malaysia and their Management* (Kuala Lumpur: Rubber Research Institute of Malaysia)
[23] Chee W C and Peng C C 2006 *Country Pasture/Forage Resource Profiles MALAYSIA* (Malaysia)
[24] Driessen P M 2001 *Lecture notes on the major soils of the world* (Food and Agriculture Organization of the United Nations)
[25] Scmid & Partner Engineering AG 2010 Dielectric Assesment Kit (DAK): High Precision Dielectric Measurements 2
[26] AG S & P E 2016 *DAK Professional Handbook V2.4*
[27] Boone R D, Grigal D F, Sollins P, Ahrens R J and Armstrong D E 1999 Soil sampling, Preparation, Archiving and Quality Control *Stand. Soil Methods Long-Term Ecol. Res.* 3–28
[28] Badawy W M 2009 Natural radioactivity of clay and sandy soils and radiation exposure doses in the Heet and Inshass regions of Egypt *Moscow Univ. Soil Sci. Bull.* **64** 105–7
[29] Johar S M, Embong Z, Azhar S and Tajudin A 2016 The gamma dose assessment and pH correlation for various soil types at Batu Pahat and Kluang districts, Johor, Malaysia *Appl. Phys. Lett* **50007**
[30] Oh M, Kim Y and Park J 2007 Factors affecting the complex permittivity spectrum of soil at a low frequency range of 1 kHz-10 MHz *Environ. Geol.* **51** 821–33
[31] Nelson S O 2008 Dielectric properties of agricultural products and some applications *Res. Agric. Eng.* **54** 104–12
[32] Liu J, Zhao S, Jiang L, Chai L and Wu F 2013 he Influence Of Organic Matter On Soil Dielectric Constant At Microwave Frequencies 1–4
[33] Maher B A 1986 Characterisation of soil mineral by magnetic measuremnet *Physicis Earth Planet.* **42**
[34] Robinson D A, Bell J P and Batchelor C H 1994 Influence of iron minerals on the determination of soil water content using dielectric techniques *J. Hydrol.* **161** 169–80
[35] Pandis N 2016 Correlation and linear regression *Am. J. Orthod. Dentofac. Orthop.* **149** 298–9

[36] Calla P N, Ranjan V, Bohra C, Naik G L, Hasan W, Sin H and Bali G 2004 Estimation of dielectric constant of soil from the given texture at microwave frequency **33** 196–200

Acknowledgments
This paper was funded by the Centre for Graduate Studies, University Tun Hussein Onn Malaysia. The authors also acknowledge for the Research centre of Applied Electromagnetic (EMCenter, UTHM) for assistance with DAK for dielectric constant measurement and Research centre for Soft Soil (RECESS, UTHM) for soil work process

IOP Conf. Series: Materials Science and Engineering **298** (2018) 012004 doi:10.1088/1757-899X/298/1/012004

The effect of gamma irradiation on chemical, morphology and optical properties of polystyrene nanosphere at various exposure time

Jibrin Alhaji Yabagi,[1,2], Mohammed Isah Kimpa[1,3], Muhammad Nmayaya Muhammad[1,2], Saiful Bin Rashid[4], Embong Zaidi[1] and Mohd Arif Agam[1*]

[1] Department of Science, Universiti Tun Hussein Onn Malaysia Pagoh Campus, 84600 Pagoh, Muar, Johor, Malaysia

[2] Department of Physics, Ibrahim Badamasi Babangida University Lapai Niger State Nigeria 286 Lapai

[3] Department of Physics, Federal University of Technology Minna Niger State Nigeria 09 Bosso

[4] Department of Physics, Faculty of Science, University of Technology Malaysia 81310 Skudai, Johor Malaysia

jibrinbida@yahoo.com

Abstract. Irradiation of polymers causes structural, chemical and the optical properties changes. Polystyrene nanosphere was drop coated to substrates and the gamma irradiation was carried out in a Cesium-137 (Cs-137) source chamber at different time (1-5 hours) with constant dose of 30 kGy. Fourier transformation infrared spectroscopy (FTIR) and Raman spectroscopy were employed to characterize the chemical properties of irradiated polystyrene while Scanning Electron Microscopy (SEM) and Atomic Force Microscopy (AFM) were used to study the surface morphological changes of the samples. The optical energy band gaps of the thin films were investigated and studied using transmittance and absorbance measurements. The results obtained revealed that as irradiation time increases the optical properties changes and polystyrene gradually undergoes crystal to carbonaceous from its amorphous state. The average particles diameter and roughness of the samples decreases with increasing irradiation time.

1. Introduction

Polymers are often exposed in radiation environments such as gamma rays [1][2] laser [3][4][5], X-rays [6] electrons beam [7][8], photonic and ions [9] which may affect the chemical and physical properties of these materials where atomic or molecular excitation and ionization occurred resulting in the changes of chemical bonds, intermolecular cross-linking, formation of free radicals and unsaturated bonds, etc. These processes cause defects and re-arrangement of polymer matrices, which are responsible for most of the changes observed in the physico-chemical properties of polymers [7]. Polymers have been extensively studied as promising materials for integrated optical device due to the broad spectrum of possible applications in data transmission and processing systems [10].

Polymers have become a very attractive for photonic device due to their easily processed; offer high flexibility, tailor made properties and have a relatively low cost, especially polystyrene, PMMA

[11]–[15]. It is known that polystyrene (PS) is an important material in both academic research and industry. PS has a simple chemical structure that consists of both unsaturated aromatic ring and saturated aliphatic chain [16]. Studies have shown that the surface modification of PS, especially the chemical reactions occurred at the unsaturated and saturated bonds. Recently, attention has been paid to modification of PS with respect to micro-fluidics, cell culture and electronic applications. In most works, the effects of gamma or laser treatments on alterations in the polystyrene have been investigated [17]–[19]. Physicochemical changes in this layer, induced by the gamma or laser radiation, have also been reported [20][21].

This work focuses on the effects of over exposure gamma irradiation of polystyrene by chemical, morphology and optical properties of polystyrene nanosphere at various time of exposure. The samples were characterized by means of Fourier Transform Infrared (FTIR), Raman spectroscopy, Scan Microscopy (SEM), Atomic Force Microscopy (AFM) and Ultra violet-visible spectrometer (UV-Vis) measurements. The optical energy band gaps and number of carbon atoms deduced from transmittance spectra are discussed.

2. Experiment

2.1. Methods and characterization

The p-type Silicon wafer was cut into 1 x 1 cm^2 and the wafers were treated in piranha solution (H_2SO_4:H_2O_2 =3:1) for 45 min at 80 ℃, the substrates were then rinsed in deionized water and dried with nitrogen before use [5]. Polystyrene nanospheres with a mean diameter of 500 nm and a concentration of 10 wt% in solution were purchased from Sigma Aldrich and was used without further purification, to achieve a large-area hexagonal close-packed nanosphere monolayer, the PS nanosphere solution was diluted to be 2 wt% with deionized water, then about 10 μL PS nanosphere suspensions with the concentration of 2 wt% was dropped on silicon wafer and allowed to dry at room temperature before gamma irradiation treatment. The gamma irradiation was performed at room temperature in air atmosphere using a Cesium-137 (Cs-137) source and irradiation times were taken as 1-5 hour.

FTIR analysis was performed using a Perkin Elmer FTIR Spectrometer LR 64912C, N3896 equipped with a universal Attenuated Total Reflectance (ATR) sample stage. The sample molecular structure was determined within the range of 4000–400 cm^{-1} with a resolution of 4 cm^{-1}. Raman spectroscopy was carried out using XploRA plus Raman microscope Horiba scientific with 532 nm line of a He-Ne laser as the excitation source. The laser power was maintained at a lower value to avoid laser induced heating of the sample. Roughness of the samples was studies using Atomic Force Microscopy (AFM) XE-100 Park systems. The measurement was carried out using non-contact mode under the room temperature conditions with scan size of 5 μm x 5 μm, high resolution 256 pixels x 256 pixels. Morphology of the samples were obtained using Scanning Electron Microscope (Hitachi SU1510) and coated with platinum before characterization. The optical energy band gap of irradiated PS and non-irradiated PS thin films were studied using the optical transmittance spectra recorded on the samples with the help of UV-Vis-NIR double beam spectrophotometer. The optical transmittance spectrums of samples were recorded in the range of 320-900 nm wavelengths.

2.2. *Data analysis*

2.2.1. *Optical energy band gap.*

The optical energy band gap can be estimated by calculating the optical absorption coefficient (α) from transmittance data obtained using the following relation:

$$\alpha = \frac{1}{d}\ln\left(\frac{1}{T}\right), \quad (1)$$

where T is the transmittance and d is the film thickness. The thin film under study has an absorption coefficient (α) obey the following relation for high photon energies ($h\nu$). The energy band gap was estimated by assuming direct and indirect bands using Tauc`s equation:

$$\alpha h\nu = B(h\nu - E_g)^n \quad (2)$$

where E_g is the band gap of the samples, $h\nu$ is the photon energy, B is band tailing parameter that depends on the electron-hole mobility and n is a constant which determines the type of transition ($n = 1/2$ and $(3/2)$ for direct allowed and forbidden transitions $n = 2$ and 3 for indirect allowed and forbidden transitions respectively) [22], [23].

2.2.2. *Number of carbon atoms* (N). The number of carbon atoms (N) in a cluster is correlated to the optical band gap (E_g) The number carbon atoms per conjugated length N was calculated by modified Tauc`s equation [24]:

$$N = \frac{2\pi\beta}{E_g} \quad (3)$$

where 2β is the energy of a pair of adjacent π sites and β is taken to be ~2.9eV as it is associated with the $\pi \rightarrow \pi^*$ optical transition in the $-C=C-$ structure [1][24].

3. Results and discussions

3.1. *Chemical properties*

The molecular structures of both irradiated and non-irradiated polystyrene with gamma irradiation are performed using FTIR and Raman spectroscopy. FTIR spectra of non-irradiated PS and irradiated PS are shown in figure 1 at various at various time of exposure (a-f). FTIR spectrum of non-irradiated PS have characteristic band at 3100 - 3000 cm^{-1} corresponds to the =C-H stretching due to the aromatic ring, peaks at 2924 and 2846 cm^{-1} is attributed to the symmetrical and asymmetrical stretching vibration of CH_2, bands at 1602, 1490 and 1447 cm^{-1} correspond to stretching vibration of the C=C bond on the benzene ring, band at 1023 cm^{-1} correspond to C-O and bands at 902 cm^{-1} is assigned to the C-H out-of-plane bending vibration of the benzene ring [25]–[27]. Irradiated PS bands are found to be affected, bands reduced in intensity and become broader compared with non-irradiated PS which indicate the structural changes as a result of gamma irradiation. It's clearly shown that intensity of

IOP Conf. Series: Materials Science and Engineering **298** (2018) 012004 doi:10.1088/1757-899X/298/1/012004

band at 1602 cm^{-1} for samples (b-f) reduced due to gamma irradiation and also the band at 2924 cm^{-1} shifted 2951 for 2hr, 2979 cm^{-1} for 3hr and 3044 cm^{-1} for 5hr.

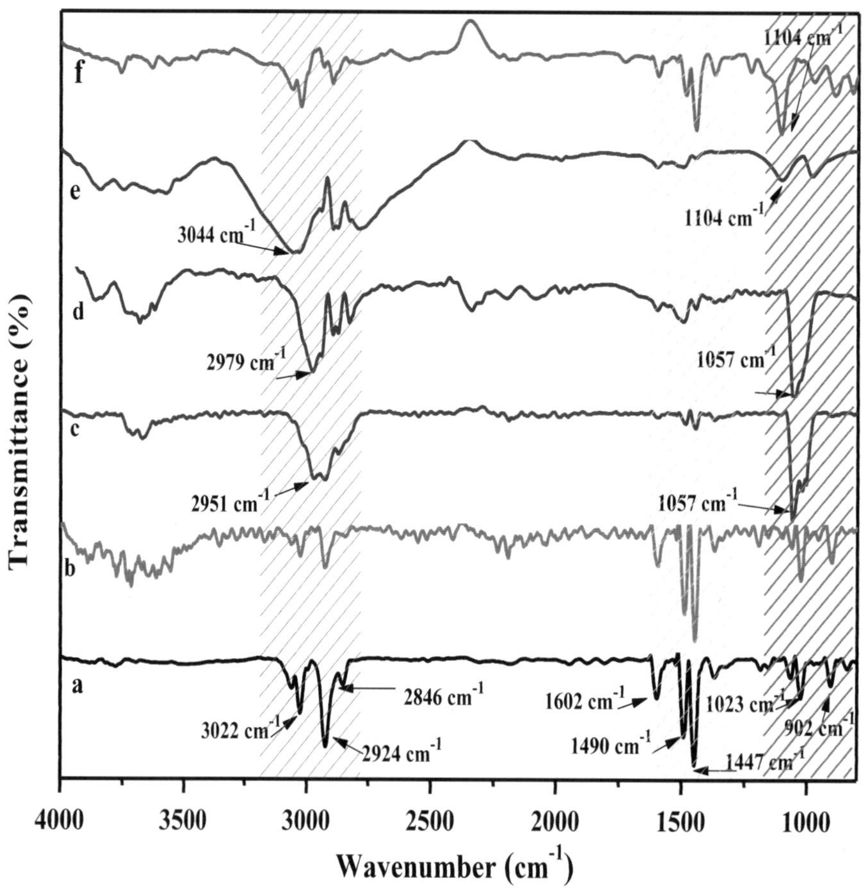

Figure 1. FTIR spectra of polystyrene at various time of exposure (a) non-irradiated PS, (b) 1hr, (c) 2hr, (d) 3hr, (e) 4hr, (f) 5hr gamma irradiation.

Raman analysis was performed using Raman micro spectrometers in backscattering geometry. Raman spectroscopy is a quick and non-destructive technique to analysis mainly carbon samples. Figure 2 (a-f) shows the Raman spectra of both non-irradiated and irradiated PS with gamma irradiation at different interval of time. Spectra peaks of polystyrene are present at 1002 cm^{-1} correspond to the aromatic breathing, peaks at 1030 cm^{-1} assigned to C-H bending, peaks at1602 cm^{-1} modes are the most intense band correspond to the C=C aromatic ring stretching and peak at 620 cm^{-1} assigned to ring deformation [28]–[32]. The non-irradiated PS spectrum reveals that peak at 1602 cm^{-1} is due to stretching of benzene rings, and increase intensity as the irradiation time increases.

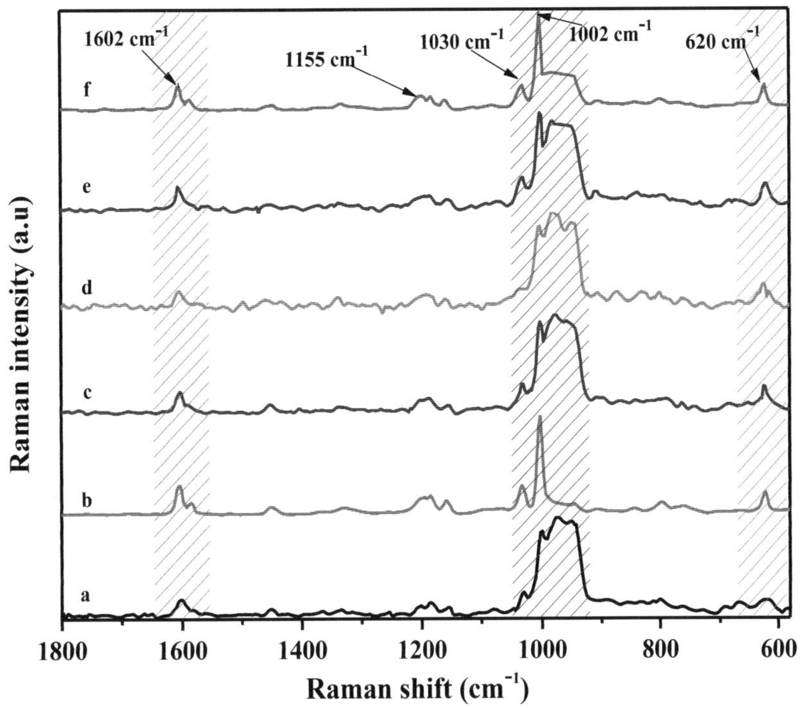

Figure 2. Raman spectra of polystyrene at various time of exposure (a) non-irradiated PS, (b) 1hr, (c) 2hr, (d) 3hr, (e) 4hr, (f) 5hr irradiation.

3.2. Morphology properties

Figure 3 (a-f) show top views of SEM images obtained for both non-irradiated and irradiated polystyrene respectively. Figure 3a shows the non-irradiated surface morphology of the PS film, it can be seen that a self-assembled monolayer of PS coated clearly formed in hexagonal shape arrangement on the silicon substrate. After irradiation at different time of exposure it revealed that the average diameter of individual PS beads reduced in size and shape, as the time of irradiation increase from 1-5 hours. However the original hexagonal lattice was maintained and become more carbonaceous material due to fragmentation formed between PS beads and also create gap in void. As reported in previous work, [4][5][33] the carbonization of polymers will produce volatile gases and the volume of polymers will decrease obviously due to loss of atoms [34]. Therefore evolution of the gap between each polystyrene bead as illustrated in figure 3(b-f).

Atomic force microscopy is a powerful tool for identify both surface morphology of the thin films and obtain information about the roughness, height asymmetries values such as the skewness and kurtosis of the films. The average roughness, root mean square (rms) and height asymmetry values are measured using a software analysis from AFM instrument and the results obtained are depicted table 1. It is appear from the table 1 that average roughness of the samples increases as the exposure time increases.

Figure 3. SEM images of polystyrene at various time of exposure (a) non-irradiated PS (b) 1hr (c) 2hr (c) 3hr (d) 4hr (e) 5hr gamma irradiation.

Figure 4 shows the AFM images obtained from both non-irradiated and irradiated polystyrene at various time of exposure. Figure 4a show that for non-irradiated PS many sharp and regular present over the whole substrate surface. The case is different for 3-5hr gamma irradiation their AFM images show significant change in topography compared to 1hr, 2hr and non-irradiated PS these could be caused by overexposure of gamma irradiation [35]. AFM results are also accorded with SEM results and optical measurements, and further confirm the enhancement of the chemical resistance of PS by gamma irradiation resulted to the formation of carbonaceous material.

Table 1: Statistical parameters of the irradiated polystyrene thin films at various time of exposure

Statistical parameters	Samples					
	Pure PS	1hr	2hr	3hr	4hr	5hr
Averege roughness (nm)	20.4	22.15	20.3	144.1	147.8	126.5
RMS roughness (nm)	23.8	30.5	27.6	162.5	166.9	153.9
Skewness S_{sk}	0.29	1.88	0.98	0.56	-0.48	0.87
Kurtosis S_{ku}	2.39	9.31	4.896	1.84	1.82	2.63

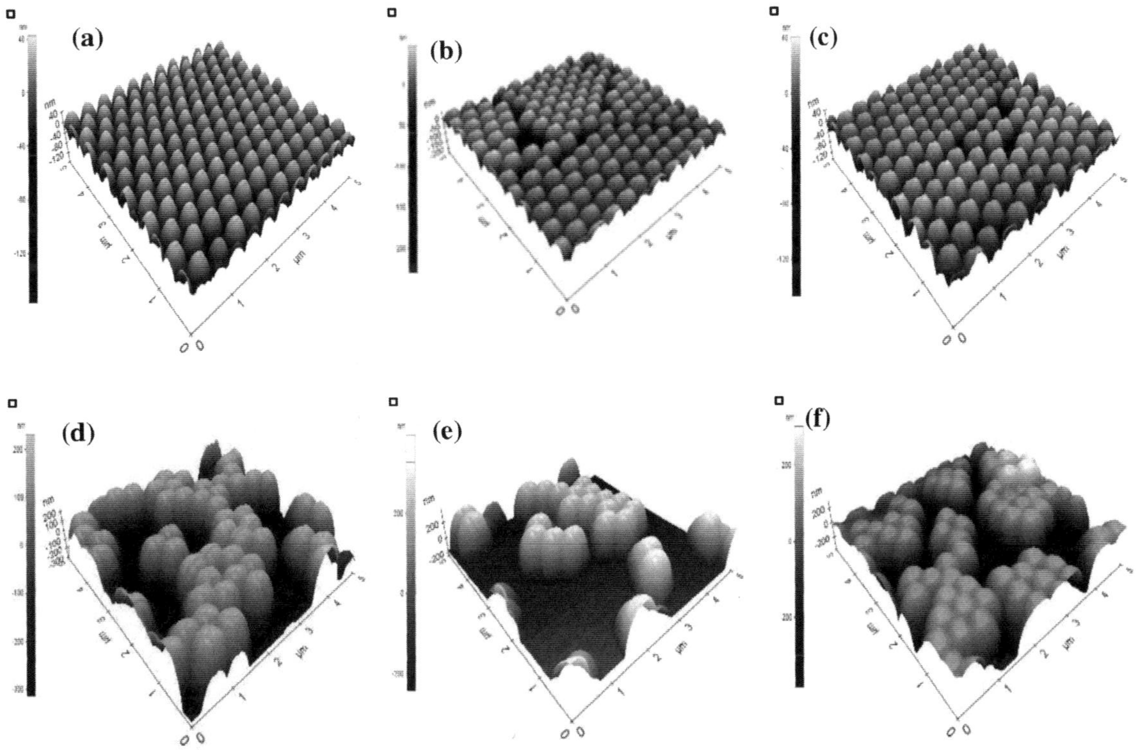

Figure 4: Atomic force microscopy images obtained from polystyrene at various time of exposure (a) topography of non-irradiated PS (b) topography of 1hr (c) topography of 2hr (c) topography of 3hr (d) topography of 4hr (e) topography of 5hr gamma irradiation of PS deposited on silicon substrate.

3.3. Energy band gap

The values of E_g at different time of exposure were estimated by extrapolation of the linear part of the plots to $(\alpha h\nu)^2 = 0$ from the curve of $(\alpha h\nu)^2$ versus $h\nu$ in the high absorption range are shown in figure 5. The values of energy band gaps (E_g) were listed in table 2. From results, the values of the optical energy band gap decreased with increase in time of exposure as observed in figure 5. This decrease in the optical energy values is due to creation of carbon enriched clusters to the partial evolution of hydrogen molecules. Also may be due enhanced diffusion rate of π-electron in the forbidden energy level of polymer formation of defects or creation of carbonaceous cluster. The results are in agreement with the results obtained by Sharma et al. [23].

Table 2: Show the values of estimated energy band gap and number of carbon atoms (N) in a cluster of both irradiated and non-irradiated polystyrene

Samples	Value of estimated energy band gap (eV)	Number of carbon atoms (N) in a cluster
PS	2.02	9
1hr	1.97	9
2hr	1.82	10
3hr	1.80	10
4hr	1.79	10
5hr	1.72	11

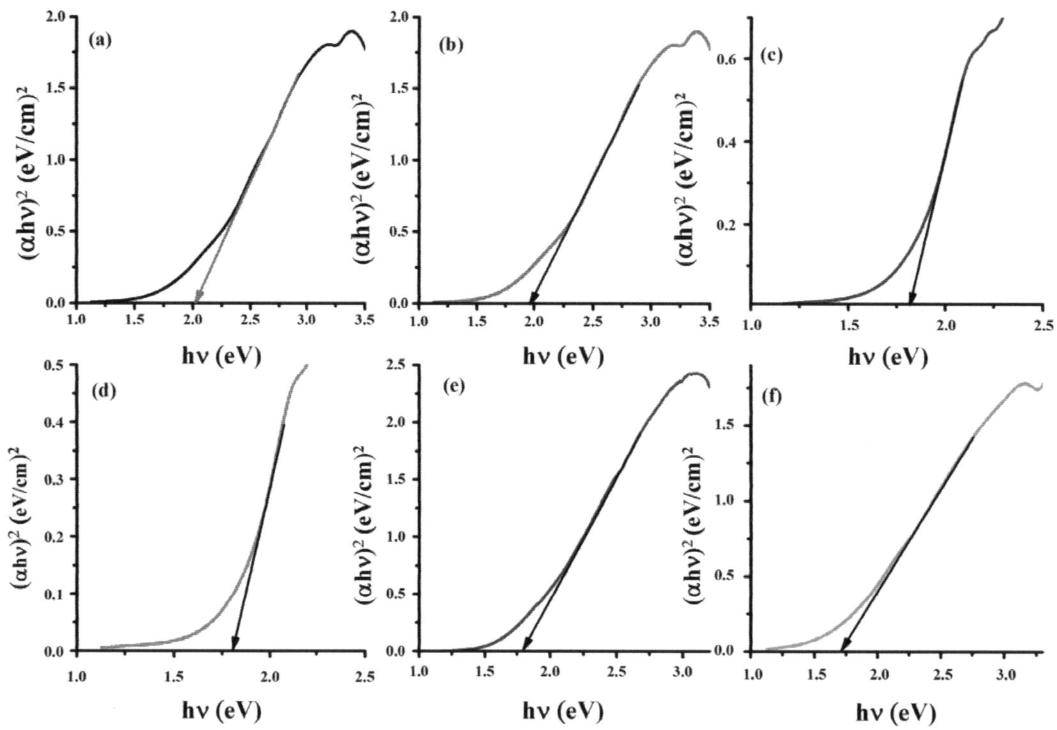

Figure 5: Energy band gap of polystyrene at various time of exposure (a) non-irradiated PS (b) 1hr (c) 2hr (c) 3hr (d) 4hr (e) 5hr gamma irradiation.

3.4. Number of the carbon atoms

The number of the carbon atoms increases with increasing irradiation time as depicted in table 2. This is attributed to the breakage of C-H bonds upon irradiation and the collapse or escape of hydrogen as hydrogen molecules, however, carbonaceous clusters are supposed to rich with charge carriers that enhance the optical properties in radiation bombarded polymers and they also influence the electrical conductivity of such materials [36].

4. Conclusions

Polystyrene nanosphere was successfully prepared by drop coating method onto substrate and treated by gamma irradiation different time (1-5 hours) with constant dose of 30 kGy. In this work, the overexposure of PS by gamma irradiation was systematically changing the morphology, chemical and optical properties of the samples. It was observed that irradiated samples reduce the average particles diameter and change the roughness of the samples as the irradiation time increases. The chemical properties revealed that the chemical resistance of PS by gamma irradiation resulted to the formation of carbonaceous material. The optical band gap of the thin films decreased with the increasing gamma irradiation time and polystyrene gradually undergoes crystallization from its amorphous state.

5. References

[1] Abdul-kader A M, Zaki M F and El-badry B A 2014 Modified the optical and electrical properties of CR- 39 by gamma ray irradiation *J. Radiat. Res. Appl. Sci.* **7** 286–91
[2] Bushra A, Hasan M A S and A A H 2012 Optical Properties of Poly Vinyl Chloride PVC Films Irradiated with Beta and Gamma – *Br. J. Sci.* **7**
[3] Mustafa M U, Agam M A, Md Juremi N R, Mohamad F, Wibawa P J and Ali A H 2011 Physical and chemical changes of polystyrene nanospheres irradiated with laser *AIP Conf. Proc.* **1347** 67–71
[4] Wibawa P J, Saim H, Agam M A and Nur H 2013 Manufacturing and morphological analysis of composite material of polystyrene nanospheres/cadmium metal nanoparticles *Bull. Chem. React. Eng. Catal.* **7** 224–32
[5] Jibrin A.Y, Isah M.K, Muhammad N.M, Kasim I.U Z E and M.A and A 2017 Structural Transformation of Polystyrene Nanosphere Produce Positive and Negative Resists by Controlled Laser Exposure *Adv. Sci. Lett.* **23** 6613–6617
[6] Fahmy A and Friedrich J 2012 Degradation behavior of thin polystyrene films on exposure to Ar plasma and its emitted radiation *J. Adhes. Sci. Technol.* **27** 324–38
[7] Mathad R D, Harish Kumar H G, Sannakki B, Ganesh S, Sarma K S S and Badiger M V. 2009 High energy electron irradiation effects on polystyrene films *Radiat. Eff. Defects Solids* **164** 656–64
[8] Onyiriuka E C 1993 Effects of high-energy radiation on the surface chemistry of polystyrene: A mechanistic study *J. Appl. Polym. Sci.* **47** 2187–94
[9] Terrasi A, Foti G, Hwu Y and Margaritondo G 1991 Valence-band states of ion-bombarded polystyrene *J. Appl. Phys.* **70** 1885
[10] Netcheva S and Bertrand P 1999 Surface topography development of thin polystyrene films under low energy ion irradiation *Nucl. Instruments Methods Phys. Res. B* **151** 129–34
[11] Becker M R, Stefani V, Correia R R B, Bubeck C, Jahja M and Forte M M C 2010 Waveguide optical properties of polystyrene doped with p-nitroaniline derivatives *Opt. Mater. (Amst).* **32** 1526–31
[12] Abdelghany A M, El-Damrawi G, ElShahawy A G and Altomy N M 2017 Structural Investigation of PVC/PS Polymer Blend Doped with Nanosilica from a Renewable Source *Silicon*
[13] Han Y, Wu Y, Shen M, Huang X and Zhu J 2013 Preparation and properties of polystyrene nanocomposites with graphite oxide and graphene as flame retardants *J. Mater.*
[14] Song X, Dai Z, Xiao X, Li W, Zheng X and Shang X 2015 Formation of carbonized polystyrene sphere/hemisphere shell arrays by ion beam irradiation and subsequent annealing or chloroform treatment *Sci. Rep.*
[15] Alsharaeh E H 2016 Polystyrene-poly(methyl methacrylate) silver nanocomposites: Significant modification of the thermal and electrical properties by microwave irradiation *Materials (Basel).* **9**
[16] Xu X, Kwok R W M and Lau W M 2006 Surface modification of polystyrene by low energy hydrogen ion beam *Thin Solid Films* **514** 182–7
[17] Wang B 2016 Surface Modification of Polystyrene by Femtosecond Laser Irradiation *J. Laser Micro/Nanoengineering* **11** 253–6
[18] Wu D, Ge X, Huang Y, Zhang Z and Ye Q 2003 Radiation synthesis of silver-polystyrene and cadmium sulfide-polystyrene nanocomposite microspheres *Mater. Lett.* **57** 3549–53
[19] Xiao T, Cai C, Peng L and Wu W 2013 Fabrication and characteristics of self-assembly nano-polystyrene films by laser induced CVD *Appl. Surf. Sci.* **282** 652–5
[20] Duan H, Zhao J, Zhang Y, Xie E and Han L 2009 Preparing patterned carbonaceous nanostructures directly by overexposure of PMMA using electron-beam lithography. *Nanotechnology* **20** 135306

[21] Bhuvana T and Kulkarni G U 2008 Polystyrene as a zwitter resist in electron beam lithography based electroless patterning of gold *Bull. Mater. Sci.* **31** 201–6
[22] Thirumavalavan S, Mani K and Sagadevan S 2015 Investigation of the Structural, Optical and Electrical Properties of Copper Selenide Thin Films *Mater. Res.* **18** 1000–7
[23] Sharma D, Sharma P and Thakur N 2009 Analysis of the optical constants of spun cast polystyrene thin film *Optoelectron. Adv. Mater. Rapid Commun.* **3** 145–8
[24] Zaki M F 2015 He–Ne laser induced changes to CN-85 polymer track detector *J. King Saud Univ. - Sci.*
[25] Eyssa, H.M. Osman, M. Kandil, S.A. Abdelrahman M M 2016 Effect of ion and electron beam irradiation on surface morphologyand optical properties of PVA *Nucl. Sci. Tech.* **26** 1–6
[26] Yoon H, Lee J, Park D W, Hong C K and Shim S E 2010 Preparation and electrorheological characteristic of CdS/Polystyrene composite particles *Colloid Polym. Sci.* **288** 613–9
[27] Song X, Dai Z, Xiao X, Li W, Zheng X, Shang X, Zhang X, Cai G, Wu W, Meng F and Jiang C 2015 Formation of Carbonized Polystyrene Sphere/hemisphere Shell Arrays by Ion Beam Irradiation and Subsequent Annealing or Chloroform Treatment *Sci. Rep.* **5** 17529
[28] Djaoued Y, Badilescu S, Balaji S, Seirafianpour N, Hajiaboli A, Kahrizi M, Truong V, Sadeghian R B, Braedley K and Bru R 2007 Micro-Raman Spectroscopy Study of Colloidal Crystal Films of Polystyrene – Gold Composites **61** 1202–10
[29] Meyer M W, Nguyen V H T and Smith E A 2013 Scanning angle Raman spectroscopy measurements of thin polymer films for thickness and composition analyses *Vib. Spectrosc.* **65** 94–100
[30] Roy D and Rae A 2013 Prospects of the Emerging Raman Scattering Tools for Surface and Nanoanalysis *Mapan - J. Metrol. Soc. India* **28** 285–97
[31] Saito R and Kataura H 2001 Optical Properties and Raman Spectroscopy of Carbon Nanotubes *Carbon Nanotub.* **80** 213–47
[32] Schrader B 1994 *Infrared and Raman Spectroscopy-Methods and Applications*
[33] Wibawa P J, Agam M A, Nur H and Saim H 2011 Changes in physical properties and molecular structure of polystyrene nanospheres exposed with solar flux *AIP Conf. Proc.* **1341** 54–61
[34] Kim E, Kyhm J, Kim J H, Lee G Y, Ko D-H, Han I K and Ko H 2013 White light emission from polystyrene under pulsed ultra violet laser irradiation. *Sci. Rep.* **3** 3253
[35] Rytlewski P and Żenkiewicz M 2009 Laser-induced surface modification of polystyrene *Appl. Surf. Sci.* **256** 857–61
[36] Abdul-Kader A M 2009 Photoluminescence and optical properties of He ion bombarded ultra-high molecular weight polyethylene *Appl. Surf. Sci.* **255** 5016–20

Acknowledgements
This work was financial supported by Centre for Graduate Studies, Graduate Research Incentive Grant (GIPS) Project Vot U302 and Faculty of Science, Technology and Human Development Universiti Tun Hussein Onn Malaysia. The authors are gratefully acknowledging the continued support from Microelectronics and Nanotechnology Shamsuddin Research Centre (MiNT-SRC) Universiti Tun Hussein Onn Malaysia and Ibrahim Badamasi Babangida University Lapai, Niger State Nigeria.

Preliminary analysis on the water quality index (WQI) of irradiated basic filter elements

Asyraf Arif Abu Bakar[1, a)], Anas Muhamad Pauzi[1, b)], Abdul Aziz Mohamed[1, c)], Syazrin Syima Sharifuddin[2, d)], Faridah Mohamad Idris[3,]

[1]College of Engineering, Universiti Tenaga Nasional, Jalan IKRAM-UNITEN, 43000 Kajang, Selangor, Malaysia
[2]National Hydraulic Research Institute of Malaysia, Kuala Lumpur, Malaysia
[3] Reactor Technology Centre, Malaysian Nuclear Agency, Bangi, 43000 Kajang, Selangor, Malaysia.

[a)]aasyraf@uniten.edu.my
[b)]anas@uniten.edu.my
[c)]azizm@uniten.edu.my
[d)]syima@nahrim.gov.my
[e)]faridah_idris@nuclearmalaysia.gov.my

Abstracts. Simple water filtration system is needed in times of extreme floods. Clean water for sanitation at evacuation centres is essential and its production is possible by using the famous simple filtration system consisting of empty bottle and filter elements (sands, gravels, cotton/coffee filter). This research intends to study the effects of irradiated filter elements on the filtration effectiveness through experiments. The filter elements will be irradiated with gamma and neutron radiation using the facilities available at Malaysia Nuclear Agency. The filtration effectiveness is measured using the water quality index (WQI) that is developed in this study to reflect the quality of filtered water. The WQI of the filtered water using the system with irradiated filter elements is then compared with that of the system with non-irradiated filter elements. This preliminary analysis only focus on filtration element of silica sand. Results shows very nominal variation in in WQI after filtered by non-irradiated, gamma and neutron filter element (silica sand), where the hypothesis could not be affirmed.

1. Introduction

Department of Environmental Malaysia (DOE) have listed a parameters that reflects the quality of water[1]. In this study, six standard main parameters tests were used in order to see the water quality of irradiated and non-irradiated filter element which are Dissolved Oxygen (DO), Biological Oxygen Demand (BOD), Chemical Oxygen Demand (COD), Total Suspended Solid (SS), Acidic and Alkaline (pH) and last but not least Ammonia Nitrogen (NH_3-N).[2] This experiment was conducted at the National Hydraulic Research Institute of Malaysia (NAHRIM), while the filter elements were radiated at the *TRIGA PUSPATI* Reactor (RTP) to obtain neuron and gamma effects and *SinaGamma* facility for gamma effects at Malaysian Nuclear Agency. This paper will discuss the comparison between WQI of water after filtered by non-irradiated and irradiated silica sand.

2. Methodology

The simple water filtration system as shown in Figure 1 is very easy to assemble. Basically, the general set-up is made out of small grain materials at the bottom such as sand, and coarse, larger rocks at the top. Contaminated water were poured from the top will go through the filtering elements, removing large impurity particles from the water.

Figure 1. Simple water filtration system set-up.

For this preliminary study, silica sand is used as the filter element. Sand is the 'biological filtration' process of water treatment. This simple system of water purification has been in continuous use since the beginning of the nineteenth century, and has proved effective under widely differing circumstances. It is simple, inexpensive and reliable and is still the chosen method of purifying water supplies for some of the major cities of the world [3]. Practically, sand may be considered to consist of small detrital fragments of rock or mineral particles liberated by mechanical disintegration of parent rock material; or biogenic particles such as shells, shell fragments or chemical precipitates occurring in nature and distinguishable by the naked eye.[4]

The methodology of obtaining the WQI follows the standard laid by the American Public Health Association (APHA)[5]. Table 1 below shows the test parameter and its reference standards.

Table 1. Analysis Standard.

Test Parameter	Reference Standards
pH	APHA 4500 –H^{+}B
Dissolved Oxygen (DO)	APHA 4500-O G
Biochemical Oxygen Demand (BOD)	In house Test Method MKA TM 04-Determination of BOD based on APHA 5210 D
Chemical Oxygen Demand (COD)	In house Test Method MKA TM 01-Determination of COD based on APHA 5220 D
Nitrogen Ammonia (NH_3-N)	In house Test Method MKA TM 03-Determination of Nitrogen(Ammonia) based on APHA 4500 –NH_3 B
Total Suspended Solid (TSS)	APHA 2540D

IOP Conf. Series: Materials Science and Engineering **298** (2018) 012005 doi:10.1088/1757-899X/298/1/012005

3. Results and discussion

There are 2 factors that influences the WQI which are geometry structure and chemical compound. It is common when a polluted water will get purified after contaminants in the water flows through the grains and micropores of filter. Simple hypothesis predicts that the greater the concentration of micropores in the sand filters, the more impurities will be trapped when water flows through. Secondly, the sand compounds will neutralize and also kills microorganism in the contaminated water. For instance, typical filtration element used to adsorb natural organic compounds, taste and odor compounds.[4]

The results for 6 parameters were recorded. Each parameters have their own classification to define the quality of water either polluted or clean. Table 2 shows the quality class for each parameter that indicate the quality of water. The range for each parameter are very important to calculate the WQI by following the best fit equations for the estimation of various sub index values as shown on table 3.[1]

Table 2. DOE Water Quality Index Classification.

Parameter	Unit	Classes				
		I	II	III	IV	V
Ammonia Nitrogen	mg/L	< 0.1	0.1 - 0.3	0.3 – 0.9	0.9 – 2.7	> 2.7
Biochemical Oxygen Demand	mg/L	< 1	1 – 3	3 – 6	6 – 12	> 12
Chemical Oxygen Demand	mg/L	< 10	10 – 25	25 – 50	50 – 100	> 100
Dissolved Oxygen	mg/L	> 7	5 – 7	3 – 5	1 – 3	< 1
pH	-	> 7	6 – 7	5 – 6	< 5	> 5
Total Suspended Solids	mg/L	< 25	25 – 50	50 – 150	150 – 300	> 300
Water Quality Index (WQI)	-	< 92.7	76.5 – 92.7	51.9 – 76.5	31.0 – 51.9	> 31.0

Table 3. Best Fit Equations for the Estimation of Various Sub Index Values.

Sub Index (SI)	Range
Dissolved Oxygen (DO) (% saturation)	
SIDO = 0	for x ≤8
SIDO = 100	for x ≤92
SIDO = -0.395 + 0.030x2 - 0.00020x3	for 8 < x < 92
Biological Oxygen Demand (BOD)	
SIDOD = 100.4 - 4.23x	for x ≤ 5
SIDOD = 108* exp(-0.055x) - 0.1x	for x > 5
Chemical Oxygen Demand (COD)	
SICOD = -1.33x + 99.1	for x ≤ 20
SICOD = 103* exp(-0.0157x) - 0.04x	for x > 20
Ammonia Nitrogen (NH_3-N)	
SIAN = 100.5 - 105x	for x ≤ 0.3
SIAN = 94* exp(-0.573x) - 5* I x - 2 I	for 0.3 < x < 4
SIAN = 0	for x ≥ 4
Suspended Solid (SS)	
SISS = 97.5* exp(-0.00676x) + 0.05x	for x ≤ 100
SISS = 71* exp(-0.0061x) + 0.015x	for 100 < x < 1000
SISS = 0	for x ≥ 1000
Acidic and Alkaline (pH)	
SlpH = 17.02 - 17.2x + 5.02x2	for x < 5.5
SlpH = -242 + 95.5x - 6.67x2	for 5.5 ≤ x < 7
SlpH = -181 + 82.4x - 6.05x2	for 7 ≤ x < 8.75
SlpH = 536 - 77.0x + 2.76x2	for x ≥ 8.75

The letter "*x*" in table 3 represents the concentration of each parameter in mg/L except for pH. After each sub index for each parameter has been calculated, the values will be substituted into equation 1 to calculate Water Quality Index, WQI.

$$WQI = (0.22 * SIDO) + (0.19 * SIBOD) + (0.16 * SICOD) + (0.15 * SIAN) + (0.16 * SISS) + (0.12 * SIpH) \quad (1)$$

where,

SIDO = Sub Index DO (% saturation)
SIBOD = Sub Index BOD
SICOD = Sub Index COD
SIAN = Sub Index NH3-N
SISS = Sub Index SS
SipH = Sub Index pH
$0 \leq WQI \leq 100$

The calculated WQI is tabulated in table 4. Silica sand were used as filtration element and the quality of water resource label as *Original*. Water were filtered through 3 types filtering elements, which are firstly the non-treatment of silica sand call as non-irradiated, secondly are the silica sand treated by the gamma and thirdly, silica sand irradiated with neutron. The filtration experiments are repeated 3 times for all gamma and neutron filters.

The results shows nominal changes in WQI between neutron-irradiated and non-irradiated silica sand of about 4% positive. Meanwhile for gamma-irradiated filters also shows nominal changes but inconsistent results of positive and negative WQI of about 4%. The variation in the results show larger percentage of uncertainties than the variation between different samples. Results from neutron-irradiated sand sample could slightly show signs of hypothesis approval.

A few factors might be contributed to the results in this lab analysis. Firstly, might be due to the water source use in this analysis. Refer to the result of original, it indicate that the water resource is already clean with WQI is 86. The value of Nitrogen Ammonia (NH_3-N), Chemical Oxygen (COD), Total Suspended Solid (TSS) are in class I (refer table 2). As the water resource initially unpolluted, might be the filtration process done without need to filter any contaminated and unnecessary particle.

Secondly, the time contact of water with the silica sand would be the one of the factor that effect the results. During the filtration process in this experiment, only half of the syringe are filled with silica sand, hence water flew through the silica sand for a short period of time. Thus, the data obtained shows very small amount of filtration.

Thirdly, it can be assumed that the treated silica sand might be have limitation in filter some of the parameter. The commercial filtration system like membrane have their own capabilities. The ultrafiltration membrane process could not filter monovalent and multivalent (Sodium chloride, magnesium sulphate and others) meanwhile reverse osmosis membrane process only allow water to pass through[6]. Hence, this experiment expected to have this kind of restraint.

Table 4. Results of WQI.

			Silica Sand						
Parameter	Unit	Original	Non irradiated	Gamma			Neutron		
				1	2	3	1	2	3
Dissolved Oxygen (DO)	mg/L	5.98	7.74	7.74	7.67	7.75	7.73	7.88	7.9
	%	74.3	94.1	94.4	93.8	93.8	94.5	95.7	96.5
Biochemical Oxygen Demand (BOD)	mg/L	2.5	0.8	2	2.5	3.1	2	2.8	2.5
Chemical Oxygen (COD)	mg/L	7.62	5.93	7.05	6.01	6.47	6.41	7.19	6.39
Nitrogen Ammonia (NH_3-N)	mg/L	0.344	0.338	0.324	0.342	0.326	0.402	0.362	0.371
Total Suspended Solid (TSS)	mg/L	10	14	10	15	13	24	13	21
pH		6.66	6.58	6.64	6.72	6.79	6.61	6.86	6.88
WQI		86	84	83	82	90	89	90	89
ΔWQI	%	0%	-2.33%	-3.49%	-4.65%	4.65%	3.49%	4.65%	3.49%

4. Conclusions

The results shows nominal changes in WQI between irradiated and non-irradiated silica sand. Irradiated silica sand does not improve the concentration of micropores or change the microstructure. Further analysis need to be conducted in order to get a better understanding of the phenomenon in this experiment. Next step will be to place more emphasize on the 3 factors discussed which were the source of water, time contact and limitation of filter. Future studies will include metallurgical analysis which are also important to obtain the amount of micro-structure changes in the silica sand due to the irradiation. X-ray Diffraction (XRD) and Scanning electron microscope (SEM) will be used for this microstructure analysis [7].

5. References

[1] Department of Environment 2007 *Malaysia Environmental Quality Report 2006* p 72.
[2] HuangY F, Ang S Y, Lee K M and Lee T S 2015 Quality of water resources in Malaysia *Research and Practices in Water Quality*
[3] Huisman L and Wood W E 1974 *Slow Sand Filtration* W. H. Organization Geneva
[4] Anuar N, Pauzi A M and Bakar A A A 2017 Methodology of water quality index (WQI) development for filtrated water using irradiated basic filter elements *AIP Conf. Proc.* **1799**
[5] Standard Methods For the Examination of Water and Wastewater (American Public Health Association 2012) **22**
[6] Anim-Mensah A 2010 *General Overview of Membranes and Applications*
[7] Bhagavannarayana G, Parthiban S and Meenakshisundaram S 2006 Enhancement of crystalline perfection by organic dopants in ZTS, ADP and KHP crystals as investigated by high-resolution XRD and SEM *J. Appl. Crystallography* **39**

Acknowledgement

Special thanks for Malaysia's Ministry of Higher Education (MOHE) for sponsoring the research project under the Fundamental Research Grant Scheme (FRGS) project number 117289-136338

IOP Conf. Series: Materials Science and Engineering **298** (2018) 012006 doi:10.1088/1757-899X/298/1/012006

The enhancement of heavy metal removal from polluted river water treatment by integrated carbon-aluminium electrodes using electrochemical method

N M Yussuf[2], Z Embong[1,2], S Abdullah[2], M I M Masirin[1,3], S A A Tajudin[1,3], S Ahmad[4], S K Sahari[6], A A Anuar[5] and O Maxwell[7]

[1]Research Centre for Soft Soil (RECESS), Office for Research, Innovation, Commercialization and Consultancy Management (ORICC) ,UniversitiTun Hussein Onn Malaysia
[2]Faculty of Science, Technology and Human Development, UniversitiTun Hussein Onn Malaysia

[3]Faculty of Civil and Environment Engineering, UniversitiTun Hussein Onn Malaysia
[4]Centre for Language Study, Universiti Tun Hussein Onn Malaysia
[5]Malaysia Nuclear Agency 42300, Kajang, Selangor
[6]Faculty of Engineering, Universiti Malaysia Sarawak (UNIMAS), 94300, Kota Samarahan, Sarawak, Malaysia

[7]College of Science and Technology, Department of Physics, Covenant University, Ota, Ogun State, Nigeria

Abstract. The heavy metal removal enhancement from polluted river water was investigated using two types of electrodes consist of integrated carbon-aluminium and a conventional aluminium plate electrode at laboratory-scale experiments. In the integrated electrode systems, the aluminium electrode surface was coated with carbon using mixed slurry containing carbon black, polyvinyl acetate and methanol. The electrochemical treatment was conducted on the parameter condition of 90V applied voltage, 3cm of electrode distance and 60 minutes of electrolysis operational time. Surface of both electrodes was investigated for pre and post electrolysis treatment by using SEM-EDX analytical technique. Comparison between both of the electrode configuration exhibits that more metals were accumulated on carbon integrated electrode surfaces for both anode and cathode, and more heavy metals were detected on the cathode. The atomic percentage of metals distributed on the cathode conventional electrode surface consist of Al (94.62%), Zn (1.19%), Mn (0.73%), Fe (2.81%) and Cu (0.64%), while on the anode contained O (12.08%), Al (87.63%) and Zn (0.29%). Meanwhile, cathode surface of integrated electrode was accumulated with more metals; O (75.40%), Al (21.06%), Zn (0.45%), Mn (0.22), Fe (0.29%), Cu (0.84%), Pb (0.47%), Na (0.94%), Cr (0.08%), Ni (0.02%) and Ag (0.22%), while on anode contain Al (3.48%), Fe (0.49 %), C (95.77%), and Pb (0.26%). According to this experiment, it was found that integrated carbon-aluminium electrodes have a great potential to accumulate more heavy metal species from polluted water compare to the conventional aluminium electrode. Here, heavy metal accumulation process obviously very significant on the cathode surface.

1. Introduction

Inorganic pollutants specifically that refer to heavy metals constitute a serious threat for the environment. Most of the metals such as copper, nickel, lead and zinc are harmful when they are discharged without treatment, because they are not biodegradable and tend to accumulate in living organisms, and many heavy metal ions are known to be toxic or carcinogenic [1]. Due to their high toxicity, these contaminated water are strictly concerned to be treated for metals removal. Various techniques have been employed for the treatment of heavy metals, including precipitation, coagulation/flocculation, adsorption, biosorption, ion-exchange, electrochemical and membrane separation [2].

Among these techniques, precipitation was said to be the most applicable technique and economical. However, this technique produces a large amount of precipitate sludge that requires further treatment [1]. Another techniques such as reverse osmosis, ion-exchange and other membrane separation were limited due to a number of disadvantages such as high material and operational cost and their operational problems [3].

Nowadays, electrochemical technologies have reached such a state that they are not only comparable with other technologies in terms of cost but also are more efficient and more compact [4]. For some situations, electrochemical technologies may be the indispencable step in treating wastewater containing refractory pollutants [5]. Electrochemical remediation involves the application of electricity to the electrodes for contaminants removal and it is used for treating water that, among others, contains COD, oil wastes or metals [6].

Hence, the development of new treatment methods for effluents bearing heavy metals is an urgent issue. Among other methods that have recently been studied, electrochemical proved very efficient in removal pollutants such as organic and inorganic matters from the industrial wastewater [7]. In addition, a number of scientific works have indicated that heavy metals in the free form can be successfully removed by electrochemical using aluminium, iron and stainless steel electrodes [8].

There are few published studies using aluminium electrodes for the treatment of metal removal, no attempt has been made to investigate metal removal by employing an aluminium electrodes integrated with activated carbon. This study primarily aims to investigate the performance of activated carbon as electrodes material in Pb, Zn, Mn and Fe removal and compare its performance with those of conventional aluminium electrodes. The removal efficiency was calculated and the metal accumulation on electrodes were characterized.

2. Materials and method

2.1. Experimental procedure

The experimental equipment schematically is shown in figure 1. It consisted mainly of a Perspex with a specific dimension 30 cm length, 10 cm width and 7 cm height, the cell and electrical circuit. The cell consisted of two aluminium electrodes. The conventional electrodes was a solid aluminium of 10 cm length, 7 cm width and 0.1 cm thick which was supported on the container wall. The integrated aluminium electrodes were fabricated by coating activated carbon slurry onto a sheet of aluminium plate using a doctor knife. Electrode slurry was prepared by mixing activated carbon powder and the polyvinyl acetate (PVAc) and methanol. PVAc was added to bind the coating to the electrode surface and methanol was added to ensure dispersion of the ingredients and allow the application of an even coating onto the electrode surface [9]. The optimum ratio of the mixture was found to be 5g: 2g: 10 mL of AC: PVAc: Methanol Then the mixture was spread on the aluminium electrodes surface, and the electrode was dried at 100°C for 2h in an oven. Final thickness of integrated electrodes formed was 0.2 cm, with 7cm width and 10 cm length. The gap between the cathode and anode was kept at 3 cm. The electrical circuit consisted of DC power supply and crocodile clippers connected to measure the cell.

2.2. Water sample characteristics

The polluted water sample was collected from *Sungai Senggarang*, Batu Pahat. It was then bring to the laboratory and preserved with nitric acid and stored in a refrigerator at 4°C for metal preservation. The experiment was conducted in a lab-scale size with two different electrode system; one with conventional aluminium electrode plates and the other one with aluminium plate integrated with activated carbon. The reactor cell was filled with 300 mL of river water while DC supply was adjusted to 90 V and the electrolysis time was set for 60 min. Both, pH and temperature were measured with a multiparameter. The multiparameter was calibrated according to the operating manual. The water samples were filtered with a 15 cm filter paper for atomic absorption spectroscopy (AAS) analysis.

IOP Conf. Series: Materials Science and Engineering **298** (2018) 012006 doi:10.1088/1757-899X/298/1/012006

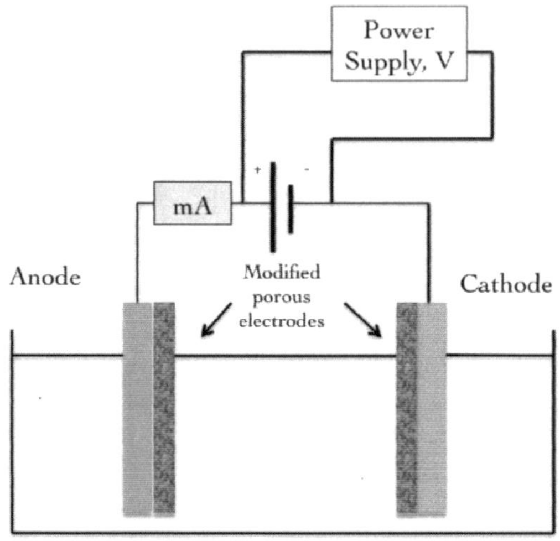

Figure 1. Schematic diagram of the experimental setup.

2.3. Measurements

At the beginning of each experiment (t=0), initial pH were measured with a Lutron pH -meter PH-222. The initial concentrations of the selected toxic metals (Mn, Fe, Zn and Pb) were, also, determined by atomic absorption spectroscopy (AAS), as mentioned above. After the 60 min electrolysis process, pH was measured for each liquid sample. Finally, 50 mL of river water samples were measured by AAS for Mn, Fe, Zn and Pb and surface of the electrodes were measured by scanning electron microscopy-energy-dispersive X-ray (SEM-EDX) for pre and post experiments. All experiments were conducted at ambient temperature.

3. Results and discussion

3.1. Removal efficiency of heavy metal

The analysis of Zn, Pb, Mn and Fe were evaluated as shown in table 1. This table gives the results of their concentration, standard deviation of concentration and relative standard deviation of concentration of the four heavy metal in water sample measured by atomic absorption spectroscopy (AAS) and comparison with United States Environmental Protection Agency (EPA) data. For the protection human health, guide lines for the presence of heavy metals in water have been set by International organization such as EPA, thus heavy metal have maximum permissible level in water specified by the organization. The maximum contaminant level (MCL) is an enforceable standard set at a numerical value with adequate margin of safety to ensure no adverse affect on human health [10].

Table 1. Concentration, standard deviation of concentration, relative standard concentration and maximum contaminant level (MCL) of Pb, Zn, Mn and Fe.

Metal	Concentration (mg/L)	SD	RSD (%)	MCL (mg/L) (EPA, 2009)
Pb	0.307	0.038	12.460	0.002
Zn	0.021	0.001	5.920	5.000
Mn	0.378	0.002	0.580	0.050
Fe	5.279	0.045	0.840	0.300

The four elements studied in this research were Pb, Zn, Mn and Fe has the maximum contamination levels of 0.015, 5.0, 0.05 and 0.3 mg/L respectively [10] while the initial pH was 3.51. From the results obtained, the concentration levels for Pb, Mn and Fe were above the specified contaminant level except for Zn. One of the aims of the present work was to check the possibility of reducing or removing the contaminant in water sample by applying electrochemical method. For this reason, an experiment was carried out and the results for metal removal efficiency (%) and final pH are presented in table 2. Figure 2 shows the evolution of the removal efficiency, calculated for operating time 60 mins, as a function of voltage used 90 V. The removal efficiency in percent, is given by:

$$\text{Removal efficiency, \%} = \frac{(C_o - C_t)\times 100}{C_o} \quad (1)$$

where C_o and C_t are respectively the concentration of metal ions before and after the treatment.

Table 2 Metal removal (%) and final pH for conventional and integrated electrodes after 60 minutes electrochemical treatment.

	Conventional electrodes	Integrated electrodes
Removal (%)		
Mn	16.14	47.88
Fe	81.61	91.80
Zn	99.00	99.50
Pb	2.20	8.47
End pH	5.94	6.02

In general, the pH of the medium raises during electrochemical as a result of the electrolysis process and the production of hydroxyl ions in the aqeous solution. According to the results, it was remarkable since all the removal rates are higher for integrated electrodes showing that in these conditions the treatment operation performs better when carbon is used. Zn removal is the highest among all as the removal efficiency is 99.50% while Fe, Mn and Pb are respectively 91.80%, 47.88% and 8.47%. For the conventional electrodes, lower removal percentage were observed for Fe, Zn and Mn (81.61%, 99.0% and 16.14% respectively).

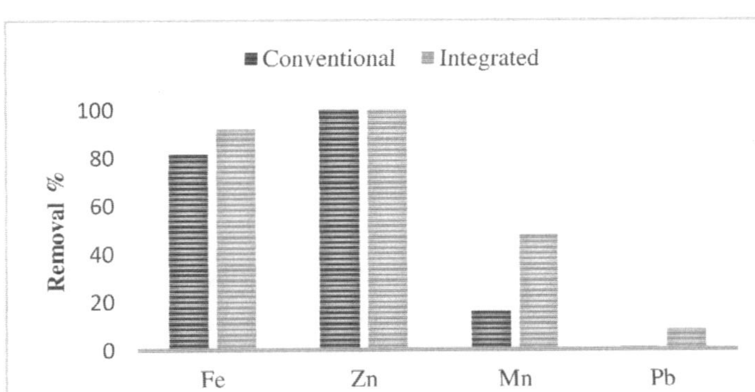

Figure 2. Comparison in metal removal efficiency between a conventional aluminium electrode and integrated electrode.

3.2. Metal accumulation on electrode plates

SEM micrographs and EDX spectra were obtained with HITACHI Horiba SU1510 scanning electron microscope equipped with a LINK analytical system. The electron energy used was 5.9 keV. EDX analysis revealed the higher metal accumulation was on carbon integrated electrode surface for both anode and cathode compared to conventional aluminium electrode, and more heavy metals were observed on the cathode. The atomic percentages of metals distributed on the cathode integrated electrode surface were O (75.40%), Al (21.06%), Zn (0.45%), Mn (0.22), Fe (0.29%), Cu (0.84%), Pb (0.47%), Na (0.94%), Cr (0.08%), Ni (0.02%) and Ag (0.22%), while on anode contain Al (3.48%), Fe (0.49 %), C (95.77%), and Pb (0.26%) (Figure 4). In contrast, less heavy metal elements were detected on the conventional electrode surface. Instead, elements that were found on conventional anode surface such as O (12.08%), Al (87.63%) and Zn (0.29%), while on cathode surface are Al (94.62%), Zn (1.19%), Mn (0.73%), Fe (2.81%) and Cu (0.64%) (figure 3).

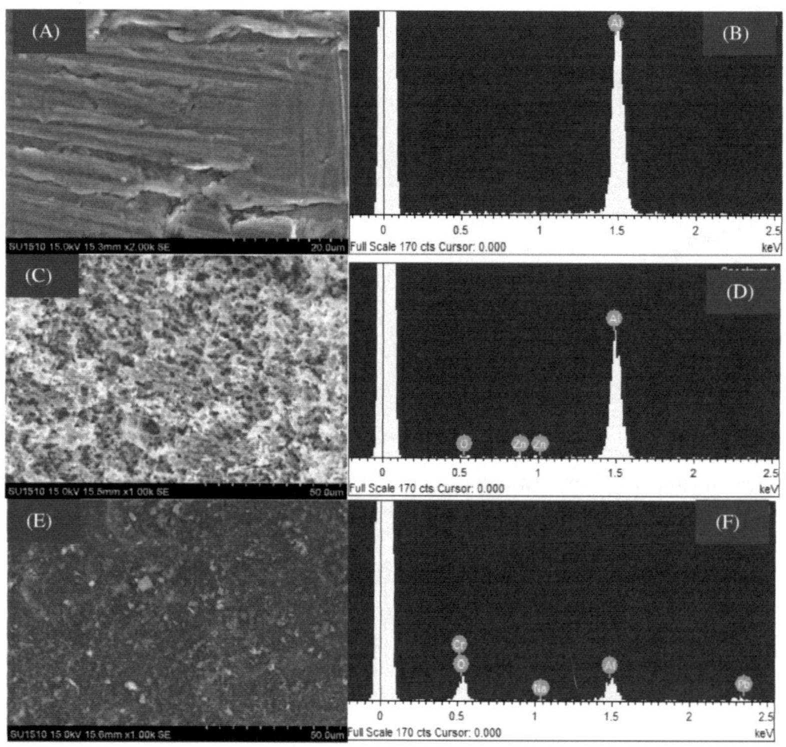

Figure 3. SEM photomicrograph of conventional aluminium electrodes (A) before treatment (C) anode surface after treatment (E) cathode surface after treatment. Corresponding EDX spectra for (A), (C) and (E) are in (B), (D) and (F), respectively.

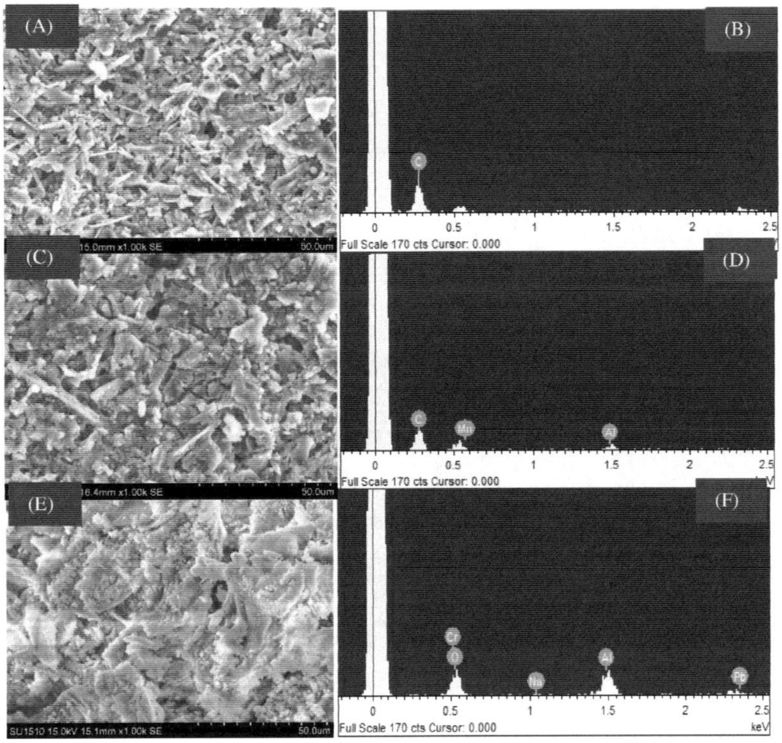

Figure 4. SEM photomicrograph of integrated aluminium electrodes (A) before treatment (C) anode surface after treatment (E) cathode surface after treatment. Corresponding EDX spectra for (A), (C) and (E) are in (B), (D) and (F), respectively.

4. Conclusion

From this work, some conclusion can be summarised as follows :

- Integrated electrode with carbon showed enhanced removal towards metal in water, particularly Fe, Mn, Pb and Zn.
- Fe and Zn showed higher removal efficiency 91.8% and 99.5% for integrated electrode; 81.6% and 99.0% for conventional electrodes, respectively while Pb and Mn removal are less than 50% for both electrodes.
- SEM-EDX analysis revealed the increment of metal accumulation especially on cathode surface.
- To conclude, it is clear that integrated carbon-aluminium electrodes is useful to improve metal removal in water.

5. References

[1] Aji B A, Yavuv Y and Koparal A S 2012 Electrocoagulation of heavy metals containing model wastewater using monopolar iron electrodes **86** 248-254.

[2] Nariyan E, Sillanpää M and Wolkersdorfer C 2017 Electrocoagulation treatment of mine water from the deepest working European metal mine – Performance, isotherm and kinetic studies. *Sep. and Purif. Tech.* **177** 363-373.

[3] Basha C A, Bhadrinarayana N S, Anantharaman N and Begum K M 2008 Heavy metal removal from copper smelting effluent using electrochemical cylindrical flow reactor. J. Hazard. Mater. **152** 71-78

[4] Chen G 2004 Electrochemical technologies in wastewater treatment *Sep. and Purif. Tech.* **38** 11-41

[5] Chen G, Chen X and Yue P L 2000 Electrocoagulation and electroflotation of restaurant wastewater *J. Environ. Eng.* **126** 858 – 863.
[6] Gatsios E, Hahladakis J N and Gidarakos E 2015 Optimization of electrocoagulation (EC) process for the purification of a real industrial wastewater from toxicmetals *J. Environ. Manage.* **154** 117–127.
[7] Rajkumar D. and Palanivelu K (2004) Electrocoagulation treatment of industrial wastewater *J. Hazard. Matter* **B113**; 123 – 129
[8] Golder A K, Samanta A N and Ray S 2007 Removal of trivalent chromium by electrocoagulation *Sep. Purif. Tech.* **53** 33 – 41.
[9] Kim K S, Kim S and Jun T H 2015 Activated carbon-coated electrode and insulating partition for Improved Dust Removal Performance in Electrostatic Precipitators *Water Air Soil Pollut.* **226** 367.
[10] United States Environmental Protection Agency (EPA) 2009 National Primary Drinking Water Regulations
[11] Hussin F, Abnisa F, Issabayeva G and Aroua M K 2017 Removal of lead by solar-photovoltaic electrocoagulation using novel perforated zinc electrode *J. of Clean. Prod.* **147** 206 – 216.
[12] Lee S Y, Jeong Y J, Chae S R, Yeon K H, Lee Y, Kim C S, Jeong N J and Park J S 2016 Porous carbon-coated graphite electrodes for energy production from salinity gradient using reverse electrodialysis *J. of Phy. and Chem. of Solids* **91** 34-40.
[13] Njiki C P N, Tchamango S R, Ngom P C, Darchen A and Ngameni E 2009 Mercury (II) removal from water by electrocoagulation using aluminium and iron electrodes *J. of Haz. Mat.* **168** 1430 – 1436.

Acknowledgement
This paper was fully sponsored by the Centre for Graduate Studies (CGS), Universiti Tun Hussein Onn Malaysia (UTHM). We also like to show our gratitude to the Research Centre of Soft Soil (RECESS), UTHM to provide us a research laboratory and thanks to Faculty of Mechanical and Manufacturing Engineering, UTHM for assistance with SEM-EDX for morphological study of the electrode surface.

IOP Conf. Series: Materials Science and Engineering **298** (2018) 012007 doi:10.1088/1757-899X/298/1/012007

Residential exposure from extremely low frequency electromagnetic field (ELF EMF) radiation

Shamesh Raj Parthasarathy[1], Roha Tukimin[1]

[1]Radiation Safety and Health Division, Malaysian Nuclear Agency, Bangi, 43000 Kajang, Selangor

shamesh@nm.gov.my

Abstract. ELF EMF radiation have received considerable attention as a potential threat to the safety and health of people living in the vicinity of high voltage transmission lines, electric distribution substations, power stations and even in close proximity to electronics and electrical household appliances. The paper highlights the study on the ELF EMF safety assessment performed at residences comprising of an owner-occupied house, a completed vacant house and an under construction condominium. The objectives of this study were to determine the ELF EMF radiation exposure level from the high voltage transmission line, electric distribution substation, power station and electrical household appliances in the residences, and to assess the potential exposure received by the occupants at the assessed locations. The results were logged in the electric and magnetic field strength with the units of volt per meter (V/m) and miliGauss (mG) respectively. The instrument setup and measurement protocols during the assessment were adopted from standard measurement method and procedures stipulated under the Institute of Electrical and Electronics Engineers (IEEE) Standard. The results were compared with the standards recommended in the International Commission on Non-Ionizing Radiation Protection (ICNIRP) guidelines.

1. Introduction

As new applications and products are discovered in parallel to the technological advancement, the exposure to electromagnetic field (EMF) frequencies become increasingly significant. Notwithstanding of the fact that the colossal benefits of using electricity in everyday life are undeniable, the general public has become increasingly concerned about the potential adverse health effects of the exposure to electric and magnetic fields at extremely low frequencies (ELF) range. The exposures are mainly from the use of electrical energy at the power frequencies of 50 Hz.

Due to public concern, a study has been conducted to assess the extremely low frequency (ELF) radiations from the high voltage transmission line, electric distribution substation, power station and electrical household appliances in residences. This paper highlights the study on the ELF EMF safety assessment performed at residences comprising of an owner-occupied house, a completed vacant house and an under construction condominium. The objectives of this study were to determine the ELF EMF radiation exposure level from the high voltage transmission line, electric distribution substation, power station and electrical household appliances in the residences, and to assess the potential exposure received by the occupants at the assessed locations. The results were then compared with the standards recommended in the International Commission on Non-Ionizing Radiation Protection (ICNIRP) guidelines.

2. Extremely low frequency (ELF) radiation

Extremely low frequency (ELF) radiation is a type of non-ionizing radiation (NIR) and is positioned at the low-energy end of the electromagnetic spectrum. Electric fields are produced by electric charges. The electric field strength is measured in units of volt per meter (V/m) or kilovolt per meter (kV/m) [1-3]. Magnetic fields are produced by moving charges and thus are proportional to electric currents in a system, irrespective of the voltage used. Their strength is measured in units of Gauss (G) and tesla (T) [1-3]. The maximum permissible exposure limits (PEL) was developed to protect workers and the general population from harmful exposure to ELF EMF. Based on scientific knowledge and information available at the moment, prolonged exposure at or below the levels recommended by these standard guidelines is considered as safe and acceptable for the purpose of protection of human health. The PEL for both workers and members of the public are excerpted from the ICNIRP and World Health Organization (WHO) guidelines. Based on these guidelines, the PEL for public, for electric and magnetic fields of ELF EMFs are 5,000 volt per meter (V/m) and 1000 miliGauss (mG) respectively for the frequency of 50Hz [1-5].

3. Methodology

The instrument setup and measurement protocols during the assessment were adopted from standard measurement method and procedures stipulated under the Institute of Electrical and Electronics Engineers (IEEE) Standard. Free body probes and ground reference instruments are the two methods used for electric and magnetic field measurement [1][4]. In this study, free body probe was used for the measurement of the fields because it does not require a known ground reference for measurements anywhere above the ground [1-4]. Measurements of the field strengths were carried out using PMM instrument Model 8053 with an attached probe Model PMM EHP-50C and EMDEX II attached with the Linear Data Acquisition System (LINDA) measurement wheel. The PMM Instrument Model EHP-50C can measure both electric and magnetic fields simultaneously and the EMDEX II attached with LINDA was used to measure the spatial distribution of the magnetic field. Readings were taken in volt per meter (V/m) for the electric fields and miliGauss (mG) for the magnetic fields.

4. Description of Assessment Site

The study comprised of ELF assessment on 3 different residential and locations specifically in an owner occupied house with primary exposure from high voltage transmission line and electrical appliances, a completed vacant house with primary exposure from electric distribution substation and an under-construction condominium with primary exposure from a power station. The 3 locations were classified as ELF1, ELF2 and ELF3 for the 3 different residential and locations. The layout of the measurement points are shown from figures 1 to 3 respectively.

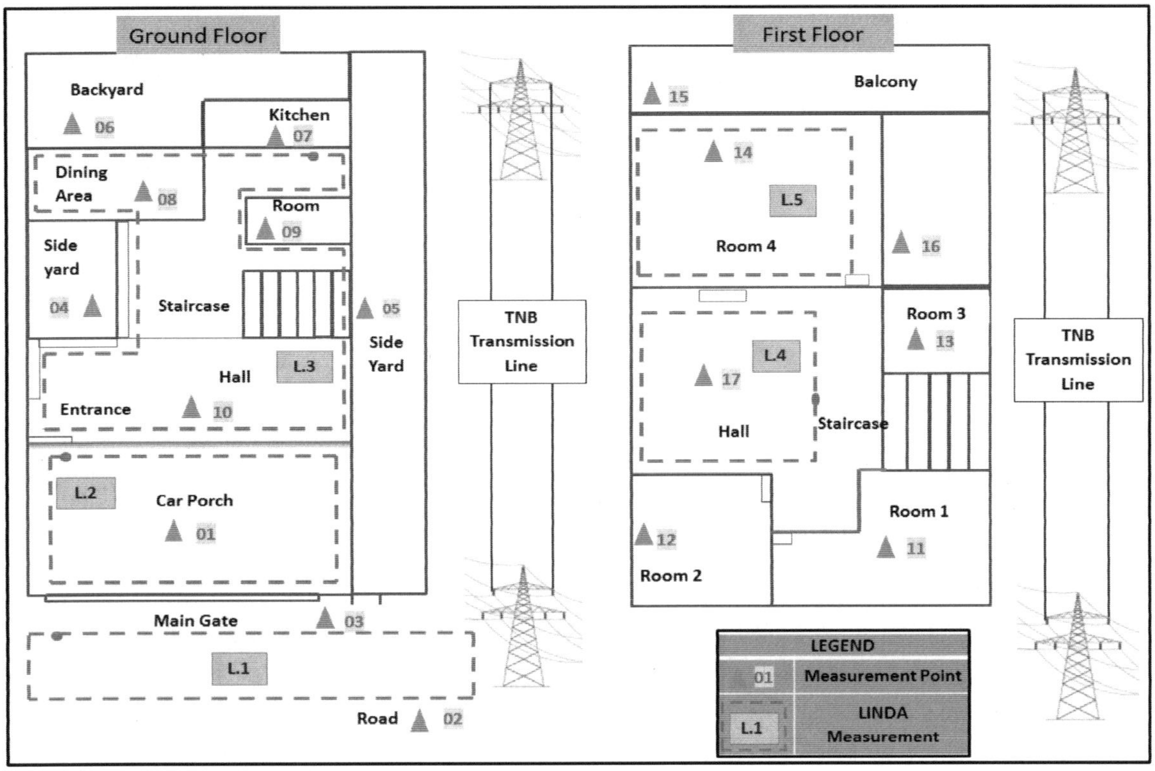

Figure 1. Layout of the measurement points for location ELF1.

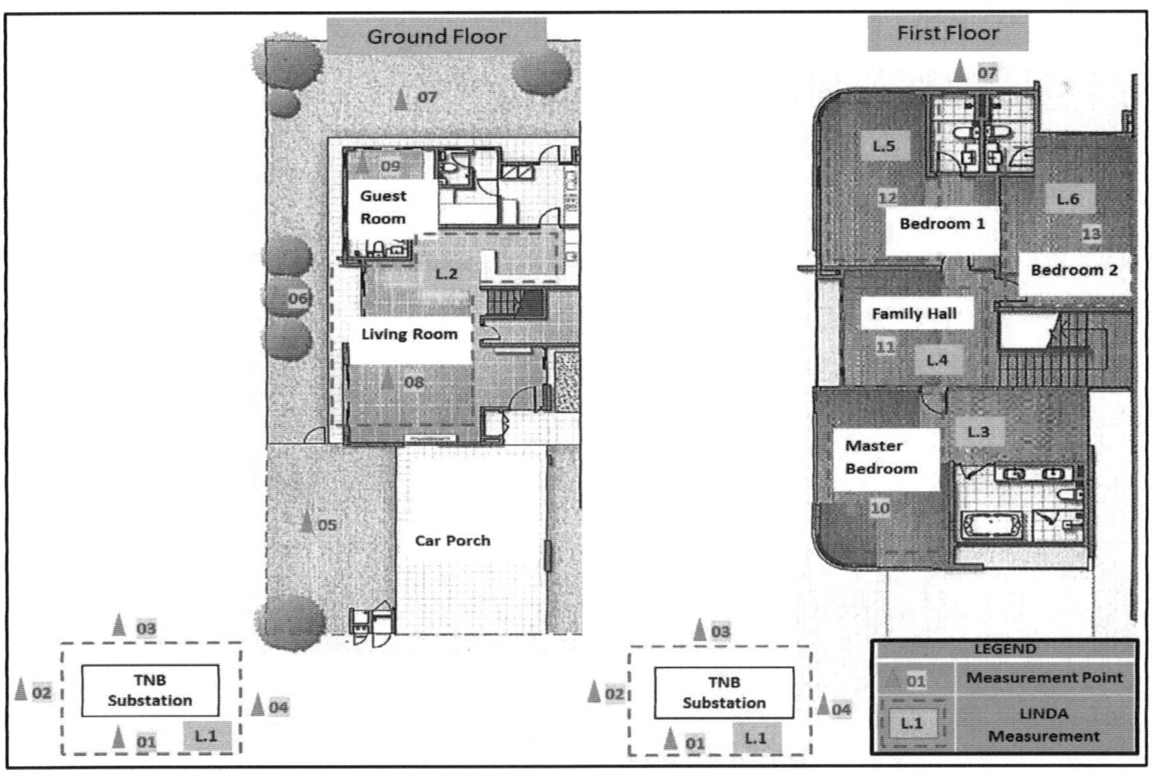

Figure 2. Layout of the measurement points for location ELF2.

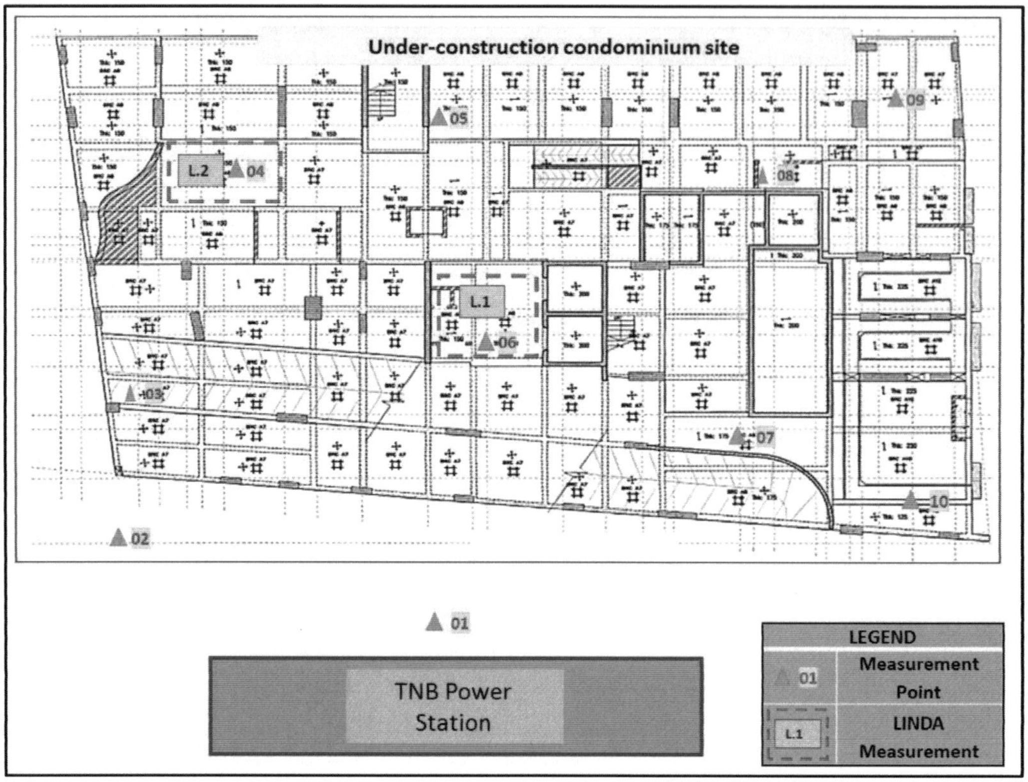

Figure 3. Layout of the measurement points for location ELF3.

5. Results and Discussion

For the first location (ELF1), the electric field strengths were found to vary from 0.100 V/m to 4.844 V/m. The highest averaged level of the electric field strengths was at 4.844 V/m measured at the car porch of the house, point 01. The value was about 0.10% or around 1032 times lower than that of the PEL recommended by ICNIRP for member of public (5000 V/m). The value is significant mainly due to the open space of the car porch without any shielding. In addition, the electrical field strength is influenced by the electrical cables and lighting points installed in that stated area. For areas inside the house, the highest averaged level of the electric field strengths was at 1.438 V/m measured at the first floor hall of the house. The value recorded at the location was about 0.03% or about 3477 times lower than that of the PEL. It should be noted that the electrical field strength value measured inside the house reduced drastically as the field strength from the transmission line was shielded by the walls of the house. For this particular condition, the electric field strength was influenced mostly by the electrical cables and electrical appliances in the house. It should be noted that at the points 03, 09, 12 and 17, the electric field strength was higher due to presence of electrical power points and electronic appliances such as television, desktop computers and in-built electronic alarm system. House wiring can produce electric fields, which are clearly strongest close to the wiring [1-5]. As for other points inside the house, the value remained constant at 0.100 V/m due to shielding by concrete walls, trees and was at a further distance from the internal and external ELF sources [1-5]. The plots of figure 4(a) indicate the electric field strength against distance and the point of measurement and their comparison with the PEL.

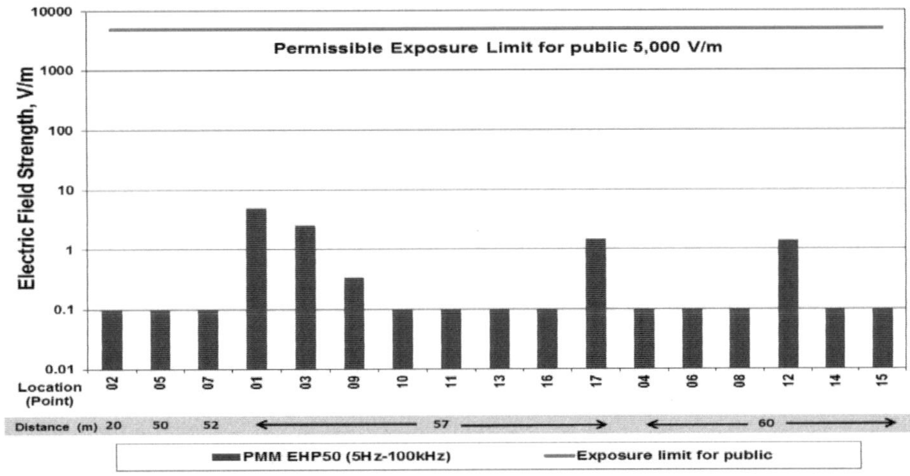

(a)

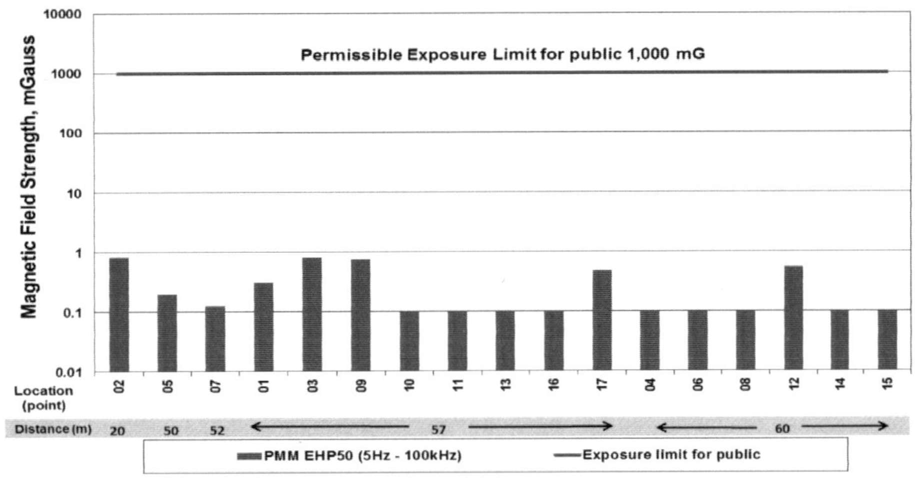

(b)

Figure 4. Plot of ELF (a) electric field strength and (b) magnetic field strength, against distance and point of measurement for location ELF1 and their comparison with the PEL for public.

The averaged magnetic field strengths were found to vary from 0.100 mG to 0.823 mG. The highest averaged value was at 0.823 mG, which was measured outside of the property at the road (Point 02), with the distance from the transmission line at 20 meters. The value was about 0.08% or around 1215 times lower than the PEL recommended by ICNIRP for members of the public (1000 mG). This value is significant as the point of measurement was the nearest compared to the other points. Another significant value was recorded at point 03 with the value recorded was 0.803 mG, which was 57 meters from the transmission line. For this situation, the magnetic field strength was more influenced by the electrical wiring from the auto gate system for the house. Magnetic field strength increases with current, hence, a stronger magnetic field would be detected near an appliance when it runs on high compared when it runs on low [1-5]. Similar with electric field, the magnetic field strength at points 03, 09, 12 and 17 was higher due to presence of electrical power points and electronic appliances. As for other areas inside the house, the averaged magnetic field strength recorded was at 0.100 mG at all the points inside the house. The value recorded inside the house was about 0.01% or about 10,000 times lower than that of the PEL. The plots of figure 4(b) indicate the magnetic field strength against distance and the point of measurement and their comparison with the PEL for public.

IOP Conf. Series: Materials Science and Engineering **298** (2018) 012007 doi:10.1088/1757-899X/298/1/012007

The spatial distribution of the magnetic field strengths around the house was captured using the LINDA system. The magnetic field strength from the contour map around the selected area indicated that the highest magnetic field strength was 0.822 mG, recorded along the road near the house, indicated as L.1 in the layout map (Figure 1). The value was at 0.08% of the PEL for public or 1216 times lower compared to the limit. As mentioned earlier, this significant value was due to the close proximity of the measurement area to the transmission line.

For the second location (ELF2), the electric field strengths were found to vary from 0.100 V/m to 0.516 V/m. The highest averaged level of the electric field strengths was at 0.516 V/m measured 1 meter in front of the TNB electrical distribution substation, point 01. The value was about 0.01% or around 9689 times lower than that of the PEL recommended by ICNIRP for members of public (5000 V/m). The value was significant compared to other points mainly due to the close proximity to the stated distribution substation and was without any shielding. The other points at 1 meter distance from the electrical distribution substation were ranging from 0.201 V/m to 0.218 V/m. There was a reduction in value as the sides and back of the substation were shielded with shrubs and trees. The electric field strength reduced drastically as the distance increased from point 5 onwards. The electric field strength inside the house was from the range of 0.130 V/m to 0.163 V/m. The value was low compared to the first location (ELF1) as there were no electrical appliances installed inside the house.

The plots of figure 5(a) indicate the electric field strength against distance and point of measurement and their comparison with the PEL. The averaged magnetic field strengths were found to vary from 0.100 mG to 3.217 mG. The highest averaged value was at 3.217 mG, which was measured at a distance of 1 meter from the front of the TNB electrical distribution substation, point 01. The value is about 0.32% or around 310 times lower than the PEL recommended by ICNIRP for members of the public (1000 mG). This value was significant as the point of measurement was the nearest compared to the other points. Another significant value was recorded from points 02 to 04 which were at 1 meter from the substation. The values ranged from 3.005 mG to 3.155 mG. It should be noted that the trees that are planted at the sides and back of the substation does not provide shielding from the magnetic field and the field strength does not differ much from point 01 which was measured without any shielding. The magnetic field strength reduced drastically as the distance increased from point 5 onwards. The values remained constant at 0.100 mG throughout the points 05 to 13. The plots of figure 5 (b) indicate the magnetic field strength against distance and point of measurement and their comparison with the PEL for public.

The spatial distribution of the magnetic field strengths around the house was captured using the LINDA system. The magnetic field strength from the contour map around the selected area indicated that the highest field strength was at 3.118 mG, recorded around the electrical distribution substation, indicated as L.1 in the layout map (Figure 2). The value was at 0.31% of the PEL for public or 320 times lower compared to the limit. This value was significant compared to other areas due to the close proximity of the measurement area to the substation.

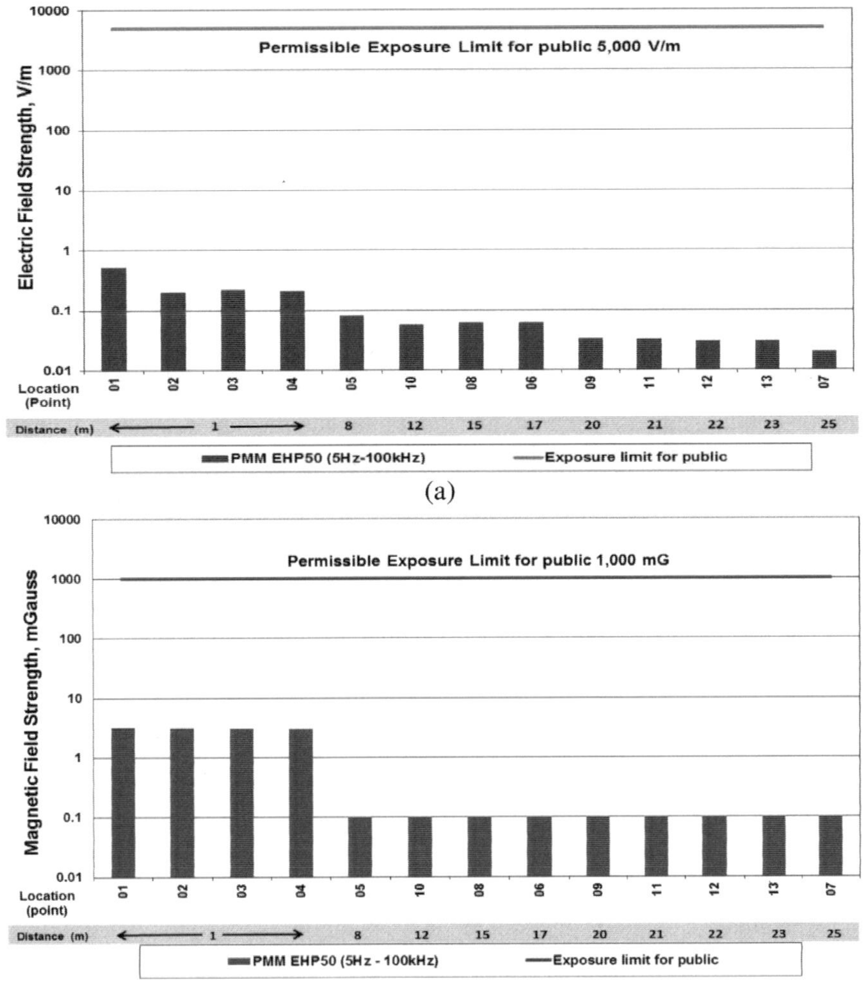

(b)

Figure 5. Plot of ELF (a) electric field strength and (b) magnetic field strength against distance and point of measurement for location ELF2 and their comparison with the PEL for public.

For the third location (ELF3), the electric field strengths were found to vary from 0.022 V/m to 0.250 V/m. The highest averaged level of the electric field strengths was at 0.250 V/m measured 120 meters in front of the TNB power station, point 01. The value is about 0.005% or around 20,000 times lower than that of the PEL recommended by ICNIRP for member of public (5000 V/m). The value is higher compared to other points of measurement mainly due to the close proximity to the stated power station. The electric field strength reduced drastically as the distance increased from point 2 onwards. The electric field strength inside the under-construction condominium was basically ranging from 0.022 V/m to 0.038 V/m. The value was low compared to the first location (ELF1) as there were no electrical wires and appliances installed inside the stated building. The plots of figure 6(a) indicate the electric field strength against distance and point of measurement and their comparison with the PEL for public. The averaged magnetic field strengths were found to vary from 0.100 mG to 0.700 mG. The highest averaged value was at 0.700 mG, which was measured 120 meters in front of the TNB power station, point 01. The value was about 0.07% or around 1428 times lower than the PEL recommended by ICNIRP for members of the public (1000 mG). This value was higher as the point of measurement was the nearest compared to the other points. The magnetic field strength reduced drastically as the distance increased from point 2 onwards. The values remained constant at 0.100 mG throughout the points 02 to 10. The plots of figure 6(b) indicate the magnetic field strength against distance and point of measurement and their comparison with the PEL for public.

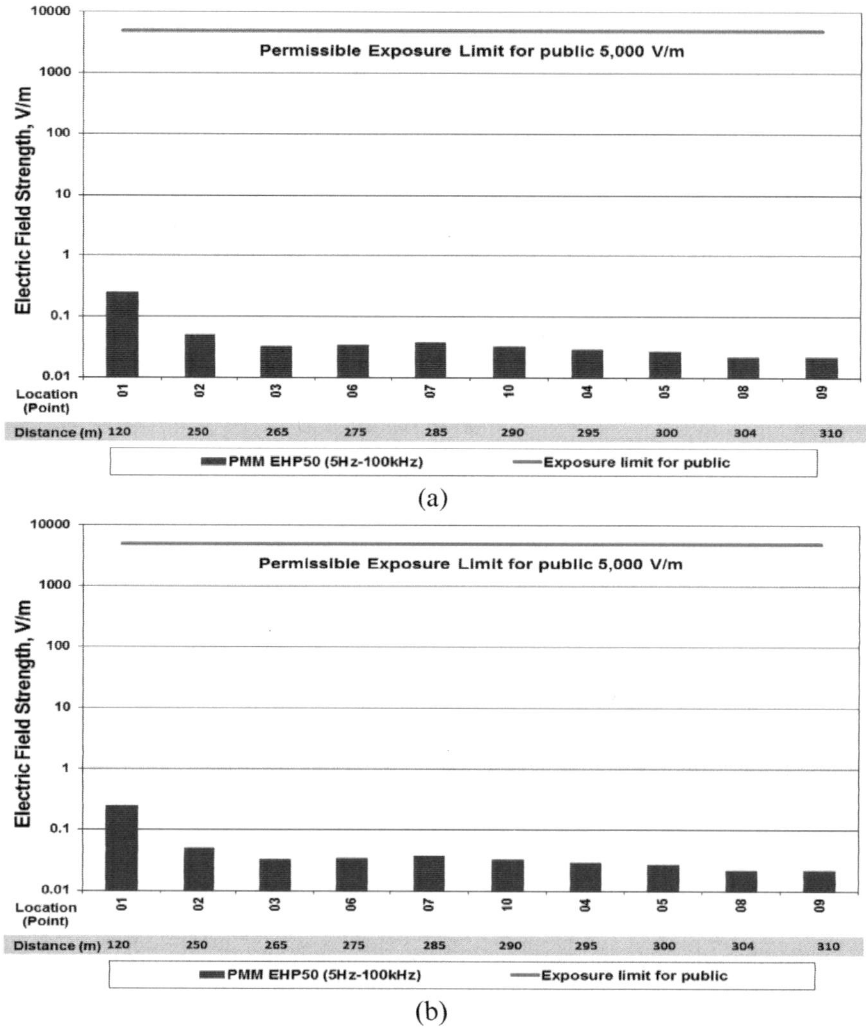

Figure 6. Plot of ELF (a) electric field strength and (b) magnetic field strength against distance and point of measurement for location ELF3 and their comparison with the PEL for public.

The spatial distribution of the magnetic field strengths around the house was captured using the LINDA system. The magnetic field strength from the contour map around the selected area indicated that the highest field strength was 0.184 mG, recorded around the electrical distribution substation, indicated as L.1 in the layout map (Figure 3). The value is 0.02% of the PEL for public or 5434 times lower compared to the limit.

6. Conclusion

The assessment of the electric and magnetic field strength has identified the presence of ELF EMF radiation at the stated locations. The highest measured exposure was at 0.10 % and 0.08 % of the PEL for the electric and magnetic field respectively. The overall results indicated that the electric and magnetic field strengths were very much lower than the maximum PEL recommended by ICNIRP and WHO, hence will not lead to any significant exposure received by members of the public.

IOP Conf. Series: Materials Science and Engineering **298** (2018) 012007 doi:10.1088/1757-899X/298/1/012007

7. References

[1] World Health Organization 2007 *Environmental Health Criteria 238, Extremely Low Frequency Fields*,

[2] International Commission on Non-Ionizing Radiation Protection (ICNIRP) 1998 Guidelines for limiting exposure to time-varying electric, magnetic and electromagnetic fields (up to 300 GHz) *Health Physics* **74**(4) 494-522

[3] International Commission on Non-Ionizing Radiation Protection (ICNIRP) 2010 Guidelines for limiting exposure to time varying Electric and Magnetic Fields (1Hz-100 kHz) 818-836

[4] Institute of Electrical and Electronics Engineers (IEEE) 2010 Recommended Practice for Measurements and Computations of Electric, Magnetic and Electromagnetic Fields with Respect to Human Exposure to Such Fields, 0Hz to 100 kHz,.

[5] IARC Working Group on the Evaluation of Carcinogenic Risks to Humans 2002 Nonionizing radiation, Part 1: Static and extremely low frequency (ELF) electric and magnetic fields *Monographs on the Evaluation of Carcinogenic Risks to Humans* **80** (Lyon IARC).

IOP Conf. Series: Materials Science and Engineering **298** (2018) 012008 doi:10.1088/1757-899X/298/1/012008

Evaluation of potential site for mineral processing plant

Muhamad Noor Izwan Ishak, Susan Maria Sipaun, Ismail Mustapha, Engku Mohd Fahmi Engku Chik, Nurliyana Abdullah, Airwan Affandi Mahmood

Plant Assessment Technology (PAT) Group, Industrial Technology Division, Malaysian Nuclear Agency, Bangi, 43000 Kajang, Malaysia.

noorizwan@nm.gov.my

Abstract. Nuclear moisture-density gauge is a type of instrument for measuring density and moisture of the material in a relatively thin zone beneath a surface of the material by using low activity of neutron and gamma radiation source. Density and moisture content data of the compacted layers are needed to determine the degree of compaction of soils, aggregate, concrete, asphalt or other materials used in civil engineering works. A gamma radiation source is mounted inside gauge housing with the source rod vertically extended to various depth positions. Direct transmission gamma radiation technique is used to obtain the count reading for the number of photons emitted before it is converted into density reading by microprocessor. This paper presents the inspection technique and results for the measurement of soil moisture and density carried out at potential site for mineral processing plant, Malaysian Nuclear Agency. Primarily, the experiment was conducted to ensure the compaction of ground is suitable for the plant construction. From the calculation, the percentages of soil wet density compaction (%WD Compact) are within acceptable limits with respect to the standard compacted wet soil density measured in the laboratory.

1. Introduction

In the construction of large structures such as roads, buildings, and levees, measurement of density and moisture content of soil is vital. Conventional and commonly used technique to determine density and moisture content of soil requires samples of soil column obtain from core sampling method and the soil from it conserved in an air-tight container. The volume of the core sampler is known and the soil weight was determined. From this measurement, the unit weight, or density of the wet soil, can be calculated. The soil must then be dried in an oven for at least twelve hours, weighed again, and the moisture content computed. This conventional method causes undue difficulties for both the contractor and the inspector at a construction site. The delay for test results is an economic liability to the contractor, while the inspector is limited by time in the number of tests he can take. Recent developments in nuclear moisture-density techniques offer a possibility for rapid, in-situ tests for soil moisture and density. Quick and efficient determination of these factors is of important concern to engineers and contractors.

2. Materials and Method

Density measurements are based on the rate of gamma photons emitted from a gamma source to be scattered or absorbed in approximate proportion to the density of the material through which they are passed. On the other hand, the process of thermalization or slowing down of fast neutrons, is the principle on which nuclear soil moisture detectors operate. Fast neutrons are thermalized by hydrogen in a soil mass. A count of thermalized neutrons produced by neutron irradiation of a soil indicates the amount of hydrogen, and therefore water, in the soil. As short irradiation times are required for both methods, complete results of nuclear moisture and density tests can be available in minutes.

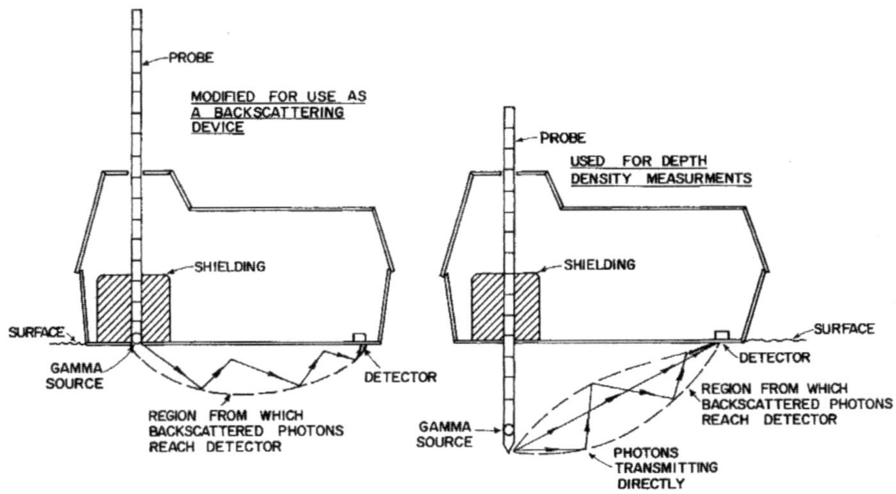

Figure 1: Gamma backscatter (left) and direct transmission (right) method.

The objective of the procedure was to evaluate the compaction value of soils on potential site for the construction of Mineral Processing Plant at Malaysian Nuclear Agency. The plant is a small scale system which is operated to ascertain the behaviour of Thorium-based nuclear process before utilizing it on a large industrial scale. A surface moisture-density gauge model Troxler 3430 was used throughout the evaluation. The instrument offers a selectable depth of measurement, ranging from back scatter position (0 mm) to 300 mm. The depth of 100 mm and counting time of one minute were used in the test. Measurement were done on flat surfaces only and a scrapper plate was used to prepare the surfaces. Care was taken not to distort the hole when removing the drill rod. The gauge was placed on the surface so that the source rod can enter the hole without disturbing any loose material around it. The source rod was lowered into the hole using the handle and trigger mechanism to 6 inches. The gauge was gently slide toward the ratemeter to make sure that the source rod was firmly in contact to the side of the hole. A set of three measurement per site was recorded. Figure 2 shows the test sites where the moisture and density measurements was taken.

2.1. Moisture Measurement

If the measurement of water content (moisture) of the soft ground, direct transmission gauge is more widely used compared to the backscatter type. Transmitted fast neutrons from the source are slowed down during interation with the nuclei of the constituent atoms in the ground. Fast neutron lose maximum energy when collide with hydrogen nucleus. The amount of fast neutrons will decrease, as more hydrogen is present in the ground. Therefore, fast neutron counting rates are roughly expressed as an exponential function of moisture content (water density), as expressed by the following equation,

$$R_m = Aexp(-B\rho_m) \tag{1}$$

where,

R_m : neutron count ratio,
A, B : calibration coefficients,
ρ_m : moisture content

3. Results and discussions

The results of the evaluation of potential site for mineral processing plant are tabulated in table 1. Figure 2 shows the sites where the moisture and density content readings are taken. Wet density, moisture, and dry density were recorded for each test site. The statistical stability test for the instrument was performed to detect any fluctuations in the readings. The analysis yields an acceptable result for the density and moisture within ideal limits. The higher wet density indicates the lesser amount of air spaces (voids) in the soil. The compaction of concrete and road asphalt also measured by dividing with standard density for the materials measured in the laboratory.

Dry density reading indicates the amount of soil density for with the absence of water (or the weight of a unit volume of a dry sample of soil, after the latter has been heated at a temperature of 103 degrees). Dry density readings measured for soil at sites 1 till 8 are within the range of 1700-1841 kg/m^3. Thus, they are within acceptable limits compared to the standard readings for compacted soil measured in laboratory (1836 kg/m^3), as indicated by percentage of wet density compaction. The lower the percentage of wet density compaction, the higher the amount of moisture contained in soil making it less suitable for any construction work. Ground on sites 3 and 4 are hard soil making it contain less moisture and higher compaction values compared with other soil regions.

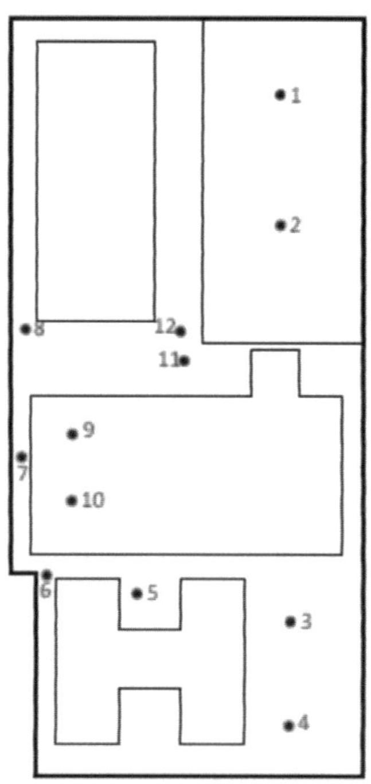

- 1, 2 : Soft Soil, Open Area
- 3, 4 : Hard Soil
- 5, 6 : Soft Soil, Barracks
- 7 : Soft Soil
- 8 : Soft Soil, Waste Storage Building, Block 33
- 9, 10 : Concrete, Block 39
- 12, 13 : Asphalt, Road

Figure 2. Site Map for Mineral Processing Plant where moisture and density content readings were taken.

Mineral Processing Plant is suggested to be located at Block 39. Sites numbered 9 and 10 are concrete ground within Block 39. The readings for the asphalt road are also taken at sites 11 and 12. The wet compaction readings for those sites indicates less than 90 percent compaction values with respect to the standard readings for concrete (2170 kg/m^3) and asphalt (2230 kg/m^3) taken in laboratory experiments.

Table 1: Test results for the percentage of wet density compaction.

Position (refer to Figure 2)	**1**	**2**	**3**	**4**	**5**	**6**	**7**	**8**
Material Measured	Soft Soil	Soft Soil	Hard Soil	Hard Soil	Soft Soil	Soft Soil	Soft Soil	Soft Soil
Wet Density (WD) (kg/m^3)	2006	1997	2032	2011	2053	2083	2028	2027
Moisture (kg/m^3)	279	297	205	170	267	260	284	230
Dry Density (kg/m^3)	1727	1700	1827	1841	1786	1823	1744	1797
Moisture (%)	14	14.8	10.3	8.5	13.4	13	14.2	11.5
WD Compact (%)	94.1	92.6	99.5	100.3	97.3	99.3	88.2	88.1

Table 2: Test results for the percentage of wet density compaction for both concrete and road asphalt

Position (refer to Figure 2)	**9**	**10**	**11**	**12**
Material Measured	Concrete	Concrete	Road Asphalt	Road Asphalt
Wet Density (WD) (kg/m^3)	2014	2001	1995	2025
Moisture (kg/m^3)	109	93	36	46
Dry Density (kg/m^3)	1908	1908	1959	1979
Moisture (%)	5.5	4.7	1.8	2.3
WD Compact (%)	87.9	87.9	87.8	88.7

IOP Conf. Series: Materials Science and Engineering **298** (2018) 012008 doi:10.1088/1757-899X/298/1/012008

4. Conclusion

The wet density measurement using nuclear gauge technique was used to evaluate the potential site for Mineral Processing Plant. From the values of percentages of wet density compaction, decisions can be made instantly by contractors whether or not the soil layer has appropriate compaction value. The evaluation result will help the contractors to determine the stability of the sites. Evaluation on compaction values for other materials like concrete and asphalts can also be made provided that the standard dry density for the materials are known before the gauge calibration is carried out. Soft soil recorded moisture percentage between 11.5 and 14.8 percent while hard soil recorded percentage between 8.5 and 10.3 percent. For concrete, moisture reading are 93 kg/m^3 and 109 kg/m^3. For road asphalt, moisture reading are 36 kg/m^3 and 46 kg/m^3.

5. References

[1] *Manual of Operation and Instruction* for *Surface Moisture-Density Gauge*, *Model 3430* Research Triangle Park, NC: Troxler Electronic Laboratories, Inc. (2015)

[2] *Density Testing And Inspection Manual* Lansing, Michigan: Manual Michigan Department of Transportation. (2010)

[3] Troxler R E and Linus W 2003 Dep, U.S. Patent No. 6,567,498 (20 May 2003)

[4] Tidwell L E, Ahlrich R C, Regan G L (1993) Evaluation of Surface Density Nuclear Gauges for Acceptance Testing of Asphalt Concrete Overlays 93-22345 Geotechnical Laboratory, Federal Highway Administration Washington.

[5] Friedenwald R L 1963 A Study Of The Troxler Nuclear Soil Density And Moisture Gauges E-27-63 Texas Transportation Institute, College Station, Texas.

[6] Ismail M A and Sinnakaruppan K 1998 The In-Situ Density Measurement for CBM at 9.57 Bound of the Kempas-Skudai Expressway MINT/T/1998/18(S) Nuclear Moisture-Density Laboratory, Malaysian Nuclear Agency, Selangor.

IOP Conf. Series: Materials Science and Engineering **298** (2018) 012009 doi:10.1088/1757-899X/298/1/012009

Potential impact of Andrassy bentonite microbial diversity in the long-term performance of a deep nuclear waste repository

M Y Mohd Tadza[1], M A Mohd Tadza[2], R Bag[3] and N S H Harith[4]
[1]Faculty of Civil Engineering & Earth Resources,Universiti Malaysia Pahang, 26300 Gambang, Pahang, Malaysia.
[2]Faculty of Science & Industrial Technology, Universiti Malaysia Pahang, 26300 Gambang, Pahang, Malaysia.
[3]Department of Civil Engineering & Environmental Engineering, Indian Institute of Technology Patna, Bihta, Patna 801106 (Bihar), India.
[4]Faculty of Engineering, Universiti Malaysia Sabah, 88400 Kota Kinabalu, Sabah Malaysia

dryuhyi@ump.edu.my

Abstract. Copper and steel canning and bentonite buffer are normally forseen as the primary containment component of a deep nuclear waste repository. Distribution of microbes in subsurface environments have been found to be extensive and directly or indirectly may exert influence on waste canister corrosion and the mobility of radionuclides. The understanding of clays and microbial interaction with radionuclides will be useful in predicting the microbial impacts on the performance of the waste repositories. The present work characterizes the culture-dependent microbial diversity of Andrassy bentonite recovered from Tawau clay deposits. The evaluation of microbial populations shows the presence of a number of cultivable microbes (e.g. *Staphylococcus, Micrococcus, Achromobacter, Bacillus, Paecilomyces, Trichoderma,* and *Fusarium*). Additionally, a pigmented yeast strain *Rhodotorula mucilaginosa* was also recovered from the formation. Both *Bacillus* and *Rhodotorula mucilaginosa* have high tolerance towards U radiation and toxicity. The presence of Rhodotorula mucilaginosa in Andrassy bentonite might be able to change the speciation of radionuclides (e.g. uranium) in a future deep repository. However, concern over the presence of Fe (III) reduction microbes such as Bacillus also found in the formation could lead to corrosion of copper steel canister and affect the overall performance of the containment system.

1. Introduction
Malaysia has interest in utilizing nuclear energy in the near future as a potential alternative for long-term energy production in peninsular Malaysia under the Economic Transformation Programme (ETP) [1][2]. The Malaysian Nuclear Power Corporation (MNPC) has been entrusted to undertake a comprehensive study to facilitate the Malaysian Government in the planning of nuclear power program including a proper disposal of high-level nuclear waste (HLW) to ensure better and safer environment [3]. It is expected that nuclear wastes generated in the nuclear fuel cycle must be safely stored for at least 100,000 years for the radiotoxicity to decrease to the levels similar to those of

natural uranium. Countries such as Sweden, Finland and France have opted for deep geological repositories (DGRs) as ultimate disposal of HLWs that is expected to be fully functional by the year 2020 [1].

The structure of DGR utilizes bentonitic clay as the buffer material. Bentonite has been selected due to its uniquely low permeability, self-sealing membrane characteristic, durability and high sorption capacity, making it a viable buffer material for HLW [4][5]. As of today, the sodium-based MX80 bentonite has been extensively researched and considered as international reference material for DGRs [5][6]. On the other hand, Deponit CA-N bentonite, from Milos formation has been considered as European standard [7][8]. Based on the these reports, various countries around the world have established their own reference material for example FEBEX for Spain, Kunigel V1 for Japan, GMZ for China to name a few [4][9][10][11]. Subsequent to national decision on nuclear power, Malaysia needs to consider its own reference material for development of its own DGR facility.

A typical bentonite is highly swelling clay derived from volcanic formation occurred several thousand years following the Pliocene age [12]. In Malaysia however, most of the bentonite used in the country are imported from neighboring country such as Pakistan and Indonesia [13]. Commercial usage of locally available bentonite was found to be scarce or nonexistent. The deposit sits in the geological heritage of Sabah and is well preserved. The first occurrence of bentonite formation in Malaysia was reported by Irawan and Samsuri [14] in the volcanic formation in Andrassy, Sabah [15]. Study of this material is crucial for the development of Malaysia's nuclear power disposal programme and may be considered as Malaysia's reference material, however, detailed characteristics and behaviour of Andrassy bentonite are scarcely documented. To date previous studies only focused on geotechnical and microbiological characterization of Andrassy bentonite and their effect on the water retention characteristics [10][16][17].

The safety of this long-term geological disposal could be compromised by physical and chemical factors, and also by the biogeochemical activity. Thus, the microbiological interactions between bentonite buffer material, host rock and copper steel canister plays a significant role in the long-term stability of the DGR. Microbiological processes can affect the geochemistry of bentonite barriers through four different mechanisms: (i) reduction of structural Fe (III) of clay minerals, (ii) alteration of mineral surfaces by the production of siderophores and small-organic acids, (iii) formation of biofilm in the clay mineral surface and (iv) to control the speciation and mobility of radionuclides [18]. For instance, previous studies have shown that the presence of *Geobactor Sp.* in MX80 bentonite has the ability to degrade radioactive material contamination [19]. On the other hand, the presence of *Rhodotula Sp.* found in Spain bentonite has the ability to precipitate Uranium based radionuclides [18]. To the best of the authors' knowledge, no attempt has been made to study the influence and potential of Andrassy bentonite microbiological diversity on the long-term performance of DGR.

The objective of the present work is twofold (i) to identify the microbiological characteristics of Andrassy bentonite and (ii) to determine the potential impact of the microbes identified on both corrosion of copper steel canister and interaction with radionuclides.

2. Material and Methods

2.1. Geographical description of the clay sample recovery sites

The bentonite samples used in this work were recovered from Andrassy in Tawau (N4° 18.97'- E 117° 57.37') shown in Figure 1. Sample was first excavated and characterized by determining the physical, chemical mineralogical and microbiological properties. The sample was excavated on site by using a hand auger up to a depth of about 5 meter. The soil sample were carefully sealed and placed in an air-tight container to avoid contamination. The geotechnical properties is presented in Table 1.

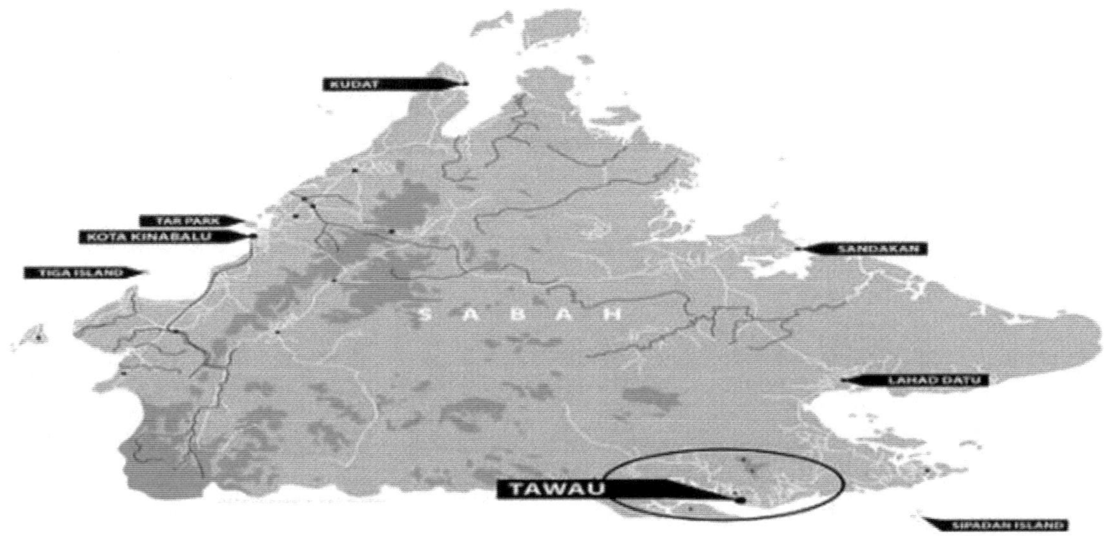

Figure 1. Map of Andrassy bentonite sample locations.

Table 1. Comparison of Geotechnical properties of Andrassy bentonite with international reference bentonites [10][17]

Properties	Andrassy	MX80	Deponit CA-N	GMZ
Specific gravity, Gs	2.78	2.80	2.84	2.66
Liquid limit (%)	129.30	437	135	276
Plastic Limit (%)	46.12	63	58	37
Shrinkage limit (%)	13.40	12.2	13.6	-
Cation exchange capacity (meq/100g)	42.15	90.31	42.78	77.3
Specific surface area (m^2/g)	734.27	676	734	570
Main minerals (%)				
Montmorillonite	63.20	75.00	81.00	75.4
Quartz	11.50	10	-	-
Vermiculite	25.30	-	-	-
Graphite	-	-	19	-
Cristobalite	-	15	-	-

2.2. Determination of microbiological characteristics

The microbiological properties of the bentonite, namely bacteria and fungus determination were carried out following plating, slide culture, streaking and isolation techniques [20]. Potato dextrose agar (PDA) was used for culturing fungi, whereas Nutrient agar (NA) was used to culture bacteria. The bentonite specimen was initially suspended in 0.9% NaCl solution to separate the microbes [21]. Identification of the specific strain of each microbes after isolation was carried out in an independent laboratory using polymerase chain reaction (PCR) protocol and referred to international microbiological characterization database.

3. Results and discussion

The bentonite in this study was found to exhibit a large surface area and high surface charge characteristics which make it ideal for soil microbes [16][17]. Based on the plating, isolation and PCR techniques, the microbes determined in Andrassy bentonite is presented in Table 2. Concurrent with the large surface area, nine microbes were successfully isolated and identified. These isolates belonged to the phyla Firmicutes, Actinobacteria, Proteobacteria, Sac Fungi, and Mycota. Based on origin of closed related sequence (i.e. where these types microbes can be abundantly be found), six of them were commonly found in soil. Surprisingly, two bacteria most commonly found in tobacco leaf and textile effluents are also present. Three species of bacteria (i.e. *Bacillus anthracis, Staphylococcus aureus, Micrococcus luteus*) and one species of fungus, a yeast strain (i.e *Rhodotorula mucilaginosa*) found in this study was similar to that of bacteria and fungi found in bentonite from Spain [18]. Another type of microbe identified was *Escherichia coli*. However, this bacteria was ruled out due to possible cross contamination during handling of the soil specimens during transport and preparation [16].

Table 2. The microbiological properties of Andrassy Bentonite

Phylum	Closest phylogenetic relative	Origin of closed related sequence
Bacteria		
Firmicutes	*Bacillus anthracis*	Soil
Firmicutes	*Staphylococcus aureus*	Tobacco leaf
Actinobacteria	*Micrococcus luteus*	Textile effluent
Proteobacteria	*Achromobacter xylosoxidans*	Soil
Proteobacteria	*Escherichia coli*	Fecal matter
Fungus		
Sac Fungi	*Paecilomyces lilacinus*	Soil
Ascomycota	*Trichoderma atroviride*	Soil
Sac Fungi	*Fusarium proliferatum*	Plant and soil
Basidiomycota (Yeast)	*Rhodotorula mucilaginosa*	Soil

3.1. Interactions of Andrassy bentonite microbial populations on radionuclides

In the case of radionuclides spillage or leakage from the DGR's gallery, the bentonite buffer would serve as buffer medium to prevent the migration of radionuclides. Furthermore, the microbes available in the buffer material would react to the movement of the radionuclides. The yeast strain *Rhodotorula mucilaginosa* found in the bentonite samples have high tolerance towards high concentration of Uranium. The yeast strain is also capable of precipitating Uranium based radionuclide into less hazardous U-phosphate minerals phases [18]. Thus, the presence of *Mucilaginosa R.* is crucial for bioremediation of radioactive contaminated sites including DGR applications. In addition, the presence of *Bacillus Sp.* have been shown to convert aerobic uranium (VI) precipitation to insoluble U phosphate phases due to the activity of acidic phosphatase which in turn assisted in immobilizing different radionuclides [22].

3.2. Effect of Andrassy bentonite microbial populations on corrosion of HLW's canister

The copper steel canister would be placed in close vicinity to the bentonite buffer material. Thus, interactions between microbes found in the bentonite buffer and the canister is expected to occur. Apart from beneficial microbes explained in Section 3.1, the presence of other Fe and Cu reducing microbes found in the same bentonite specimen would cause corrosion problems to the copper steel canisters used for storing HLWs. For instance, the presence of *Mucilaginosa R.*has also been reported to cause damages to aircrafts [23]. They can cause degradation of aircraft components due to their ability to uptake Cu [24]. Similarly, the existence of *Luteus M., Aureus S., Xyloxidan A.* and *Lilacinus P.* may also contribute to the degradation of the canister under long exposures. Studies revealed that these microbes have the ability to cause corrosion in medical devices, steel pipelines and transmission towers [25][26][27]. Due to their unique ability, these microbes has long been used as remediation agents in remediation of contaminated sites and contaminated water all over the world [28][29][30]. However, in the perspective of construction of DGRs, in the long-term these microbes may degrade and affect the stability of the copper steel canister and causes leakage of radionuclides into the environment.

4. Conclusion

The present study describes the culture-dependent microbial diversity of Andrassy bentonite formations. The bentonite were carefully collected and characterised. Based on the microbiological analyses, it was noted that the bentonite contained beneficial microbes (i.e. *Mucilaginosa R.* and *Bacillus Sp.*) for DGR application as buffer material in restricting the migration of radionuclides. On the contrary the bentonite also consists of microbes (i.e. *Luteus M., Aureus S., Xyloxidan A.* and *Lilacinus P.)* that have to potential to induce corrosion to copper steel canister used for safe long-term storage of the HLWs.

5. References

[1] Jaafar M Z, Nazaruddin N H and Lye J T T 2017 *AIP Conference Proceedings* **1799**(1) 020001
[2] Economic Planning Unit (EPU) 2017 Strategy Paper 17: Sustainable Usage of Energy to Support Growth
[3] Pusch R 2009 *Geological storage of highly radioactive waste: Current concepts and plans for radioactive waste disposal*. Springer Science & Business Media.
[4] Madsen F T 1998 *Clay Minerals*, **33**(1) 109-129
[5] Tadza M Y M 2011 *Soil-water characteristic curves and shrinkage behaviour of highly plastic clays: an experimental investigation.* Cardiff Universiti PhD Thesis
[6] Tripathy S, Tadza M Y M and Thomas H R 2014 *Canadian Geotechnical Journal* **51**(8) 869-883
[7] Koch D 2008 *Science & Technology Series* **334** 23-30
[8] Mancuso C, Jommi C and D'Onza F 2012 *Unsaturated soils: research and applications* Springer
[9] Ye W M, He Y, Chen Y G, Chen B and Cui Y J 2016 *Environmental Earth Sciences* **75**(10) 906
[10] Tadza M Y M, Azmi N S M, Mustapha R, Desa N D and Samuding K 2017 *AIP Conference Proceedings* **1799**(1) 030002
[11] Villar M V and Lloret A 2007 *Physics and Chemistry of the Earth, Parts A/B/C* **32**(8) 716-729
[12] Reeves G M, Sims I and Cripps J C 2006 *Clay materials used in construction*. Geological Society of London
[13] Lim S C, Gomes C, Kadir M Z A A and Abidin M 2013 *International Journal of Electrochemical Science* **8**(9) 11429-11447
[14] Irawan S and Samsuri A 2007 *Petrochemical Catalysis Technology*
[15] Tahir S, Musta B and Rahim I A 2010 *Geol. Soc. Of Malaysia* **56** 79-85
[16] Tadza M Y M, Asras M F F, Tadza, M A M, Ismail J, Mahazam N and Azmi, N S M 2016 Jurnal Teknologi **78**(8-5) 11-16.
[17] Tadza M Y M, Mahazam N and Tripathy S 2017 *International Journal Of Geomate* **12**(32) 30-36
[18] López-Fernández M, Fernández-Sanfrancisco O, Moreno-García A, Martín-Sánchez I, Sánchez-

Castro I and Merroun M L 2014 *Applied geochemistry* **49** 77-86
[19] De Canniere P and Meleshyn A 2013 *Proceedings of EUROSAFE Forum 2013* Cologne Germany
[20] Benson H J 2001 *Microbiological Applications: A Laboratory Manual in General Microbiology* McGraw-Hill Science
[21] Zborowski M, Malchesky P S, Jan T and Hall G S 1992 *J. of General Microbiology* **138**, 63-68
[22] Zhang G, Senko M J, Kelly D S, Tan H, Kemner M K and Burgos D W 2009 *Geochim Cosmochim Acta* **73** 3523–3538
[23] Hagenauer A, Hilpert R and Hack T 1994 *Werkst Korros* **45** 355–360
[24] Leticia V, Beatriz L and Mauricio C 2012 *International Biodeterioration & Biodegradation* **69** 28–37
[25] Sette L D, Rodrigo M, Passarini Z, Rodrigues A, Leal R R, Christina K and Pagnocca F C 2010 *Fungal diversity* **44**(1) 53–63
[26] Wan Y Z, Xiong G Y, Liang H, Raman S, He F and Huang Y 2007 *Applied Surface Science*, **253**(24) 9426–9429
[27] Romero J M, Angeles-Chavez C and Amaya M 2004 *Corrosion Engineering, Science and Technology* **39**(3) 261–264
[28] Nakajima A, Yasuda M, Yokoyama H, Ohya-Nishiguchi H and Kamada H 2001 *World J Microb Biot* **17**, 343-347
[29] Tunali S, Cabuk A and Akar T 2006 *Chem Eng J* **115** 203-211
[30] Siddiquee S, Rovina K, Azad S, Al Naher L, Suryani S and Chaikaew P 2015 *Microbial & Biochemical Technology* **7**(6) 384–393

IOP Conf. Series: Materials Science and Engineering **298** (2018) 012010 doi:10.1088/1757-899X/298/1/012010

Measurement of 238U and 232Th radionuclides in ilmenite and synthetic rutile

M I Idris[1,a)], K K Siong[1] and S M Fadzil[1]

[1]School of Applied Physics, Faculty of Science and Technology, Universiti Kebangsaan Malaysia, 43600 UKM, Bangi, Selangor Darul Ehsan

[a)]aidzat@ukm.edu.my

Abstract. The only factory that currently processes ilmenite to produce synthetic rutile is Tor Minerals in Ipoh, Perak, Malaysia. These two minerals contain radioactive elements such as uranium and thorium. Furthermore, this factory was built close to the residential areas. Thus, the primary issues are radiation exposure attributed to the decay of the radionuclides. Hence, the objectives of this study are to measure the dose and to evaluate activity levels of uranium and thorium. Dose rates from surrounding area of factory indicate the normal range for both on the surface and 1 meter above the ground (0.3-0.7 μSv/hr) lower than the global range of 0.5-1.3 μSv/hr set by UNSCEAR. The mean activity levels of uranium and thorium for ilmenite are 235 Bq/kg and 503 Bq/kg while for synthetic rutile are 980 Bq/kg and 401 Bq/kg, respectively. The result shows that uranium activity levels of synthetic rutile is 4 times higher than ilmenite but it is still lower than the regulatory exemption limit of 1000 Bq/kg set by IAEA Basic Safety Standards. Even though the dose rates at the factory and the activity levels are within safe limits, safety precautions must be followed by the factory management to prevent any unwanted accident to occur.

1. Introduction

Tor Minerals International, was founded by Benelite Corporation of America in 1973. "Benelite Process" was developed by Benelite and patented for producing synthetic rutile. Synthetic rutile is the raw material used in producing HITOX TiO_2 pigment. Benelite licensed and helped to build several plants around the world, which have utilized this process, including Tor Minerals in Ipoh, Malaysia [1].

Synthetic rutile can be produced by reducing the iron oxide in ilmenite to metallic iron using carbon monoxide, followed by reoxidation and separation from the TiO_2 rich fraction (Becher process) or leaching with hydrochloric acid (Benelite process) [2]. Both synthetic rutile and ilmenite contain radioactive elements such as Uranium-238 and Thorium-232. The probability of exposure to radiation due to decay of radioactive products is high depending on the dose rate and the activity levels during processing of these materials. Hence, a safety measure should be taken to ensure that radiation exposure to the public and workers does not exceed the limit. Therefore, the objectives of this study are to measure the dose rates surrounding the factory area which is located near the residential area and to evaluate the activity levels of uranium and thorium in the ilmenite and synthetic rutile samples.

IOP Conf. Series: Materials Science and Engineering **298** (2018) 012010 doi:10.1088/1757-899X/298/1/012010

2. Methodology

2.1. Study area

The study areas were located around Tor Minerals factory (Figure 1 and 2). The Tor Minerals factory is in the Kinta district in the state of Perak (Jalan Bukit Merah). However, it was built near residential area.

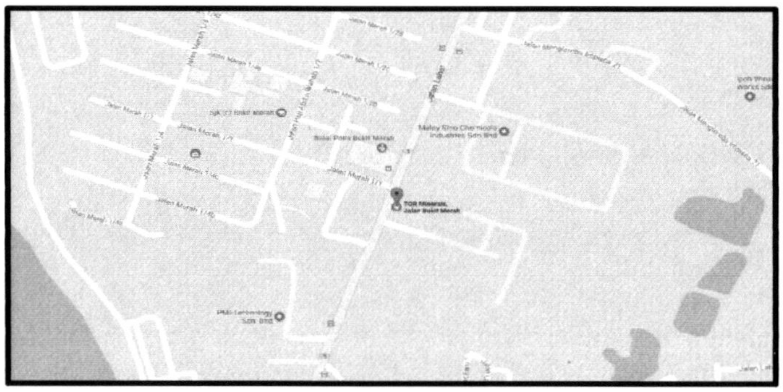

Source: Google Map, 2017

Figure 1. Location of Tor Minerals, Ipoh.

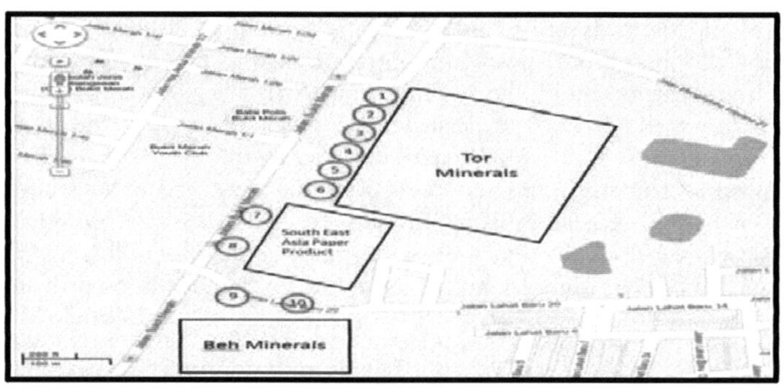

Source: Google Map, 2017

Figure 2. Sampling Points at Tor Minerals factory

2.2. Dose rates measurement

"LUDLIM Rate Meter" (Geiger-Mueller Counter) was used to measure the radiation dose around the study area. At each sampling point, two radiation dosage readings were recorded, one at the soil surface and the other at one meter above the soil surface [3]. Coordinates of each location were recorded using Silva Multi-Navigator equipment.

2.3. Radioactivity measurement

Samples of ilmenite and synthetic rutile were taken directly from the factory's stockpile. Each sample was weighed, sealed and left for a month to achieve a secular equilibrium [4]. The samples were measured using a gamma spectrometry system with a High-Purity Germanium (HPGe) detector and analyzed with Genie 2000 software. The HPGe detector is enclosed inside CANBERRA 747 shielding with 10 cm thickness lead coated with 1 mm and 1.6 mm of tin and copper respectively, in order to reduce the background radiation from building and cosmic rays. Energy resolution for the 1332.5 keV energy peak was 1.80 keV and the relative efficiency of the detector was 30%. The energy calibration of gamma spectrometry system was done using a mixture of ^{22}Na, ^{51}Cr, ^{57}Co, ^{60}Co, ^{85}Sr, ^{88}Y, ^{109}Cd, ^{113}Sn, ^{137}Cs, ^{123}mTe, and ^{241}Am radionuclides. The efficiency calibration was done using an in-house preparation silica standard spiked with mixture of ^{51}Cr, ^{57}Co, ^{60}Co, ^{85}Sr, ^{88}Y, ^{109}Cd, ^{113}Sn, ^{137}Cs, ^{123}mTe, ^{210}Pb and ^{241}Am radionuclides. Activity levels of Thorium (^{232}Th) were determined through the ^{228}Ac (338.32, 911.20, and 968.97 keV) and ^{208}Tl (583.19 and 2614.51 keV) [5,6].

2.4. Determination of specific activities of radionuclides

The specific activity of radionuclide was determined using comparative method with the following equation [7]:

$$W_s = \left(\frac{M_p \times A_s}{M_s \times A_p}\right) \times W_p \tag{1}$$

where,

W_s : Specific activities in samples (Bq/kg).
W_p: Specific activities in certified reference material (Bq/kg).
M_s: Mass sample (g).
M_p: Mass certified reference material (g).
A_s: Activities of samples (cps)
A_p:Activities of certified reference material (cps)

3. Results and discussion

The surface dose rates from surrounding areas of Tor Minerals factory show normal range reading for both, on the surface (0.3-0.7 μSv/hr) and 1 meter above the ground (0.3-0.7 μSv/hr).

The results indicate that sampling points number 3, 4, 9 and 10 show higher dose rates than the other locations because they are located near ilmenite storage and Beh Minerals factory which has monozite storage. However, the dose rates from surrounding areas are below the global range of 0.5-1.3 μSv/hr [4]. In general, dose rate on soil surface depends on the weather. Based on Assessment of radiation doses to the population in the environment of a nuclear power plant, calculations may statistically consider the variability of weather-dependent dispersion conditions as well as any seasonal differences in radiation doses. For example in dry weather, radionuclides are deposited on the surface of the soil. However, the radionuclides quickly penetrate into the first centimeter of soil after rainfall. Therefore, if there is no further rainfall, the radiation will return to normal within a few hours or days as the soil saturation disappears. This will influence the radiation measurement at the surface of the soil. One meter from the soil surface is generally used as reference for surface radiation dose measurement [4].

Table 1. Coordinates of sampling points and dose rates measurement.

Sampling point	Location		Surface Dose Rate (μSv/hr)	
	North	**South**	**On the surface**	**1 m above**
1	4°32.800	101°2.441	0.3	0.3
2	4°32.805	101°2.561	0.4	0.3
3	4°32.794	101°2.532	0.6	0.5
4	4°32.801	101°2.549	0.7	0.6
5	4°32.772	101°2.555	0.3	0.3
6	4°32.801	101°2.548	0.4	0.3
7	4°32.751	101°2.537	0.4	0.3
8	4°32.701	101°2.505	0.5	0.4
9	4°32.662	101°2.506	0.7	0.6
10	4°32.681	101°2.514	0.7	0.7
		Range	0.3-0.7	0.3-0.7
		Mean	0.5	0.43

The results indicate that sampling points number 3, 4, 9 and 10 show higher dose rates than the other locations because they are located near ilmenite storage and Beh Minerals factory which has monozite storage. However, the dose rates from surrounding areas are below the global range of 0.5-1.3 μSv/hr [4]. In general, dose rate on soil surface depends on the weather. Based on Assessment of radiation doses to the population in the environment of a nuclear power plant, calculations may statistically consider the variability of weather-dependent dispersion conditions as well as any seasonal differences in radiation doses. For example, in dry weather, radionuclides are deposited on the surface of the soil. However, the radionuclides quickly penetrate into the first centimeter of soil after rainfall. Therefore, if there is no further rainfall, the radiation will return to normal within a few hours or days as the soil saturation disappears. This will influence the radiation measurement at the surface of the soil. One meter from the soil surface is generally used as reference for surface radiation dose measurement [4].

The mean activity levels of uranium and thorium for ilmenite are 235 Bq/kg and 503 Bq/kg while for synthetic rutile are 980 Bq/kg and 401 Bq/kg, respectively as shown on Figure 3.

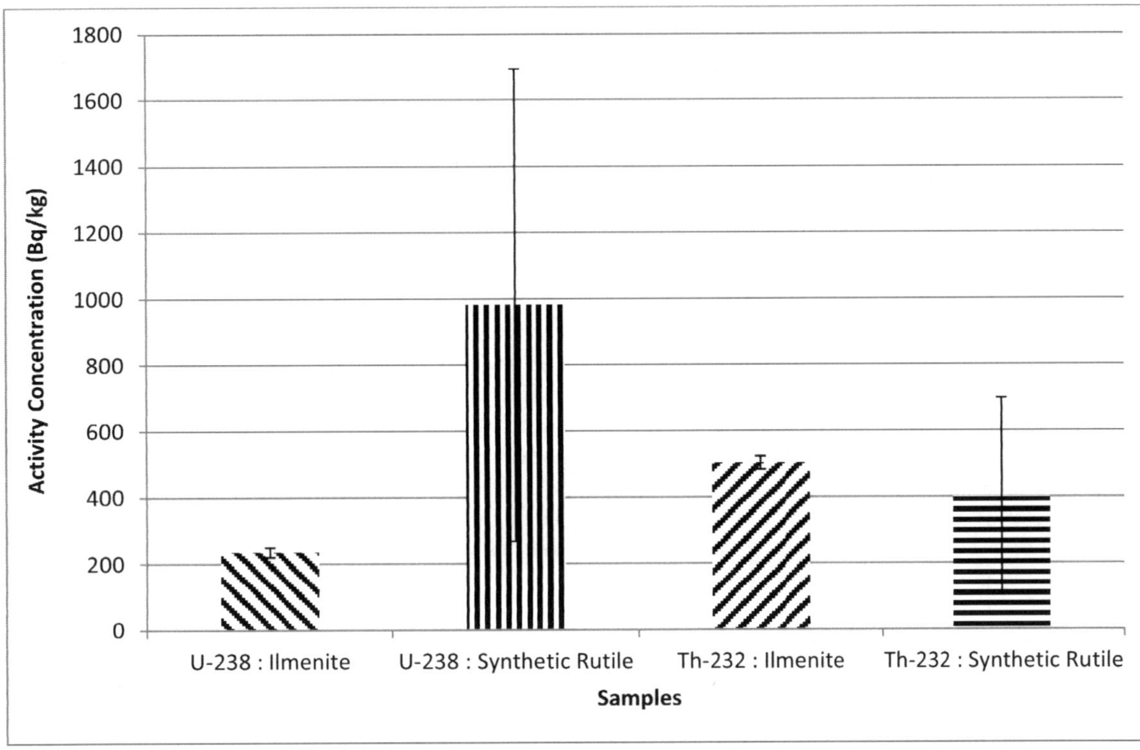

Figure 3: Activity levels for ^{238}U and ^{232}Th in ilmenite and synthetic rutile samples.

From figure 3, the result shows that uranium activity level of synthetic rutile is 4 times higher than activity level of ilmenite but it is still below than the limit of 1000 Bq/kg set by IAEA Basic Safety Standard [8]. Comparing with Chinnaekksaki *et al.* [9], the mean activity levels of Uranium in synthetic rutile taken from production plant in Tamil Nadu, India is very low (38.23 Bq/kg). Additionally, Cooper [10] showed that synthetic rutile processed in Australia also has low activity level of Uranium at 210 Bq/kg and it is lower than uranium activity in its corresponding ilmenite (300 Bq/kg). In this study, the activity level of Uranium in synthetic rutile is shown higher than ilmenite may be caused by the incomplete Uranium extraction from ilmenite. Based on Iluka's Synthetic Rutile Production Reports [11], all ilmenites consist low levels of Uranium and Thorium. However, in some mineral sands deposits these levels can be elevated near 500ppm U plus Th. The elements tend to remain with the TiO_2 during the synthetic rutile process. Therefore, they are found in even higher levels in the synthetic rutile product.

In the ilmenite chemical processing, it has been reported that 77% of the initial Thorium activity end up in the iron oxide waste and 18% will go into synthetic rutile [12]. Apparently from figure 3, the activity level of thorium in synthetic rutile is not 18% of the initial but it is still lower than ilmenite sample.

4. Conclusion

It is observed that the dose rates surrounding Tor Minerals factory are higher than the background dose due to the presence of thorium and uranium in ilmenite and synthetic rutile. Moreover, it is observed that activity level of uranium in synthetic rutile is higher than activity level of thorium. It is concluded that the residential area surrounding Tor Minerals factory are safe in the context of public health with respect to radioactivity. Future studies shall also include the interior of the factory in order to assess the radiation risks to the plant workers. Radon level reading shall also be included.

5. References

[1] Tor Minerals International 2007 Tor Minerals History (http://www.torminerals.com/history.html. 10 May 2012).
[2] Mcnulty G S 2007 Production of titanium dioxide (IAEA TECDOC -NORM-V Vienna IAEA)
[3] UNSCEAR 2000 Sources and effects of ionizing radiation (New York United Nation).
[4] Ismail B, Nasirian M and Pauzi A 2007 Radioactivity and radiological risk associated with effluent sediment containing technologically enhanced naturally occurring radioactive materials in amang (tin tailings) processing industry *J. of Environmental Radioactivity* **95** 161-170.
[5] Hassan A M, Abdel W M, Nada A, Walley E N and Khazbak A 1997 Determination of Uranium and Thorium in Egyptian monazite by gamma-ray spectrometry *Appl. Radiation Isotope* **48** 149-152.
[6] Aznan F I, Yasir M S, Amran, Ab M, Ismail B, Redzuwan Y and Irman, Abd R 2009 Radiological studies of naturally occurring radioactive materials in some Malaysia's sand used in building construction *The Malaysian J. of Analytical Sciences* **13** 29-35.
[7] Yasir M S, Majid A Ab, Ibrahim F, Tap M S Q and Abidin M R Z 2006 Analisis ^{238}U, ^{232}Th, ^{226}Ra dan ^{40}K dalam sampel amang, tanah dan air di Dengkil, Selangor menggunakan spektrometri sinar gama *Malaysia J. of Analytical Sciences* **10 (1)** 35- 40.
[8] IAEA 1996 Basic Safety Standards Safety Series **115** (Vienna: IAEA)
[9] Chinnaesakki S, Bara S V, Sartandel S J, Tripathi R M and Puranik V D 2010 Radiological characterisation of synthetic rutile using HPGe gamma spectrometry *Rad. Protection Dosimetry* **140(4)** 378-382.
[10] Cooper M B 2005 Naturally Occurring Radioactive Materials (NORM) in Australian Industries-Review of Current Inventories and Future Generation (Radiation Health and Safety Advisory Council).
[11] Iluka's Synthetic Rutile Production Reports 2012 (Product and Technical Development in Australia).
[12] Haridasan P P, Pillai P M B, Tripathi R M and Puranik V D 2007 Thorium in ilmenite and its radiological implications in the production of Titanium Dioxide *Rad. Protection Dosimetry* **49** 220-227.

Acknowledgement
The authors wish to thank to Ministry of Higher Education for the research grant given to this project (reference number UKM-ST-07-FRGS0235-2010) also to the technical support from Nuclear Science Programme, UKM.

IOP Conf. Series: Materials Science and Engineering **298** (2018) 012011 doi:10.1088/1757-899X/298/1/012011

Radioactive release during nuclear accidents in Chernobyl and Fukushima

Siti Nur Ain Sulaiman[1(a)], Faizal Mohamed[2], Ahmad Nabil Ab Rahim[3]

[1,2] Nuclear Science Programme, School of Applied Physics, Faculty of Science & Technology, National University of Malaysia (UKM)
[3] Reactor Technology Centre, Technical Support Division, Malaysian Nuclear Agency

sitinurain@siswa.ukm.edu.my

Abstract. Nuclear accidents that occurred in Chernobyl and Fukushima have initiated many research interests to understand the cause and mechanism of radioactive release within reactor compound and to the environment. Common types of radionuclide release are the fission products from the irradiated fuel rod itself. In case of nuclear accident, the focus of monitoring will be mostly on the release of noble gases, I-131 and Cs-137. As these are the only accidents have been rated within International Nuclear Events Scale (INES) Level 7, the radioactive release to the environment was one of the critical insights to be monitored. It was estimated that the release of radioactive material to the atmosphere due to Fukushima accident was approximately 10% of the Chernobyl accident. By referring to the previous reports using computational code systems to model the release rate, the release activity of I-131 and Cs-137 in Chernobyl was significantly higher compare to Fukushima. The simulation code also showed that Chernobyl had higher release rate of both radionuclides on the day of accident. Other factors affecting the radioactive release for Fukushima and Chernobyl accidents such as the current reactor technology and safety measures are also compared for discussion.

1. Introduction

Nuclear accident does not commonly occur, but when it happens, it will raise worldwide interest especially on the safety aspect. There are many factors affecting the plant operation and leads to failure such as human error, reactor design, and natural disaster. The age-profile of the NPP may also indicate whether the safety evaluation is satisfactory and in-line to protect from the latest potential threat.

The fission products released to the environment or source term is originated from the irradiated fuel rod in the reactor core. Depending on the half-life of radionuclide, this factor will determine how long it will stay in the radioactive plume and may dispersed towards the public area. However, some short half-lived radionuclides may be hazardous as well, for example; Iodine-131 (I-131) because iodine is naturally absorbed by thyroid and food chain. For long half-lived radionuclides, this may be a great concern as it will require additional effort such as decontamination and relocation for a longer term.

The Chernobyl and Fukushima nuclear accidents are now remained the most catastrophic accident in the nuclear history. Although the cause of the accident and the effects are different from each other, it is still trigger the question whether having a nuclear power plant in one's country is a good source of energy when the safety issues are still a concern. Both accidents were rated Level 7 by International Nuclear Events Scale (INES) which describes the accident to be serious enough to require a country to implement countermeasures to protect the public from the health and environmental effects of radiation, but it does not mean that these effects have occurred.

Radionuclides I-131 and Cs-137 were chosen as the primary release in this work. The radioactive release caused by the accidents was analyzed by referring to the previous reports which have used computational code systems to model the release rate for both accidents. While comparing the results as both showed different aftermath accident consequences, discussions are made to enlighten the issue.

1.1. Chernobyl Nuclear Accident

On 26th April 1986, operators were preparing to test the backup cooling system in reactor Unit 4. This test allows power to be provided in the event of a station blackout. However, the routine safety drill went horribly wrong. During the test, almost all of the control rods were removed. The reactor became unstable due to large positive coefficient of reactivity under certain conditions and the faulty actions of the operators for switching off the emergency safety systems of the reactor [1].

Technicians began to lose control of coolant flow as the lower transitions joints that link the zirconium fuel channels in the core to the steel pipes that carry the inlet cooling water weakened, causing the temperature in the reactor core to be extremely increased. Fuel rods started to melt, exposing zirconium (Zr) to react with hydrogen in steam (H¬2O), thus resulted in an explosion. It blew up the roof top of the building including the top of the reactor core vessel, spilling out eight tons (8 Tons) of radioactive debris.

Information on the details of accident including the radioactive release on that time is available in a meeting report by IAEA [2]. The radioactive release from this accident occurs continuously for over 10 days. Although the release rates recorded were varying, the highest release rate was on the day of the explosion, which bared the inside of reactor directly to the atmosphere. After 10 days, the release decreased abruptly, indicating the end of intense release period [3]. The radioactive release was affecting greatly to the reactor neighbouring regions in European part of the former Soviet Union.

1.2. Fukushima Nuclear Accident

A great earthquake with a magnitude of 9.0 in East Japan awakened a 15-metre tsunami and causing three Fukushima reactors to be disabled. When the earthquake hit the coast, the seismic sensors triggered the insertion of the control roads to stop the fission reaction. However, the off-site power supply was lost, activated the diesel-generator supplying electricity to a backup system needed for core cooling. The tsunami waves arrived less than an hour after the earthquake and went over the sea wall, thus flooding the lower parts of building. The emergency diesel generator had stopped functioning which then resulted in heat in the core to highly increase.

The three reactor units were unable to maintain their cooling system and water circulation functions. Despite all the hard work from engineers to bring back the backup system online, the heat trapped in the core vessel started to cause the water evaporated into steam and the fuel rod began to melt after reaching 2,300 °C. Just like Chernobyl case, the zirconium from the cladding of the fuel rod started to react with hydrogen resulted in the explosions in reactor unit 1 and 3. For reactor unit 4, sources of flammable gas remained unknown as the spent fuel in the spent fuel pool was still covered with water. Explosion occurred in reactor unit 2 was reported due to lower suppression area.

The core meltdown scenario was almost the same for all three units, except that it happened on different dates. Temperatures above 2000°C in reactor core may not only cause melting of UO_2, but also the zircaloy cladding and steel [4]. The melted fuel rod when mixed with other materials in the core vessel will form a quenched meat, named corium. Some volatile fission products such as Cs and I were released during melting process, but the less volatile fission product and the actinides were still remained in the corium as it cools down. Therefore, the corium may still contains the radionuclides and leads to a long-term radiotoxicity in the reactor core [5].

2. Methodology

In this paper, the results from the previous reports using System for Prediction of Environment Emergency Dose Information (SPEEDI) were referred. It is a type of computer-based decision support system developed by Japanese Atomic Energy Research Institute (JAERI) for real-time dose assessment [6] and predictions of environmental radioactivity [7] in radiological emergencies. SPEEDI is used for domestic local range accidents and involved in a national scale emergency response programme. However, Chernobyl accident necessitated the expansion of the prediction area of SPEEDI to evaluate the long-range transport of airborne radioactivity due to a severe nuclear accident in a foreign country.

The worldwide version of SPEEDI, better known as WSPEEDI was then developed to predict the radiological impact on Japanese people in case of nuclear accident occurs in neighbouring country. As detailed explanation on both systems are already reported in other literatures [6][8][9], a summary of description for SPEEDI and WSPEEDI has also been outlined by [10]. For comparison purpose, we will look into the results using SPEEDI only for both Chernobyl and Fukushima accidents.

SPEEDI flow of computational codes is described in figure 1. The first step in using this SPEEDI is to insert the information such as the location of nuclear accident site. This system has characteristic data of nuclear reactor sites in Japan such as longitude, latitude and stack height. Then, it will predict the meteorological condition automatically, whenever the observed data are made input. The WEADUS2 code will forecast the wind speed and wind directions at the measuring points. The output data file generated by this code has both observed data and forecasted data. Other codes involve in main calculations are WIND04, PRWDA and CIDE. WIND04 code calculates the wind fields by using a three-dimensional mass-consistent model that is influenced by terrain. PRWDA code is used to simulate the diffusion of airborne and CIDE to evaluate the dose [11].

When SPEEDI was used in 1986 to predict the release rate from Chernobyl accident, the system had a limitation in terms of topographical data. SPEEDI was only equipped with topographical data of Japanese nuclear power plant. According to the report [7], the application of SPEEDI was continued in this case by assuming a flat terrain, as the objective region is almost flat. The radioactive release was assumed to start at 25th April 1986 with a height of 1000 m for 12 hours from the start of the release and 200 m after the first 12 hours. During Fukushima accidents, there were some difficulties in estimating the time of the release. Therefore, the explosive sound around the suppression chamber of Unit 2 was reported and marked as the starting point to record the release rate [10]. This was also supported by the extremely increased dose rate measured at the main gate of the plant.

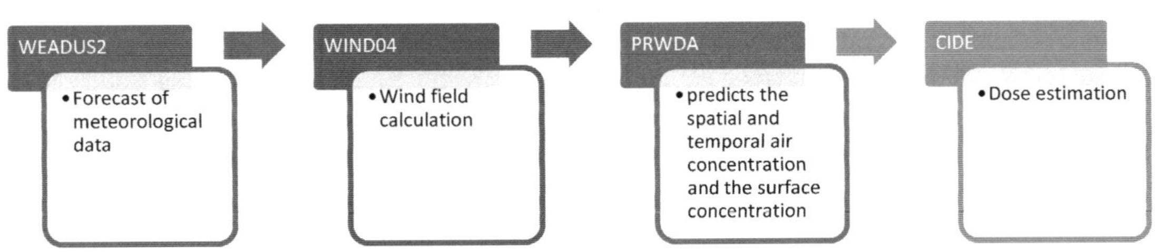

Figure 1. The flow of computational codes in SPEEDI.

3. Results and Discussion

The activity for the release of Cs-137 and I-131 release during Chernobyl are 2.3×10^6 Ci and 50.0×10^6 Ci respectively are shown in figure 2. These values were significantly higher than that of Fukushima which are 0.41×10^6 Ci and 4.3×10^6 Ci respectively. In Chernobyl, two explosions were reported occurred during the accident. The first one was the initial steam explosion followed by a second explosion happened after two to three seconds, possibly caused by the build-up hydrogen due to zirconium-steam reactions. A number of fires started as a result from the ejected fuel, moderator and structural materials. The destroyed core was exposed and lead to the radioactive release directly to atmosphere.

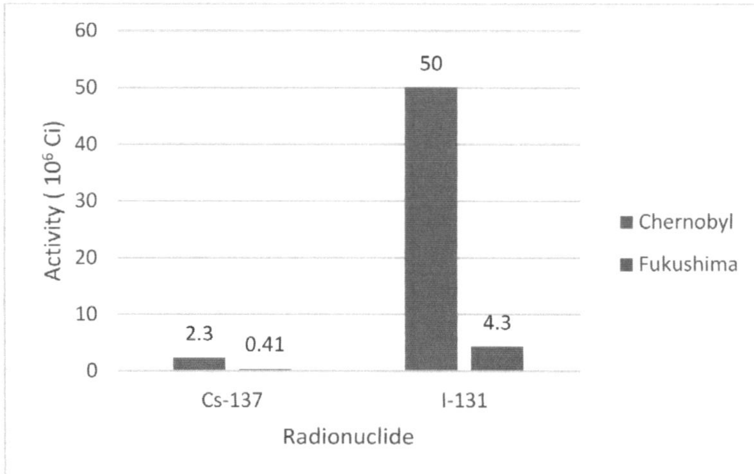

Figure 2. The release activity of radionuclide Cs-137 and I-131 during Chernobyl and Fukushima accidents [13].

Most of the released were made up of short physical half-lives, making the long-lived radionuclides in smaller group. Hence, short-lived radionuclides released by the accident have already decayed. However, I-131 with half-life of eight days was raising concern towards the public. Current estimation of I-131 deposition was required to improve thyroid dose reconstruction [12]. After the initial period, Cs-137 also became a problem as it is in the form of salts, making it highly soluble with water thus increase the probability of water contamination.

However, in Fukushima, most of the radioactivity released from the reactor was captured in the water inside the reactor building [13]. The explosions from three reactors were signaling that a loss of coolant had occurred. This had caused the fuel temperature increasing up to 1,000 °C and the zirconium alloy cladding to oxidize when reacted with steam. The oxidation process created the hydrogen gas which then leaked out to the ceiling of the building, mixed with air and then came the blast. Even though the ceiling of the containment blew off, the reactor vessel was still remained intact.

The estimated release rate of Cs-137 and I-131 in table 1 was reported by using SPEEDI. The simulation process was done by the same researchers for both accident but on different reports, therefore they had credible experience in handling the model. Although there was a large time gap in between both accidents, the release rate will be discussed in terms of technical issues occurred.

Table 1. The estimated release rate of Cs-137 and I-131 on the day of accident [10][14].

Radionuclide	Release rate on the day of accident (10^3 Ci / hr)	
	Chernobyl (26 April 1986)	Fukushima (12 March 2011)
Cs-137	5.6	-
I-131	2000	0.621

Chernobyl was reported to have a release rate of Cs-137 and I-131 which were 5.6×10^3 Ci and 0.02×10^3 Ci respectively. Within the first 24 hours from the explosion started, half of the noble gas inventory has been discharged [7]. The higher release of Chernobyl when compared to Fukushima was still mainly caused by the major exposure of the reactor core. The explosion also caused the core debris to be dispersed and started multiple fires on the roofs of the reactor building and the machine hall [3]. These areas were covered with highly flammable tar. However, it is still unclear the fires were originating from the reactor cavity during the first 20 hours after the explosion.

In Fukushima, there was no detection of Cs-137 but only for I-131 which was 0.621×10^3 Ci. Cs-137 was only detected on March 15th, the third day after the first explosion. It was reported that the cause of undetected Cs-137 was the dust sampling missed to capture the plumes [10]. The release of I-131 was probably occurred during primary containment vessel venting [15] in order to reduce the pressure inside the reactor. The steam vented was believed to contain small amount of radioactive material. The first hydrogen explosion which happened a few hours later at Unit 1 might as well had cause the release to be increasing.

The results of the radioactive release during Chernobyl and Fukushima accidents can also be used to initiate various further studies. These data are crucial in deposition studies, estimating atmospheric dispersion locally and globally and exposure pathways. As both accidents were rated Level 7 by INES, a detailed research can also be a guideline for other countries with NPP to prepare with an updated and comprehensive emergency plan. Although nuclear accidents may reach to the catastrophic level if it goes beyond human control, it is still recorded as low number of accidents in the industry.

4. Conclusion

This paper is aimed at discussing on the radioactivity release during nuclear accidents in Chernobyl and Fukushima. The release rates focusing on Cs-137 and I-131 were estimated by referring to the previous reports using SPEEDI computer code system. The radioactive release and the release rate from Chernobyl accident were significantly higher than that of Fukushima accident. Both accidents were rated level 7 by INES scale for the major release of radioactive material with widespread health and environmental effects. Chernobyl accident which mainly caused by human error can be avoided if only the operators did not turn off their emergency backup system. The exposed core had resulted in great radioactive release and consequently, higher exposure dose to public. Although Fukushima accident was not prophesized to be this worst as it was affected by high-magnitude earthquake and tsunami, the backup system did not last long enough to keep the core cooled. It was very unfortunate for such accidents to occur but the lesson learned should be taken to improvise the current reactor technology towards better nuclear power generation.

5. References

[1] Gerasko V N, Kluchnikov A A, Korneev A A. *et al.* 1997Sarcophagus *Object. History, State and Prospects* (A.A. Kluchnikov, ed.). Kiev

[2] International Nuclear Safety Advisory Group 1986 Summary Report on the Post-accident Review Meeting on the Chernobyl Accident, INSAG-1, IAEA, Vienna

[3] UNSCEAR 2000 Exposures and effects of the Chernobyl accident *Annex J.* 451-566

[4] Grambow B and Poinssot C 2012 Interactions between Nuclear Fuel and Water at the Fukushima Daiichi Reactors *Elements* **8(3)** 213-219

[5] Kinoshita N, Sueki K, Sasa K, Kitagawa J, Ikarashi S, Nishimura T, Wong Y S, Satou Y, Handa K, Takahashi T, Sato M and Yamagata T 2011 Assessment of Individual Radionuclide Distributions from the Fukushima Nuclear Accident Covering Central-East Japan, *Proc. of the National Academy of Sciences of The United States of America* **108(49)** 19526-19529

[6] Chino M, Ishikawa H, Yamazawa H 1993 SPEEDI and WSPEEDI: Japanese Emergency Response Systems to Predict Radiological Impacts in Local and Workplace Areas due to a Nuclear Accident *Radiation Protection Dosimetry* **50(2-3)** 145-152

[7] Chino M, Ishikawa H, Yamazawa H and Moriuchi S 1986 Application of SPEEDI System to The Chernobyl Reactor Accident, JAERI-M 86-142, Japanese Atomic Energy Research Institute (JAERI)

[8] Nagai H, Chino M, and Yamazawa H 1999)Development of scheme for predicting atmospheric dispersion of radionuclides during nuclear emergency by using atmospheric dynamic model, *Nippon Genshiryoku Gakkai-Shi* **41(7)** 777-785

[9] Terada H. and Chino M 2008 Development of an Atmospheric Dispersion Model for Accidental Discharge of Radionuclides with the Function of Simultaneous Prediction for Multiple Domains and its Evaluation by Application to the Chernobyl Nuclear Accident, *J. of Nuclear Science and Technology* **45(9)** 920-931

[10] Chino M, Nakayama H, Nagai H, Terada H, Katata G and Yamazawa H 2011 Preliminary Estimation of Release Amounts of 131I and 137Cs Accidentally Discharged from the Fukushima Daiichi Nuclear Power Plant into the Atmosphere *J. of Nuclear Science and Technology* **48(7)** 1129-1134

[11] Imai K, Chino M. and Ishikawa H. 1985 SPEEDI: A computer code system for the real-time prediction of radiation dose to the public due to an accidental release (JAERI-1297)

[12] IAEA 2006 Environmental Consequences of the Chernobyl Accident and their Remediation: Twenty Years of Experience *Radiological Assessment Report Series* Vienna

[13] von Hippel F N 2011 The Radiological and Psychological Consequences of the Fukushima Daiichi Accident *Bulletin of Atomic Scientist* **67(5)** 27-36

[14] Katata G, Ota M, Terada H, Chino M and Nagai H 2012 Atmospheric discharge and dispersion of radionuclides during the Fukushima Dai-ichi Nuclear Power Plant accident. Part I: Source term estimation and local-scale atmospheric dispersion in early phase of the accident, *J. of Environmental Radioactivity* **109** 103-113

[15] TEPCO 2012 Estimation of Radioactive Material Released to the Atmosphere during the Fukushima Daiichi NPS Accident, www.tepco.co.jp/en/press/corp-com/release/betu12_e/images/120524e0205.pdf

IOP Conf. Series: Materials Science and Engineering **298** (2018) 012012 doi:10.1088/1757-899X/298/1/012012

Surface study of graphene ink for fine solid lines printed on BOPP Substrate in micro-flexographic printing using XPS analysis technique

S Hassan[1,a]**, M S Yusof**[1,b]**, Z Embong**[2,c]**, S Ding**[3,d] **and M I Maksud**[1,e]

[1] Faculty of Mechanical and Manufacturing Engineering, Universiti Tun Hussein Onn Malaysia, 86400 Batu Pahat, Johor, Malaysia.
[2] Faculty of Science, Technology and Human Development, Universiti Tun Hussein Onn Malaysia, 86400 Batu Pahat, Johor, Malaysia.
[3] School of Aerospace, Mechanical and Manufacturing Engineering, RMIT University East Campus, Melbourne, VIC 3083, Australia

[a] suhaimihas@uthm.edu.my
[b] mdsalleh@uthm.edu.my
[c] zaidi@uthm.edu.my
[d] songlin.ding@rmit.edu.au
[e] midris1973@gmail.com

Abstract. Micro-flexographic printing is a combination of flexography and micro-contact printing technique. It is a new printing method for fine solid lines printing purpose. Graphene material has been used as depositing agent or printing ink in other printing technique like inkjet printing. This graphene ink is printed on biaxially oriented polypropylene (BOPP) by using Micro-flexographic printing technique. The choose of graphene as a printing ink is due to its wide application in producing electronic and micro-electronic devices such as Radio-frequency identification (RFID) and printed circuit board. The graphene printed on the surface of BOPP substrate was analyzed using X-Ray Photoelectron Spectroscopy (XPS). The positions for each synthetic component in the narrow scan are referred to the electron binding energy (eV). This research is focused on two narrow scan regions which are C 1s and O 1s. Further discussion of the narrow scan spectrum will be explained in detail. From the narrow scan analysis, it is proposed that from the surface adhesive properties of graphene, it is suitable as an alternative printing ink medium for Micro-flexographic printing technique in printing multiple fine solid lines at micro to nano scale feature.

1. Introduction

Surface adhesion could be described as the molecular force of attraction betweeen dissimilar materials for example graphene with plastic substrate and lanthanum with glass substrate. The attraction strenght of the surface adhesive was determined by the surface energy of the material which known as binding energy (eV). The higher surface binding energy, the molecular attraction will be greater [1]. A simple particle intercalation method was reported to characterize graphene adhesion which was essential in designing graphene based electronic devices and in gaughing its realibility [2].

IOP Conf. Series: Materials Science and Engineering **298** (2018) 012012 doi:10.1088/1757-899X/298/1/012012

Micro-flexographic printing technique was the combination of flexography and micro-contact printing (μCP) techniques. It was a new printing technique in roll to roll flexography printing group. The combination of both printing techniques were expanded in the new era of high speed printing machine that could print fine solid lines image in micro to nano scale feature [3]. The achievement was very useful in printing industry especially in micro-electronics industry [4]. By using micro-flexographic printing technique, the authors managed to print fine solid lines in micro measurement which was 3 μm line width and 3 μm line gaps. The achievement was the lowest fine solid lines printed by a flexography printing technique. Graphene ink was used as printing ink medium in this research which was printed on biaxially oriented polypropylene (BOPP) thin film surface as substrate.

From the latest printing development process, the printing components played important roles in producing micro to nano scale printing image with a good surface adhesive property. Examples of printing components were ink, printing plate, substrate, roller engagement and others. Ink and substrate properties which constitute some of the main components in printing also played the main role in achieving the best quality of image or fine solid lines printing. The ink properties could be ink chemistry, viscosity, rheological behavior, solvent evaporation rate and drying [5].

Previous research had indicated that the higher electricity generation with graphene particle type was attributed to higher electrical conductivity in thickness direction, large surface area, catalytic activities of oxygen reduction and the open structure into that facilitates mass transfer of fuels and ions [6]. Graphene material also exhibited better electrochemical performance and lower impedance compared with activated carbon (C) or unmodified electrode [6]. The study had investigated two types of graphene which were regular graphene sheet and crumpled graphene particle in modifying anode and cathode electrodes in microbial fuel cell (MFC). In particular, the regular graphene particle performed even better than regular graphene sheet.

The attraction and chemical bonding between graphene metal printed and surface was due to a chemisorption and physisorption process. On previous research, the study of graphene attraction could be referred to the example of Lanthanum Oxide (La_2O_3) adhesion property on Silica, Si substrate which had been successful when O 1s binding energy could be seen from La_2O_3 and SiO_2 components. There were four components that contributed to O 1s photoelectron which were La_2O_3, SiO_2, O bulk and water vapour [1].

In fine solid lines printing, Yusof [7] managed to print out 50 μm line width and 50 μm line gap by using carbon graphic inks as printing inks. The author used web press industrial machine as a printing method. This technique used photopolymer as a mold to transfer the ink from plate roller to substrate. While, in micro-contact printing (μCP), Perl [8] showed that μCP could produce fine solid lines below 1 μm [8] which was smaller than flexography printing technique. This research used heavy-weight dendritic thioethers as printing inks and octadecanethiol as backfilling agent.

By combining both printing techniques, it was clearly seen that with the advantages and disadvantages of these techniques, a new era of printing technology could be explored in producing micro to nano scale printing image [3]. The knowledge gap in contact mechanism, printing plate deformation, ink spreading mechanism and other important factors were still under further development.

Graphene was the example of material that could be used as printing ink for Micro-flexographic printing process. Graphene ink was a step to move forward in order to achieve high speed printing in electronic with simple, rapid, low cost method, less waste and roll-to-roll capability. This type of ink was practically used in electronic printing industries that aimed on printing multiple micro to nano fine solid lines feature [9].

Graphene characteristics were the most important thing that must be understood whether it was suitable to be used as a printing ink for combination of flexography printing and μCP which known as Micro-flexographic. This study was investigated on the surface adhesion analysis of using graphene ink in Micro-flexographic printing technique to achieve the printed fine solid lines image with good surface adhesion on BOPP substrate.

2. Research Methodology

The Micro-flexographic printing process was initiated from the preparation of printing plate [10]. A pattern of fine solid lines image was produced on printing plate [11] which it was attached at plate cylinder. Micro-flexographic printing process was almost same as flexography printing process [12]. Ink was transferred to the printing plate by using an engraved cylinder known as an anilox roll as shown in figure 1. The anilox roll was the primary control of the quantity of ink transferred to the plate and subsequently the print. The ink characteristic was studied by referring to its morphology, size and depth of engrave cells. Here, the thin film analysis of printed graphene ink had been performed the surface characterization using X-ray Photoelectron Spectroscopy (XPS).

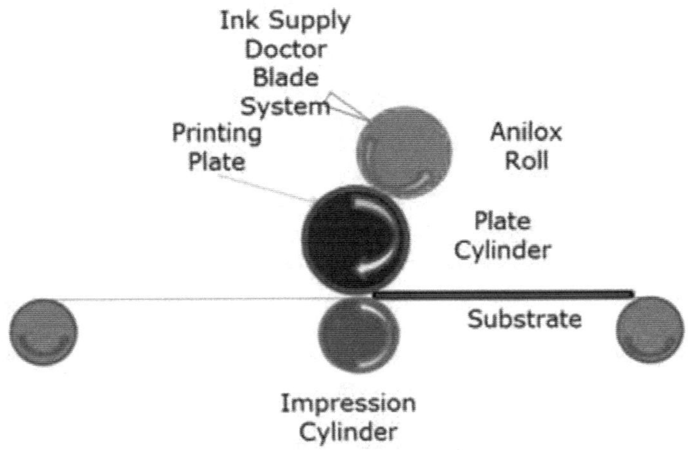

Figure 1. Schematic of the Micro-flexographic processing

2.1. Graphene Characteristics: In General

Generally, graphene had chemical and electronic properties that meet the requirements for applications in various fields such as transistor channel, printed circuit board and etc. The base material for the samples used in this area was normally commercially available graphene ink. Graphene was a pure carbon in the form of a very thin in micro to nano range. It was nearly transparent sheet and one atom thick. The average thickness of the graphene sheet with four layers was approximately 1.3 nm [13]. It was also remarkably strong for its very low weight and it conducts heat and electricity with a great efficiency. The material had high thermal conductivity too and because it was a two-dimensional (2D) material that was almost transparent, it interacted in other interesting and useful ways with light and with other materials. Graphene could be used in many applications but it stilled not commercially used.

Graphene material consisted of a two-dimensional (2D) sheet of covalently bonded carbon atoms, forms the basis of both 3D graphite and 1D carbon nanotubes. Previous experiment established graphene as the strongest material ever measured and showed that atomically perfect nanoscale materials could be mechanically tested to deform well beyond the linear regime [14].

2.2. Surface Analysis Using XPS Study

X-ray Photoelectron Spectroscopy (XPS) was an advanced surface analytical technique that was used to characterize the elemental composition of the graphene printing ink. The method was favourable for several nanometer thickness oxide or deposition [15][16]. XPS can be used to detect elements present within approximately the outermost 1 to 3 nm of the nano particle surface. XPS analysis technique also capable to perform surface chemical composition [17]. XPS spectrum were obtained using Thermo Scientific K-Alpha equipped X-ray source energy from monochromated aluminium Kα at 1486.7 eV and operated under vacuum chamber condition as low as 1 x 10^{-9} mbar. The XPS system was equipped with a flood gun that will allow for surface sputtering etched. Etching could be performed using an Ar^{+} ion gun.

XPS spectrum analysis was deconvoluted using commercial software known as Casa XPS. This software was capable to manipulate the spectrum iteration for photoelectron and background signal. The synthetic spectrum for the photoelectron peaks and its background signal was optimized using a specific model such as linear, Shirley and Tougaard method. In this work Shirley method had been used for background model, while the synthetic photoelectron signal was assigned using mixed of Gaussian-Lorentzian method. The ratio for both Gaussian and Lorentzian had been assigned using area in the ratio of 70% and 30% respectively. In this work, the data deconvolution was performed for element of interest on the substrate by referring to the narrow scan. There were two regions of interest which were carbon, C 1s and oxygen, O1s. Further discussion for this narrow scan deconvolution will be discussed in the result and analysis section.

3. Result and Analysis

The Micro-flexographic experimental printing trial had be done successfully on biaxially oriented polypropylene (BOPP) substrate by using graphene ink. The result from Scanning Electron Microscope (SEM) analysis was shown in figure 2. The printed lines contained narrow and wider. The narrow line showed the BOPP substrate with 1 μm gap while the wider line showed the graphene ink with 3 μm width.

The printed fine solid lines image of graphene ink on the top surface of BOPP substrate was analyzed by using X-ray Photoelectron Spectroscopy (XPS).

The position for each synthetic component in the narrow scan of C 1s and O 1s were referred to the electron binding energy (eV). The spectrum analysis was deconvoluted by Casa XPS software.

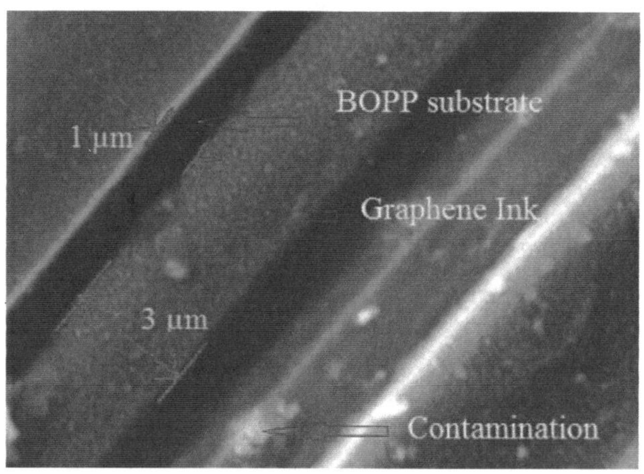

Figure 2. Printed fine solid lines image of graphene ink on BOPP substrate

The deconvolution of Carbon, C 1s photoelectron spectrum for graphene were involved by C-O, C sp^3 and C sp^2 species with binding energy position of 286.1, 284.4 and 283.4 eV respectively as shown in figure 3. The deconvolution of oxygen, O 1s photoelectron spectrum for graphene indicated that the oxygen species was contributed by photoelectron peaks at 530.3, 532.3 and 533.8 eV associated with O (Graphene), O (Graphite) and OH^- as illustrated in figure 4. As shown in figure 4, it is obviously seen that the surface mainly dominated by three species of oxide which were came from the bulk of the carbon. On the other hand, there existence of hydroxyl component, OH^- could be due to water vapour that covers the sample surface and its came from the ambient environment.

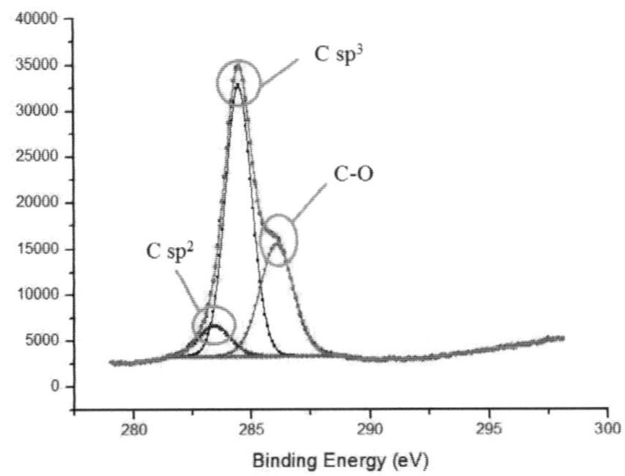

Figure 3. XPS spectrum deconvolution for C 1s for graphene on BOPP substrate

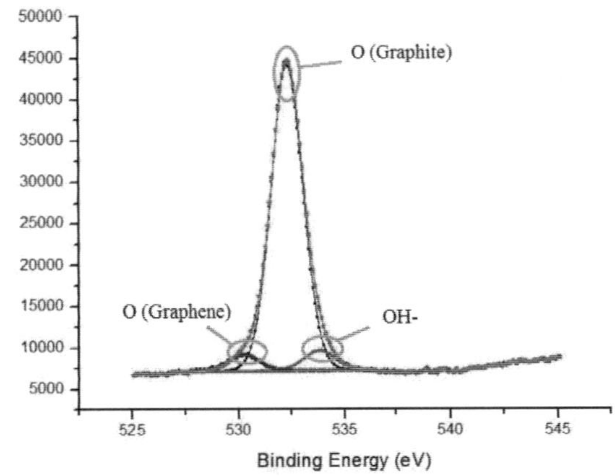

Figure 4. XPS spectrum deconvolution for O 1s for graphene on BOPP substrate

Graphene functionalized had a complex C 1s spectrum contained with sp^2 component. While sp^3 component came from graphite, the C sp^3 peaks must be 1 eV to the higher binding energy side of C sp^2 component. From the analysis result, Carbon sp^2 in fine solid lines graphene ink was seen at 283.4 eV which was 1 eV lower than Carbon sp^3 binding energy. This indicated the surface may be dominated by two types of carbon species which were graphene and graphite. The XPS analysis result has proven the graphene ink functionality in fine solid lines printed by Micro-flexographic printing method.

Overall result indicated that the graphene C 1s and O 1s narrow scan revealed the carbon and oxygen species of this particular ink printed that were associated with BOPP substrate. The carbon species, C from C 1s narrow scan indicated there were three species of carbon bonding with the C 1s signal which were contributed by C-O, C sp^3 and C sp^2. The information of oxygen species, O^{2-} component from O 1s narrow scan indicated there was three species of oxygen associated with the O 1s signal which was contributed by oxygen (Graphene), O (Graphite) and OH^- components. The photoelectron signal from both species had become a clear evidence to support this finding. It means that most of the graphene printed was residual at the top layer of the substrate surface.

Therefore, it could be concluded that the surface adhesion property of the graphene ink fine solid lines printed was influenced by the biaxially oriented polypropylene (BOPP) substrate. The formation of these fine solid lines printed or bonding was respected to a good adhesion between ink printed and substrate due to the anion and cation interaction under a thermodynamic influenced.

4. Conclusions

Surface adhesion study of graphene ink prints on BOPP substrate is successfully done. From this work, several conclusions can be summarised as follows:

- There are two types of species on the substrate which are carbon and oxygen.
- The components that contribute to O 1s photoelectron are oxygen (Graphene), O (Graphite) and OH^- (water vapour).
- The components that contribute to C 1s photoelectron are C-O, C sp^3 and C sp^2.
- The chemical bonding formation between the fine solid lines graphene ink printed and the BOPP substrate surface is due to a chemisorption and physisorption process.
- A successful observation of graphene fine solid lines adhesion property on BOPP substrate is clearly seen from the C 1s binding energy from C-O, C sp^3 and C sp^2 component.
- Micro-flexographic printing process is good candidate for printing electronic with graphene ink property, substrates and process parameters are main role to success the implementation.

5. References

[1] Hassan S, Yusof M S, Embong Z and Maksud M I 2016 Angle resolved x-ray photoelectron spectroscopy (ARXPS) analysis of lanthanum oxide for micro-flexography printing *AIP Conf. Proc.* **1704** 040002.

[2] Zong Z, Chen C L, Dokmeci M R and Wan K T 2010 Direct measurement of graphene adhesion on silicon surface by intercalation of nanoparticles *J. of Applied Phys* **107** 026104.

[3] Hassan S, Yusof M, Embong Z and Maksud M 2017 Surface adhesion study of La2O3 thin film on Si and glass substrate for micro-flexography printing *AIP Conf. Proc.* AIP Publishing pp 040006.

[4] Hassan S, Yusof M S, Maksud M I and Nodin M N 2017 A Study of Micro to Nano Structure Fine Solid Line Printing for Graphene Ink by Micro-flexographic Printing Method *Int. J. of Mechanical Engineering* **1** 109-113.

[5] Hrehorova E, Pekarovicova A and Fleming P D 2006 Gravure printability of conducting polymer inks, *NIP & Digital Fabrication Conference*, Society for Imaging Science and Technology pp 107-110.

[6] Xiao L, Damien J, Luo J, Jang H D, Huang J and He Z 2012 Crumpled graphene particles for microbial fuel cell electrodes *J. of Power Sources* **208** 187-192.

[7] Yusof M S 2011 Printing Fine Solid Lines in Flexographic Printing Process Swansea University Swansea 201 pp.

[8] Perl A, Reinhoudt D N and Huskens J 2009 Microcontact printing: limitations and achievements *Advanced Materials* 21(22): 2257-2268.

[9] Hassan S, Yusof M S, Maksud M I, Nodin M N and Rejab N A 2015 A Feasibility Study of Roll to Roll Printing on Graphene *Applied Mechanics and Materials* 799-800 402-406.

[10] Hassan S, Yusof M S, Maksud M I and Nodin M N 2016 Investigation of Micro-flexographic Printing Method in Fine Solid Lines Printing Structure by Graphene Ink *Int. J. of Materials* **3** 98-103.

[11] Hassan S, Yusof M S, Maksud M, Nodin M, Rejab N A and Mamat K 2015 A study of nano structure by roll to roll imprint lithography, *Int. Symp. on Technology Management and Emerging Technologies (ISTMET)* IEEE pp 132-135.

[12] Deganello D, Cherry J A, Gethin D T and Claypole T C 2010 Patterning of micro-scale conductive networks using reel-to-reel flexographic printing *Thin Solid Films* **518**(21) 6113-6116.

[13] Liang X, Fu Z and Chou S Y 2007 Graphene transistors fabricated via transfer-printing in device active-areas on large wafer *Nano Letters* **7**(12) 3840-3844.

[14] Lee C, Wei X, Kysar J W and Hone J 2008 Measurement of the elastic properties and intrinsic strength of monolayer graphene *Science* **321**(5887) 385-388.

[15] Venkata Krishnan R, Mittal V K, Babu R, Senapati A, Bera S and Nagarajan K 2011 Heat capacity measurements and XPS studies on uranium–lanthanum mixed oxides *J. of Alloys and Compounds* **509**(7) 3229-3237.

[16] Embong Z 2011 XPS, AES and Laser Raman Spectroscopy: A Fingerprint for A Materials Surface Characterisation *J. Sains Nuklear Malaysia* **23**(2) pp. 26-45.

[17] Murdock R C, Braydich-Stolle L, Schrand A M, Schlager J J and Hussain S M 2008 Characterization of nanomaterial dispersion in solution prior to in vitro exposure using dynamic light scattering technique *Toxicological sciences* **101**(2) 239-253.

Acknowledgments

This project was supported by 'Skim Latihan Akademik Bumiputera' (SLAB) from Ministry of Education Malaysia.

IOP Conf. Series: Materials Science and Engineering **298** (2018) 012013 doi:10.1088/1757-899X/298/1/012013

Investigations of Au-198 as radiotracer in laboratory porous media using gamma camera: a preliminary study

N Othman[1], W H B Wan Kamal[2], N H Yusof[1], E M F Engku Chik[1], M A S Yunos[1], M A K Adnan[1], M R Shari[1]

[1]Plant Assessment Technology, Industrial Technology Division, Malaysian Nuclear Agency, 43000 Kajang, Selangor
[2]Medical Technology Division, Malaysian Nuclear Agency, 43000 Kajang, Selangor

noraishah@nm.gov.my

Abstract. Preliminary experiment has been carried out using irradiated Au-198 as radiotracer inside the laboratory porous media. The objectives are to check the compatibility of Au-198 as the radiotracer inside the porous media as well as to provide insights of fluid hydrodynamics inside the media using gamma camera.198Au is gamma emitter isotope with half-life of 2.7 days and energy of 0.41 MeV (99%). The porous media consists of fine sandstone with grain size 850μm, lubricant as the mimic of original oil in plant (OOIP) or trapped oil and a layer of cement on top of the rig as the bed rock. Gamma camera is arranged next to the porous media in order to capture the movement of radiotracer which has been set to 1minute per frame. Initially, the gold wire which has isotope of 197Au was irradiated inside the rotary rack of Reactor Triga PUSPATI (RTP) to produce 198Au. RTP is located in Nuclear Malaysia, Bangi has energy of 750kW and neutron flux of 5 x 102 n/cm2/s. 198Au, which is in liquid form, is injected inside the porous media and monitored and recorded by gamma camera. The gamma camera gives a quantitative determination of local fluid saturations over the area of observation.

1. Introduction

Crude oil development and production of oil reservoirs can include up to three distinct phases: primary, secondary, and tertiary recovery. During primary recovery, the natural pressure of the reservoir or gravity drives oil into the wellbores, combined with artificial lift techniques which bring the oil to the surface [1]. The expected initial extraction rate of available hydrocarbons reservoir's original oil in place is around 10-15% typically produced during primary recovery. This will leave around 85% of hydrocarbons in the reservoir. Secondary recovery techniques extend a field's productive life generally by injecting water or gas to displace oil and drive it to a production wellbore [2]. Pump jacks for water flooding and initial gas injection can increase the recovery up to 25-30% of the original oil in place. However, producers have attempted enhanced oil recovery (EOR) techniques, techniques that offer prospects for ultimately extracting another 10-15% of the initially available hydrocarbons or producing 30 to 60 % or more, of the reservoir's original oil in place [3].

The petroleum potential in Malaysia is mainly fractured basement rock reservoir located offshore and characterized by a fracture system of very high heterogeneity in porosity and permeability. The complexity of the reservoir structure introduces a considerable uncertainty in the reservoir model and fluid flow simulation. The extraction of residual oil requires modification in reservoir characteristics in which water flooding is most commonly used for enhanced oil recovery. In this operation, the water is injected through the residual oil under high pressure to extract the oil. This situation calls for more application of tracer technique and challenges to tracer technology. In view of rising energy consumption, the use of advanced technologies to boost oil recovery from mature oil fields in Malaysia is most welcomed [4]. In addition, the innovative application of the gamma-camera-imaging system as a quantitative method for investigating the physics of oil displacement within a thin slab of porous medium has been successfully carried out using radiotracing medium [5]. The gamma camera can provide the behavior of the water and oil movement in the porous media during water flooding. In this paper, preliminary result is shown on the intervention of radiotracer in porous media which mimics the reservoir with gamma camera as a tool in quantitative determination of local fluid saturations over the area of observation. Thus, the objectives are to investigate the compatibility of Au-198 as radioactive tracer in porous media and to visualize the water-oil movement inside the porous media using gamma camera. The selection of Au-198 is made based upon assumption that the particle of irradiated Au-198 is tagged with hydrocarbon inside the laboratory porous media. The radiation from Au-198, in this case gamma ray emission is detected and captured by NaI scintillation detectors and provide data monitoring and profile of oil movement.

2. Materials and Methodology

About 2 mg of gold wire with 99% purity from Goodfellow Cambridge Limited is sent to research reactor for irradiation and produce about 1mCi of activity. *Aqua regia* solution with ratio of 3:1 (HCl: NHO_3) is prepared beforehand to dissolve the irradiated Au-198 to liquid form. The volume of 10ml gold solution is prepared for this study. All preparations of gold radiotracer are carried out inside the radioactive fume hood for safety. The survey meter is always available to monitor the dose rate level is at the permissible level which is 10 μSv/hr or 1 mR/h at all time. Gamma camera Hamamatsu model BHP6602 is used for imaging and monitoring the flow of fluid inside the porous media.

A lab-scale rig for porous media consist of water reservoir which can accommodate about 2L of water, is attached to porous media container using Teflon uniform tubing. In this preliminary study, tap water is used to supply continuous water as source of water flooding. The water reservoir is located about 50 cm above the oil reservoir and gravity is used to push the water flow from water reservoir to the oil reservoir respectively. The outlet from the porous media is then connected to separator as the outlet and also to assess the oil recovery as well as the water as shown in figure 1. The porous media consists of fine sandstone with grain size 850μm, 20ml lubricant as the mimic of original oil in plant (OOIP) or trapped oil and a layer of cement on top of the rig as the bed rock. The dimension of oil reservoir is 20 cm in diameter and height respectively.

The water from the reservoir will be pushed inside the porous media using the gravity pull for fluid displacement activity. Prior to that, the gold radiotracer is mixed inside the water reservoir and this is called as tracer dilution technique. This means that radiotracer is further diluted and this may affect the sharpness of the image. Gamma camera is set to capture image of one frame per minute and once the gamma camera is ready for dynamic study, the experiment is on. The half-life of gold is 2.7 days and emits gamma ray with energy of 0.41MeV (99%).

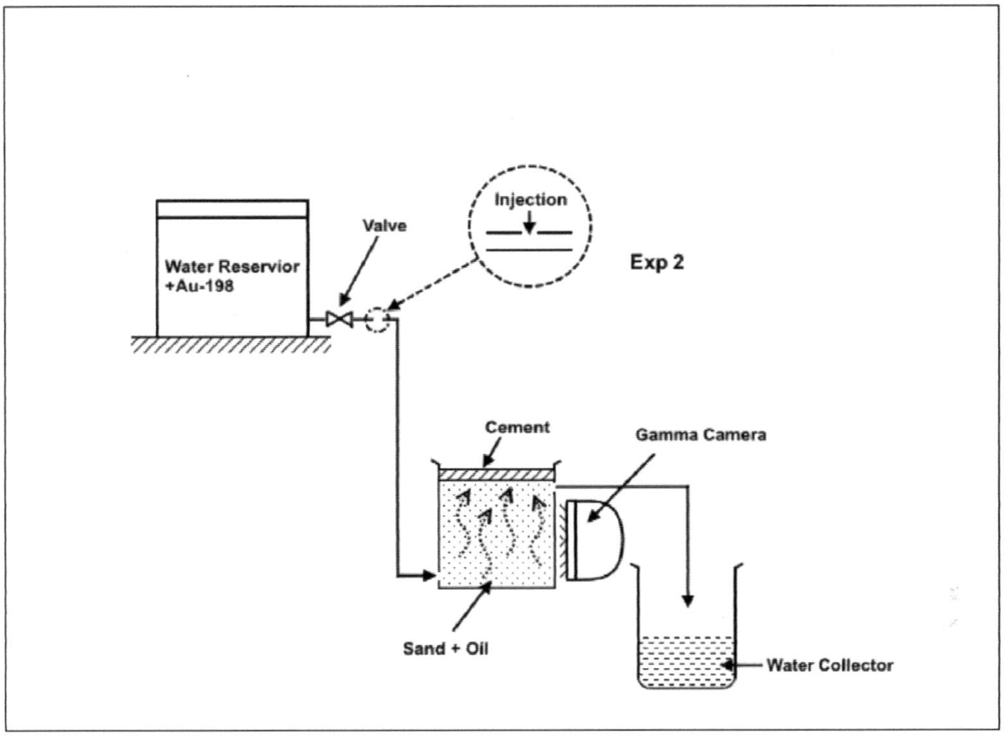

Figure 1. Schematic diagram of experimental set up.

3. Results and Discussions

3.1. Determination of actual activity

The dose rate recorded was 100μSv/h at surface of uncollimated liquid gold and 3μSv/h at 1m distance. Therefore, actual activity is calculated using the equation below for 1meter distant:

$$Dose\ rate\ \left(\frac{mR}{h}\right) = 0.55\ x\ Energy\ (MeV) x\ Activity\ (mCi) \tag{1}$$

$$Activity\ (mCi) = \frac{\frac{0.3mR}{h}}{0.55\ x\ 0.41} = 1.3\ mci/48.1MBq \tag{2}$$

Thus, the actual activity for radiotracer injection is 1.3mci or 48.1MBq throughout the study.

3.2 Flow visualization using gamma camera

Figure 2a shows the hotspot of radioisotope Au-198 inside the vial. The bright red and yellow colors indicate the high intensity of highly concentrated Au-198 which is measured as 1,825,192 counts for 10ml gold solution before injection inside the rig. It shows that activity of 48.1MBq is sufficient and capable for image detection using gamma camera. Figure 2b displays the image intensity of gold solution as soon as water is added to the concentrated gold solution for dilution of radiotracer. The intensity of Au-198 is gradually decreased as depicted in blue greenish color.

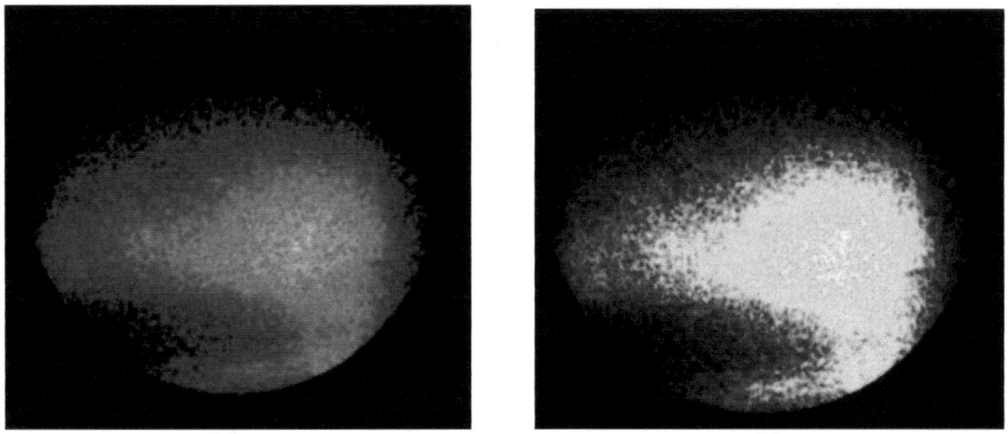

(a) (b)

Figure 2. The images of Au-198 intensity (a) 10ml gold solution, (b) diluted gold solution.

The hydrodynamics of fluid inside the oil reservoir can be observed once Au-198 is injected inside the porous media as shown in figure 3 where the intense green colour is recorded and monitored. At earlier minutes (t=2-3min), it seems that the liquid radiotracer is dispersed inside the media where continuous water is pushing the radiotracer inside the oil reservoir. It seems that the water containing radiotracer is occupying the porous media which contains oil and therefore, oil recovery is expected at the outlet. This is due to the fact that water reservoir is meant to push the trapped oil inside the media as water flooding mechanism. Nevertheless, after 20 minutes, no oil was collected at the collector pot. Figure 3 shows that the accumulation of Au-198 occurs inside the media and the reading are around 2200-2300 counts and remain plateau. There is no significant tracer breakthrough at the outlet as expected although additional of water is carried out. Besides, the media's dose rate is recorded as 2mR/h which is equivalent to 0.8mCi or 29.6MBq. This indicates higher intensity of radiotracer is trapped inside the media. These preliminary results show that Au-198 is adsorbed onto the sand and does not behave as water tracer.

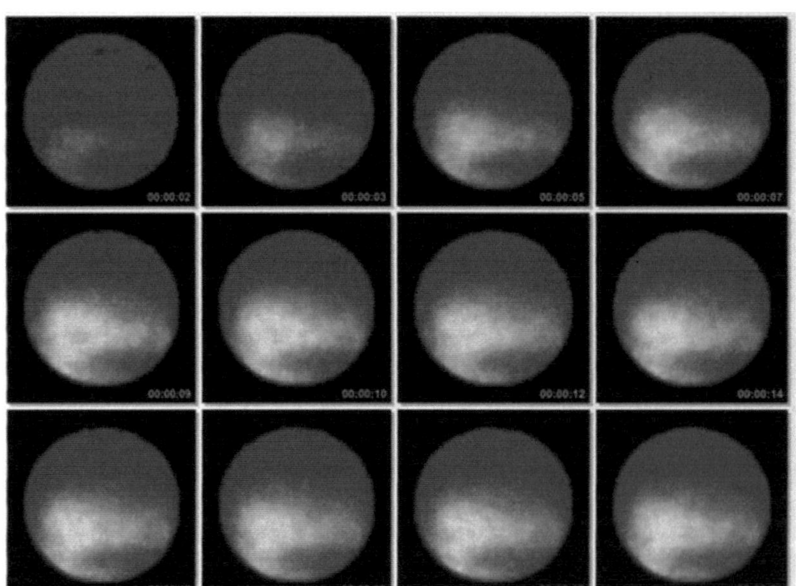

Figure 3. Image of dynamic flow of Au-198 recorded at 1frame per minute for Experiment 1.

In order to rebut with the previous experiment, second experiment is carried out in which the modification is done to the tracer injection mechanism. Instead of injecting the radiotracer Au-198 inside the water reservoir which is causing higher dilution of radiotracer, the radiotracer which is 3ml in volume (concentrated) is injected to the system by Dirac pulse. Prior injection, the dose rate of radiotracer was measured using survey meter and it was recorded as 100μSv/hr and the flow rate of water supplied was 180ml/min. The monitoring of the process took about an hour. (OOIP) or trapped oil and a layer of cement on top of the rig as the bed rock. The dimension of oil reservoir is 20 cm in diameter and height respectively.Figure 4 shows the flow of radiotracer inside the reservoir and it can be seen clearly that the impact of concentrated radiotracer without dilution provides more distinctive image of injection process. It shows that, once the injection starts, the detector captures small green area (1,1) to indicate the tracer has already entered the reservoir and the injection has just commenced. Only small green area is observed which shows the Dirac pulse method of injection is implemented. The blue area is the reservoir or porous media which has not been touched by Au-198 yet. From coordinate (1.2) to (1,8), the distribution of radiotracer covers about two-third of the overall circle. Thus, it can be deduced that the dispersion of Au-198 as radiotracer is not homogenous and still remain and stuck inside the sand although continuous water injection is exerted and sufficient time of water flooding is given. To improve the hiccup of this experiment, the researcher should monitor and record the dose rate reading of the collected water to confirm the presence of Au-198 at the outlet if any.

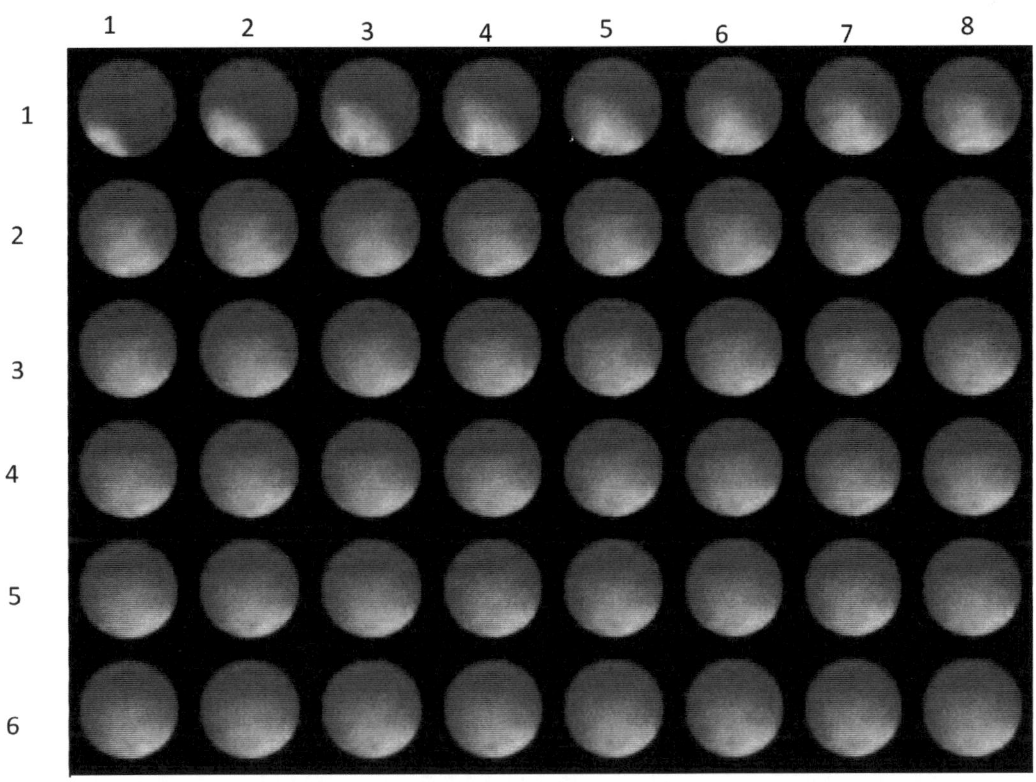

Figure 4. Image of dynamic flow of Au-198 recorded at 1 bit per minute for Experiment 2.

4. Conclusions

From this preliminary result, a conclusion can be made that Au-198 is not recommended for oil-tagging tracer but more for sand tracing application. This is because, both results from different set of experiments show that most Au-198 tracer is trapped inside the oil-laden sand and the rest is distributed on the sand particles although water flooding is introduced continuously. Nevertheless, gamma camera can be used to study about the hydrodynamics of oil inside the Laboratory Porous Media provided the selection of oil tracer is correct. This is because, for lab study, the movement of water or oil can be traced accordingly (static or dynamic) and monitor respectively.

5. References

[1] Aluka Dinesh R 2013 *Int. J. Sci.Res.* IJSR(2013)ISSN: 2319-7064
[2] Jain P 2003 Tracer Application in Oilfield Investigations (IAEA, Vienna)
[3] Tunio S Q, Tunio A H, Ghirano N A, El Adawy Z M 2011 *Int. J. Appl. Sci. Technol* **1** 5
[4] http://etp.pemandu.gov.my/Oil,Gas_and_Energy/
[5] Yi B H, Carl C G 1988 JPT *J. of Petroleum Technology* **40** 1355

Acknowledgements

The authors would like to express their special thanks to Dr Julia Abdul Karim from Reactor, Pn Nazaratul Asyifa from ACA/BAS, Reactor and KFK staff for assisting us throughout this study. Not to forget IAEA and Nuclear Malaysia in making the research fruitful.

IOP Conf. Series: Materials Science and Engineering **298** (2018) 012014 doi:10.1088/1757-899X/298/1/012014

Preliminary study of tin slag concrete mixture

Mohd Jamil Hashim[1,a], Ishak Mansor[2,b], Mohamad Pauzi Ismail[2], Suhairy Sani[2], Azhar Azmi[2], Shaharudin Sayuti[2],Mohd Zaidi Ibrahim[3,c], Abul Adli Anuar[2] and Abdul Adha Abdul Rahim[1]

Agency Nuclear Malaysia: Engineering Division[1], Industry Technology Division[2]and Waste Technology Development Centre[3]

[a)]Jamil@nm.gov.my
[b)]ishak_mansor@nm.gov.my,
[c)]mohd_zaidi@nm.gov.my

Abstract. The study focuses on practices to facilitate tin smelting industry to reduce radioactive waste product (Tin Slag) by diluting its radioactivity to a safe level and turning it to a safer infrastructural building product. In the process the concrete mix which include Portland cement, sand, tin slag, water and plasticizer are used to produce interlocking brick pavements, piles and other infrastructural products.The mixing method follows DOE (UK) standard method of mixing targeted at in selected compressive strength suitable for its function and durability. A batching machine is used in the mixing and six test cubes are produced for the test. The testing equipment used are a compressional machine, ultrasonic measurement and a Geiger Muller counter to evaluate of the concrete mix to find the lowest emission of radiation surface dose without compromising the strength of concrete mix. The result obtained indicated the radioactivity of tin slag in the mixing process has reduced to background level that is 0.5μSv/h while the strength and workability of the concrete has not been severely affected. In conclusion, the concrete mix with tin slag has shown the potential it can be turned into a safe beneficial infrastructural product with good strength.

1. The first section in your paper

Malaysia is one of the major world's tin-producing countries. Tin ingot for export is produced from smelting plants. In smelting process, tin ore concentrates which contain significant amount of radioactive minerals such as monazite, zircon and ilmenite were used as raw materials. In the process, cassiterite ($SnO2$) is reduced to tin metal by heating with carbon at 1200°C to 1300°C.Reverberatory furnaces are used to smelt tin concentrate and to re-smelt slag (the waste left after the ore has been smelted) for additional tin recovery. A furnace charge consists of cassiterite, a carbon reducing agent, and limestone and silica fluxes. The molten batch is tapped into a settler from which the slag overflows into water, where the molten slag is granulated and solidifies to glass-like material[20].

Tin slag contains lime, iron, silica, small amount of heavy metals and natural occurring radioactive materials such as uranium series and thorium series [20][16]. The highest reading, recorded while travelling near monazite and zircon mineral dumps, was 13 times the mean environmental radiation level of in Malaysia. It is evident that the radioactive material dumps on the roadsides can influence the radiation level on the road [15]. High background dose rates were detected in area where tin slag is used as sea defense [10].

There are significant needs to be addressed in using tin slag for commercial used indoor or outdoor. Concrete mixture was considered due to its multiple usages as outdoor and indoor or buried products to isolate waste product such as tin slag. The paper reviews some preliminary works on tin slag concrete mixing which is prepared in according to DOE method of mixing is in accordance to British Standard.

Tin slag contains elevated concentrations of radionuclides from both the uranium and the thorium decay series and is sometimes used as a source of niobium and tantalum. Table 1 shows the activity concentrations of radionuclides in tin slag according to Omar *et al.* [16].

Table 1. The activity concentrations of radionuclides in tin slag (Omar, 2008).

Sample	^{238}U (Bq/g)	^{226}Ra (Bq/g)	^{232}Th (Bq/g)	^{228}Ra (Bq/g)
Tin Slag 1	3.6	2.4	5.1	1.0
Tin Slag 2	6.7	3.4	7.3	1.2
Tin Slag 3	0.9	0.5	0.8	1.0

The use of slag might need to be restricted, especially in building materials where it could give rise to high indoor radon or thoron concentrations and gamma radiation [17]. Therefore the isolation of tin slag in building material should not be indoor but buried or as infrastructure material. Table 2 shows Ra-226 and Th-232 contents in various building materials.

Table 2. Ra-226 and Th-232 contents in some German building materials and wastes (Winter and Wicke, 1993).

Material	^{226}Ra ($Bq\ Kg^{-1}$)	^{232}Th ($Bq\ Kg^{-1}$)
Granite	30-500	17-311
Brick	10-200	12-200
Limestone	4-41	2-20
Gypsum	2-70	2-100
Concrete	7-92	4-71
Light-weight concrete	6-80	1-60
Old copper slag	861-2100	18-78
Tin slag	1000-1200	230-340
Brown coal ash	4-200	6-150

Proper management of tin slag are important to ensure the public and environment are safe, current and in the future. Malaysia implemented 1 Bq g -1 as the clearance limit for naturally occurring radionuclide from series of uranium and thorium, as stipulated in the Atomic Energy Licensing (Radioactive Waste Management) Regulations 2011. Processing of mineral disposal of waste containing naturally occurring radionuclides below the activity concentration of 1 Bq g -1 were exempted from regulatory control (AELR, 2011). Table 1 and 2 show that total activity concentration in tin slag always higher than 1 $Bq\ g^{-1}$. Because most of tin slag containing higher activity concentration than the clearance limit, processing and disposal of tin slag need approval by the Atomic Energy Licensing Board (AELB). From the table 2 building material shows tin slag give Radium 226 a concentration of 1~1.2 Bq g-1. Thorium 232 of 0.23~0.34 Bq g-1. The annual limit public dose is 1 mSv. This 1mSv/year is equal to 0.11 μSv/hr. Public exposure is subjective to time of occupancy and distance from radiation source. The net public dose must not above annul limit. Since the concrete is

IOP Conf. Series: Materials Science and Engineering **298** (2018) 012014 doi:10.1088/1757-899X/298/1/012014

exposed or embedded to the infrastructure. The exposure is calculated from occupancy factor of 0.8 for indoor and 0.2 for outdoor [18].

2. Methodology

Concrete mixing procedure BRE Mix Design Method [8] (formerly known as DoE method of mixing) which is in accordance to British Standard is adopted in this work. Basically it is a weight proportion method where all its constituents are measure by weighing with the electric mass balance. All the concrete mixtures, which are, coarse aggregates, fine aggregates, cement and water is placed in the batching machine. The concrete mixture is tested for slump before making cubes for strength test. There are three types of concrete mixtures. They are as follows normal concrete M30, 100% tin slag replacing sand and 20% tin slag replacing sand. Table 3 shows the mix proportion used in the experiment.

Table 3. The mix proportion (* Cube volume 0.01 m^3).

Mix	Proportion	W/C	Cement (kg)	Fine aggregates	Coarse aggregates
				River sand (kg)	Tin Slag (kg)
Normal concrete M30	1:0.5:2.7:1.48	0.5	5.04	13.8	0
100% slag M30	1:0.54:2.17:2.4	0.54	3.62	0	7.85
20 % slag M30	1:0.5:1.8:2.32	0.5	4.2	6.36	1.59

Material type and grade used are portland cement grade 43, fine aggregates (River sand) of 600μ sieve passes 22%, clean water with cement to water ratio, tin slag granular with particle 600μ sieve passes 11%, coarse aggregates (granite) of 20 mm diameter. Mix design process in this experiment, for the mix design process the following factors have to been taken into consideration are minimum compressive strength required, water to cement ratio, type of cement, durability, workability and water content, choice of aggregate, cement content and aggregate content.

The concrete containing tin slag cubes were prepared based on British DOE method sample preparation (BS 1881: part 108, 1983). The block sizes were 150mm x 150mm x 150mm and the workability is medium. The samples were cured by immersion into water for 28 days. The samples were tested at the age of 7 days, 14 days and 28 days. Testing procedure for destructive and non-destructive testing is shown in table 4 below.

Table 4. List of tests.

Destructive testing	Non-destructive testing
Compression cube test	Sieve testing of sand for 600 μ
	Slump test
	Weight test
	ultrasonic pulse velocity (UPV) testing
	Surface radiation dose test (MS EN 12504-4: 2013)

A dual dial gauge hydraulic compression machine is used for compression cube test in accordance to BS 1881 Part: 116, 1983. Firstly the cube is chosen with flat surfaces and is placed against the machine contact surface at the centroid. The dial marker is set to zero with the dial gauge. The compression machine is loaded by lowering its load cell. The marker will indicate its result as the compression decreases or reaches its peak.

The slump test equipment is shown in figure 1 and procedure in accordance to BS1881 Part: 102, [2]. The mix concrete is placed in the cone (bottom diameter is bigger) in three layers and 25 times the steel rod is stamped in every layer to the top of the cone. The cone is pull up and the slump is measured by putting the cone beside the concrete mass and rod is placed perpendicular to the top surface. The gap between the concrete mass and the rod is the slump measurement.

Figure 1.Test apparatus for concrete workability.

The UPV test instrument used is TICO as shown in figure 2 (a) and (b) or PUNDIT or their equivalent company approved instrument in accordance to BS 1881 part: 203, 1986. The probes shall be in general selected from PUNDIT or equivalent with frequency about 50kHz Calibration Block or reference bar as provided by the equipment manufacturer with known and calibrated transit time. Conventional couplant – typical couplants are petroleum jelly, grease, soft soap and kaolin/glycerol paste. No couplant is required for exponential tip probe.Surface transmission should be only used in cases i) when one face of the concrete is accessible, ii) when the depth of a surface crack is to be determined or iii) when the quality of the surface concrete relative to the overall quality is of interest.

Table 5. The general quality of concrete based on UPV (Malhotra V.M, 1975)

Pulse velocity (m/s)	Quality
Above 4500	Excellent
3600 – 4500	Generally good
3000 – 3600	Questionable
2100 – 3000	Generally poor
below 2100	Very poor

Common usage is in the determination of the variability and quality of concrete and in the estimation of crack depth. Using a variety of transmission techniques, the extent of such defects such as voids, honeycombing, cracks and segregation can be determined. The method can also be used in the integrity testing of repairs and construction joints, particularly in water retaining structures. This technique is also useful when examining fire damaged concrete. It may also be used to detect presence and degree of microcracking in steel. Transducers of different frequencies are needed for different types of materials. Ultrasonic pulse velocity is the most recommended NDT used for assessing concrete strength (Malhotra, 1975).

The equipment used is a Geiger Muller counter as shown in figure 2 (b). The procedure is based on MS EN 12504-4: 2013. The surface radiation levels were measured in unit μSv/hr. Each surface of the cubic sample was measured with the counter and the average value is calculated.

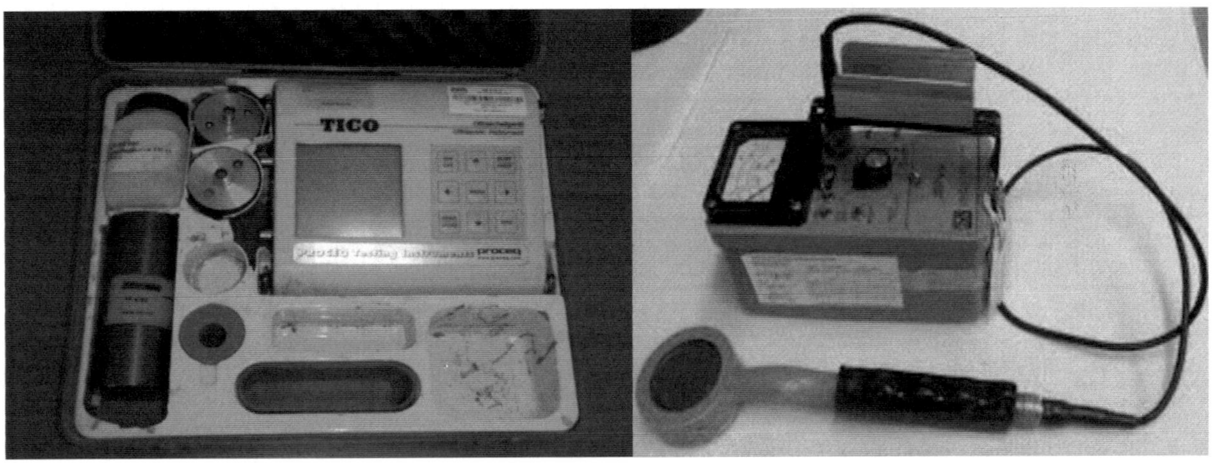

(a) (b)

Figure 2. (a) Brand: Proceq, Model: TICO, Serial No. : 21.1759, (b) Geiger counter for surface radiation measurement.

3. Results and Discussion

The main purpose of this experiment is to find the lowest emission of radiation surface dose rate in modified tin slag concrete mixture without compromising strength of concrete. The first information of the mineral contain in tin slag was measured which obtain from Malaysian Smelting Corporation Berhad. It was analyzed using neutron activation analysis and the result shown in table 6 with the Certificate reference number NAA/4D/17 (23).

Based on UPV result, the general condition of concrete mixture type is graded excellent as shown in table 7. It can be seen that concrete mixture with 20% tin slag has highest velocity value compared to concrete mixture with 100% tin slag within 7 – 14 days but similar in strength. Meanwhile, both concrete mixtures have shared the same value of velocity and strength after 28 days.

During the curing stage concrete strength generation shows increment of strength for 7 days and gradually low generation for 14 days and achieved higher strength at 28 days shown in table 7. Results of the experiment are tabulated in table 8. It is found that the concrete density and slump (workability) are increased with the introduction of tin slag in the mix. However, the radiation level is also increased slightly when the tin slag is introduced.

Table 6. Elements Analysis in tin slag sample.

No.	Element	Slag MSC – 1 (Concentration)	Slag MSC – 1 (Concentration)
1.	Sm	414 μg/g	401 μg/g
2.	U	116 μg/g	115 μg/g
3.	Yb	127 μg/g	121 μg/g
4.	As	162 μg/g	43.8 μg/g
5.	Sb	3.2 μg/g	3.3 μg/g
6.	Br	<0.5 μg/g	5.0 μg/g
7.	La	0.29 %	0.27 %
8.	Ce	0.44 %	0.43 %
9.	Lu	11.8 μg/g	11.1 μg/g
10.	Th	0.13 %	0.13 %
11.	Cr	0.18 %	0.18 %
12.	Hf	370 μg/g	358 μg/g
13.	Ba	0.46 %	0.38 %
14.	Nd	0.14 %	0.13 %
15.	Cs	9.9 μg/g	15.2 μg/g
16.	Tb	31.4 μg/g	29.2 μg/g
17.	Sc	102 μg/g	94.8 μg/g
18.	Rb	419 μg/g	292 μg/g
19.	Fe	12.40 %	12.42 %
20.	Zn	0.26 %	0.22 %
21.	Co	24.7 μg/g	27.5 μg/g
22.	Eu	9.5 μg/g	9.6 μg/g

Table 7. Results experiment of UPV.

Concrete Mixture type	concrete cube age in pulse velocity (m/s)			Strength
	7 days	14 days	28 days	
20 % slag	5430	5554	5610	Excellent
100 % slag	5332	5497	5610	Excellent
Normal M30 concrete	5340	5408	5508	Excellent

Table 8. Results of the experiment.

Concrete Mixture type	Density kg/m^3	Concrete grade	600μ sieve	Radiation μSv/hr	Slump (mm)	Ultrasonic pulse velocity (m/s)	Compressive test at 28 days (N/mm^2)
Normal	2.192	M30	22%	0.5	10	5508	20.0
20% tin slag	2.488	M30	24%	0.8	20	5610	24.8
100% tin slag	2.488	M30	11%	1.5	25	5610	20.8

The test result in table 8 shows the increase of density with percentage of tin slag. The 20% tin slag mixture gives the best result with 0.8 μSv/hr surface dose and 24.8 N/mm^2 in strength. The compressive strength is good for 7 days. This shows good early strength formation. In overall the 20% tin slag shows surface radiation below dose limit 1 μSv/hr and 100% tin slag above the s.. In short, it indicates that percentage of tin slag is slightly effect the concrete strength mixture.

The surface radiation dose 0.8μSv/hr was obtained from 20% tin slag mixture compared with background dose of normal natural occurring radiation material in concrete 0.5μSv/hr. Therefore the net surface radiation is:

$$0.8\mu Sv/hr - 0.5\mu Sv/hr = 0.3\ \mu Sv/hr \qquad (1)$$

$$\text{Total limit dose } 0.3\mu Sv/hr \times 0.2 \text{ (occupancy factor)} \times 24 \text{ (hr/day)} \times 365 \text{ (day/year)} = 525\ \mu Sv/year\ (1) \qquad (2)$$

Therefore surface dose limit of 0.525mSv/year is less than 1 mSv/year which is the annual limit for public member. The result suggest 30% and 40% of tin slag mixture may be used as long as the public dose limit for radiation is not breach and the concrete strength is maintain.

4. Conclusion

The factor effecting surface radiation is tin slag percentage in the concrete mixture. The isolation of tin slag is suc-cessful in concrete mixture with 20% sand replacement formulation which only emits the best result of 0.8 μSv/hr surface dose. Where net total limit dose of 525 μSv/year is about 50% lower than 1 mSv/year annual limit dose for public exposure for tin slag concrete mixture and 20% tin slag mixture give the best strength development for 28 days as 24.8 N/mm2. Preliminary result conclude that tin slag additional is considered able to suit in concrete envi-ronment and replaced sand in normal concrete at optimum ratio of 20%.

5. Further works

The future work in this project explores the new mix design for outdoor application such as a cement brick mortar for brick laying, pole for agricultural or livestock fencing in rural areas project in helping the rural communities and underground civil foundation structure. The research includes studies on new mix designs that would enable the incorporation of the idea of altering water-cement ratio mixture for better improvement.

The facts that tin slag has a high density composition makes it possible for making bricks or wall that can be used as a radiation shielding. The tin slag collected from tin smelting industry tested with NAA method (table 6) has shown to contain twenty two elements. However, to reduce the effect of radiation some elements may be removed by special extraction processes.

The amount of the tin slag used in these mix design is less compared to the tin slag waste from the tin smelting industry found in Malaysia. The waste, in large amount, is threatening to the environment due to the fact that it needs much bigger area for safe dumping purpose in safe manner. Typically in tin smelting industry the sighting of piling up waste is common. This is not healthy for long term exposure to human being as the small particles from the tin slag may be inhaled into the body. It is much more crucial since some of the particles are radioactive. Hopefully, through this project as a start some proposals will be given due attention and the appropriate actions may be carried out accordingly in the next task of plan. Thus this would pave much more interesting and beneficial code of conduct and practices as the Nuclear Malaysia and local government struggle to find ways and procedures to solve the pile of tin slag.

6. References

[1] AELR 2011 Atomic Energy Licensing (Radioactive Waste Management) Regulations
[2] British Standard Institution 1983 *Method for determination of slump* (London BSI Part 102 BS 1881)

IOP Conf. Series: Materials Science and Engineering **298** (2018) 012014 doi:10.1088/1757-899X/298/1/012014

[3] British Standard Institution 1983*Method of making test cube from fresh concrete* (London, BSI Part 108 BS 1881)

[4] British Standard Institution 1983 *Method for determination of the compressive strength of concrete cube* (London BSI, Part 116 BS 1881)

[5] British Standard Institution 1986 *Recommendations for measurement of velocity of ultrasonic pulses in concrete* (London BSI Part 203 BS 1881)

[6] British Standard Institution 1997*Guide to Specifying Concrete* (London BSI BS 5328)

[7] British Standard Institution 1985 *The Structural use of Concrete* (London BSI BS 8110)

[8] Building Research Establishment 1997 *BRE Mix Design Method* (formerly DoE method)

[9] Building Research Establishment 1988 *Design of Normal Cooncrete Mixes* (ed Marsh B.K 2nd ed.)

[10] Gilmore J C and Jackson R G 1992 *Radiological Hazards from Deposit of Tin-Smelting Slag and the Problems of Site Clearance and Disposal*

[11] IAEA 2002 Training course series no. 17 *Guidebook on Non-destructive Testing of Concrete Structures* (IAEA Vienna)

[12] Malhotra V M, ACI 1975 monograph no. 9 *Testing of hardened concrete: Nondestructive methods*, (American Concrete Institute

[13] Mohamad Pauzi Ismail *et. al*, 2015 *Ujian Tanpa Musnah terhadap konkrit struktur* (DBP, Kuala Lumpur)

[14] Omar M 1998 The Status of Activities involving NORM in Malaysia (RCA Expert Advisory Group Meeting to Review and Develop Radiation Protection Guidance for Naturally Occurring Radioactive Materials in the Oil and Gas and other Minerals Extraction and Processing Industries, ANSTO, Sydney, Australia, 16-20 Mar. 1998)

[15] Omar, M., Hassan, A., and I. Sulaiman. Radiation exposure during travelling in Malaysia. *Radiation Protection Dosimetry*, 121 (4), 456–460, (2006)

[16] Omar M, Hamzah M S, and A K Wood 2008 Radioactivity disequilibrium and total Activity Concentration of NORM Waste *J. of Nuclear and Related Technologies* **3(2)** 47 – 56

[17] Ramachandran T V 2009 *Indian Journal of Physics*, Thoron (220 Rn) in the indoor environment and work places **8** pp1079-1098

[18] United Nations Scientific Committee on the Effects of Atomic Radiation (UNSCEAR) 2000 Report to the General Assembly with scientific annexes

[19] Winter M and Wicke A 1993 Environmental radiation ecology, radiation effects *Progress in Radiation Protection* (Band I., Verlag TUV. Rheinland, In German)

[20] Yusof M A W 2005 Investigating the Potential for Incorporating Tin Slag in Road Pavements (PhD Thesis, University of Nottingham)

IOP Conf. Series: Materials Science and Engineering **298** (2018) 012015 doi:10.1088/1757-899X/298/1/012015

Large-scale synthesis of Tellurium nanostructures via galvanic displacement of metals

Kuan-Ying Kok[1,a), Thye-Foo, Choo[1,b), Nur Ubaidah Saidin[1,c) and Che Zuraini Che Ab Rahman[2,d)

[1]Malaysian Nuclear Agency, Bangi, 43000 Kajang, Selangor D.E., Malaysia
[2]Institute of Microengineering and Nanoelectronics (IMEN), Universiti kebangsaan Malaysia, 43600 UKM Bangi, Selangor D.E., Malaysia

[a)]kyk1000@nm.gov.my
[b)]ctfoo@nm.gov.my
[c)]ubaidah@nm.gov.my
[d)]iniaruz_s@yahoo.com

Abstract. Tellurium (Te) is an attractive semiconductor material for a wide range of applications in various functional devices including, radiation dosimeters, optical storage materials, thermoelectric or piezoelectric generators. In this work, large scale synthesis of tellurium (Te) nanostructures have been successfully carried out in different concentrations of aqueous solutions containing TeO_2 and NaOH, by galvanic displacements of Zn and Al which served as the sacrificial materials. Galvanic displacement process is cost-effective and it requires no template or surfactant for the synthesis of nanostructures. By varying the concentrations of TeO_2 and NaOH, etching temperatures and etching times, Te nanostructures of various forms of nanostructures were successfully obtained, ranging from one-dimensional needles and rod-like structures to more complex hierarchical structures. Microscopy examinations on the nanostructures obtained have shown that both the diameters and lengths of the Te nanostructures increased with increasing etching temperature and etching time.

1. Introduction

Functional nanostructures are critical to the ever-increasing complexity of technologies due to their capabilities in achieving the desired materials properties and functions for the state-of-art performances. It is important to have a cost-effective materials processing technique that produces nanostructures in large quantity with reproducible structures and properties to make them commercially viable and competitive. Trigonal-Tellurium (t-Te), for example, is a useful p-type semiconductor material with narrow bandgap of 0.35 eV at room temperature. It has been used in various technological applications due to its unique physical and chemical properties such as piezoelectricity, thermoelectricity, photoconductivity and photoelectricity [1]. It is a good candidate material for sensors, radiation dosimeters, optical storage materials, thermoelectric or piezoelectric generators. Te nanowires have been synthesized by thermal evaporation of Te powders [2], laser ablation [3], direct thermal oxidation at ambient pressure in a flow of O_2 without catalyst [4], hydrolysis of tellurium isopropoxide in the presence of tetra alkyl ammonium bromide solution [5], sol-gel process [6] and hydrothermal techniques [7]. These methods usually involve high temperature or pressure that limits their manufacturability and scalability. Nevertheless, reports on the large-scale synthesis of Te nanostructures, in particular, the formation of one-dimensional (1-D) Te nanostructures by galvanic displacement (GD) have rarely been reported in spite of their wide applications.

We present, here in, an account of the large-scale synthesis of Te nanostructures via the process of galvanic displacements on commercially available zinc and aluminum foils as the sacrificial substrates. The method is facile, allows the use of flexible substrates and requires no template or surfactant. The effects of electrolyte concentration, reaction time and temperature on the morphology, dimension and crystallinity on the Te nanowires obtained were systematically analyzed to gain an insight into the growth mechanisms involved. X-ray diffractometry (XRD) and scanning electron microscopy (SEM) were used to characterize the microstructures, morphologies, phases and dimensionalities of the nanostructures obtained.

2. Materials and Methodology

Commercially available Zn and Al foils were cut into sizes measuring approximately 1 x 1 cm, chemically cleaned using standard procedures and dried in an oven. The substrates were then immersed into an alkaline Te bath containing 0.5 ml aqueous solution of x mM TeO_2 and 1 M NaOH in a Teflon beaker. The composition of the x mM TeO_2 was systematically varied between x = 1, 10, 100 mM and the reaction times for etching ranged between 30 - 60 minutes with the etching temperatures varied from room temperature (RT) to 50°C. The as-prepared samples obtained were subsequently centrifuged for about 3 minutes before soaking in 0.5 ml of 1 M H_2SO_4 and rinsing thoroughly with DI water to reveal the Te nanostructures. The nanostructures were characterized by using an XRD (PANAlytical XPert Pro MPD) and SEM (FEI QUANTA 400).

3. Results and Discussion

Figure 1. Te nanostructures formed from: (a-c) Zn and (d-f) Al substrates in electrolytes containing different concentrations of TeO_2 and 1 M NaOH at room temperature for 30 min.

The SEM micrographs in figure 1 reveal the various nanostructures of Te formed using different concentrations of TeO_2 with Zn and Al as the sacrificial materials respectively. The Te nanostructures formed exhibited different morphologies; (a) uniform granular structure, (b) pine leave structure of high aspect ratios with needle-shaped tips, (c) nanoflower structure, (d) nanofiber-like structure, (e) pine leave structure with pencil-shaped tips and (f) micron-sized pine leave structures with pencil-shaped tips. Lower tellurium ion concentrations (1 mM) were found to favor the kinetically-controlled 1-D nanostructures while higher tellurium ion concentrations favored the hierarchically-branched structures. More uniform nanostructures were obtained from the Zn substrates.

As shown in the SEM micrographs in figure 2, finer Te nanostructures were more apparent at higher reaction temperatures. With increasing reaction times for etching, the diameters of the nanostructures formed increase from 150 to 350 nm while their lengths increase steadily from 2 to15 μm.

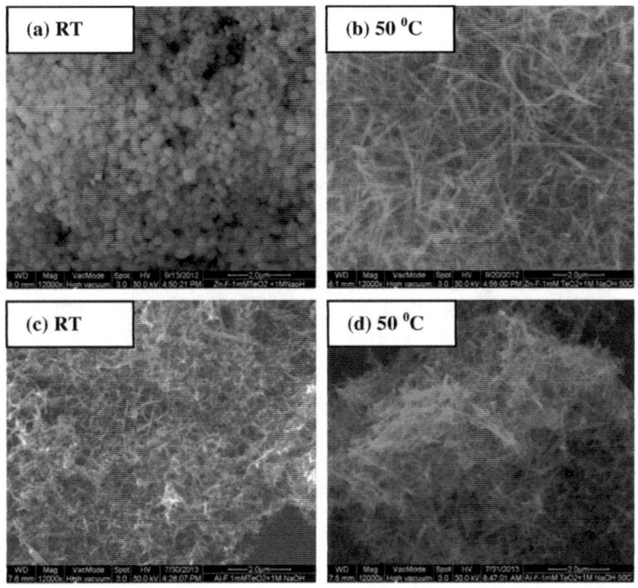

Figure 2. The effects of reaction temperatures on the Te nanostructures formed when (a-b) Zn and (c-d) Al substrates were etched for 30 min in 1 mM TeO_2 and 1 M NaOH.

Figure 3. Effect of reaction time on the Te nanostructures synthesized from Zn (a-b) and Al substrates (c-d) in 10 mM TeO_2 and 1 M NaOH at room temperature.

In figure 4, the X-ray diffractogram for the Te nanostructures obtained from 1 mM TeO_2 and 1 M NaOH at room temperature on Zn showed no prominent diffraction peak which is indicative of the amorphous phase of Te. In contrast, the Te nanostructures obtained from 10 mM TeO_2 and 1 M NaOH on Al at room temperature are crystalline with all peaks indexed to the space group P3121 (152) hexagonal crystal structure of Te (the corresponding SEM micrographs are shown in Figs. 1 (e) and 3 (c).

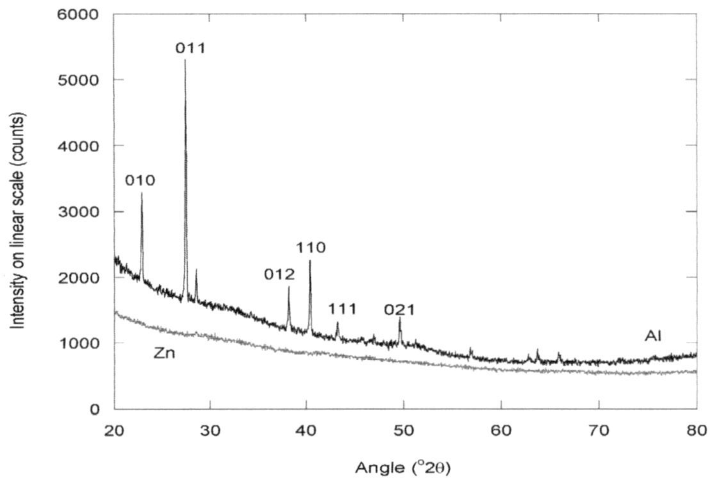

Figure 4. X-ray diffractograms of the Te nanostructures synthesized using: (bottom) 1 mM TeO_2, 1 M NaOH and Zn substrate and (top) 10 mM TeO_2 and 1 M NaOH and Al substrate, both at room temperature

In general, the growth mechanisms for the nanostructures involved basically simple redox reactions where zinc and aluminum were oxidized while Te ions were reduced as shown in Fig. 5. Zn or Al provided the negative redox potentials to reduce the TeO_3^{-2} ions to Te (0) nanostructures on the surface of the metal foil [8]. This generated an electric current to support the continued growth of Te, yielding Te hierarchical structures, such as nanoflowers or pine leave structures, depending on the growth conditions used. Different electrolyte concentrations affected the potential difference and current directionality, resulting in the formation of the Te nanostructures with different morphologies and large variations in length and diameter of the nanostructures. Galvanic displacement (GD) is a spontaneous electrochemical reaction driven by the difference in the redox potential between a solid substrate and the ions of a source material, leading to the deposition of a more noble material at the expense of a less noble material [9]. Since in a GD reaction, the substrate acts as the reducing agent, the synthesized product is of high purity. Besides capable of producing materials in large quantity, another advantage of GD is that it does not require any electrical connection. In this way, materials can be grown or deposited on any isolate part in a system.

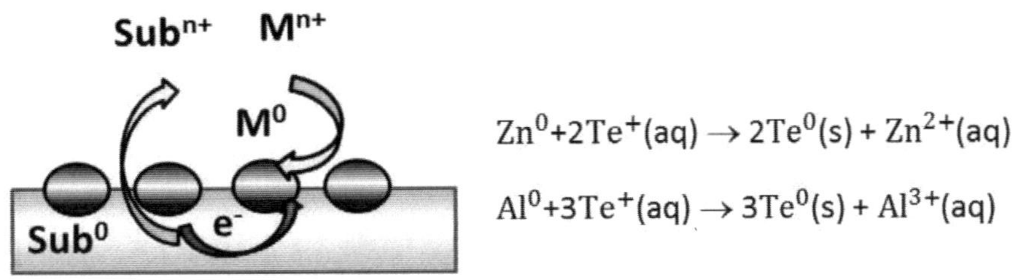

Figure 5. Schematics showing the reaction paths involved in a galvanic displacement reaction of Zn and Al.

4. Conclusion

The present study has successfully demonstrated the large-scale formation of Te nanostructures ranging from 1-D nanostructures to hierarchical branched structures via galvanic displacement. The morphologies of the Te nanostructures could be precisely tuned by adjusting the electrolyte concentration, etching temperature and etching time. This ability to tune the morphologies with different growth parameters is critical for achieving optimum Te nanostructures. In the formation of uniform and finer Te nanostructures, Zn acted as a better candidate material for galvanic displacement reactions compared to Al.

5. References

[1] Gautam U K and Rao C N R J 2004 *Mater. Chem.* **14** 2530–2535.
[2] Jiang Z Y, Xie Z X,. Zhang X H, Xie S Y, Huang R B, Zheng L S 2004 Synthesis of α-tellurium dioxide nanorods form elemental tellurium by laser ablation *Inorg. Chem. Commun* **7** 179-181.
[3] Liu Z, Yamazaki T, Shen Y, Kikuta T, Nakatani N 2008Synthesis and characterization of TeO2 nanowires *Jpn. J. Appl. Phys.* **47** 771-774.
[4] Siciliano T, Tepore A, Micocci G, Genga A, Siciliano M, Filippo E 2009 Transition formn- to p-type electrical conductivity induced by ethanol adsorption on α-tellurium dioaxide nanowires *Sens. Actuators B* **138** 207-213.
[5] Daniele Huriet S, Hubert-Pfalzgraf L G 2005 Effect of titanium additives on the growth of tellurium dioxide crystals in a sol-gel process *Mater. Lett.* **59** 2379-2382.
[6] Corte S, Thomas P, Lecomte A, Champarnaud-Mesjard J C, Guinebretiee R and Dauger 2002A *Phys. Chem. Glasses* **43C** 405.
[7] Zhu H, Zhang H, Liang J, Rao G, Li J, Liu G, Du Z, Fan H and Luo J 2011 *J. of Physical Chemistry C* **115** 6375 – 6380.
[8] Wu T, Zhang M, Lee K H, Kim S, Choa Y and Myung N V 2014 *Electrochimica Acta* **150** 298 – 307.
[9] Wu T, Myung L Y, Zhang M, Lee K H, Lee Y L, Lim H R, Kim B S, Choa Y H and Myung N V 2015 *Electrochimica Acta* **176** 1382 – 1392.

Acknowledgment

The authors would like to acknowledge Prof Dr Nosang Myung from University of California-Riverside for his technical support.

Hydrothermal growth of ZnO: a substrate-dependent study on nanostructures formation

N U Saidin[a], T F Choo[b] and K Y Kok[c]

Materials Technology Group, Industrial Technology Division,
Malaysian Nuclear Agency, Bangi 43000 Kajang, Selangor, Malaysia

[a]ubaidah@nm.gov.my
[b]ctfoo@nm.gov.my
[c]kyk1000@nm.gov.my

Abstract. ZnO has received lots of attention in recent years because of its unique properties, rendering it a good candidate material for various industrial applications. Amongst the methods used for synthesizing ZnO structures, hydrothermal technique offers a simple, cost-effective and yet environment friendly synthesis route for the materials. To investigate the effect of substrate on the formation of hydrothermally grown ZnO, several types of substrates were used and the structures of the ZnO were systematically studied by field emission scanning electron microscopy (FESEM). It was demonstrated that the type of substrates used had a great influence on the morphologies, density and alignment of the ZnO formed. Pre-coated substrate with Au yielded highly aligned ZnO nanorod arrays compared to bare Si and Al substrates. Possible mechanisms for the variation of morphology obtained are discussed.

1. Introduction

Zinc oxide (ZnO) possesses unique chemical and physical properties such as wide band gap (3.37 eV at room temperature) and high-exciton binding energy. This II-VI semiconductor is of great interest for a wide range of applications including sensors [1], light emitting diodes [2], solar cells [3], photocatalysts [4] and others.

Previous research has shown that the properties of ZnO depend strongly on it microstructures. To date, many forms of ZnO nanostructures have been reported, for instance tubes, spheres, wires, belts, helices, plates, islands, boxes, walls, rods, and flakes. These structures are affected by processing parameters like the synthesis method used, type of substrate, synthesis temperature and others. Various methods on the synthesis of ZnO have been reported in the literatures: chemical vapor deposition [5], magnetron sputtering deposition [6], pulsed laser deposition [7], electrochemical deposition [8] and hydrothermal method [9]. Hydrothermal method has become one of the favorite techniques in the growth of ZnO because, besides requiring moderate synthesis temperature, it is low-cost, simple and environmental friendly [10, 11].

Normally for hydrothermal method, a seed layer of ZnO nanocrystals is required to provide the homoepitaxial nucleation sites. This seed layer is generally prepared by sol-gel, sputtering, spin coating or physical vapor deposition techniques. This is followed by annealing process to ensure that the seed particles are successfully formed and adhered to the substrate [12]. However, the use of high temperature for annealing is detrimental to hydrothermal growth. The seedless approach, on the other hand, requires expensive substrates such as GaN or external electrical field to enhance ZnO growth [11]. To overcome this, the growth of ZnO can be performed via seedless hydrothermal technique assisted by galvanic process on different types of substrates. Zheng et al. (2013) demonstrated that the growth process could be performed on different conducting materials substrates [11]. In this work, we adopted this method for the synthesis of ZnO on different substrates, namely silicon (Si) wafer, gold (Au)-coated Si wafer, aluminum (Al) wire mesh and Au-coated Al wire mesh. A systematic study on the morphological changes of ZnO structures on these substrates was carried out.

2. Experimental

All chemicals in this work were of analytical grade and used as received without further purification. The synthesis of ZnO by the galvanic hydrothermal technique has been reported in our previous work [13]. In this work, the substrates used were Si wafer, Au-coated Si wafer, Al wire mesh and Au-coated Al wire mesh. The Si wafers were cut into a dimension measuring1 cm × 2 cm while the Al wire meshes were cut into the size of 2 cm × 2 cm. All substrates were first rinsed ultrasonically in acetone, ethanol, isopropanol and distilled water for 15 min successively. The edges of the substrates were then wrapped around with Al foils. An aqueous solution containing 25 mM zinc nitrate hexahydrate ($Zn(NO_3)_2.6H_2O$) and 25 mM hexamethylenetetramine (HTMA $C_6H_{12}N_4$) was used as the electrolyte. The solution was maintained at 75 °C on a hotplate and the growth time for ZnO was 4 h. After reaction, all the samples were taken out of the electrolyte and rinsed with deionized water for several times. The morphologies and microstructures of the ZnO were characterized by a field emission scanning electron microscopy (FESEM, GeminiSEM 500, Carl Zeiss). Figure 1 shows the schematics of the experimental set-up.

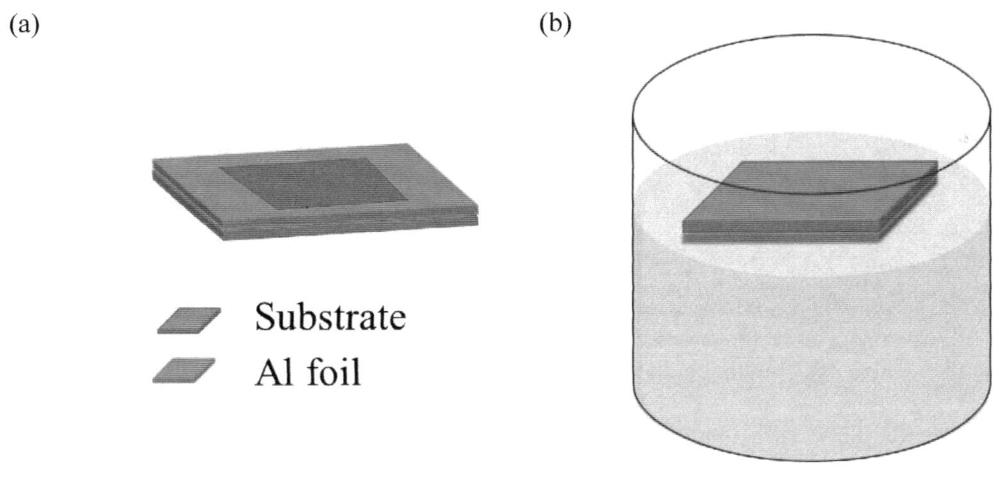

Figure 1. Schematics of the experimental set-up: (a) edges of the substrate were covered with an Al foil and (b) the substrate was floated down-face on the solution

3. Results and discussion

In this study, zinc nitrate was used as the Zn^{2+} precursor and HMTA as a weak base and pH buffer. ZnO was synthesized based on the following chemical reactions [10, 12]:

Decomposition reaction:

$$(CH_2)_6N_4 + 6H_2O \rightarrow 6HCHO + 4NH_3 \quad (1)$$

Hydroxyl supply reaction:

$$NH_3 + H_2O \rightarrow NH^{4+} + OH^- \quad (2)$$

Supersaturation and ZnO growth reaction:

$$Zn^{2+} + OH^- \rightarrow Zn(OH)_2 \rightarrow ZnO + H_2O \quad (3)$$

It is generally accepted that HMTA decomposes slowly in a heated aqueous solution to yield ammonia and formaldehyde as the initial reactants, which in turn resulting in the formation of hydroxide ions (Eq. 1-2). The hydroxide ions further reacts with the Zn^{2+} to form zinc complexes which finally transforms into ZnO.

For galvanic-based hydrothermal method, the growth mechanism is similar to that of the electrochemical deposition, except that a galvanic cell is employed instead of using the external power source. Difference in the work functions between Al (sacrificing anode) and the substrate (cathode) provides the bias to drive electrons from Al to the substrate. In this method, Al atoms lose electrons to develop positive charges and the electrons are transferred to the substrate that acts as a cathode. At the cathode, reduction process occurs in which oxygen is dissolved. This is followed by the formation of zinc hydroxide, Zn(OH)2 which then undergoes dehydration to form ZnO as shown in Eq. 4-6 [13]:

$$O_2 + 2H_2O + 4e^- \rightarrow 4OH^- \quad (4)$$

$$Zn^{2+} + 2OH^- \rightarrow Zn(OH)_2 \quad (5)$$

$$Zn(OH)_2 \rightarrow ZnO + H_2O \quad (6)$$

Figures 2-5 display the FESEM images of the samples. Different substrates used produced different morphologies of ZnO structures. When bare Si wafer was used as the substrate, flower-like ZnO morphology was obtained. These flower-like structures were assembled from many rod structures with average length of 2.9 μm. On close examinations, the structures reveal that each single rod was hexagonal in shape with average diameter of ~598 nm in size (Figure 2a-b). However, when Au-coated Si wafer was used, aligned ZnO nanorods of pencil-like structure with average diameter of 1.8 μm were grown perpendicularly to the substrate surface (Figure 3a-b).

Figure 2. (a) FESEM images of ZnO nanorods grown on bare Si wafer (a) side-view and (b) top-view. Inset shows that the rod is hexagonal in shape

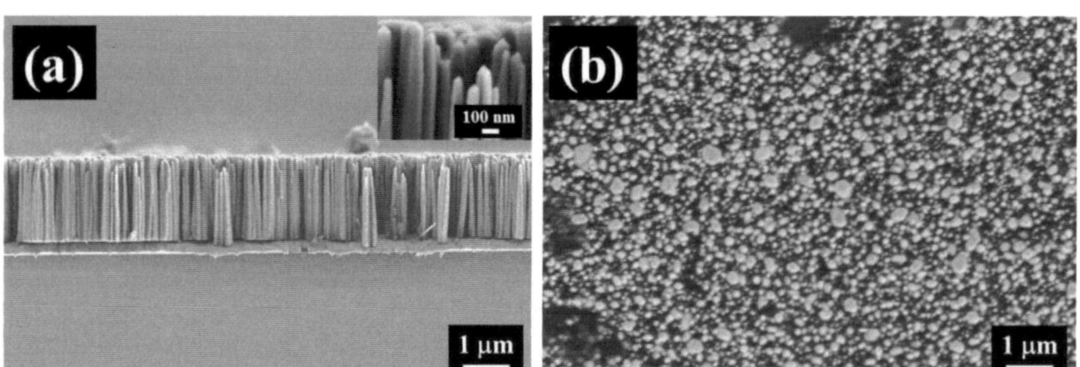

IOP Conf. Series: Materials Science and Engineering **298** (2018) 012016 doi:10.1088/1757-899X/298/1/012016

Figure 3. (a) FESEM images of ZnO nanorods grown on Au-coated Si wafer (a) side-view and (b) top-view. Inset shows that the rods are pencil-like in shape

When bare Al wire mesh was used as the substrate, randomly interconnected ZnO microflakes covered with nanorods were obtained (Figure 4). In contrast, the Au-coated Al wire mesh substrate produced uniform arrays of ZnO nanoflowers. This suggests that galvanic reaction helps to promote ZnO growth in nanorod shapes (Figure 5 a - b).

Figure 4. (a) FESEM images of ZnO nanorods grown on bare Al wire mesh (a) side-view and (b) top-view

Figure 5. (a) FESEM images of ZnO nanorods grown on Au-coated Al wire mesh (a) side-view and (b) top-view

Table 1 simplifies various growth structures of the ZnO obtained with different types of substrates:

Table 1. Types of ZnO growth structures

Substrate	Structures

IOP Conf. Series: Materials Science and Engineering **298** (2018) 012016 doi:10.1088/1757-899X/298/1/012016

Si wafer	Hexagonal flower-like	
Au-coated Si wafer	Aligned pencil-shape	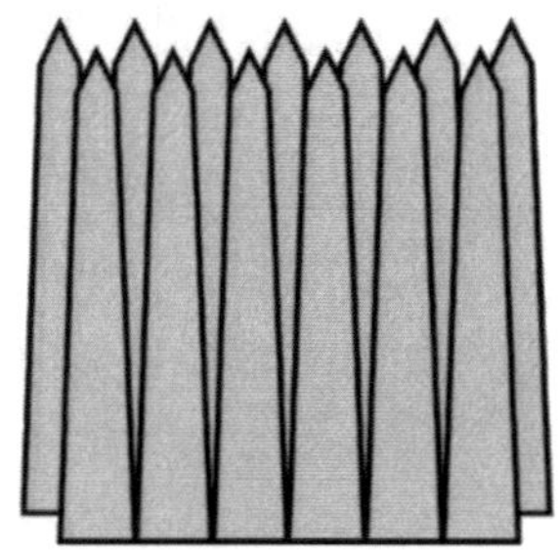
Al wire mesh	Hexagonal flower-like + flakes	
Au-coated Al wire mesh	Hexagonal flower-like	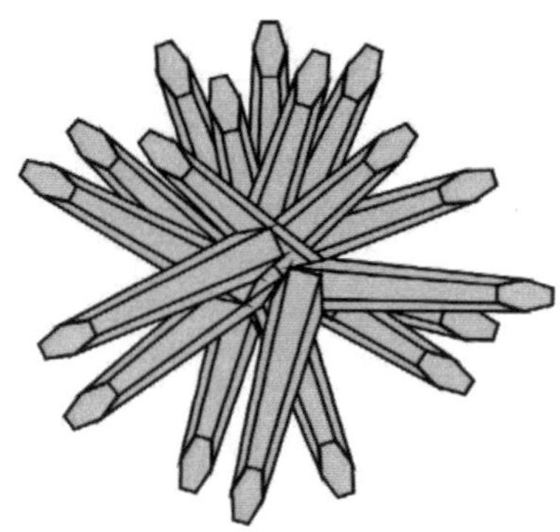

4. Conclusions

Distinct morphologies for the ZnO growth can be achieved on different substrates using seedless galvanic assisted hydrothermal method. ZnO growth on Au-coated Si wafer exhibits well-aligned pencil-shaped rod-like structures while others produced flower-shaped rod structures.

IOP Conf. Series: Materials Science and Engineering **298** (2018) 012016 doi:10.1088/1757-899X/298/1/012016

5. References

[1] Y. H. Navale, S. T. Navale, N. S. Ramgir, F. J. Stadler, S. K. Gupta, D. K. Aswal and V. B. Patil, Sensors and Actuators B: Chemicalhttp://dx.doi.org/10.1016/j.snb.2017.05.085.

[2] S. Zhi-Feng, Z. Yuan-Tao, C. Xi-Jun, Z. Shi-Wei, W. Bin, J. Jun-Yan, C. Xian-Wei, D. Xin, Z. Bao-Lin and D. Guo-Tong, CrystEngComm 17, 40-49 (2015).

[3] Z. Mei-Lin, J. Feng, Z. Mei-Ling, L. Jie, Z. Zhen-Sheng and D. Xuan-Ming, RSC Adv. 4, 10462-10466 (2014).

[4] P. Manoj and Y. Zahira, Applied Surface Science 292, 520-530 (2014).

[5] M. Madhusudhana Reddy, G. Ramanjaneya Reddy,K. Chennakesavulu, E. Sundaravadivel, S. S. Prasath, A. M. Rabe and J. Sreeramulu, Journal of Porous Mater, DOI 10.1007/s10934-016-0247-3.

[6] M. Parvathy Venu, B. V. Shrisha, K. M. Balakrishna and K. Gopalakrishna Naik, AIP Conference Proceedings 1831, 080041(2017).

[7] X. W. Sun and H. S. Kwok, J. Appl. Phys. 86, 408-411 (1999).

[8] Y. Dali, H. ming, l. Shenyu, L. Jiran, W. Yaqiao and M. Shuagyun, Electrochemica Acta 115, 297-305 (2014).

[9] H. Amir, B. Nicoleta, A. C. John and B. Pablo, Materials and Design 119, 464-469 (2017).

[10] Z. Shibu, C. Xiangnan, Z. Feibiao, J. Man, Z. Zuowan and H David, Journal of Solid State Chemistry 197, 69-74 (2013).

[11] Z. Zhaoke, S. L. Zhi, P. Yuan, Y. Lu, C. Lang and W. Junling, Scientific Reports 3, 1-5 (2013).

[12] Zhang. Yangyang, K. R. Manoj, K. S. Elias and D. Y. Goswami, Journal of Nanomaterials,624520, 1-22 (2012).

[13] K. Y. Kok, I. K. Ng, N. U. Saidin, T. F. Choo, B. S. Wee and Y. Abdullah, Materials Science Forum **888**, 304-308 (2017).

Acknowledgement

This work is financially supported by the Ministry of Science, Technology and Innovation, Malaysia and the Malaysian Nuclear Agency.

Gamma irradiation induced method in preparation of $Gd_2O_2S:Eu^{3+}$ phosphors: the effect of dose towards luminescent properties

S Rahim[1], M H Hasim[1], M T M Ayob[1], I A Rahman[1] & S Radiman[1*]

[1]Nuclear Technology Research Centre, Faculty Science and Technology, Universiti Kebangsaan Malaysia, 43600, Bangi, Selangor, Malaysia

shahidan@edu.ukm.my

Abstract. A novel gamma irradiation induced synthesis method of $Gd_2O_2S:Eu^{3+}$ phosphors was investigated in the presence of cetyltrimethylammonium bromide (CTAB). The effect of irradiation doses (50-150kGy) on structural and morphology analysis as well as luminescence properties were characterized by X-ray diffraction (XRD), field emission scanning microscopy (FESEM) and photoluminescence spectrometer (PL). The results show that gamma radiation is potentially induced formation of $Gd_2O_2S:Eu^{3+}$ phosphors from radiation reduction and/or precipitation of insoluble compounds as the hexagonal phase structure was formed without any impurities as proven in XRD pattern. The morphologies were observed that the obtained $Gd_2O_2S:Eu^{3+}$ phosphors possess sphere structure with smooth surface at 100 kGy irradiated dose. PL spectroscopy reveals that the strongest red emission peaks is located at 626 nm under 325 nm light excitation, which corresponds to $^5D_0 \rightarrow {}^7F_2$ transition of Eu^{3+} ions. An optimized dose for excellent luminescent was observed at 100 kGy. The results suggested that the $Gd_2O_2S:Eu^{3+}$ phosphors may have a beneficial approach in field of imaging device or media.

1. Introduction

Trivalent europium activated gadolinium oxysulfide ($Gd_2O_2S:Eu^{3+}$) phosphors is an important luminescent material especially for imaging display either in medical or industrial applications. The material featured high density and refractive index thus perform high light absorption and energy transfer efficiency with a broader excitation band and shows red emission. Radiation-induced processes have some advantages over conventional chemical methods:

- Temperature independent
- High purity of yield materials
- Narrow size distribution of particles
- Simple method

In this work, we have investigated the potential of gamma irradiation method to synthesize $Gd_2O_2S:Eu^{3+}$ phosphors as well as the effect on luminescent properties.

2. Experimental details

Gadolinium (III) nitrate hexahydrate ($Gd(NO_3)_3.6H_2O$, 99.9%), europium (III) nitrate ($Eu(NO_3)_3.5H_2O$, 99.9%) and cetyltrimethylammonium bromide (CTAB, ≥ 98%) were purchased from Sigma-Aldrich. Ammonium sulfate ($(NH_4)_2SO_4$) was purchased from Merck. All reagents were analytical grade and used without further purification. Deionized water (18Ω) was used as solvent throughout the experiment.

In this study, gadolinium nitrate, $Gd(NO_3)_3$ 0.5 M, $Eu(NO_3)_3$ 0.05 M and ammonium sulfate, $(NH_4)_2SO_4$ were weighed in suitable stoichiometric ratio and dissolved in deionized water as described detail in our previous work [1]. 1% mol of CTAB was added into the solution and stirred with a magnetic stirrer continuously for 2 hours before being portioned into four samples and irradiated in ^{60}Co gamma cell (1.45 kGy/h) with irradiation doses at 50kGy, 100kGy and 150kGy. The precipitate samples obtained were centrifuged and washed with ethanol and deionized water for several times. Afterwards, the solid precipitates were dried overnight in air at ambient temperature and hydrogenation treatment was carried out towards dried precipitates at 900°C for 2 hours.

2.1. Characterization

The gamma source of ^{60}Cobalt with a dose rate of 1.45 kGy/h model MDS NORDION / Gamma Cell 220 Excel was used under atmospheric pressure and room temperature. The X-ray diffraction (XRD) measurements of $Gd_2O_2S:Eu^{3+}$ phosphors samples were carried out in the reflection mode with a Bruker X-ray diffractometer that operated at 40 kV voltages and a current of 40 mA with Cu Kα radiation (λ=0.15406 nm). The morphology and size of the samples were inspected by the field emission scanning electron microscopy (FESEM-Carl Zeiss, Supra 35VP) imaging. Meanwhile, the photoluminescence spectra of the samples were gathered over a range between 300 to 800 nm of wavelength via FLSP920 Edinburgh spectrometer.

3. Results and Discussion

The X-ray diffraction pattern of $Gd_2O_2S:Eu^{3+}$ phosphors at different irradiation doses are shown in FIGURE 1. All samples show same diffraction patterns including none-irradiated sample (control) and these well-defined diffraction peaks can be indexed to pure hexagonal $Gd_2O_2S:Eu^{3+}$ phase according to JCPDS No: 00-026-1422 without any impurity peaks.

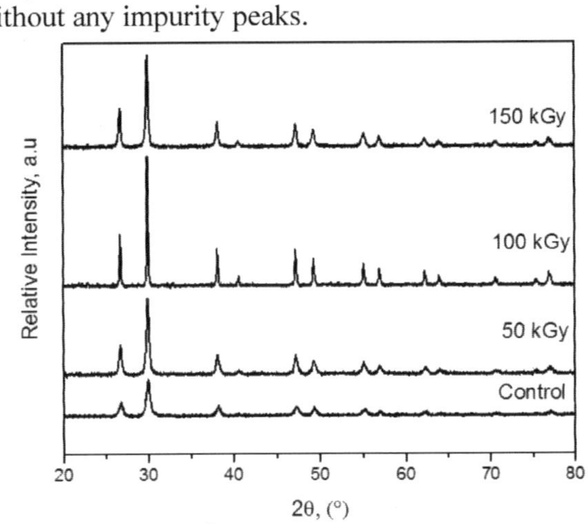

FIGURE. 1: XRD patterns of $Gd_2O_2S:Eu^{3+}$ phosphors of none-irradiated sample (control), 50 kGy, 100 kGy and 150 kGy irradiation dose.

The FESEM images of the produced $Gd_2O_2S:Eu^{3+}$ phosphors at different gamma irradiation doses were shown in FIGURE 2. In FIGURE 2(a), the none-irradiated samples exhibit agglomerated and rough surface of the particles. As the irradiation dose increases, the particle morphologies transformed into uniform spherical shape and the agglomeration also reduced. Therefore, the gamma irradiation could also produce defined structure as well as high homogeneity. The sample with 100 kGy irradiation dose shows high homogeneity and smooth surface. After 150 kGy irradiation dose, the spherical shape becomes ruptured and slightly porous due to molecular damage of CTAB thus unable to protect the particle surface. In this case, CTAB played an important role in controlling the morphologies and sizes of any nanomaterials [2]. As expected, the low concentration of CTAB used in this work tends to formed sphere structures as the surfactant associates into micelles and controlled particles surface thus slow down the nucleation process which also avoided from the aggregation process.

FIGURE 2.: FESEM images of $Gd_2O_2S:Eu^{3+}$ phosphors of non-irradiated sampel/control (a) and by different irradiation doses b) 50 kGy, c) 100 kGy and d) 150 kGy.

FIGURE 3 shows the emission spectra of the $Gd_2O_2S:Eu^{3+}$ phosphors at different gamma irradiation doses. Upon the excitation at 325 nm, the strongest red emission peaks which splits into two peaks at 616 nm and 626 nm arises from the forced electric-dipole $^5D_0 \rightarrow {}^7F_2$ transition of the Eu^{3+} ions. In this process, trivalent Gd^{3+} ions, as sensitizer, absorb ultraviolet excitation light and subsequently transfer energy to the neighboring Eu^{3+} ions act as activator, resulting in the overall red emission of Eu^{3+} ions [3]. An optimized dose that demonstrated excellent luminescent was observed at 100 kGy. It was found that the smooth surface of particles showed higher photoluminescence intensity [4].The slight decrease in luminescence intensity after reaching a certain crystallite size was caused by the large size of the crystallite, which tends to scatter light in particle excitation, leading to a reduction in intensity in luminescence measurements (FIGURE 3(b) 7).

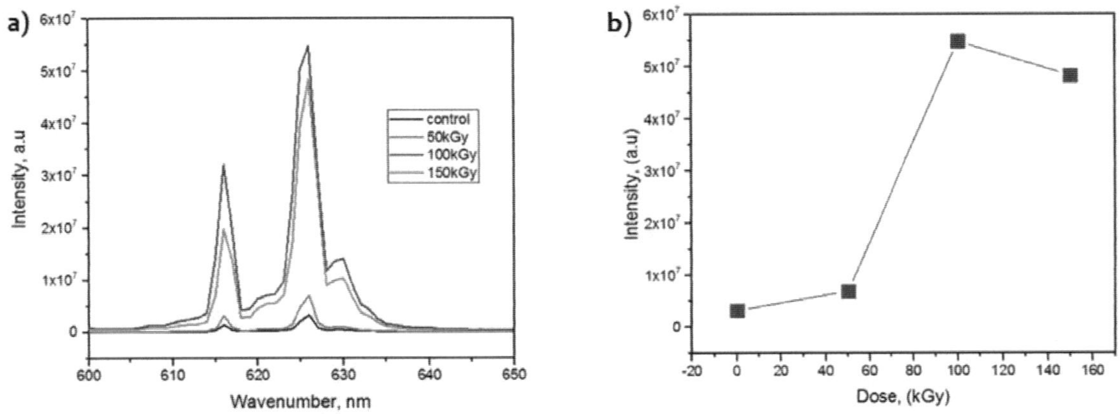

FIGURE. 3: (a) Photoluminescence emission spectra of $Gd_2O_2S:Eu^{3+}$ phosphors at different radiation doses and control (none irradiated sample) and (b) relation between irradiation dose and luminescent intensity.

4. Conclusion

In summary, $Gd_2O_2S:Eu^{3+}$ phosphors have been successfully synthesized by gamma irradiation induced method combined with hydrogenation process. The CTAB played important roles in the formation of the spherical morphology as well as in producing smooth surface of particles. It was found that the optimized dose for excellent luminescent observed at 100 kGy and the strongest red emission peaks is located at 626 nm under 325nm light excitation, which corresponds to $^5D_0 \rightarrow {}^7F_2$ transition of Eu^{3+} ions. The results obtained suggest that the $Gd_2O_2S:Eu^{3+}$ phosphors may have a beneficial approach in field of imaging device or media.

5. Acknowledgment

The present work was support by the Ministry of Education (MOE), Malaysia (UKM (AP-2015-006) and Public Service Department of Malaysia by giving scholarship to Sapizah Rahim

6. References

[1] Hasim M H, Rahman I A, Rahim S, Ayob M T M, Sharin S & Radiman S 2017 *J. Nanomater.* **2017** 7431432
[2] Chen F, Chen G, Liu T, Zhang N, Liu X, Luo H, Li J, Chen L, Ma R & Qiu G 2015 *Sci. Rep.* **5** 17934
[3] Huang J, Song Y, Sheng Y, Zheng K, Li H, Zhang H, Huo Q, Xu X & Zou H 2012 *J. Alloys Compd.* ***532*** 34
[4] Wang W N, Widiyastuti W, Ogi T, Lenggoro I W & Okuyama K 2007 *Chem. Mater.* ***19*** 1723.

Neutron shielding behavior of thermoplastic natural rubber/boron carbide composites

Nurazila Mat Zali, Hafizal Yazid and Megat Harun Al Rashid Megat Ahmad

Malaysian Nuclear Agency, 43000 Kajang, Bangi Selangor, Malaysia

nurazila@nuclearmalaysia.gov.my

Abstract. Many shielding materials have been designed against the harm of different types of radiation to the human body. Today, polymer-based lightweight composites have been chosen by the radiation protection industry. In the present study, thermoplastic natural rubber (TPNR) composites with different weight percent of boron carbide (B_4C) fillers (0% to 30%) were fabricated as neutron shielding through melt blending method. Neutron attenuation properties of TPNR/B_4C composites have been investigated. The macroscopic cross section (Σ), half value layer (HVL) and mean free path length (λ) of the composites have been calculated and the transmission curves have been plotted. The obtained results show that Σ, HVL and λ greatly depend on the B_4C content. Addition of B_4C fillers into TPNR matrix were found to enhance the macroscopic cross section values thus decrease the mean free path length (λ) and half value layer (HVL) of the composites. The transmission curves exhibited that the neutron transmission of the composites decreased with increasing shielding thickness. These results showed that TPNR/B_4C composites have high potential for neutron shielding applications.

1. Introduction

Exposure to high energy or ionizing radiation may result in cell mutation, cancer and death. Therefore, the radiation must be attenuated enough to protect the personnel from the harmful effects caused by it. Radiation shielding is one of the method to control radiation exposure. To design an appropriate radiation shielding, it is necessary to have known the neutron attenuation properties of the shielding materials. Since radiation interacts with matter, its intensity will decrease as it travels through a materials. The attenuation properties of radiation will affect how much shielding is necessary and how much dosage one receives. Attenuation coefficient is the most important factor in determining the effectiveness of shielding material used [1]. This shielding parameter also known as macroscopic cross section (Σ) for neutron is related to half value layer (HVL) and mean free path length (λ). Many researchers have determined the values of different shielding parameters in various ways to know the shielding effectiveness of the shielding materials developed [2-5].

Neutrons shielding unlike other forms of radiations introduces some complications because of the wide range of energy that must be considered [6]. High speed neutrons are more difficult to shield against because absorption cross sections are much lower at higher energies. Thus, it is first necessary to moderate high energy neutrons through elastic or inelastic scattering interactions. It is widely known that hydrogenous materials are used as neutron shielding for their effectiveness of neutron moderation. Materials such as water, polyethylene, polyester with higher hydrogen content have greater chance of shielding neutrons [7]. Hydrogenous materials can undergo elastic scattering with fast and intermediate energy neutrons. When boron is added to hydrogenous materials, it promotes the absorption of neutrons and reduces secondary gamma radiation. Boron containing polymer composites are generally used since boron compounds have excellent thermal neutrons absorption ability and are relatively cheap [8-10]. In this work, thermoplastic natural rubber (TPNR) with various amount of boron carbide (B_4C) fillers were fabricated as neutron shielding and the attenuation properties of the composites were investigated.

2. Materials and method

Composites with various filler loadings were prepared by melt mixing method. Natural rubber (NR), liquid natural rubber (LNR), high density polyethylene (HDPE) and boron carbide (B_4C) (0-30 wt%) were mixed in Haake internal mixer for 15 minutes at 135°C. The composite compound was then heated press at 135 °C to form a slab.

To investigate the attenuation behavior of the composites against neutron, transmission tests were performed by measuring the ratio of incident and transmitted neutron fluxes. Samples with various thicknesses were exposed to thermal neutron for equal interval of time using the thermal neutron beam from Malaysian Nuclear Agency's 1 MW TRIGA reactor. Samples were installed at the beam port located 719 cm from the source. The transmitted beam was collected using a fission chamber detector (Model 3053, LND Inc.).

3. Results and discussion

3.1. Macroscopic cross section, half value layer and mean free path

The capability of neutron shielding material lies on its attenuation coefficient value which is known as macroscopic cross section, Σ. The macroscopic cross section is applicable to a thick samples that often contain a mixture of elements [11]. Macroscopic cross section can be determined by measuring I and I_o of a shielding material of known thickness (x). When a neutron beam traverses a shielding material, the intensity of the beam will be attenuated according to the Beer Lambert's law:

$$I = I_o e^{-\Sigma x} \tag{1}$$

where I_o is the initial intensity of neutrons, I is the intensity of neutrons after traversing the medium, Σ is the macroscopic cross section and x is the thickness of the medium.

Figure 1 presents the macroscopic cross section of TPNR composites as a function of B_4C filler loading. It is clear that macroscopic cross section of the composites with B_4C addition is greater than TPNR matrix. The graph reveals a significant increase in macroscopic cross section as the B_4C filler loading increase which means that the probability of interaction increases as B_4C content became higher. The macroscopic cross section is the probability that a neutron will undergo a reaction per unit path length travelled in the material. Therefore, the greater the macroscopic cross section values, the higher the number of neutron that can be absorbed by shielding material. The macroscopic cross section reaches up to 343% upon raising the B_4C loading to 30 wt%. Increasing of macroscopic cross section value of TPNR composites with B_4C loading is due to the high thermal neutron absorption of ^{10}B isotope in B_4C. Natural boron consists primarily of two stable isotopes, ^{11}B (80%) and ^{10}B (20%) and its (n, α) reaction cross section for thermal neutrons is about 3840 barns (for 0.025 eV neutron) [12].

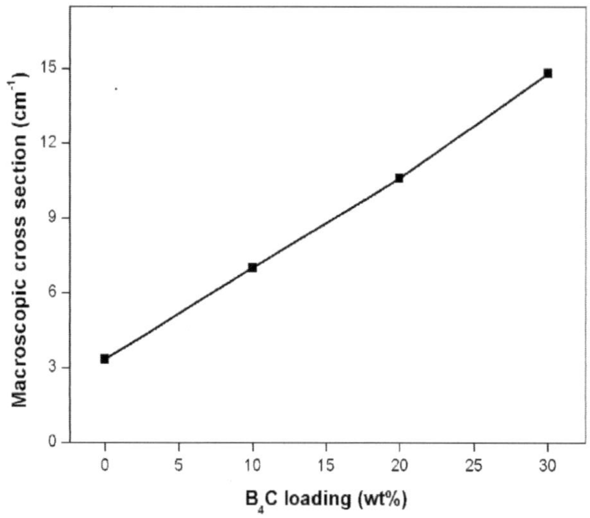

Figure 1. Macroscopic cross section of TPNR composites as a function of B_4C loading

The calculated macroscopic cross section, half value layer and mean free path length values of TPNR/B_4C composites have been shown in table 1. An increase in B_4C loading decreases the half value layer and mean free path length of the neutron. Half value layer (HVL) is an important parameter in designing any radiation shielding since half value layer indicates the required thickness of an absorber to reduce the radiation level to half of its initial value [13]. HVL is calculated by the following equation:

$$HVL = \frac{ln\,2}{\Sigma} = \frac{0.693}{\Sigma} \quad (2)$$

where Σ is the macroscopic cross section of the absorber. From the table, it is seen that the half value layer decreases from 0.207 cm to 0.047 cm with a weight percent of B_4C increase from 0% to 30% indicating an improvement of the attenuation properties of the composites. This results attributed to the increase of the macroscopic cross section values with the increase of the filler content. The results also indicate that composite with 30 wt% of B_4C loading requires only 0.047 cm of the thickness to attenuate 50% of the neutron intensity thus making it a better shielding material than others.

Similarly, decreasing trend of mean free path length (λ) values can be seen in table 1. The mean free path length (λ) is the average distance between two successive interactions. Neutrons travel on certain mean free path before undergoing a collision. The mean free path appears in the exponential and for this reason it is sometimes called the "relaxation length". This parameter is equivalent to the reciprocal of macroscopic cross section which is:

$$\lambda = \frac{1}{\Sigma} \quad (3)$$

where Σ is the macroscopic cross section of the absorber. The mean free path length depends on both the type of material and the energy of the neutron. After each collision, the energy is decreased and the mean free path length is affected accordingly [11]. By studying this parameter, the shielding properties of the materials can be easily compared. As shown in table 1, raising the loading of B_4C filler to 30 wt% result in reduction of mean free path length to 77% from the starting value. It also can be concluded that high B_4C content can attenuate neutron in a short distance while low amount of B_4C need a longer distance to attenuate the neutron.

Table 1. Macroscopic cross section (Σ), half value layer (HVL) and mean free path length (λ) values of TPNR composites with various B_4C loadings.

B_4C loading (wt%)	Σ (cm^{-1})	HVL (cm)	λ (cm)
0	3.343	0.207	0.299
10	7.014	0.098	0.143
20	10.604	0.065	0.094
30	14.803	0.047	0.068

3.2. Neutron transmission

Neutron transmission test was conducted for different thicknesses of the composites and the variation of neutron transmission factor (I/I_o) with shielding thickness for TPNR composites with 0-30 wt% of B_4C are depicted in Figure 2. The shape of the graph are approximately the same for all types of composites. Neutron transmission factor (I/I_o) decreases with increasing shielding thickness which follows the exponential law (equation 1) for all composites. From this figure, it can be observed that composite with 30 wt% of B_4C has the lowest neutron transmission factor due to its high macroscopic cross section value. So at this composition, it could be say that the shielding performance against neutron is much better than others. From this figure, it is also shows that higher thickness of shielding material gives a better protection against radiation. Transmission factor at 0.3 cm thickness of 10 wt%, 20 wt% and 30 wt% composites were found to be 0.106, 0.05 and 0.025 respectively. Whereas composite without B_4C addition has the highest transmission value of 0.369.

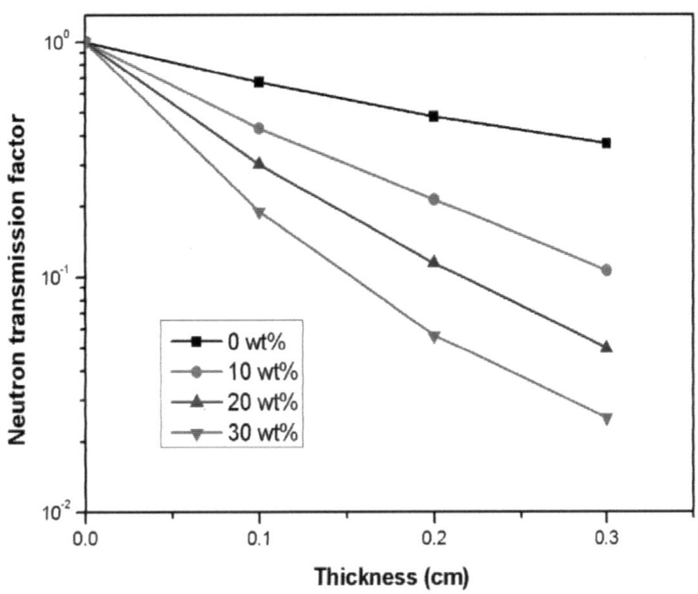

Figure 2. Comparison of neutron transmission factor of composites at different filler loadings

4. Conclusion

Neutron attenuation properties of TPNR filled with various filler loadings of B_4C were investigated in this study. It has been found that the addition of B_4C into TPNR matrix greatly influenced the shielding effectiveness of the composites. The macroscopic cross section increases whereas half value layer and mean free path length decrease by the increase of B_4C filler in the composites. Meanwhile, increasing shielding thickness gives a better transmission against neutron radiation.

5. References

[1] Shultis J K and Faw R E 2002 *Fundamentals of Nuclear Science and Engineering, Second Edition* (CRC Press)
[2] Singh V P and Badiger N M 2014 *Investigation of gamma and neutron shielding parameters for Borate Glasses containing NiO and PbO.* Hindawi Publishing Corporation, Physics Research International. Article ID 954958
[3] Madhusudhan Rao A S, Narender K, Gopal Kishan Rao K, Krishna G N and Murthy R K 2013 *Res. J. Phy. Sci.* **1(6)** 11-16
[4] El-Khayatt A M and Akkurt I 2013 *Annals of Nuclear Energy* **60** 8-14
[5] Elmahroug Y, Tellili B and Souga C 2013 *Int. J. Phys. Re.* **3** 33-40
[6] Igwesi D I and Thomas O S 2014 *Journal of Multidisciplinary Engineering Science and Technology* **1** 281-284
[7] Mensah W O, Fletcher J J and Danso K A 2012 *Int. J. Sci. Technol.* **5** 455-461
[8] Kharita M H, Yousef S and AlNassar M 2011 *Prog. Nucl. Energy. Prog.* **53** 207-211
[9] Jumpee C and Wongsawaeng D 2015 *J. Phys. Conf. Ser.* **611** 012019
[10] Yao F,Wu Q, Lei Y and Xu Y 2008 *Ind. Crops. Prod.* **28** 66-72
[11] Rinard P 1991 Neutron Interactions with Matter in *Passive Nondestructive Assay of Nuclear Materials* (Nuclear Regulatory Commission, NUREG/CR-5550) pp 357-377
[12] Vikas M 2015 *Surface Modification of Nanoparticle and Natural Fiber Fillers*, (Wiley-VCH Germany)
[13] Biswas R, Sahadath H and Mollah A S 2016 *J. Radiat. Res. Appl. Sci.* **9** 26-34

IOP Conf. Series: Materials Science and Engineering **298** (2018) 012019 doi:10.1088/1757-899X/298/1/012019

The optimization of heating temperature for carbon extraction from peat soil

FahrulRazi Md Jamien[1,a], Zaidi Embong[1,2,b], Saiful Azhar Ahmad Tajudin[1,3], Salawati Ahmad[4], Azwan Mat Lazim[5]

[1] Faculty of Applied Science and Technology, Universiti Tun Hussein Onn Malaysia, Pagoh Education Hub, KM 1 Jalan Panchor, 86400 Muar, Johor, Malayisa
[2] Research Centre for Soft Soil (RECESS), Office for Research, Innovation, Commercialization and Consultancy Management (ORICC), Universiti Tun Hussein Onn Malaysia
[3] Faculty of Civil and Environment Engineering, Universiti Tun Hussein Onn Malaysia
[4] Centre for Language Studies, Universiti Tun Hussein Onn Malaysia
[5] School of Chemical Sciences and Food Technology, Universiti Kebangsaan Malaysia

[a]fahrulrazimdjamien@gmail.com
[b]zaidi@uthm.edu.my

Abstract. This study aims to find the optimised heating temperature for carbon extraction from peat soil. Ccarbon from peat soil was extracted by the pyrolisation process at temperature, T = 200, 300, 400, 500, 600 and 700°C for 5 hours. The carbon, C and silica, Si content extracted from peat soil at various heating temperatures were measured by using an Energy Dispersive X-Ray analyzer (EDX) at 3 points on the sample surface. High atomic percentage of carbon (95.44 %) was measured at T = 400°C, while the highest atomic percentage of silicon was recorded at T = 700°C (30.79 %). The surface morphology of peat soil was analysed using Scanning Electron Microscope (SEM) at 800 magnifying power, and it was clearly seen that peat fragments were pyrolised at high heating temperature with pores enhancement. The molecular structure parameter of the extracted carbon were identified using Laser Raman analysis. The peak positions of D-band and G-band for raw peat soil were observed at Raman shift of 1379.62 and 1549.02 cm^{-1}, respectively. Meanwhile, the D-band for peat soil heated at temperature of 400°C was 1391.56 cm^{-1} whereas the G-band has peak position at 1562.16 cm^{-1}. From the Raman spectra of the optimum heating temperature at 400°C, it was revealed that the carbon molecular structure from the peat soil is mainly attributed by graphite and diamond structure.

1. Introduction

Land surface in the world comprises of peat soil for about 400 million ha or equivalent to 3% of the land surface in the world [1]. According to Spedding [2], a significant amount of reserve peat soil have been discovered in Indonesia, Brazil and other subtropical region such as Malaysia. Peat is an organic-rich and fibrous material that is generated from partially decayed organic complex material with lignin and cellulose as major constituent [3] . Organic material found in the peat soil generally have a high content of carbon. The exhaustion of peat soil from the burning process generates a huge quantity of carbon. There has been a report on the extraction of carbon by using peat as a precursor [4]. Large amount of carbon obtained from peat soil could be activated physically or by using a chemical reagent such as concentrated acid or alkaline [5]. Activated carbon is very useful in various areas in science, especially in environmental remediation. In order to extract the carbon, peat soil needs to be heated at various temperature. Thus, peat soil can be used as a precursor for the production of activated carbon. Since peat soil has a high amount of carbon content, it is normally composed of a significant amount of graphite-like structure or sp^2 dominant depending on the heating procedure in which it consist of high temperature thermal treatment [5] . However, appropriate heating temperature needs to be determined. Thus, the changes of carbon molecular structure in the sample will be observed and the temperature that contributes in extracting the highest content of carbon will be investigated in this study.

2. Methods

2.1 Sample Preparations

In this work, peat soil sample was collected at the depth of 0 – 30 cm from the surface at Parit Nipah Darat, Parit Raja, Johor. In order to prevent cross contamination and to keep moisture of the peat sample, polyethylene bag was used to keep the sample. The soil was then transported to the Research Centre for Soft Soil (RECESS), UTHM. The peat soil was taken out from the polyethylene bag and loaded into a steel loader for drying process.. The drying process was performed in the oven at 110°C for 24 hours. After the drying process, the samples were cooled at room temperature at 28°C. The samples then underwent pulverization with grinding machine and were sieved at 62 micron size particle. The process of sieving is important in order to obtain uniform size particle of samples. The sieved sample was loaded into crucible with lid and then proceed with pyrolisation process. The pyrolisation process undergoes no chemical reagents and an electric furnace was used (WiseTherm Laboratory Instrument) at various temperatures. The baking temperatures used in this study were 200, 300, 400, 500, 600 and 700°C, and the baking process for each temperature were performed within 5 hours period. The pyrolised samples were cooled at room temperature and kept in a dessicator for the next procedure.

2.2 SEM-EDX Analysis

Micrographs of the raw peat soil and pyrolysed peat surface area were acquired by using JEOL JSM-6380LA scanning electron microscope (SEM) facilited with EDX system. The samples were loaded onto carbon adhesive tape, then via into SEM coating system. A thin layer of gold was sputtered on the samples by using Fison Instrument SEM coating system in order to dissipate charge during scanning. The coated samples were then transferred into the specimen stage and viewed under SEM operating with an accelerating voltage of 10kV, and under low pressure at 8 Pa. The elemental analysis of carbon, C and silicon, Si were performed using SEM-EDX Analysis Station software.

2.3 Laser Raman Spectroscopy

Laser Raman spectroscopy was used to determine the molecular structure of pyrolised samples. Potassium bromide, KBr technique with impregnation ratio 1:100 was used to prepare the pellet. The pellet from KBr impregnation was then mounted on the sample stage of the spectroscopy instrument for analysis. The measurement was in the intensity. The analysis was carried out in a wave number range of 250-3000 cm^{-1}. The wavelength and laser spot size were 532nm (green) and ~30μm of diameter. In this research, the laser exposure time for the sample was 15 seconds per sample. The deconvolution of Raman spectra to determine the D-band and G-band were conducted by using a Lorentzian curve-fitting procedure using Origin 8.0 software.

3. Results and Discussions

3.1 SEM-EDX Analysis of Carbon Content at Various Baking Temperature.

Figure 1 shows the changing of C and Si atomic percentage at various baking temperature. The concentration of carbon was gradually increased from 200°C to 400°C. Further baking at 400°C, exhibits maximum concentration of carbon, C was obtained at 95.44 at %. The maximum percentage of carbon, C was achieved due to the organic ,atter degradation from peat itself within the range of temperatures (optimum pyrolisation) at atmospheric pressure [6] . However, the concentration of carbon, C gradually decreases while the concentration of silicon, Si increases with respect to a higher temperature. At the maximum baking temperature, carbon content in the peat has the lowest atomic percentage for about 25.55 at %. The decline of carbon content at the higher temperature was due to the possible contamination of oxygen from the oven ambient [7] . Carbon content loss beginning at temperature of 500°C to 700°C were due to oxidation of carbon to form carbon dioxide (CO2) and carbon monoxide (CO) [9]. However, carbon content in untreated condition (raw peat) is slightly higher compared to the carbon content at temperature of 200°C (85.69 at %) with atomic percentage of 91.34 at %. The silicon, Si content in untreated peat has a least concentration for only 0.77 at % compared to other various baking temperature. The concentration of silicon increases with respect to temperature from room temperature (raw peat) to 700°C. At 500°C, silicon concentration shows rapid increase to a higher temperature. At 700°, silicon shows the maximum concentration for about 30.79 at %.

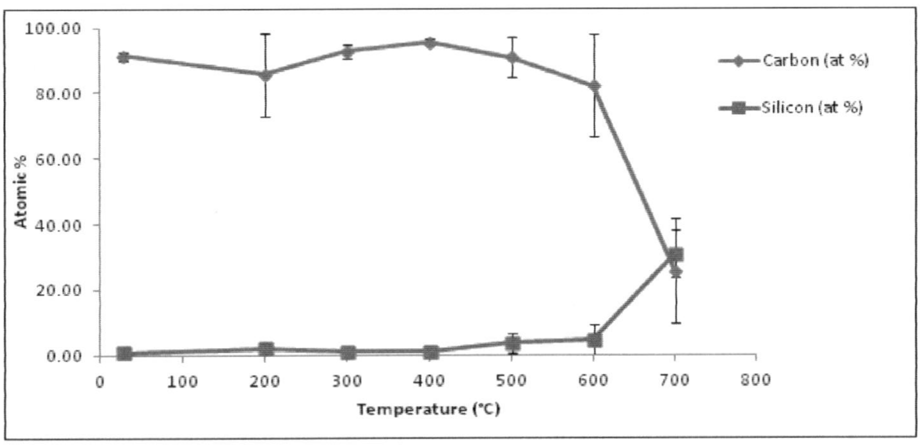

Figure 1. C and Si elemental composition (at%) changes at baking temperature between 200°C to 700°C.

Figure 2 (a-b) shows the morphology structure of the sample from SEM untreated and treated at 400°C baking temperature. Figure 2(a) shows the surface of untreated peat soil that has a rough surface area. This could be from the peat fragments such as wood, bark or overground plant components[10]. However, the morphology of the sample for 400°C treated sample exhibits much fine surface and there is no deposits sticking on its surface. Those peat fragments were distinguished as pyrolised at high baking temperature beginning from 400°C.

Figure 2(b) also proves that there is some enhancement of pores in the treated sample surface. Here, the enhancement of pores could be due to high temperature. High temperature causes the decomposition and softening of some peat fragment to form an intermediate melt in the carbon structure. Some additional pores were developed due to the releases of volatile organic matter. This is in agreement with the surface study of activated carbon from pistachio–nut shell that had been reported by Lua *et al.*[10].

IOP Conf. Series: Materials Science and Engineering **298** (2018) 012019 doi:10.1088/1757-899X/298/1/012019

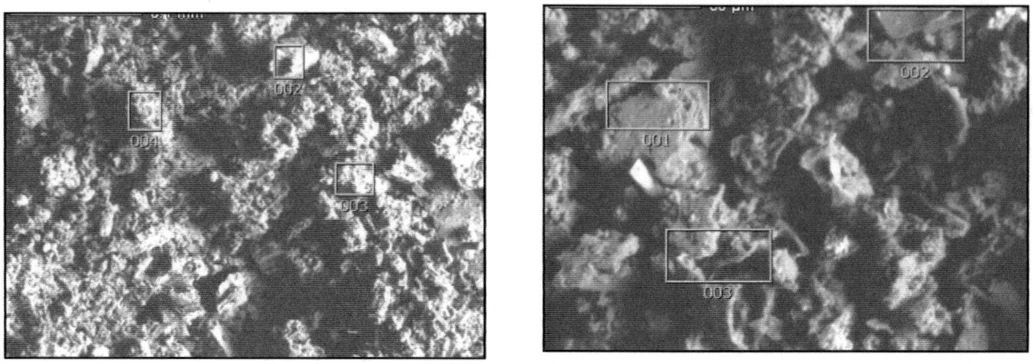

(a) (b)

Figure 2. Surface morphology comparison between (a) untreated raw peat and (b) heat treated at baking temperature, T = 400°C.

3.2 Molecular Structure Analysis for Carbon Compound at Optimum Heating Temperature Using Laser Raman.

The raw peat and the treated samples at an optimum temperature, 400°C were analysed by using Laser Raman spectroscopy. Laser Raman spectroscopy was used in order to determine the molecular structure of carbon compound for treated sample. Figure 3 shows the Raman spectroscopy for untreated peat soil and heat treated peat at T = 400°C, respectively.

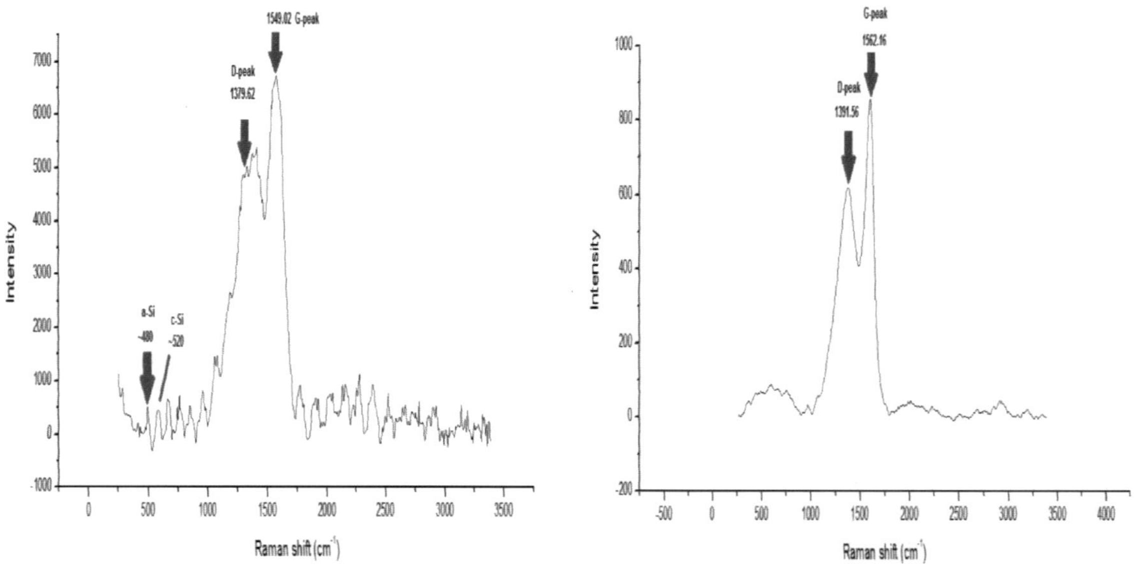

(a) (b)

Figure 3. Laser Raman spectrum for (a) untreated peat soil and (b) heat treated at baking temperature, T = 400°C

Figure 3(a) shows Raman spectra proves the presence of amorphous silicon, a-Si where the Raman shift is clearly observed at ~480 cm^{-1}, and the peak intensity of crystalline silicon, c-Si was found at a wavenumber around ~520 cm^{-1}. However, the saturation point of concentration can be observed from figure 1 which lies on approaching 700°C. Figure 4 shows Laser Raman spectrum deconvolution for untreated peat soil and heat treated at baking temperature 400°C.

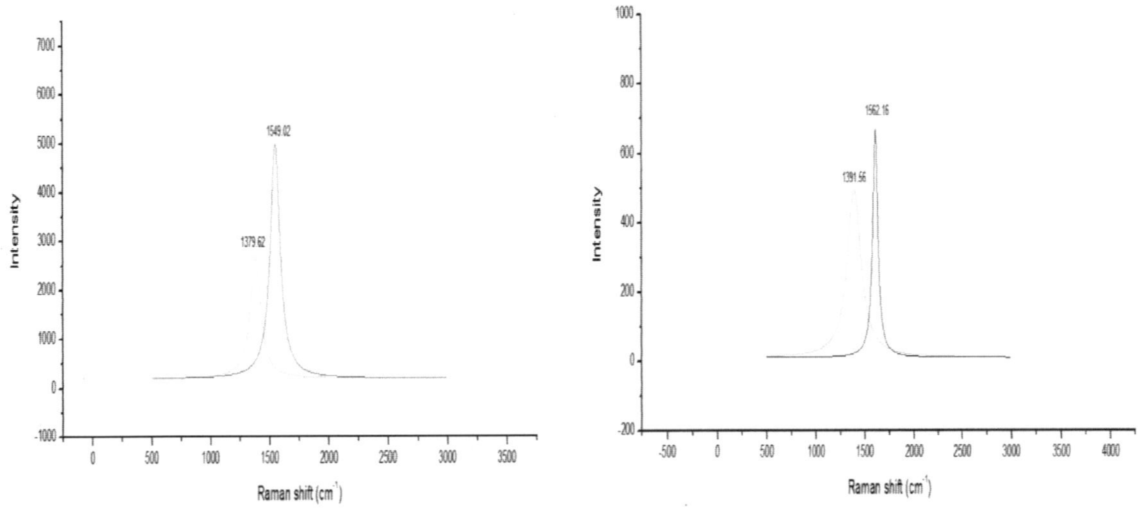

(a) (b)

Figure 4. Laser Raman spectrum deconvolution for (a) untreated peat soil and (b) heat treated at baking temperature, T = 400°C.

In figure 4, the baking process affects the position of the Raman spectra is displayed. The spectrum deconvolution needs to perform in order to recover the features of the original spectrum although peaks are broadened. According to Qi Wang et al [11], the spectrum deconvolution is significance to monitor how each Raman shift changes when materials are heated. The D-bands were slightly changed from 1379.62 cm^{-1} to 1391.56 cm^{-1}. The position of the D-band shifts to a higher wavenumber as the temperature increases. This is due to the changes of structure of carbon materials. The D-bands were broad spectrum indicating that carbon in both spectrums was consistent. But, there was a slight Raman shift observed between both, as shown in figure 4. However, D-band in Raman spectrum on peat at T=400°C was stiffer than the raw peat spectrum as shown at the base of spectrum. The stiffening reduction of band indicates carbon structure were somewhat similar [13]. Diamond-like (sp^3) carbon network was identified in both Raman spectra which indicates structure defects [14]. The G-band shows the graphitic structure of carbon organisation. It can also be seen that there are some changes from 1549.02 cm^{-1} to 1562.16 cm^{-1}. The G-band was also shifted to a higher wavenumber which means, more graphitisation (carbon, C arrangement into graphite formed) after being baked [15]. Graphitic-like (sp^2) carbon network exist in both Raman spectrum.

The spectra of raw peat and treated peat soil sample at T=400°C were contained into two main bands called as the graphitic (G) band and the disordered (D) band. G-band occurs at ~1500 cm^{-1} while D-band occurs at ~1300 cm^{-1}. From the laser Raman spectrum, the Raman shift position for both samples agrees with [16] the shift position for D-band and G-band occur in the range of 1100 – 1800 cm^{-1}. The maturity of sp^2 carbonaceous material or graphitic characteristic of peat can be measured by the value of intensity ratio, ($R = I_G/I_D$), where I_D is diamond-peak intensity, and I_G is graphitic-peak intensity. The value of peak intensity steeply decreases over the temperature of 400°C. The values of R in this study for untreated peat and treated peat soil sample at T = 400°C were 1.24 and 1.08, respectively. According to Khadiran *et al.* [17], the samples have reasonable graphitic content because the values of R more than 1.0.

4. Conclusion

Carbon, C from peat was extracted from peat. Highest carbon content was obtained at heating temperature of 400°C (95.44 at %). Carbon concentration increases with respect to high thermal energy but begins to decrease at temperature of 500°C, and carbon, C was expected to react with oxygen at high thermal energy forming carbon monoxide (CO) and carbon dioxide (CO_2). Silicon obtains the highest concentration at maximum baking temperature of 700°C. Viewing of peat samples via SEM describes the surface morphology and mineral composition, thus verifies the degree of peat decomposition. The result of Raman spectra shows that two structures of graphitic-like (sp^2) and diamond-like structure (sp^3) were dominant in peat. The structural changes of carbon have been interpreted by the considerable changes of intensity ratio.

5. References

[1] Strack M. Peatlands and climate change. University of Calgary; 2008.
[2] Spedding PJ. Peat. *Fuel*. 1988 Jul 1; **67**(7):883–900
[3] Brown P., Gill S., Allen S. Metal removal from wastewater using peat. *Elsevier*. 2000; **34**(16): 3907-3916
[4] Uraki Y, Tamai Y, Ogawa M, Gaman S, Tokura S. Preparation of activated carbon from peat. *BioResources*. 2008; **4**(1):205–13
[5] Dias JM, Alvim-Ferraz MCM, Almeida MF, Rivera-Utrilla J, Sánchez-Polo M. Waste materials for activated carbon preparation and its use in aqueous-phase treatment: A review. *J Environ Manage*. 2007; **85**(4): 833–846
[6] Lazzarini A, Piovano A, Pellegrini R, Agostini G, Rudić S, Lamberti C, et al. Graphitization of Activated Carbons: A Molecular-level Investigation by INS, DRIFT, XRD and Raman Techniques. *Phys Procedia*. 2016; **85**:20–26
[7] Petsch ST. Carbon (Organic, Degradation). In *Springer Netherlands*; 2011. p. 234–238
[8] Chung CK, Wu BH. Reaction of carbon and silicon at high temperature deposition. In: 2008 3rd *IEEE International Conference on Nano/Micro Engineered and Molecular Systems. IEEE*; 2008; 136–139
[9] Maahs HG. Oxidation of carbon at high temperatures: or transport control reaction-rate control. Washington.D.C; 1972.
[10] Ryedelek P. Application of Scanning Electron Microscope (SEM) in Peat Studies. *Polish J Environ Study*. 2006;**15**(5):117–121
[11] Lua AC, Yang T, Guo J. Effects of pyrolysis conditions on the properties of activated carbons prepared from pistachio-nut shells. *J Anal Appl Pyrolysis*. 2004;**72**(2):279–287
[12] Qi Wang D, D.D. A, L.V. K. Deconvolution of the Raman spectrum of amorphous carbon. *J Raman Spectrosc*. 1995; **26**(1):1039–1043
[13] Beyssac O. Raman spectra of carbonaceous material in metasediments : a new geothermometer. *J Metamorph Geol*. 2002; **20**(9):859–871
[14] Beyssac O, Goffé B, Petitet J-P, Froigneux E, Moreau M, Rouzaud J-N. On the characterization of disordered and heterogeneous carbonaceous materials by Raman spectroscopy. *Spectrochim Acta Part A Mol Biomol Spectrosc*. 2003; **59**(10):2267–2276
[15] Pimenta, Dresselhaus, Dresselhaus, Cancado, Jorio, Saito. Studying disorder in grpahitic-based systems by Raman spectroscopy. *Phys Chem*. 2007; **9**:1276–1291
[16] Roh J-S. Structural Study of the Activated Carbon Fiber using Laser Raman Spectroscopy. *Carbon Lett*. 2008; **9**(2):127–30
[17] Khadiran T, Hussein MZ, Zainal Z, Rusli R. Activated carbon derived from peat soil as a framework for the preparation of shape-stabilized phase change material. *Energy*. 2015; **82**:468–78

Acknowledgements
This paper was fully sponsored by the Centre for Graduate Studies (CGS), Universiti Tun Hussein Onn Malaysia (UTHM). We would also like to show our gratitude to the Research Centre of Soft Soil (RECESS), UTHM for providing us the electric furnace and also sampling tools and thanks to Faculty of Mechanical and Manufacturing Engineering, UTHM for assistance with SEM-EDX for morphological study and elemental contents on peats. Also, we would like to thank Microelectronics and Nanotechnology – Shamsuddin Research Centre (MiNT-SRC) for great assistance with Laser Raman spectroscopy for determining molecular structure on peats. We are also immensely grateful to all contributors, directly and indirectly helping us to complete this study.

IOP Conf. Series: Materials Science and Engineering **298** (2018) 012020 doi:10.1088/1757-899X/298/1/012020

Mass attenuation coefficients of several bio-adhesive based oil palm particleboards at 16.59-25.26 keV photon energies

U A Abdu Mustapa[1], M F Mohd Yusof[2], P N K Abd Hamid[4], R Hashim[3], M Z Ahmad[1], M Z Abd Aziz[1]

[1]Advanced Medical and Dental Institute, Universiti Sains Malaysia, 13200 Penang, Malaysia
[2]School of Health Sciences, Universiti Sains Malaysia, 16150 Kota Bharu, Kelantan, Malaysia
[3]School of Industrial Technology, Universiti Sains Malaysia, 11800 Penang, Malaysia
[4]School of Physics, Universiti Sains Malaysia, 11800 Penang, Malaysia

umiatulasma@student.usm.my

Abstract. Particleboards made of oil palm with addition of polylactic acid (PLA), starch, and fish oil were fabricated with target density of 1.0 g/cm^3. The mass attenuation coefficients of the particleboards were measured using x-ray fluorescence (XRF) configuration in conjunction with niobium, molybdenum, palladium and tin metal plates that provided $K_{\alpha 1}$ photon energies between 16.59 and 25.26 keV. The results were compared to the calculated value of water using XCOM. The results showed that all particleboards having mass attenuation coefficients near to the value of water with the mass attenuation coefficient different less than 0.25. The method of fabrication did not give significant different to the mass attenuation coefficients of the particleboards. The results had indicated the potential of bio-adhesive based palm oil particleboards to be developed as phantoms for low energy photons.

1. Introduction

Increasing demand for wood products in many countries is creating certain amount problem for raw material supply. Declining of timbers from the natural forests has also pressured the wood-based industry to find alternative sources for consistent raw material supply. An alternative to solid wood products, various types of panel such as laminated veneer lumber, plywood, medium density fiberboard and particleboard have been developed in past. Oil palm (*Elaeis guineensis*) which is native species from Africa has the largest plantation land among other agricultural crops in Malaysia. It has a significant commercial value in the form of oil that can be produced from the mesocarp of the fruit. Currently Malaysia is one of the largest producers of palm oil in the world. Increasing land area of oil palm plantation has been leaving substantial amount of residue in harvesting sites.

The potential use of raw and natural materials in the development of binderless particleboard production successfully discovered by many researchers [6][7][8][9]. Binderless particleboard offer a significant effect towards ecological problem. Thus, in this study, the mass attenuation coefficient of such type of board was determined to be used as phantom material. Phantom is a simulated system that has been used in medical field for quality control (QC), quality assurance (QA), radiation protection, calibration and testing purpose. Polylactic acid (PLA) is biodegradable thermoplastic aliphatic polyester derived from renewable resources such as corn starch, tapioca roots, chips, starch or sugarcane. PLA basic chemical formula is $(C_3H_4O_2)n$. Starch is a carbohydrate abundantly present in various plant species, such as potato, wheat, sago, rice and corn. It is soft, white and tasteless powder that is insoluble in cold water, alcohol, or other solvents. The basic chemical formula of the starch molecule is $(C_6H_{10}O_5)n$.

Starch is a polysaccharide comprising glucose monomers joined in α 1,4 linkages. The simplest form of starch is the linear polymer amylose; amylopectin is the branched form.

Fish oil is in the form of liquid at room temperature but generally solidify below 150-100 C [13]. Fish oil supplements are now the third most frequently used supplements following multivitamin-minerals and calcium-containing supplements. Throughout the world, medical scientists are encouraging people to eat more fish as fish oil is the only major source of the polyunsaturated omega-3 fatty acids, eicosapentaenoic acids (EPA) and docosahexaenoic acids (DHA). In medical, fish oil is not only being used as medicine but also be used in different methods such as a markers in MRI. The objective of this study was to fabricate oil palm trunk particleboards added Polylactic acid (PLA), starch and fish oil respectively with a target density of 1.0 g/cm^3, which were similar with the density of water. The mass attenuation coefficients of the oil palm trunk particleboards produced were determined at energy range of 16.59, 17.47, 21.21 and 25.26 keV.

2. Material and Methods

2.1. Sample Preparation

Oil palm trunks were chipped using a hammer mill machine to convert from wood chip into wood particle. In the next step, the wood particles were grounded to fine material with particle size ranges of 149 - 250 μm in length in average. The moisture content of the wood particle has been measure and should reach a value less than 10%. The binderless particleboard was manually formed and pre-pressed into a mold of dimension 20.5 cm x 20.5 cm with target thickness of 0.5 cm. Polylactic acid (PLA) and starch, in the form of powder were added at rates of 10% each and mixed manually to formed the different particleboard. As for fish oil, in the formed of liquid, four different particleboard were formed which were divided at rates of 10 %, 20 %, 30% and 50% treatment levels. The particleboards were fabricated using hot pressing machine at of 160°C pressing temperature for 20 minutes and target density of 1.0 g/cm^3. The density of the particleboards were measured based on the external dimensions of the particleboard samples.

2.2. Measurement of Mass Attenuation Coefficients

The mass attenuation coefficient was measured using X-ray fluorescent (XRF) configuration based on the study by Marashdeh *et al.* [11] as shown in figure 1. An annular ^{241}Am source was used in conjunction with Niobium (Nb), Molybdenum (Mo), Palladium (Pd) and Tin (Sn) metal plate that provided $K_{\alpha 1}$ photons of 16.59 keV, 17.47 keV, 21.21 keV and 25.26 keV energies respectively. A low-energy Germanium detector (LEGe) was used to measured the transmitted photons from the sample. Multichannel analyzer (MCA) was used to determined the signal collected from spectroscopy amplifier corresponding to the peak energy. The mass attenuation coefficients of the particleboard samples were measured based on the transmitted photons according to the Beer-Lambert law given by the equation:

$$I = I_\circ e^{-\mu x} \tag{1}$$

with $I_\circ$ and I is the intensity of initial and transmitted photons across the thickness, μ is the linear attenuation coefficient of sample in cm^{-1} and x is the thickness of the material in cm. . The value of linear attenuation coefficient is calculated using the equation

$$\mu = \frac{1}{x}\ln\left(\frac{I_\circ}{I}\right) \tag{2}$$

The mass attenuation coefficients, μ/ρ of the particleboard samples were measured by dividing the value of linear attenuation coefficients to the density of the particleboard samples.The values of mass attenuation coefficients of the samples were compared to the calculated value of water using photon cross section database (XCOM) [10].

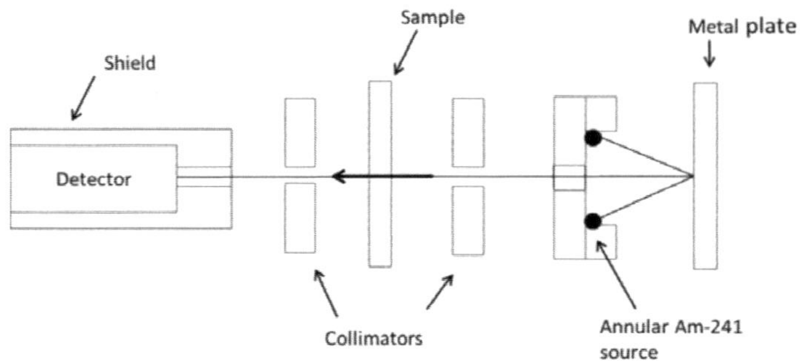

Figure 1. The experimental set-up used for the linear attenuation coefficient measurement of the fabricated oil palm trunk particleboards using an X-ray fluorescence (XRF) configuration.

3. Results and Discussions

The measured density of the oil palm particleboards using gravimetric method is presented in table 1. The results showed that the density of fabricated oil palm particleboards were close to the values of water

Table 1. The measured density of oil palm trunk particleboards measured using gravimetric methods based on the external dimensions of the particleboards.

Sample	Density (g/cm^3)			% difference to water
	Average	Max	Min	
A	0.978	1.099	0.884	2.2
B	1.050	1.103	1.004	5.0
C	1.075	1.094	1.035	7.5
D	1.214	1.233	1.189	21.4
E	1.080	1.098	1.057	8.0
F	1.082	1.113	1.057	8.2
G	0.997	1.049	0.987	0.3

(A) binderless oil palm trunk particleboard, (B) PLA added oil palm trunk particleboard, (C) starch added oil palm trunk particleboard, (D) 10% fish oil added oil palm trunk particleboard, (E) 20% fish oil added oil palm trunk particleboard, (F) 30% fish oil added oil palm trunk particleboard and (G) 50% fish oil added oil palm trunk particleboard.

The measured mass attenuation coefficients of oil palm trunk particleboards measured using the XRF configuration in comparison to the XCOM values of water is illustrated in figure 2. The detail of linear and mass attenuation coefficients of the particleboards is presented in table 2. The results showed that the mass attenuation coefficients of all oil palm trunk particleboards were close to the values of water at all XRF photon energies. The results were in good agreement to the study by Baskaran *et al.* [7] and the previous works on the fabrication of particleboards as phantom [11][14][2][1]. A paired sample t-test was performed to compare the mass attenuation coefficients of water to the oil palm trunk particleboards and presented in table 3. The results showed that the method of fabrication of the oil palm trunk particleboards did not significantly influenced the mass attenuation coefficients of the particleboards in comparison to water. The results were in good agreement to the study by Abuarra *et al.* [2]. The overall results had indicated the similarity of attenuation coefficients of oil palm trunk particleboards to water indicating their potential use as phantoms for low energy photons.

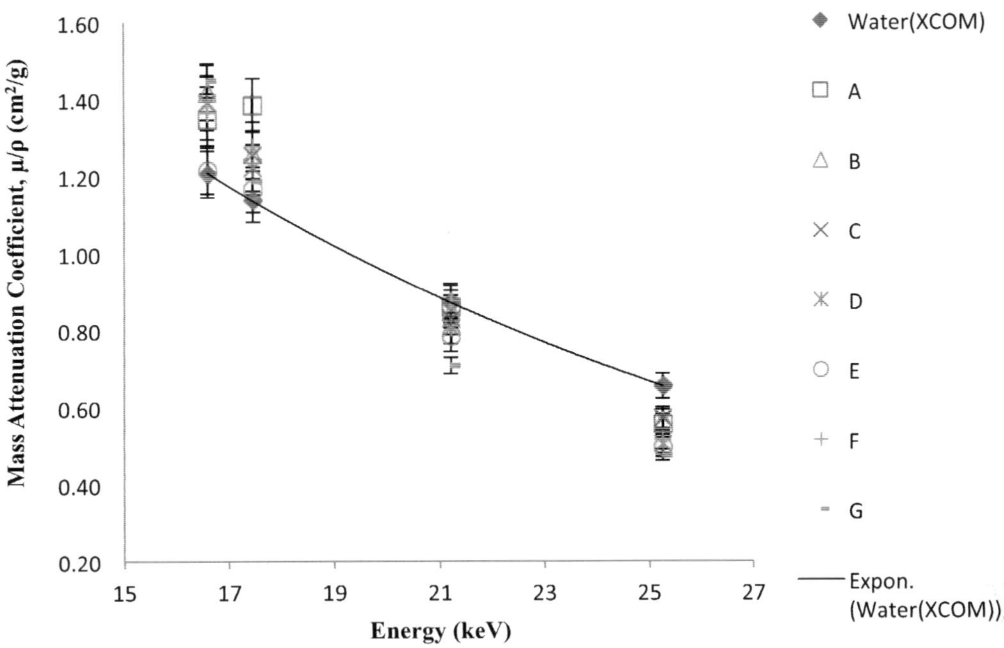

Figure 2. The mass attenuation coefficients of oil palm trunk particleboards in comparison to water (XCOM) at XRF photon energies between 16.59 and 25.26 keV.

Table 2. The mass attenuation coefficients of oil palm trunk particleboards measured using X-ray fluorescence (XRF) configuration.

Sample	**16.59keV**		**17.46 keV**		**21.21 keV**		**25.26 keV**	
	μ(cm^{-1})	μ/ρ (cm^{2}g^{-1})	μ(cm^{-1})	μ/ρ (cm^{2}g^{-1})	μ(cm^{-1})	μ/ρ (cm^{2}g^{-1})	μ(cm^{-1})	μ/ρ (cm^{2}g^{-1})
W	-	1.205	-	1.138	-	0.872	-	0.654
A	1.317	1.347	1.354	1.384	0.841	0.860	0.542	0.554
B	1.488	1.417	1.319	1.256	0.890	0.848	0.594	0.566
C	1.493	1.389	1.349	1.255	0.891	0.829	0.615	0.572
D	1.689	1.391	1.482	1.221	0.976	0.804	0.611	0.503
E	1.311	1.214	1.258	1.165	0.845	0.783	0.534	0.495
F	1.476	1.364	1.382	1.277	0.948	0.876	0.557	0.515
G	1.443	1.448	1.183	1.187	0.705	0.707	0.473	0.474

(A) binderless oil palm trunk particleboard, (B) PLA added oil palm trunk particleboard, (C) starch added oil palm trunk particleboard, (D) 10% fish oil added oil palm trunk particleboard, (E) 20% fish oil added oil palm trunk particleboard, (F) 30% fish oil added oil palm trunk particleboard and (G) 50% fish oil added oil palm trunk particleboard.

Table 3. The paired sample t-test of mass attenuation coefficients between water and oil palm trunk particleboards.

Standard	*p*-value of sample-standard pair						
	A	B	C	D	E	F	G
Water-sample	0.2178	0.2401	0.2689	0.4364	0.1557	0.2974	0.4530

4. Conclusion

The density of fabricated oil palm trunk particleboard with 50% of fish oil were nearly achieved the target density of water of 1.0 g cm^{-3} with the density value of 0.987 g cm^{-3} and differentation percentage density of 0.3 percent. Therefore, sample with 50% of fish oil is suitable as a water equivalent. The mass attenuation coefficient of the oil palm trunk particleboard is successfully determine. The mass attenuation coefficient of binderless, PLA, starch and fish oil added oil palm trunk particleboards were found closer to the value of calculated water by XCOM in the range of photon energy 16.59 keV – 25.26 keV. Therefore, it can be conclude that oil palm trunk binderless particleboard, added PLA,starch and fish oil particleboards were appropriate to be use as water equivalent material at 16.59 keV – 25.26 keV photon energy and developed as phantom of low energy photon.

5. References

[1] Ababneh B A 2016 Characterisation of solid wood and almond gum bonded Rhizophoras pp. particleboard as breast phantom for MRI and CT *PhD Thesis*, Universiti Sains Malaysia, Malaysia.

[2] Abuarra A, Bauk S, Hashim R, Kandaiya S, Tousi E. T and Ababneh B A 2014 XRF technique for the evaluation of gum arabic bonded Rhizophora spp. particleboards as tissue-equivalent material *Int. J. Appl. Phys. and Math.* **4(3)** 201-204.

[3] Ahmad Z, Saman H M and Tahir P M 2010 Oil palm trunk fiber as a bio-waste resource for concrete reinforcement *Int. J. Mech. and Mater.* **5(2)** 199-207.

[4] Bailey R L, Gahche J J, Miller P E, Thomas P R and Dwyer J T 2013 Why US adults use dietary supplements *JAMA Internal Medicine* **173(5)** 355-361.

[5] Baskaran M, Hashim R, Said N, Mohamed Raffi S, Balakrishnan K, Sudeh K, Sulaiman O, Arai T, Kosugi A, Mori Y, Sugimoto T and Sato M 2012 Properties of binderless particleboard from oil palm trunk with additional of polyhydroxyalkanoates *Composites: Part B* **43** 1109-1116.

[6] Baskaran M, Hashim R., Mohd Yusoff M Y, Bauk S, Sulaiman O, Sato M and Kumar S 2015 Mass attenuation coefficients of binderless and polylactic acid added oil palm trunk particleboard in the diagnostic energy rang. *Int. J. Adv. Sci. Engi .nformation Tech.* **5** 355-357.

[7] Baskaran M, Hashim R, Sulaiman O, Hiziroglu S, Sato M and Sugimoto T 2015 Optimization of press temperature and time for binderless particleboard manufactured from oil palm trunk biomass at different thickness levels *Materials Today Comm.* **3** 87-95.

[8] Hashim R., Wan Nadhari W N A, Sulaiman O, Sato M, Hiziroglu S, Kawamura F, Sugimoto T, Tay Guan Seng and Tanaka R 2012 Properties of Binderless Particleboard Panels Manufactured from Oil Plam Biomass *BioResources* **7(1)** 1352-1365.

[9] Hashim R, Wan Nadhari W N A, Sulaiman O, Kawamura F, Hiziroglu S, Sato M, Sugimoto T, Tay Guan Seng and Tanaka R 2011 Characterization of raw materials and manufactured binderless particleboard from oil palm biomass *Materials and Design* **32** 246-254.

[10] Hubbell J H 1982 Photon mass attenuation and energy-absorption coefficients *Int. J. Appl. Radiat. Is.* **33** 1269-1290.

[11] Marashdeh M W, Bauk S, Tajuddin A A and Hashim R 2012 Measurement of mass attenuation coefficients of *Rhizophora spp.* binderless particleboards in the 16.59–25.26 keV photon energy range and their density profile using x-ray computed tomography *Appl. Radiat. Is.* **70** 656-662.

[12] Ngu K T, Bauk S, Hashim R, Tajuddin A A and Shukri A 2015 Fabrication of formaldehyde-based *Rhizophora* spp. particleboards and their mass attenuation coefficients at 15.77, 17.48, 21.18 and 25.27 keV photon energies *J. Physical Sci.* **26(1)** 27-33.
[13] Pike H I and Jackson A 2010 Fish oil: Production and use now and in the future *Lipid Technology* **22** 50-61.
[14] Tousi E T, Bauk S, Hashim R, Jaafar M S, Abuarra A, Aldroob, K S A and Al-Jarrah A M 2014 Measurement of mass attenuation coefficients of *Eremurus-Rhizophora* spp. particleboards for X-ray in the 16.63-25.30 keV energy range *Radiat. Phy.s Chem.* **103** 119-12

Acknowledgement
The authors acknowledge the financial support provided from the Research University grant number RU1001/CIPPT/8011001 by Universiti Sains Malaysia.

Phase formation and microstructure of gamma irradiated Bi-2223 Superconductor

Zaahidah 'Atiqah Mohiju, Natasha Alieya Adnan, Nasri A. Hamid [1, a] and Yusof Abdullah[2]

[1]Nuclear Energy Center, College of Engineering, Universiti Tenaga Nasional, Jalan IKRAM-UNITEN, 43000 Kajang, Selangor, Malaysia

[2]Materials Technology Group, Industrial Technology Division, Malaysian Nuclear Agency, Bangi, 43000 Kajang, Selangor, Malaysia

Nasri@uniten.edu.my

Abstract. The Bi-2223 superconductor has been synthesized using the conventional solid state reaction method. The effect of gamma irradiation on phase formation and microstructure of high-temperature Bi-2223 superconductor ceramic was investigated. The bulk samples sample were palletized with 7 tons pressure of hydraulic press machine and sintered at 840° C for 48 hours. The gamma irradiation was performed at the Nuclear Malaysian Agency with dose of 50 kGray at room temperature. Structure characterization using X-ray diffraction (XRD) showed that the patterns for all the samples demonstrate well-defined peaks all of which could be indexed on the basis of a Bi-2223 phase structure. However, for irradiated sample, it showed reduction in the peak intensity indicating a decrease in the content of the Bi-2223 superconducting phase. The effect of gamma (γ) irradiation on surface morphology and its composites has also been investigated by scanning electron microscope (SEM) and the micrograph shows that the grains are distributed randomly with poorly connected inter and intra-grain microstructure. This shows that the morphology of the Bi-2223 superconductor is very sensitive to gamma irradiation. The effect on the phase formation and microstructure of non-irradiated and gamma irradiated of Bi-2223 superconductor is compared and evaluated.

1. Introduction

Highly textured bulk superconductors have a high potential for electric and magnetic applications such as transmission cables, transformers, current limiters, motors and generators. This is due to its ability to carry high transport critical current densities, J_c. However, the performance of Bi-2223 bulk superconductor is influenced by the density and homogeneity of the microstructure [1-3]. Low electrical current densities, J_c in superconductor are usually associated with grain boundaries weak coupling problem. Since grain boundaries of these materials act as weak links, it is important to produce a single phase material of high transition temperature, T_c (high-T_c) oxide superconductors. Nevertheless due to the high complexity of the reaction and to the small difference in their thermodynamic stabilities, it will be intensely difficult to prepare the (Bi, Pb)-2223 phase as a single phase since it usually intergrowths with the (Bi, Pb)-2212 phase.

IOP Conf. Series: Materials Science and Engineering **298** (2018) 012021 doi:10.1088/1757-899X/298/1/012021

There are several methods used to synthesize the high-quality Bi-2223 precursor powder. The most recommended methods are solid-state reaction, oxalate co-precipitation, spray drying and spray pyrolysis [4-6].

It is believed that the formation of Bi-2223 phase is significantly improved through the partial substitution of Bi by Pb. Thus, it is proven that this elemental substitution is effective in stabilizing the formation process of a high-T_c Bi-2223 phase [7-10].

The purpose of this work is to focus on understanding the phase formation and microstructure of gamma irradiated Bi-2223 superconductor. The phase development, microstructure and mechanical properties of gamma (γ) irradiated Bi-2223 sample are reported in this paper.

2. Experimental Procedure

The high-T_c superconducting samples of composition $(Bi_{1.6}Pb_{0.4})$ $Sr_2Ca_2Cu_2O_x$ were prepared by the conventional solid-state reaction method. After the initial grounding process, the compound went through annealing process at temperature of 800°C for 36 hours. In order to develop the homogeneity of powder, the resulting oxide powder was reground in agate mortar before pelletized into bulk form. The powder was then pressed into pellets under hydrostatic pressure around 7 tons or 70 000 psi. The pellets were then sintered at 840°C for 48 hours and then furnace-cooled down to room temperature.

Gamma irradiation on the samples was carried out using JS10000 (IR-219) irradiator with radiation dose of 50 kGray. The irradiation was conducted at SINAGAMA Plant of the Malaysian Nuclear Agency. Structural investigations of non-irradiated and irradiated samples were conducted using the Bruker D8 Advanced X-Ray Diffractometer (XRD) to verify their phase formation. Microstructure investigation was carried out using the Hitachi S3400N Scanning Electron Microscope (SEM). Compression tests on the samples were done at room temperature in order to obtain the plot of stress-strain curve, and thus the mechanical properties of the superconductor ceramics.

3. Results and Discussion

The structure of samples was characterized by X-ray diffraction. In figure 1, it shows that both samples are multiphase where Bi-2223 phase co-existed with Bi-2212 phase. Well resolved sharp peaks corresponding to the Bi-2223 phase are observed and indicates a high fraction of textured Bi-2223 phase. The grain alignment is indicated by the predominance of the three main Bi-2223 peaks, while the other reflections of this phase are less obvious. A few peaks that correspond to secondary phase and other impurities are also observed. After the sample is exposed to gamma irradiation, the volume fraction of Bi-2223 phase reduced. This indicates that gamma irradiation diminished the presence of Bi-2223 phase in the superconductor ceramics.

Figure 2(a) shows a well-distributed crystalline structure with flake-type grains and some needle-type grains of different sizes. The grains are distributed randomly and poorly connected, and oriented anisotropically. This growing feature promotes a volume increase, thus the bulk density which resulting in the expansion of the sample.

Meanwhile gamma irradiated sample in figure 2(b) shows that the loss of uniformity and fractured pattern with flaky disposition of the grains regularly disappeared with decreasing grain size, and thus resulted in increasing porosity. These grains are disconnected due to the presence of minor impurities and caused by the energy deposition and thus validates to the decreasing superconductor properties of the ceramics. The microstructure exposes a significant amount of small crystals in the order of 1–2 μm when comparing the morphology of the samples before and after gamma irradiation.

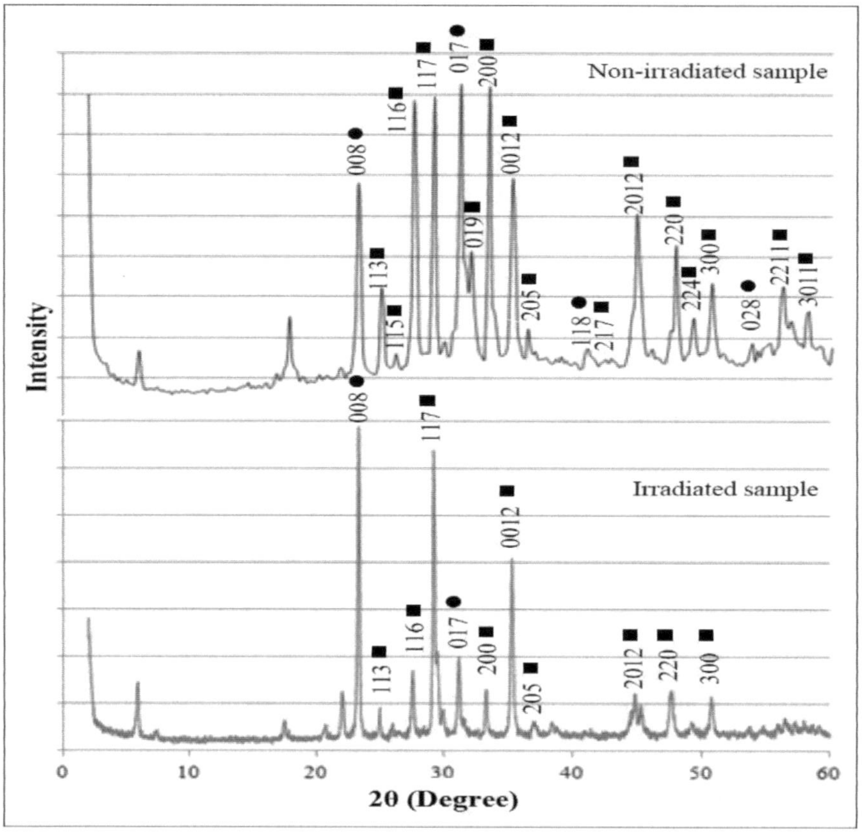

Figure 1. XRD patterns of Bi-2223 for non-irradiated and gamma irradiated sample exposed to 50 kGray.

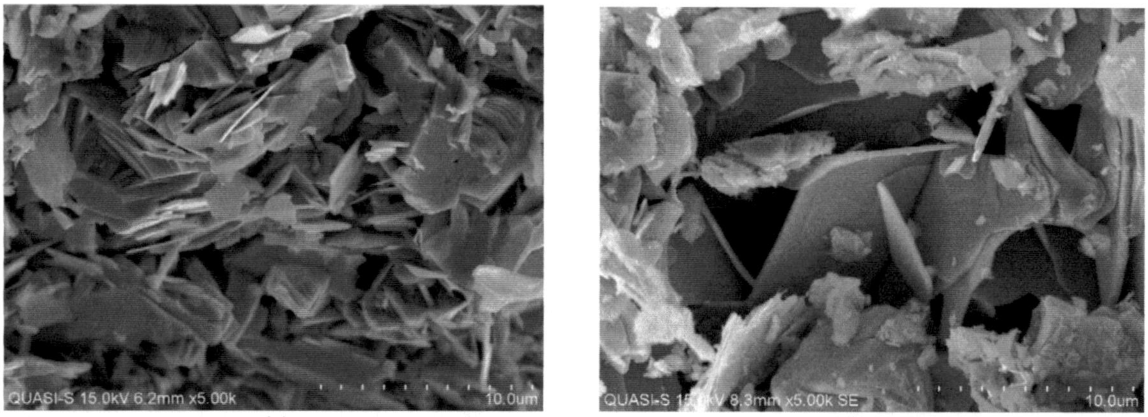

(a) (b)

Figure 2. SEM micrographs of (a) non-irradiated samples (b) gamma irradiated samples with 50 kGray dose.

The energy deposition to local position such as lattice and/or atoms at lattice sites may lead to this rare occurrence of small crystals since the dose rate was not high enough to heat up the samples. Decreasing in grain size would resulted in a decrease in T_c, because of discretization of the energy levels that has been shown to decrease the effective density of states. This phenomena is in accordance with the Bardeen-Cooper-Schrieffer (BCS) relation. For the whole concentration, the transition temperature of the sample decreases proportionally with increasing irradiation dose. Nevertheless further investigation is needed to understand the mechanism better.

4. Conclusion

Irradiation of 50 kGy of gamma ray does not change the phase formation of Bi-2223. However, the intensity of peaks in irradiated sample of Bi-2223 superconductor decreases, and thus reduces the flux pinning and superconducting properties of the sample. SEM analysis was used to view sample's surface topography and composition by creating various images. The non-irradiated micrograph showed a randomly distributed crystalline structure with flake-type grains mixed with needle type-grains. However, the gamma irradiated micrograph showed that the uniformity has lost and the structure is fractured with flaky disposition of the grains that regularly disappeared with decreasing grain size. This resulted in increasing porosity of the microstructure.

5. References

[1] Kurihama T, Yamashita M, Tsuruta K and Izumi T 1994 Physica C vol 219 (Amsterdam: North-Holland/Elsevier) p 57
[2] Leyva A, Mora M, Martin G and Martinez A 1995 Supercond. Sci. & Technol. vol 8 (Bristol: IOP Publishing) p 816
[3] Bohandy J, Suter J, Kim B F, Moorjani K and Adrian F J 1987 Appl. Phys. Lett. vol 51 (College Park: American Institute of Physics) p 2161
[4] Maeda H, Tanaka Y, Fukautomi M and Asano T 1988 Japanese Journal of Applied Physics vol 27(2) (Tokyo: Japan Society of Applied Physics) p 303
[5] Hau Chen D, Shei C Y, Sheen S R and Chang C T 1991 Japanese Journal of Applied Physics vol 30(6) (Tokyo: Japan Society of Applied Physics) p 1198
[6] Yoo J, Kim S, Ko J W and Kim Y K 2004 Supercond. Sci. Technol. vol 17(9) (Bristol: IOP Publishing) p S538
[7] Sunshine A, Siegrist T, Schneemeyer L F, Murphy D W, Cava R J, Batlogg B, Van Dover R B, Fleming R M, Glareem S H, Nakahara S, Farrow R, Rajewski J J, Zahurck S M, Waszezak J V, Marshall J H, Marsh P, Rupp Jr L W and Peek W F 1988 Phy. Rev. B vol 38, (College Park: American Physical Society) p 893
[8] Green S M, Jiang C, Meiyu Luo H L and C. Politis C 1988 Phy. Rev. B vol 38 (College Park: American Physical Society) p 5016
[9] Chavira E, Escudero R, Rios Jara D and Leon L M 1988 Phy. Rev. B vol 38 (College Park: American Physical Society) p 9272
[10] Khan M N, Memon A and S. Al-Dallal S 1993 Int J. Mod. Phys. B vol 7 (London: World Scientific Publishing) p 1687

Acknowledgement

This project is funded by the UNITEN Bold Research Grant with Project Code 10289176/B/9/2017/32, and conducted under collaboration with Malaysian Nuclear Agency.

IOP Conf. Series: Materials Science and Engineering **298** (2018) 012022 doi:10.1088/1757-899X/298/1/012022

Verification of relative output factor measurement using Gafchromic films for small-field radiosurgery photon beams.

R Abdullah[1,4], M S Sapeei[1], N S A Shukor[1], M F M Yusof[1], M Mohamed[2,3], N R N Idris[2,3], A L Yusoff[2,3], A Zakaria[1], and M Z A Aziz[4]

[1]School of Health Sciences, Health Campus USM, Universiti Sains Malaysia, Kelantan, Malaysia
[2]Department of Nuclear Medicine, Radiotherapy, and Oncology, School of Medical Sciences, Health Campus USM, Universiti Sains Malaysia, Kelantan, Malaysia

[3]Department of Nuclear Medicine, Radiotherapy, and Oncology, Hospital Universiti Sains Malaysia, Kelantan, Malaysia.

[4]Advanced Medical and Dental Institute, Universiti Sains Malaysia, Bertam, Pulau Pinang, Malaysia.

reduan@usm.my

Abstract. Accurate dosimetry of small-field photon beams has always been difficult to achieve, due to the steep dose gradient and absence of lateral electronic equilibrium. The purpose of this study was to verify the measurement of relative output factor (ROF), which is one of the dosimetric parameter required for stereotactic radiosurgery (SRS) treatment planning. The ROFs for Radionics circular cone collimators with diameter in the range of 10.0 to 45.0 mm were measured using Gafchromic EBT2 and EBT3 films. The measurements were then compared with the ROFs obtained using a PinPoint ionisation chamber and Monte Carlo (MC) simulation. From the results, the ROFs measured by the ionisation chamber, EBT2, and EBT3 films were good agreement with the MC simulation, with deviations of less than 1.5, 2.6, and 5.0 % respectively. Based on the film dosimetry, the EBT2 film showed in a more reliable measurement for field size ranging from 15.0 to 45.0 mm, compared with EBT3. As a conclusion, based on the special characteristic of the small-field photon beams, ROF measurement using PinPoint ionisation chamber are being favoured, due to its accuracy. However, the EBT2 film can be used as an alternative, when high spatial resolution is required.

1. Introduction

In order for a stereotactic radiosurgery (SRS) treatment planning system (TPS) to be commissioned to work with a medical linear accelerator (LINAC), a large number of measurements using the LINAC will be required, to be entered into the system. One of the parameters to be measured is the relative output factor (ROF). The ROF is a simple way of comparing the output of a LINAC under desired settings, to the output of the same LINAC under reference condition (i.e. field size of 10 x 10 cm^2 at the reference depth inside a water phantom, and source to surface distance (SSD) of 100 cm), with known absolute dose. The ROFs need to be accurately measured for very small field size, as it can be used to tell the TPS how the LINAC output varies as the field size reduces, allowing a more accurate calculation of the dose.

The accuracy of the ROF directly affects the accuracy of the dose delivered in a radiosurgical treatment [1]. Accurate measurement of the ROF for small-field photon beams ($\leq 4 \times 4\ cm^2$) used in SRS can be made difficult due to the presence of lateral electronic disequilibrium and steep dose gradients. The ionisation chamber is a standard dosimeter used for measuring the dosimetric parameters of photon beams, such as percentage depth dose (PDD) beam profiles, and ROF. However, for small-field photon beams, several challenges are present, as discussed in the published literatures [2-4]. Various dosimeters can be used as alternative to the ionisation chamber, e.g., diode, diamond detector, thermoluminiscent detector (TLD), metal-oxide-semiconductor field effect transistor (MOSFET), glass rod dosimeter (GLD), radiographic, and radiochromic films. Each dosimeter has its own pros and cons when it is used for the small-field photon beams, as elaborated in the published literatures [5-10]. To date, there is no single detector which can be used on its own, a narrow circular field (with diameter of < 20 mm) [11]. This led to many relying on the Monte Carlo (MC) simulation. Although the MC calculation is still considered as an experimental method, it is often used as a detector-independence reference. The relative performance of each detector can therefore be quantified, as deviation from the MC simulation results.

Therefore, the ROFs of small-filed photon beams can be obtained, through careful comparison between the ROFs estimated by the small volume ionisation chamber, film dosimetry, and MC simulation. This study was conducted to compare the measured ROF using two different types of Gafchromic film (i.e. EBT2 and EBT3) and a PinPoint ionisation chamber, and validated using MC simulation.

2. Material and methods

All irradiations were performed with 6 MV photon beam, using a LINAC (PRIMUS, Siemens Healthineers, Erlangen, Germany) equipped with a multi-leaf collimator (MLC). The LINAC output was calibrated based on the IAEA TRS-398 protocol [12]. The ROF for circular cone collimators (Radionics, Integra Radionics, Massachusetts, USA) with diameters of 10.0, 15.0, 20.0, 22.5, 27.5, 32.5, and 45.0 mm, were measured in this study.

2.1. Measurement of ROF using PinPoint Ionisation Chamber

The measurement of ROF using an ionisation chamber (PinPoint®, PTW, Freiburg, Germany) with vented sensitive volume of $0.015cm^3$, was conducted by placing it inside a water scanning system (MP3, PTW, Freiburg, Germany). The ionisation chamber was placed at the centre of a 45.0 mm circular field size, depth of maximum dose (dmax) of 1.5 cm and SSD of 100 cm. the procedure was repeated by replacing the circular cone collimator with that of a different diameter. The ROF for each circular cone collimator was calculated, by calculating the ratio of the charge obtained at dmax (Cw, ic, A), to the charge ontained at similar depth with 10 x 10 cm^2 or reference field size (Cw, ic, ref). The ROFs were calculated using Equation 1, as follows:

$$ROF\ (A) = \frac{Cw{,}ic{,}A}{Cw{,}ic{,}ref} \tag{1}$$

2.2. Film calibration

Gafchromic EBT2 films (Ashland Inc., Kentucky, USA) were irradiated with 6 MV photon beam, by placing it inside a 30 x 30 x 20 cm3 solid water phantom (RW3, PTW, Freiburg, Germany), at the centre of a 10 x 10 cm2 field size, dmax of 1.5 cm and SSD of 100 cm. Film calibration was carried out by placing the film perpendicular to the beam, for dose ranging from 50 to 500 cGy. Unexposed film was also developed, to obtain the background reading. After 24 hours, the calibrated films were scanned using a flatbed scanner (Epson Expression 10000XL, Seiko Epson Corporation, Nagano, Japan) in landscape orientation. The ratio of film response to the dose delivered was determined in pixel value, using a patient plan verification software (Verisoft®, PTW, Freiburg, Germany). Equation 2 was used to convert pixel value to the net optical density (OD). Similar procedure was repeated for the EBT3 film (Ashland Inc., Kentucky, USA). Finally, a calibration curve (dose versus net OD) was generated for each film type, i.e. EBT2 and EBT3 films.

$$Net\ OD = Log\ 10\ \frac{\text{pixel value non irradiate film}}{\text{pixel value irradiate film}} \quad (2)$$

2.3. Measurement of ROF using Gafchromic EBT2 and EBT3 films
Gafchromic films (i.e. EBT2 and EBT3) taken from the same batch as the calibrated films, were irradiated with 6 MV photon beam, by placing it inside the 30 x 30 x 20 cm^3 solid water phantom, at the centre of a 45.0 mm circular field size, dmax of 1.5 cm and SSD of 100 cm. the procedure was repeated by replacing the circular cone collimator with that of a different diameter. The irradiated films were scanned using the EPSON flatbed scanner, at 24 hours post-irradiation and analysed using Verisoft software. The net OD was converted to dose, based on calibration curve. The ROF for each circular cone collimator was calculated, by calculating the dose obtained at dmax, to the dose obtained at similar depth with 10 x 10 cm^2 field size.

2.4. Modelling and simulation
The ROF for the Radionics circular cone collimators were calculated using BEAMnrc and DOSXYZ nrc software components, which are included in the EGSnrc source code (national Research Council Canada, NRCC). The code was installed on a personal desktop computer. The BEAmnrc is a general purpose MC simulation system for modelling of radiotherapic source, while DOSXYZnrc is used to calculate the dose in three dimensional (3D) medium [13]. The simulations were performed in three stages. For the first stage, the BEAMnrc was used to simulate the PRIMUS LINAC head, to obtain the phase space data for the first 20 cm distance after target (right after the projection mirror). This phase space file was used for the subsequent circular and square (10 x10 cm^2) field sizes. In this stage, 50 million particle histories were used.

For the second stage, all circular cone collimators used in this study were simulated. The collimators were simulated as stainless steel materials with a density of 8.03g/cm^3. The geometry of each collimator was simulated based manual measurement, since technical drawing of the component was unavailable. The radionics assembly holder and circular cone collimators were simulated using the BEAMnrc's flattening filter (FLATFILT) component module. The 50 x 50 mm^2 (phase space file was used for second stage) and space files were generated for all cone collimators, at SSD of 80 cm. The energy cut-off values for photon (PCUT) and electron (ECUT) transport were set to 0.01 and 0.70 MeV, respectively. The phase space files generated during the second stage of the simulations were used in the final stage, to calculate the 3D dose distributions in water phantom using the DOSXYZnrc software.

3. Results and Discussion

Figure 1 shows the calibration curves of the EBT2 and EBT3 films, following irradiation with 6 MV photon beam. Both films showed linear response to the dose. Based on the R^2 values, EBT3 showed a more linear response to the dose ranging from 50 to 250 cGy, compared to the EBT2 film. The differences between the films were smaller than ~ 4 %, with almost similar sensitometric curve pattern. In general, the results showed no significant variation in terms of dose response, for both films. However, the variations in sensitivity (net OD per unit dose) may be present, when different batches of film were used. This has been proven in the study done by Butson et al. (2010), where it was concluded that there were variations in the sensitivity between different batches of each film (i.e. EBT2 and EBT3), although both batches possess similar thickness of active layer (~ 28 μm) [5]. Hence, the EBT2 and EBT3 films must be calibrated, every time a set of measurement needs to be performed using films from different batches.

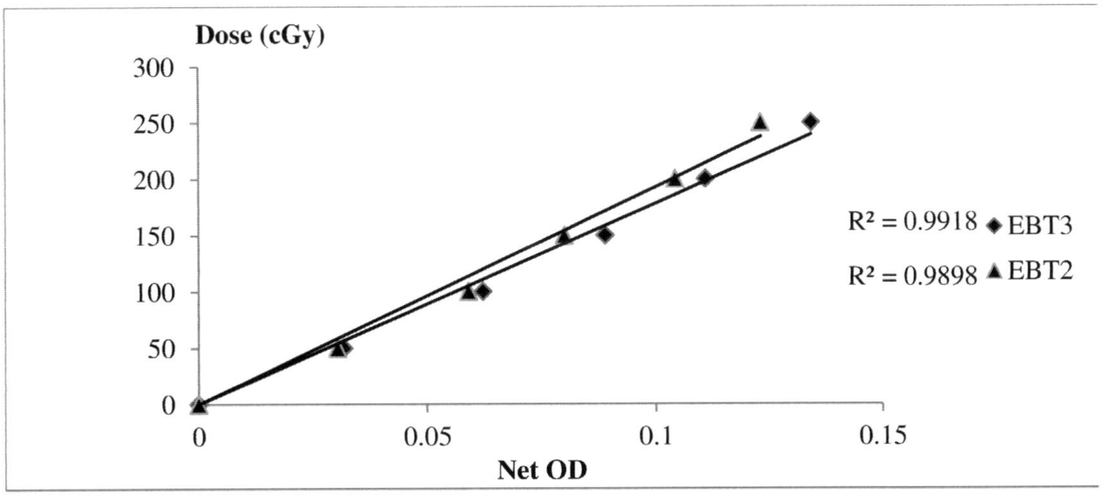

Figure 1. Dose (cGy) versus net OD for Gafchromic EBT2 and EBT3 films.

The comparison between the ROFs measured using the PinPoint ionisation chamber, EBT2 and EBT3 films with respect to the MC calculated ROF, are shown in Table 1 and 2. The ROFs measured using the ionisation chamber were in excellent agreement with the MC calculated ROFs (with less than 1.5 % deviation, except for the 10 mm circular collimator. The accuracy of the ionisation chamber in small-field photon beam measurement was dependent on the resolution of the ionisation chamber. A smaller sensitive volume will contribute to higher resolution. The ROFs measured using the EBT2 and EBT3 films were also found to be in good agreement with the MC-calculated ROFs (with deviations of less than 2.6 and 5.0 %, respectively), except for 10 mm circular collimator. The results were in parallel with the study done by Underwood (2013), which reported the inherent variations in dose measurement of around 2 to 5 % (for both EBT2 and EBT3 films), even with an optimised dosimetric process [14].

Table 1. The ROFs obtained using the films (EBT2 and EBT3), ionization chamber (PP), and MC simulation, for different circular cone diameters.

Cone (mm)	ROF			
	EBT2	EBT3	PP	MC
10.0	0.616	0.625	0.733	0.770
15.0	0.829	0.833	0.827	0.824
20.0	0.859	0.843	0.873	0.879
22.5	0.867	0.850	0.887	0.886
27.5	0.892	0.850	0.896	0.895
32.5	0.887	0.868	0.902	0.902
45.0	0.938	0.905	0.928	0.915

Table 2. The percentage deviations between the ROFs obtained using the films (EBT2 and EBT3) and ionization chamber (PP), relative to the MC simulation, for different circular cone diameters.

Cone (mm)	Deviations (%)		
	EBT2/MC	EBT3/MC	PP/MC
10.0	-20.03	-18.86	-14.73
15.0	0.62	1.10	0.37
20.0	-2.24	-4.06	-0.65
22.5	-2.14	-4.06	0.11
27.5	-0.34	-5.03	0.11
32.5	-1.71	-3.82	-0.05
45.0	2.56	-1.05	1.47

Unlike the ionisation chamber, the self-developed Gafchromic films record doses through indirect measurement. Thus, the films were required to be 'rested' for the first 24 hours following irradiation, before they can be read and analysed. Still, systematic error may be introduced during the measurement, as well as the scanning process. This is because, the flatbed scanner was not specifically made for film dosimetry, and thus, hardware issues may be present, which could results in a dose fluctuation. Human error may also be present during the setup of the experiment, the scanning process and analysis of the films using the dosimetric software. The energy dependence of the films may be another contributor to the variations. Nonetheless, for the 6 MV photon beams used in this study, a relatively small difference in the films response of < 10 % were demonstrated, from 60 KeV to MeV range [15].

4. Conclusion

Gafchromic EBT2 and EBT3 films were shown to be capable of measuring the ROFs for small-field photon beams, due to their features, i.e. lack in lateral electronic equilibrium, high in spatial resolution, and near tissue equivalent. Both films are inexpensive and provide good option for smaller radiotherapy centre that do not own the equipment or personnel, to perform a more complex small-field relative dosimetry. However, the films are limited such that, the read out process needs careful attention, to avoid measurement uncertainties. Therefore, based on the special characteristics of small-field photon beams, the PinPoint ionisation chamber was suggested due to its accuracy. However, the EBT2 film can be used as an alternative, when high spatial resolution is required.

5. References

[1] Rah J E, Shin D O, Jang J S, Kim M C, Yoon S C and Suh T S 2008 Application of a glass rod detector for the output factor measurement in the CyberKnife *Appl. Radiat. Isot.* **66** 1980-85

[2] Khelashvili G, Chu J, Diaz A and Turian J 2012 Dosimetric characteristics of the small diameter BrainLab (TM) cones used for stereotactic radiosurgery *J. Appl. Clin. Med. Phys.* **13** 4-13

[3] Sharma S D, Kumar S, Dagaonkar S S, Bisht G, Dayanand S, Devi R, Deshpande S S, Chaudhary S, Bhatt B C and Kannan S 2007 Dosimetric comparison of linear accelerator-based stereotactic radiosurgery systems *J. Med. Phys.* **32** 18-23

[4] R Abdullah, N R N Idris, A Zakaria, A L Yusof, N I Mohsin and M Mohamed 2015 Measurement of dosimetric parameters and dose verification in stereotactic radiosurgery *Malaysian J. Health Science* **13** 39-9

[5] Butson M J, Yu P K N, Cheung T and Alnawaf H 2010 Energy response of the new EBT2 radiochromic film to x-ray radiation *Radiat. Meas.* **45** 836-9
[6] Sharpe M B, Jaffray D A, Battista J J and Munro P 1995 Extrafocal radiation: A unified approach to the prediction of beam penumbra and output factors for megavoltage x-ray beams *Med. Phys.* **22** 2065-74
[7] Zhu T C, Bjarngard B E and Shackford H 1995 X-Ray Source and the Output Factor *Med. Phys.* **22** 793-8
[8] R Abdullah, N I Mohsin, M Mohamed, N R N Idris, A Zakaria and A L Yusof 2016 Verification of Relative Output Factor measurement for radiosurgery small photon beams *Health Environm. J.* **7** 20-31
[9] Alalawi A I, Jafari S M, Najem M A, Alsaleh W, Clark C H, Nisbet A, Abolaban F, Hugtenburg R P, Hussein M, Alzimami K S, Bradley D A and Spyrou N M 2014 Preliminary investigations of two types of silica-based dosimeter for small-field radiotherapy *Radiat.Phys. Chem.* **104** 139-44
[10] Ralston A, Liu P, Warrener K, McKenzie D and Suchowerska N 2012 Small field diode correction factors derived using an air core fibre optic scintillation dosimeter and EBT2 film *Phys. Med. Biol.* **57**
[11] Fan J, Paskalev K, Wang L, Jin L, Li J, Eldeeb A and Ma C 2009 Determination of output factors for stereotactic radiosurgery beams *Med. Phys.* **36** 5292-300
[12] Andreo P, Huq M S, Westermark M, Song H J, Tilikidis A, DeWerd L and Shortt K 2002 Protocols for the dosimetry of high-energy photon and electron beams: a comparison of the IAEA TRS-398 and previous international Codes of Practice *Phys. Med. Biol.* **47** 3033-53
[13] Aziz M Z A, Yusoff A L, Osman N D, Abdullah R, Rabaie N A and Salikin M S 2015 Monte carlo dose calculation in dental amalgam phantom. *J. Med. Phys.* **40** 150-5
[14] R J Underwood 2013 *Small field dose calibrations with Gafchromic film* MSc. Thesis Georgia Institute of Technology
[15] Muench P J, Meigooni A S, Nath R and Mclaughlin W L 1991 Photon Energy-Dependence of the sensitivity of Radiochromic film and comparison with Silver-Halide film and Lif Tlds used for brachytherapy dosimetry. *Med. Phys.* **18** 769-75

Acknowledgments
This study was supported financially by the Short-term Grant (304/PPSK/61313048) of Universiti Sains Malaysia. This work was made possible with the cooperation of the Nuclear Medicine, Radiotherapy, and Oncology Department, School of Medical Sciences, Hospital Universiti Sains Malaysia, and School of Health Sciences, Universiti Sains Malaysia (USM).

IOP Conf. Series: Materials Science and Engineering **298** (2018) 012023 doi:10.1088/1757-899X/298/1/012023

Dosimetry of Al_2O_3 optically stimulated luminescent dosimeter at high energy photons and electrons

M F Mohd Yusof[1], N A Joohari[1], R Abdullah[1], N S Abd Shukor[1], A B Abd Kadir[2], N Mohd Isa[2]

[1] School of Health Sciences, Universiti Sains Malaysia, 16150 Kota Bharu, Kelantan, Malaysia
[2] Medical Physics and Secondary Standard Dosimatry Laboratory, Malaysian Nuclear Agency, 43000 Bangi, Selangor, Malaysia

mfahmi@usm.my

Abstract. The linearity of Al_2O_3 OSL dosimeters (OSLD) were evaluated for dosimetry works in clinical photons and electrons. The measurements were made at a reference depth of Z_{ref} according to IAEA TRS 398:2000 codes of practice at 6 and 10 MV photons and 6 and 9 MeV electrons. The measured dose was compared to the thermoluminescence dosimeters (TLD) and ionization chamber commonly used for dosimetry works for higher energy photons and electrons. The results showed that the measured dose in OSL dosimeters were in good agreement with the reported by the ionization chamber in both high energy photons and electrons. A reproducibility test also reported excellent consistency of readings with the OSL at similar energy levels. The overall results confirmed the suitability of OSL dosimeters for dosimetry works involving high energy photons and electrons in radiotherapy.

1. Introduction

The use of OSLD has been increasing and has recently begun replacing the thermoluminescent dosimeter (TLD) for personnel monitoring. Typical OSLDs are constructed using aluminum oxide powder doped with carbon Al_2O_3:C. The chip-shape also known as nanodot, has rendered OSLDs capable of measuring radiation doses from every direction and position. This device can be exposed and read repeatedly, with good reproducibility rate of detection. OSLD provides a wide range of energy detection with 5 kV-10 MV photons and electron energy beyond 250 keV. OSLDs also reported excellent energy dependence of 10% over a diagnostic energy range of 70-140 kVp; 5% for photon and electron from 5 - 20 MeV [1]. To date, studies concerning the characteristics of dosimetry involve the sensitivity, reproducibility and energy dependence of OSL towards low and high energy photons [2-8]. A previous study by Yahya [9] suggested that OSLDs has excellent energy dependence and dose linearity at low and high energy X-rays compared to the thermoluminescent dosimeters.

Studies on the application of OSLDs for particulate ionizing radiations such as electrons on the other hand, are limited. A readily available dosimetry protocols such as the IAEA TRS 398:2000suggested a reference depth Z_{ref} as a dosimetric parameters for electrons. The Z_{ref} was determined to be R_{80} of the electron depth dose, which is known as the therapeutic range of the electrons. The R_{80} is defined as the depth on which the target volume receives 80% of the prescribed dose. The Z_{ref} for photons, on the other hand, is measured at an approximate depth of 5 cm for 6 and 10 MV photons. This study focused on the dose measurements at Z_{ref} of high energy photons and electrons measured using the OSLDs.

IOP Conf. Series: Materials Science and Engineering **298** (2018) 012023 doi:10.1088/1757-899X/298/1/012023

2. Methodology

2.1. Preparation of Optically Stimulated Luminescent Dosimeters (OSLD) and Themoluminescent Dosimeters (TLD100)

The OSL dosimeters were annealed in a secondary standard dosimetry laboratory (SSDL) in the Malaysian Nuclear Agency (Nuclear Malaysia). Annealing was carried out by exposing the OSL to high intensity light to remove trapped charge within the OSL. The initial readings of the OSL were recorded prior to the measurement. The readings of the OSL can be calculated using the equation:

$$Dose_{OSL} = D_y - D_x \tag{1}$$

with D_x and D_y being the initial and final readings of the OSL, respectively, measured in milliSievert (mSv).

The TLD100 was selected to obtain a group of TLD chips with the lowest standard deviation values (SD) of TL signals within the group. The TLD100 chips were annealed using an annealing oven at 400°C for an hour followed by 100°C for 3 hours as per recommendations by the manufacturer [10]. Both OSL and TLD100 were calibrated at the SSDL in Nuclear Malaysia using 1 mGy of ^{137}Cs gamma energy at 1 m distance for ~1.31 minutes.

2.2. Dose Measurements in High Energy Photons and Electrons

The absorbed dose at Z_{ref} was measured at 6 and 10 MV photons, and 6 and 9 MeV electrons using Varian Clinax IX as the source of the photons and electrons. Solid water phantoms were used as a medium and the OSLDs and TLD100 were placed at the Z_{ref}, as shown in figure 1. The Z_{ref} for photons were measured at a depth of 5 cm for 6 and 10 MV photons while the Z_{ref} for electrons were measured at depths of 1.34 and 2.06 cm for 6 and 9 MeV electrons, respectively, based on previous calibration measurements. Bolus was added onto the solid water phantoms to provide depth and minimize air gaps created while positioning the OSLDs and TLD100. The OSLDs and TLD100 were exposed to photons and electrons at 100 cm source to surface distance (SSD), 10 cm x 10 cm field size on the surface and 100 monitor unit (MU) based on the calibration condition recommended by IAEA TRS 398:2000.

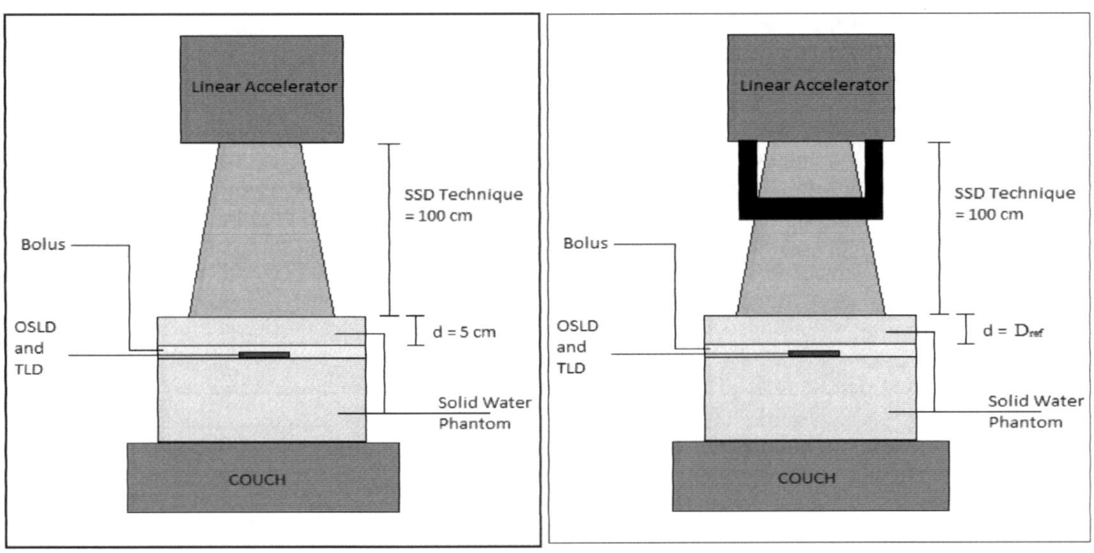

Figure 1. The experimental set up for dose measurement at (a) photons and (b) electrons.

The doses by OSLDs were measured using a portable OSL reader type MicroStar. The OSL dosimeter was analyzed using a stimulated element aluminum oxide, (Al_2O_3) at a wavelength of 532 nm (green light). Radiation charges from X-ray or high energy photon of Themoluminescent dosimeter (TLD-100) was read by the TLD reader model Harshaw 4500 using a WinREMS software. The measurement error, σ for both OSLDs and TLD100 was determined using equation 2:

$$\text{Error}, \sigma = \frac{\text{dose}_{\text{max}} - \text{dose}_{\text{min}}}{2} \tag{2}$$

The absorbed dose by OSLDs and TLD100 was compared to the value by ionization chamber, acting as a standard dosimeter for calibration purposes. The chi-square χ^2 calculation was used to determine the closeness readings by OSLDs and TLD100 to ionization chamber, as per equation 3:

$$\chi^2 = \left(\frac{D_{std} - D_{exp}}{\sigma}\right)^2 \tag{3}$$

with D_{std} and D_{exp} being the dose measured using ionization chamber and OSLDs/TLD100, respectively, and σ is the measurement error of the dosimeter measured using equation 2.

3. Results and Discussions

The measured absorbed dose at Z_{ref} for 6 and 10 MV photons compared to the readings of the ionization chamber is tabulated in Table 1. The results showed that the absorbed dose by OSLDs was closer to the value reported by the ionization chamber in the case of both 6 and 10 MV photons. The absorbed dose by TLD100, on the other han,d was significantly lower than that of the ionization chamber and OSLDs. The results were in good agreement to the study of absorbed dose by OSLDs, at a depth of maximum dose d_{max} at high energy photons, reported by Yahya [9]. The reproducibility of OSLDs, and TLD100 were better when the measurements were done at lower energy photons, as confirmed by the lower σ values. The reproducibility at higher energy photons was significantly higher for both OSLDs and TLD100, which is consistent with Yahya [9].

Table 1. The detailed value of absorbed dose by OSLDs and TLD100 compared to the ionization chamber for 6 and 10 MV photons

Energy (MV)	Absorbed dose (cGy)				
	Ionization chamber	**OSLD**	σ_{OSLD}	**TLD100**	σ_{TLD}
6 MV	874.54	815.41	4.93	481.07	3.10
10 MV	937.76	874.61	10.69	473.07	15.96

The absorbed dose at Z_{ref} in 6 and 9 MeV electrons measured using OSLDs and TLD100 compared to the ionization chamber is shown in Table 2. The results showed that the readings by the OSLDs were closer to that of the ionization chamber at 6 and 9 MeV electrons compared to the TLD100. The readings by TLD100 were significantly lower than that of the ionization chamber and OSLDs, similar to the results obtained in the photons measured earlier. The OSLDs reported better reproducibility compared to TLD100 at both 6 and 9 MeV electrons, as evidenced by its lower σ values.

Table 2. The detailed value of absorbed dose by OSLDs and TLD100 in comparison to ionization chamber for 6 and 9 MeV electrons

Energy (MV)	Absorbed dose (cGy)				
	Ionization chamber	**OSLD**	σ_{OSL}	**TLD100**	σ_{TLD}
6 MeV	997.998	892.12	1.33	522.309	21.18
9 MeV	1012.725	866.22	11.69	548.179	42.23

The χ^2 values of OSLDs and TLD100 to ionization chamber at both photons and electrons is shown in Table 3. The results showed that the dosimetry readings by OSLDs were closer to the value ionization chamber at all measured photons and electrons compared to that of the TLD100 as per its lower χ^2 values. The results are consistence with a previous study comparing OSLDs and TLD100 signal towards ionizing radiations by Yahya [9]. The overall results confirmed the suitability of

OSLDs as a dosimeter for quality assurance and dosimetry works involving high energy photons and electrons.

Table 3. The χ^2 values of OSLDs and TLD100 to ionization chamber for 6 and 10 MV photons and 6 and 9 MeV electrons

Energy	OSLD	TLD100
6 MV	2.068 x 10^4	1.611 x 10^6
10 MV	34.54	8.472 x 10^2
6 MeV	25.99	5.044 x 10^2
9 MeV	20.84	1.210 x 10^2

4. Conclusions

Dose measurement using OSLDs were almost similar to that of reported by the ionization chamber compared to TLD100H at all high energy photons and electrons. The OSLDs also reported good reproducibility of readings as per by the low value measurement errors. The dose measured using TLD100 was significantly lower than that reported by the ionization chamber and OSLDs. The results confirmed the suitability of OSLDs as a dosimeter for high energy photons and electrons.

5. References

[1] Tien C J, Ebeling R, Hiatt J R, Curran B and Sternick E 2012 Optically stimulated luminescent dosimetry for high dose rate brachytherapy, *Front. Oncol.* 91(2) 1-7

[2] Viamonte L, da Rosa A R, Buckley L A, Cherpark A, and Cygler J E 2008 Radiotherapy dosimetry using a commercial OSL system, *Med. Phys.* **35(4)** 1261-1266

[3] Perks C, Yahnke C, and Million M 2008 Medical dosimetry using optically stimulated luminescence dots and microStar® readers, *12th Int. Congress of the Int. Radiation Protection Association*, IRPA12

[4] Reft C S 2009 The energy dependence and dose response of a commercial optically stimulated luminescent detector for kilovoltage photon, megavoltage photon, and electron, proton, and carbon beams *Med. Phys.* **36(5)** 1690-1699

[5] Bäck S J, Mattsson, Kjaer-Kristoffersen F and Medin J 2004 Real-time optical-fibre luminescence dosimetry for radiotherapy: physical characteristics and applications in photon beams, *Phys. Med. Biol.* **49(9)** 1655-1699

[6] Jursinic P A 2007 Characterization of optically stimulated luminescent dosimeters, OSLDs, for clinical dosimetric measurements, *Med. Phys.* **3436(37)** 1690-1699

[7] Dunn L, Lye J, Kenny J, Lehmann J, Williams I and Kron T 2013 Commissioning of optically stimulated luminescence dosimeters for use in radiotherapy, *Radiat. Meas.* **52** 31-39

[8] Knežević Ž, Stolarczyk L, Bessieres I, Bordy J M, Miljanić S and Olko P 2013 Photon dosimetry methods outside the target volume in radiation therapy: Optically stimulated luminescence (OSL), thermoluminescence (TL) and radiophotoluminescence (RPL) dosimetry, *Radiat Meas.* **57** 9–18

[9] Yahya M H 2015 Energy dependence of *Optically Stimulated Luminescent (OSL) dosimeter at low and high energy X-rays*, B.Sc Dissertation, Universiti Sains Malaysia.Aznar M C, Andersen C E, Bøtter-Jensen L,

[10] Tobergte D R and Curtis S 2013 Low dose evaluation with TLD100 using Numerical Stabilization, *J. Chem. Inf. Model.* **53(9)** 1689–1699

Acknowledgements

The authors also would like to acknowledge the Oncology, Radiotherapy and Nuclear Medicine Department, Universiti Sains Malaysia for the use of the linear accelerator in this study.

IOP Conf. Series: Materials Science and Engineering **298** (2018) 012024 doi:10.1088/1757-899X/298/1/012024

Study on the irradiation effect towards water filtration element

Asyraf Arif Abu Bakar[1,a),]**Anas Muhamad Pauzi**[1,b)], **M.Mujaini**[1,c)] **Abdul Aziz Mohamed**[1,d)], **Faridah Mohamad Idris**[2,e)]

[1]College of Engineering, Universiti Tenaga Nasional, Jalan IKRAM-UNITEN, 43000 Kajang, Selangor, Malaysia
[2]Reactor Technology Centre, Malaysian Nuclear Agency, Bangi, 43000 Kajang, Selangor, Malaysia.

[a)] aasyraf@uniten.edu.my
[b)]anas@uniten.edu.my
[c)]madihah@unite.edu.my
[d)]azizm@uniten.edu.my
[e)]faridah_idris@nuclearmalaysia.gov.my

Abstract. Gamma and neutron irradiation effect on material had been studied intensively. It was proven that there are strong correlation between neutron fluence and physical and mechanical damage of materials. The paper is part of the project to study the use of neutron and gamma irradiation to increase the effectiveness of simple water filtration element which are mostly sands and paper or pulp. Studies on irradiation effect on these element are compiled and analyze its changes from atomic structure to physical properties. On the atomic level, the change in atomic structure are relatively similar to the effect of irradiation on any solid, but with slight difference due to different in material composition and grain structure. Change in chemical properties might be less prominent but the change in grain size and its ability to filter water may be different. List of publication on previous experiments conducted on irradiating filtering element are compiled to provide basic idea on the parameters needed to provide significant changes. As a conclusion, the paper provides a preliminary correlation between the fluence and energy level of irradiation with the physical effects towards these filtration element, hence will be used to analyze events leads to change in filtering effectiveness upon irradiation

1. Introduction

Fundamentally, ionising radiations such as neutron and gamma can affect the properties through interaction with matter such as silica and paper. These two materials are the main components to be studied in this paper. Filtration system in this study uses sand and paper to filter water for domestic users. Interaction of sand and paper with neutron and gamma will affect the porosity of silica and paper due to chemical changes of these two materials after irradiation with neutron and gamma. Flow rate, heavy metals, nutrients in water and colour after filtration would be affected when irradiated materials such as sand and paper were use in filtration system. In this paper literature review were undertaken to study levels of radiation and its effect on chemical properties of sand, paper and water. Most significant changes due to irradiation occurs in high density fluence around structural material of Silicon dioxide, SiO_2 showing radiation damage processes in structures, including negative charge (electron) to the oxide and consequently reduce the positive charge trapping after irradiation with neutron as shown in figure 1.

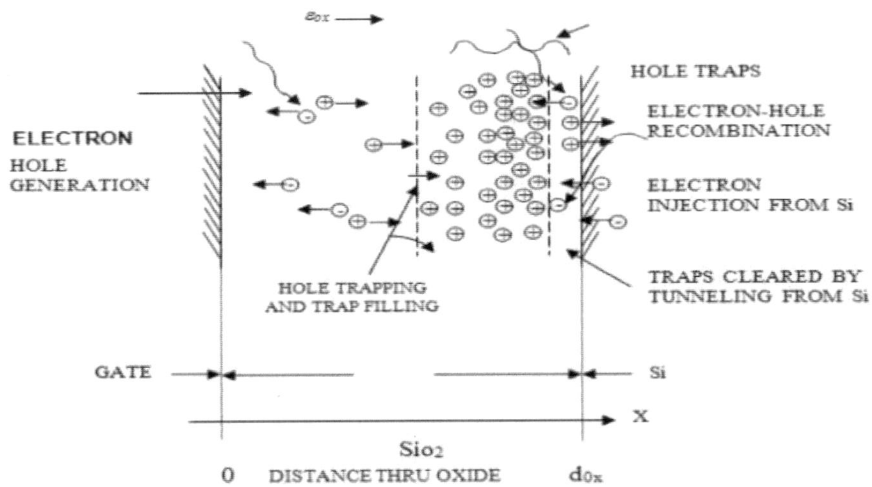

Figure 1. Oxide changes in silicon dioxide (MOS) showing radiation damage in the structure after exposure to neutrons [14].

Neutron and gamma are the ionization radiation that could modify the atomic structure of material if the material expose in a significant time. The change of atomic structure would tend to have a permanent defect on the material. Neutron is neutral in electrical charge and can pass through electron clouds and interact directly with the nucleus while gamma have high energy in form of electromagnetic radiation can ionise the atom. Previous studies had been done intensively on steels and alloys while not many were done on sand and paper. The damage events would likely to be similar with other materials since the basic interaction of radiation towards a material in an atomic level would be similar.

In this paper, the study will look into the effect of radiation of neutron and gamma towards basic water filtration element, which are mainly sand and papers or pulp, especially on the mechanism and damage events that lead to the changes in microstructure of filtering element, hence, changes the ability of these element to function to purify water.

2. The radiation damage event

A primary knock-on atom (PKA), is a displacement of the atom from its lattice site cause by radiation, and further on creates additional knock-on atom until eventually a cascade of displaced atoms are created. The PKA would stop at an interstitial location. Comparing with the atomic structure of sand, which is mostly Silicon Oxide (SiO_2), the effect of radiation are similar to effect on any other solids as shown in figure 2.The radiation damage event is the transfer of energy from the radiation toward the solid target where it consists of a few processes, starting from interaction of radiation with an atom in the lattice atom of the atoms in the sand and paper.

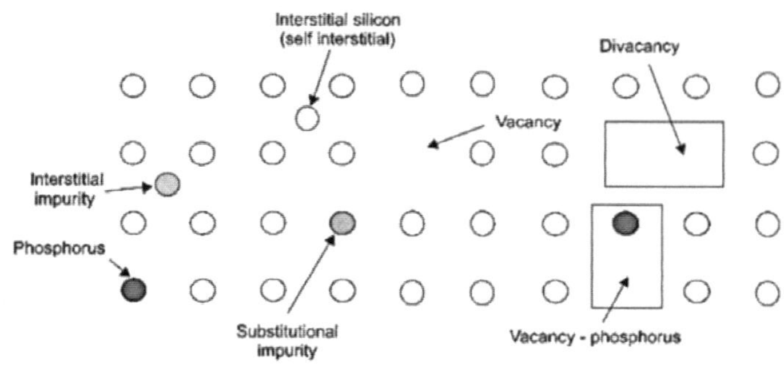

Figure 2. Diagram of some defects in the n-type silicon crystal lattice due to point defect[1].

2.1. Effect of Fast Neutron Radiation

High energy of neutron or fast neutron (generally between 0.5MeV and 20 MeV) can disturb the arrangement of atom and crystalline structure of materials. During an exposure of material with fast energy neutron, the structure of material may damage and cause temporary localized. The structural damage more significant for metals because of their relative immunity to damage by ionizing radiation.

2.2.2. Effect of Thermal Neutron Radiation. Thermal neutron have less important effect on the structural damage as they can be captured by nuclei of the irradiated material, which will become radioactive. Normally, after the neutron captured by the nucleus, the nucleus is in an excited or high energy state and excess energy is released by emitting high energy gamma rays, resulting the emitting nucleus recoils and is displaced.

2.2.3. Effect of Gamma on Material. Gamma rays are high energy photon produced from the decay of unstable nucleus. The energy of gamma rays may achieve from a few keV to several MeV. There were three ways of interaction between matter which are photoelectric effect, pair production and the Compton effect. Unlike fast neutron, the probability of an atom being displaced by a direct interaction between gamma rays and atomic nuclei in crystal is very small.

3. Discussion on the other paper

Above explanation proved that interaction of ionization radiation have a significant contribution on the defect of material depends on the energy it have. In this chapter will discuss the previous study that relates to the irradiation of gamma ray and neutron with the elements that use in simple water filtration. The main factor that will be discussed are the neutron fluence and gamma dose rate on the structure of material. Table 1 below shows a few recent study on the effect of irradiation towards the material.

Table 1. Fluence and effect of neutron and gamma radiation on various sample.

No	Fluence	Sample	Radiation effect	Outcome Measure
1.	10 to 50 kGy	Styrene Acrylic Esther (SAE) cement mortar	Different doses of gamma rays on the physio-mechanical properties of polymer modified white sand cement mortar.	Total porosity and water absorption values decreased with increasing irradiation dose, compressive strength values for polymer modified cement mortars increased with increasing irradiation dose while[2]
2.	50 – 300kGy	Quartz (SiO_2)	High gamma-irradiation doses on natural quartz crystals by ESR technique	The structure of quartz were not change when were irradiated by gamma-doses [3]
3.	10 to 50 kGy	Ordinary Portland cement with 5% of active rice husk ash (RHA)	Ordinary Portland cement with 5% of active rice husk ash (RHA) are impregnated with unsaturated polyester resin (UPE) followed by gamma rays ranging from 10 to 50 kGy	Polymer-impregnated blended cement (OPC–RHA–UPE) paste irradiated at a dose of 30 kGy has a good resistance towards sulfate and seawater attack as compared to the neat blended cement (OPC–RHA) paste[4]
4.	10^{22}n/m^2 or 12MGy	KU1 fused silica	Concentration of E'-type defects and oxygen-related defects produced in KU1 samples irradiated with high neutron fluences	Highest fluence generate a much higher concentration of defect than gamma irradiations at highest dose [5]
5.	12MGy	Fused Silica	Relevance of the manufacturing process (defects and impurities) present in the material before irradiation.	Impurities and defects produced during the manufacturing process of the silica are defect at least up to 12MGy doses[6]
6.	Neutron 10^{16} cm^{-2}, gamma & proton	Epitaxial silicon material	Radiation induced defects and their differences observed in various kinds of epitaxial silicon material	Neutron Irradiation introduces highest clustered regions compared to all other performed irradiation experiments with reduced generation of point defects[7]
7		Silicon with impurities oxygen and carbon	Defect parameters on high resistivity silicon with different concentrations of the impurities oxygen and carbon irradiated with neutrons and charged particles.	Irradiation with neutrons mainly create clusters single isolated displacements is enhanced due to Coulomb scattering.[8]
8	10 – 50kGy	Polymethyl methacrylate alite composite	Effect of gamma exposure dose on the physico-chemical and mechanical properties of polymer-alite composit.	Polymer loading, compressive strength and bulk density increase with increasing absorbed dose. The formation of new products in the pores appeared after impregnation.[9]
9	1.332 -1.173 MeV	Basaltic Rock fragment	Gamma-ray irradiation on water-rock reactions are investigated by comparing leaching experiments with and without irradiation	Water-rock interaction is promoted by gamma-ray irradiation in most cases of major basalt constituents, the absorbed dose are contained more in the feldspars[10]
10		Silica	High-energy radiation towards silica-core optical fiber	Fast-neutron irradiation induces substantially higher optical losses in the pure-silica core of the CeramOptec fiber than in the F-dopedsilica cladding[11]
11		Synthetic silica glasses	Defect of E' centers by irradiation of gamma.	Gamma irradiation creates defect pairs on E' center and an oxygen hole center[12]
12	50-300kGy	Paper	Filtration properties of four different filter papers on basis of the results of fibers swelling determination	Gamma irradiation of the filter paper induces the swelling restrictions of cellulose fibers, with progressive effect on prolongation of gamma-irradiation [13]

Based on the related literatures compiled in table 1, the materials could be categorized into three groups which are Polymer, papers and silicon. Table 1 shows the effect of neutron and gamma fluence on various types of samples with constituents related to sand and paper. Irradiation of gamma on Quartz (SiO_2) between 50kGy to 300 kGy does not change the quartz structure but irradiation of neutron with 10^{22}n/m^2 or 12MGy on KU1 silica sand causes highest concentration of defect. The optimum range on energy that needed for neutron to have its effect with material is between 10^{21}n/m^2 and 10^{22}n/m^2 neutron fluence meanwhile for gamma is 12MGy doses. Irradiation of gamma on paper using 50 to 300kGy causes induces the swelling restrictions of cellulose fibers, with progressive effect on prolongation of gamma-irradiation.

3.1. Effect of radiation on polymer

Based on table 1 above, row number 1, 3 and 8, the gamma radiation causes both cross-linking and degradation during treatment. Cross-linking is the most important effect of polymer irradiation because it is usually accompanied with an increase of the number of the polymeric chains under the effect of irradiation dose, leading to the formation of a network structure. Several studies indicated that, during irradiation polymerization an interaction between calcium silicate hydrate formed during the hydration reaction of the blended cement and the polymer presents in the pores takes place.4

3.2. Effect of radiation on paper

Gamma irradiation of the filter paper as in row number 12 in table 1, induces the swelling restrictions of cellulose fibers. The effects of paper simultaneously and progressively with prolongation of gamma-irradiation.

3.3. Effect of radiation on silica

Based on table 1, row number 2, 4, 5, 6, 7, 10 and 11 shows the effect of irradiating various types of radiation on silicon especially used for fusion and fission reactor environment. Overall, it shows that a higher gamma dose causes higher changes in microstructure such as pairs defects on E' center and an oxygen hole center. But for pure SiO_2 or quarts, radiation does not induce any changes on the microstructure. Meanwhile, irradiation with neutrons mainly causes clusters are create while using charged particles the generation of single isolated displacements is enhanced due to Coulomb scattering.

4. Conclusion

The defects of materials especially silica and paper depended on neutron and gamma energies.For the filtration system to be studied we propose to use 12MGy gamma doses and 10^{22}n/m^2 neutron fluence on the silica and paper in the system. The results from previous experiment can be a benchmark for the further study on comparing the quality index of filtrated water using gamma and neutron irradiated basic filter element.

5. References

[1] Gianluigi C 1998 The effect of hadro irradiation on the electrical properties of particle detectors made from various silicon materials Universite Joseph Fourier-Grenoble

[2] Khattab M M 2014 Effect of gamma irradiation on polymer modified white sand cement mortar composites *J. of Industrial and Engineering Chemistry* **20**

[3] Insiripong S, Kedkaew C, Thamaphat K, Chantima N, Limsuwan P and J. Kaewkhao 2012 Irradiation effect on natural quartz from Zambia *Procedia Engineering* **32**

[4] Khattab M M and Younes M M 2011 Durability of gamma irradiated polymer-impregnated blended cement pastes *Construction and Building Materials* **25**

[5] Bravo D, Lagomacini J C, León M, Martín P, Martín A andLópez F J 2009 Comparison of neutron and gamma irradiation effects on KU1 fused silica monitored by electron paramagnetic resonance, *Fusion Engineering and Design* **84**

[6] León M, Martín P, Ibarra A and Hodgson E R 2009 Gamma irradiation induced defects in different types of fused silica *J. of Nuclear Materials* **386-388**

[7] Honniger F 2007 *Radiation damage in silicon* Department of Physic, University of Hamburg

[8] Kuhnke E F M and Lindstroem G 2001 Defect generation in crystalline silicon irradiated with high energy particles *Nuc. Inst. and Meth. in Phys. Research Section B: Beam Interactionswith Materials and Atoms* **186**

[9] EL-Fass M M, Ismai M R, Abd El-Rahman H A and El-Miligy A A 1999 Physico-chemical studies on polymethyl methacrylate alite composite *J. of Radioanalytical and Nuclear Chemistry* **240**

[10] Yonezawa C, Tanaka T and Kamioka H 1996 Water-rock reactions during gamma-ray irradiation *Appl Geochemistry* **11**.

[11] Griscom D L, Gingerich M E, Friebele E J, Putnam M and Unruh W 1994 Fast-neutron radiation effects in a silica-core optical fiber studied by a CCD-camera spectrometer *Appl. Optics* **33**

[12] Imai H, Isoya J, Hosono H, Abe Y and Imagawa H 1993 Generation of E' centers and oxygen hole centers in synthetic silica glasses by y irradiation *The American Physical Society* **48**

[13] Balcerzak R C C and Dabrowski J 1985 Effects of gamma-irradiation on filtration properties of paper *Nukleonika* **30**
[14] Chee F P, Haider F, Amir A, Husin A H, Salleh S and Alias A 2015 Radiation damage in the silicon dioxide (SiO_2) *J. of Nuclear And Related Technologies* **12**

Acknowledgment
Special thanks to Malaysia's Ministry of Higher Education (MOHE) for sponsoring the research project under the Fundamental Research Grant Scheme (FRGS) project number 117289-136338.

IOP Conf. Series: Materials Science and Engineering **298** (2018) 012025 doi:10.1088/1757-899X/298/1/012025

Ultrasonic inspection of fake gold jewelry

Mohamad Pauzi Ismail[1, a)] Suhairy Sani[1], Faris Syazwan bin Mohd Shofri[2], Mohd. Harun[3] and Norlaili Binti Omar[4]

[1]Agensi Nuklear Malaysia, Bangi

[2]Universiti Putra Malaysia

[3]Pensioner, Bandar Baru Bangi

[4]Institut Penyelidikan dan Perundingan YaPEIM (INPUT), Kuala Lumpur.

[a)] pauzi2010@gmail.com

Abstract. Hollow jewelry made from combination of gold and other material was found in the market. At the outside it is made of gold and the inside layer is made of other material. X-ray fluorescent method cannot detect the inside material that was covered by gold. This paper explained the experimental result of ultrasonic inspection of fake gold used for jewelry. The ultrasonic pulse echo contact method was used to measure longitudinal wave velocity in the gold jewelry. The results of measurements are explained and discussed.

1. Introduction

X-ray fluorescence (XRF) testing is one of the common methods to detect fake gold. It is however applicable to surface and near surface detection. If the metal is covered with adequate thickness of gold layer then the X-ray fluorescence cannot detect the main metal behind the gold. In other cases, fake gold bar is produced by replicating gold bar with dimensions and weight, and with enough gold plating to fool X-ray fluorescence testing using tungsten. Tungsten is very nearly the same density as gold and it can be used as a core for counterfeit bars and coins that can be made to appear identical to the real item, and also keep the correct weight. However there is one physical property that cannot be confused between tungsten and gold and that is the speed at which sound travels through metal. The sound velocity for gold is 3240 m/s and for tungsten it is 5180 m/s. The velocity of sound for any particular metal can be measured by applying ultrasonic pulses and measuring how much time it takes for the pulses to travel through the metal. This is why ultrasonic testing has become known as the best method to detect fake gold bullion bars and coins (1).

It is not difficult to detect fake gold bar and coins using ultrasonic but for complicated geometry like jewelry the technique becomes more challenging.

2. Technical background

In ultrasonic testing sound energy is transmitted through a solid material by a series of small displacement within the materials. There are three basic types of mode of sound propagation in materials, i.e. longitudinal, transverse and surface) wave. A longitudinal wave is a wave formed by individual particles oscillating in the direction of propagation. This mode of transmission is the one most often used in ultrasonic testing. This wave is the most easily generated and detected. Almost all of-the sound energy used in ultrasonic testing originates as longitudinal sound and then may be converted to the other modes for special applications.

The velocity of sound propagation varies from one material to another. It depends on the elastic property and density of the material. It is given by,

$$c = \sqrt{\frac{E}{\rho}} \qquad (1)$$

where c is the wave velocity, E elastic modulus and ρ density. Sound velocity in various materials is given in table 1.

When the ultrasonic waves passing through a material reach the interface between that material and a material with different acoustic properties, part of the ultrasonic waves are partly passed through it and the other partly reflected at the interface. There are many kinds of interface material or boundary involved in ultrasonic testing, for example, the interface between the transducer and the test surface in a contact test, that between water and the test surface in an immersion test and that between flaws or discontinuities and the test material.

When the ultrasonic wave or sound beam strikes a plane separating two materials, some of the wave is transmitted forward and the remainder reflected backward (as shown in **FIGURE 1**). The relative amounts of reflected and transmitted beam intensities are expressed by the reflection and transmission coefficients, defined as follows,

$$\text{Transmission coefficient, } T = \frac{4Z_1\, Z_2}{(Z_1+Z_2)^2} \qquad (2)$$

where Z_1 and Z_2 are the acoustic impedance of the two materials. Z is a multiplication of density and velocity of sound in materials.

$$\text{Reflection coefficient, } R = 1 - T \qquad 1)$$

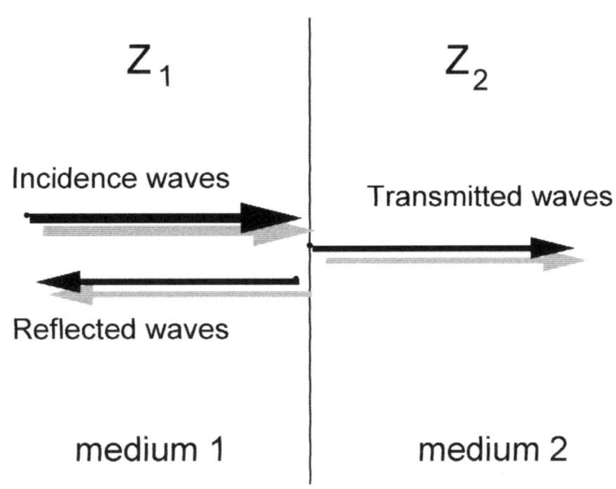

Figure 1. Reflection and Transmission at a plane surface, normal incidence

The coefficient may be expressed either as percentages or as decreases in the number of decibels. The equation shows that the transmission coefficient approaches unity and the reflection coefficient tend to zero when Z_1 and Z_2 have approximately similar value. The materials are then said to be well matched or coupled. On the other hand, when the two materials have substantially dissimilar characteristic impedance, e.g. for a solid or liquid in contact with a gas, the transmission and reflection coefficients tend to zero and 100 per cent, respectively. The materials are then said to be mismatched or poorly coupled.

Figure 2 illustrates the principle of ultrasonic flaw detection.

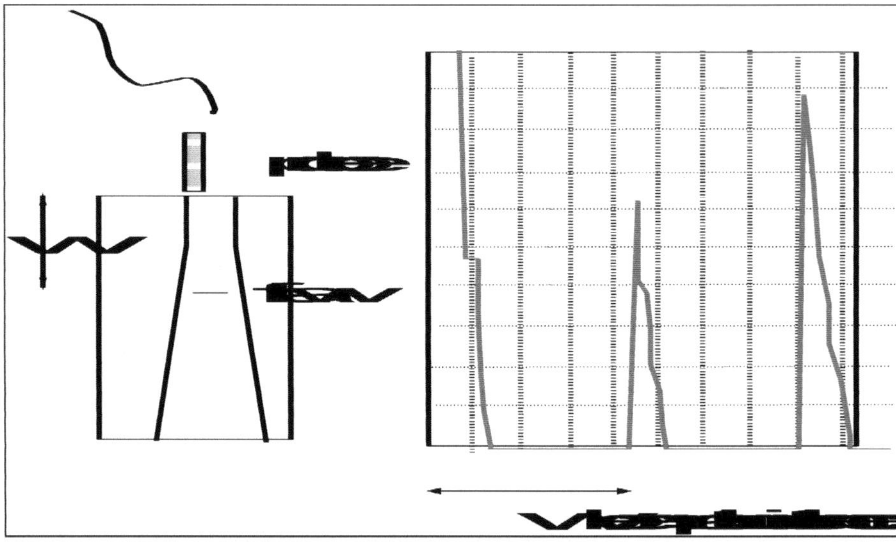

Figure 2. Principle of ultrasonic flaw detection

3. Experimental Method

The fake gold jewelleries received from YaPIEM are as in figure 3. The measurements were made by immersion method using 5 MHz focused probe and SIUI CTS-9009 Plus Ultrasonic Flaw Detector (figure 4). Engine oil was used as a couplant between probe and specimen.

Bracelet GLG-1

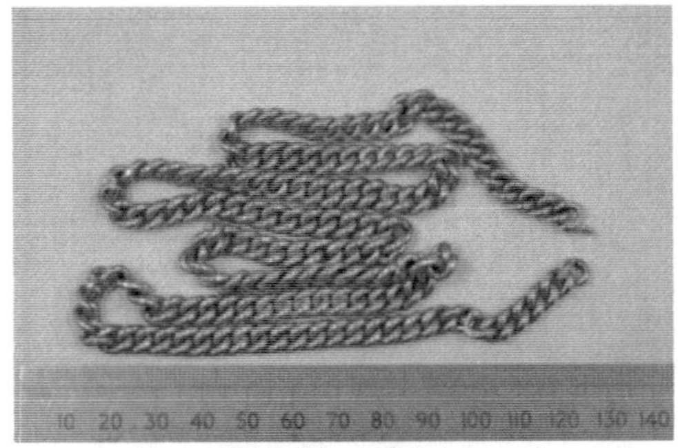
Necklace RL1

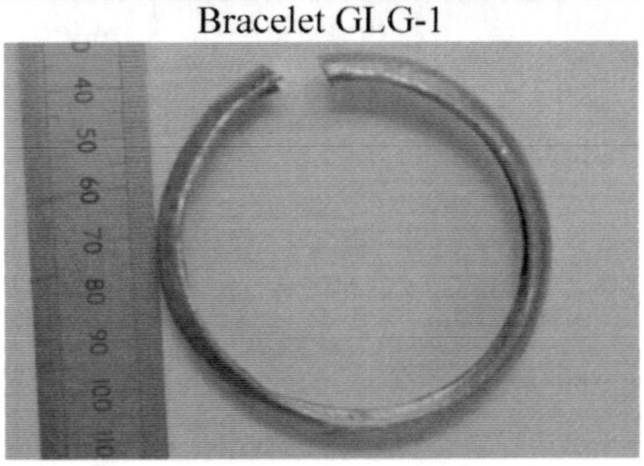
Bracelet GLG-2

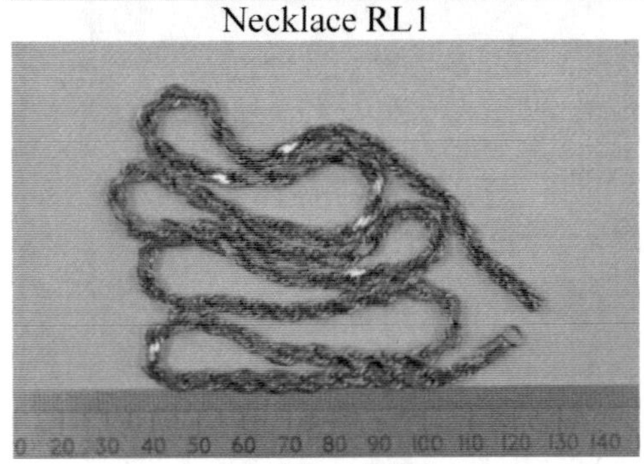
Necklace RL6

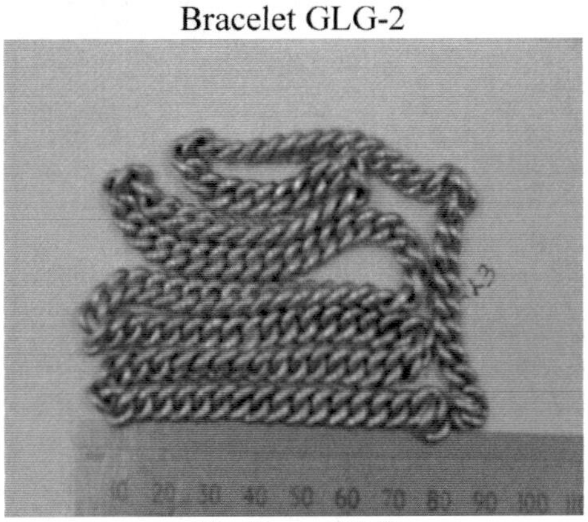
Necklace RL3

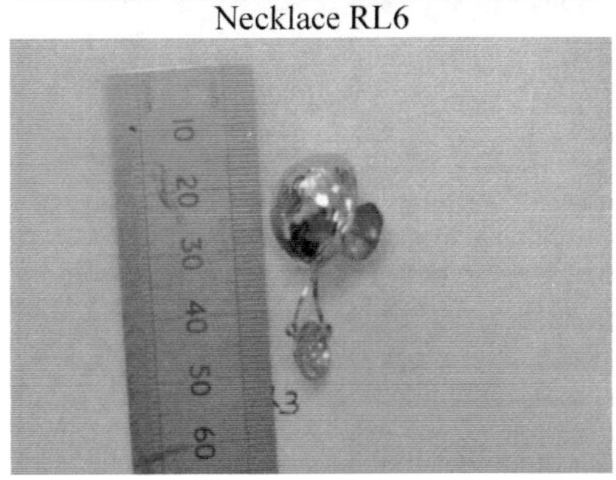
Ear ring

Figure 3. Fake gold jewelleries

IOP Conf. Series: Materials Science and Engineering **298** (2018) 012025 doi:10.1088/1757-899X/298/1/012025

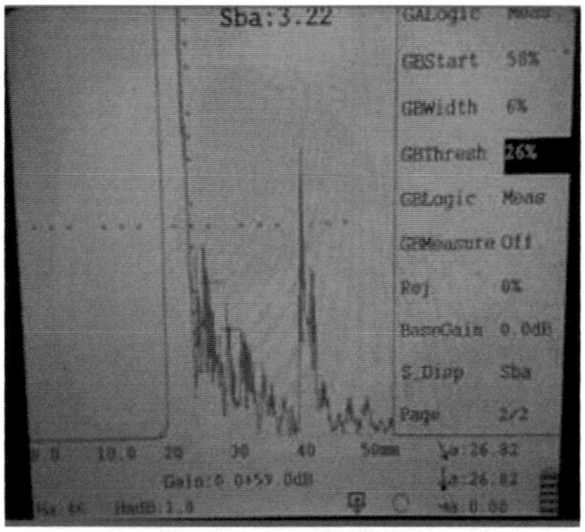

a) Bracelet GLG-1

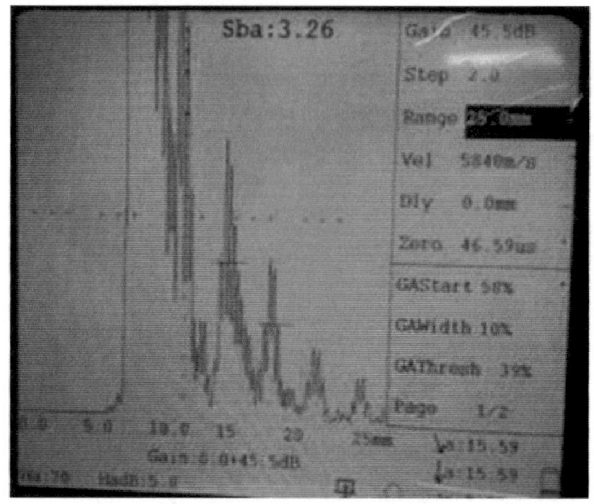

b) Bracelet GLG-2

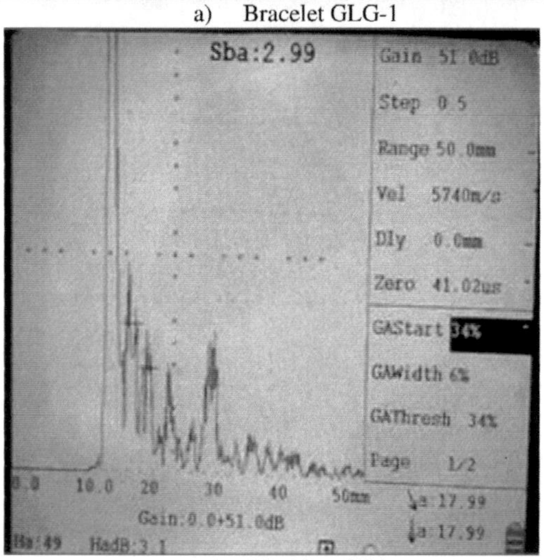

c) Necklace RL1

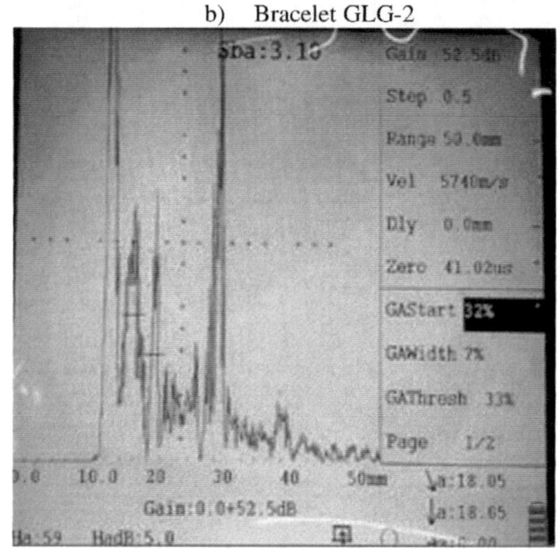

d) Necklace RL3

Figure 4. Typical ultrasonic signals from bracelet and necklace

4. Results and discussion

Figure 4 shows typical ultrasonic signals from the fake gold jewelleries. The echo signal from bracelet and pendant are quite clear and easy to interpret. However the echo from necklace is not clear and difficult to interpret due to complex geometry of the sample.

Experimental results are tabulated in table 1. The measured sound velocities showed that the values are much different from actual sound velocity in pure gold, i.e. 3240m/s.

Table 1. Results of ultrasonic measurement as compared to XRF (2) and SEM (2)

Sample identification	Contents	Density (g/cc)	Sound velocity (m/s)	XRF	SEM-EDX (% weight)			Measured sound velocity (m/s)
				(%weight)	Surface	Sub-surface	Internal surface	
	Fe	6.95-7.86	3500-6000	0				4023
	Co			0				3931
	Ni	8.3-8.8	5300-5900	0				3740
	Cu	8.4-8.9	3500-4700	1.737	2.68	4.66		4443
	Zn			0				4327
	Ru			0				4592
GLG1	Rh			0				
	Pd			1.143				
	Ag	8.4-10.5	**3600-4760**	63.849	3.52	6.26	100	
	Os			0.015				
	Ir			0				
	Pt			0.215				
	Au	19.3	3240	33.04	93.8	89.08		
	Fe	6.95-7.86	3500-6000	0				4023
	Co			0				3931
	Ni	8.3-8.8	5300-5900	0				3740
	Cu	8.4-8.9	3500-4700	0				3852
	Zn			0				3981
	Ru			0				3736
GLG2	Rh			0				
	Pd			0				
	Ag	8.4-10.5	**3600-4760**	1.882	3.57	3.93	100	
	Os			0				
	Ir			0				
	Pt			0.488				
	Au	19.3	3240	97.63	96.43	96.07		
	Fe	6.95-7.86	3500-6000	0.122				4223
	Co			0				4400
	Ni	8.3-8.8	5300-5900	47.746	7.77	19.09		4059
	Cu	8.4-8.9	**3500-4700**	49.093	0.76	2.9	94.97	
	Zn			0				
	Ru			0.053				
RL6	Rh			0				
	Pd			0.194				
	Ag	8.4-10.5	3600-4760	0.471			5.03	
	Os			0.012				
	Ir			0				
	Pt			0				
	Au	19.3	3240	2.31	91.47	78.01		
	Fe	6.95-7.86	3500-6000	0.137				3097
	Co			0				3331
	Ni	8.3-8.8	5300-5900	0				3448
	Cu	8.4-8.9	**3500-4700**	84.224	14.71	24.7	100	3981
	Zn			1.463				4009
	Ru			0.004				4342
RL3	Rh			0				
	Pd			0				
	Ag	8.4-10.5	3600-4760	0.235				
	Os			0.09				
	Ir			0				
	Pt			0.027				
	Au	19.3	3240	13.759	85.29	75.3		
	Fe	6.95-7.86	3500-6000					7014
	Co							6300
	Ni	8.3-8.8	5300-5900					
	Cu	8.4-8.9	3500-4700	6.081	3.16	6.03		
	Zn							
	Ru							
Ear ring	Rh							
	Pd							
	Ag	8.4-10.5	**3600-4760**	2.019	1.11	3.56	100	
	Os							
	Ir							
	Pt							
	Au	19.3	3240	91.9	95.73	90.41		

Measurement is also compared with 916 gold pendant which give sound velocity values of 3304 – 3454 m/s.

5. Conclusion

Ultrasonic testing can be used as an additional NDT method for checking fake gold. It is quite reliable for gold bar or coin. For jewelry with complicated shape and thin material, XRF and SEM are adequate but sound velocity measurement can be used as a complementary method.

6. References

[1] The Application of Ultrasonic Testing for Silver Bars, Bullion and Coins (5 June 2017). https://www.scribd.com/document/267911050/The-Application-of-Ultrasonic-Testing-for-Silver-Bars

[2] M. Harun, M. P. Ismail, A. A. Abas and S. Sani, 2013, Teknologi ultrasonik dalam pengujian ketulinan fizikal barang kemas berongga, Seminar Mineral Berharga: Pengaruh Pasaran Dan Penawarannya di Dunia, Hotel Grand Season, Kuala Lumpur, 13-14 Mei 2013.

Acknowledgments

The first author would like to thank to NDT team of Nuclear Malaysia for their helps in collecting NDT data. They are Suhairy Sani (SKM UT 2) and Amry Amin Abbas (SKM UT 3). Thanks also given to Ms Laili binti Omar from Institut Penyelidikan dan Perundingan YaPEIM (INPUT) for providing test samples and information related to fake gold.

IOP Conf. Series: Materials Science and Engineering **298** (2018) 012026 doi:10.1088/1757-899X/298/1/012026

Composite material characterisation using an advanced small angle x-ray (SAXS) technique

[1]**Hafizal Yazid,** [1]**Julie A. Murshidi,** [1]**Rafhayudi Jamro,** [1]**Megat Harun M. A.,** [2]**Abdul Aziz Mohamed**

[1] Malaysian Nuclear Agency, Bangi 43000 Kajang, Selangor, Malaysia

[2] Universiti Tenaga Nasional, Jalan Ikram-Uniten, 43000 Kajang, Selangor, Malaysia

Abstract. Materials development in the field of composite material spurs the use of advanced characterization technique. As the fillers become in the nanoscale range in size, the effect of agglomeration become apparent and cannot be avoided. The use of Small Angle X-Ray (SAXS) Scattering technique revealed the information on agglomeration based on the value of specific surface (m^2/g). Thermoplastic natural rubber composite was found isotropic based on 2D saxs scattering pattern. As the amount of fillers increased from 2-10% wt., the value of specific surface dropped accordingly. This indicated the higher the amount of filler used, the higher the degree of agglomeration. The SAXS system was also tested by Alumina (BAM) powder and yield result which was in good agreement with BET technique.

1. Introduction

The sample is a blend of natural rubber and high density polyethylene or known as NR/HDPE. Both phases are compatibilised with liquid natural rubber (LNR). This LNR fills in the gap between the dispersed phase of HDPE and the matrix phase of NR [1]. Fillers are added in the blend to form composite system which act as radiation shielding material. However, the fillers which in the nanometer size range are very sensitive to moisture and prone to form agglomerates. The fillers were therefore treated with surfactant to minimize agglomeration. The degree of agglomeration of filler inside the blend was then characterized by using SAXS technique.

SAXS is an analytical method to determine the structure of particle systems in terms of averaged particle sizes or shapes. It is a non-destructive method for investigating nanostructures from typically 1 nm to 100 nm in size. In this context, the nanostructures are regarded as the object of interest which scatter the incoming x-rays. The scattered x-rays are captured by the detector (1 or 2 Dimensional) behind the object as shown in figure 1. Information of the object (scatterer in the nano range) is obtained by performing data analysis on the scattered data signal. Comparison of SAXS technique with the other techniques is given in figure 2. SAXS instrument used in the study is shown in figure 3 (a). Differences between SAXS and XRD is shown schematically in figure 3 (b).

IOP Conf. Series: Materials Science and Engineering **298** (2018) 012026 doi:10.1088/1757-899X/298/1/012026

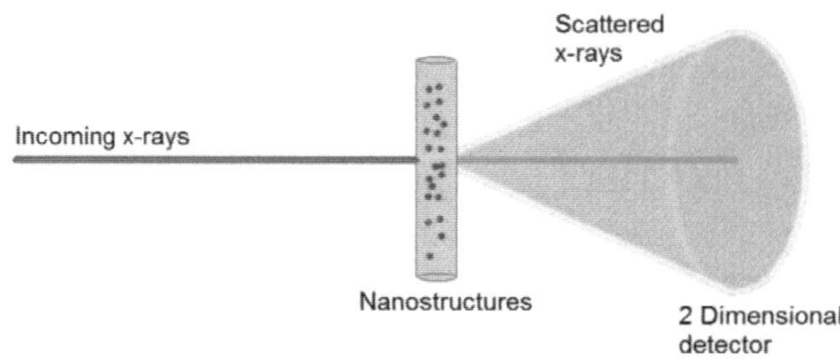

Figure 1. X-rays source, nanostructures and detector configuration.

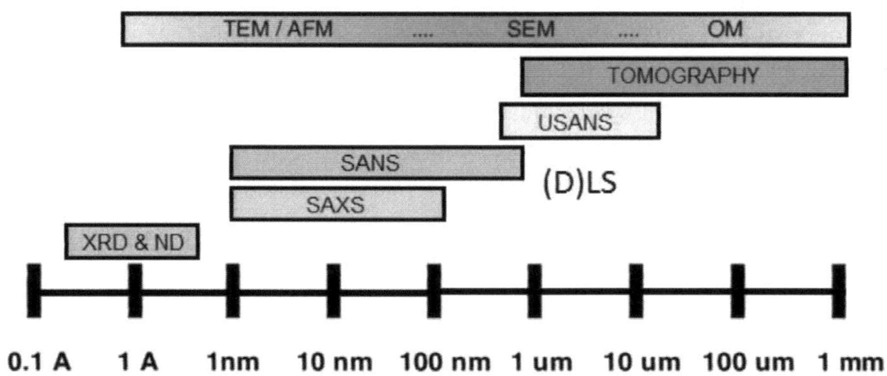

Figure 2. Comparison of SAXS technique with other techniques

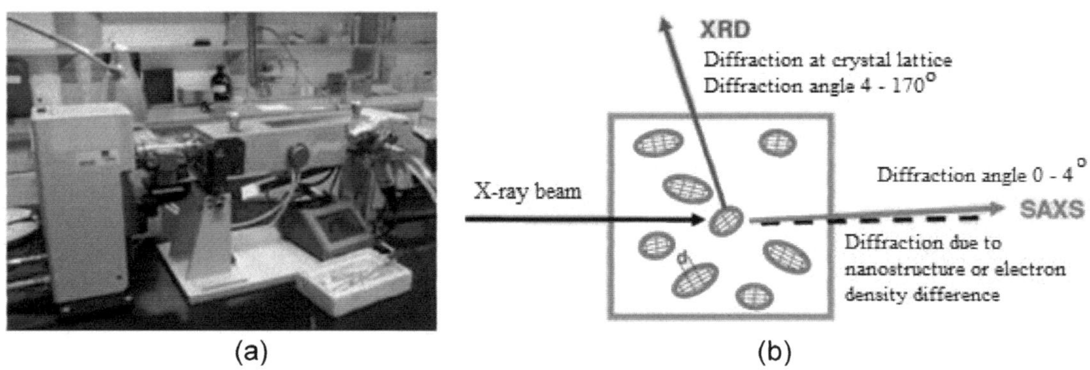

Figure 3. (a) SAXS instrument (b) Schematic diagram of differences between SAXS and XRD

2. Materials and method

2.1. Sample preparation

Sample of polymer blend consists of 50% natural rubber (Nr), 10% liquid natural rubber (Lnr) and 40% high density polyethylene (HDPE). Nr (SMR L grade) was supplied by Malaysian rubber research institute, RRI (Malaysia). The density of Nr is 0.95g/cm^3. HDPE was supplied by Titan Chemicals Corp. Bhd. (Malaysia) with a density of 0.91g/cm^3. Lnr was produced from Nr with weight average molecular weight relative to polystyrene as Mw = 40 x 10^4. Nano fillers B_4C with the density of 2.52g/cm^3 was used and added to the blend starting from 2%wt until 10%wt. Before the fillers were added to the blend, they are treated with surfactant and observed under scanning electron microscope.

2.2. Experimental work

SAXS experiments were performed using Hecus SWAX instrument, operated at 50kV and 1mA with the point collimation geometry. The radiation used was a Ni filtered CuKα radiation of wavelength 0.154nm (Seifert, Ahrensburg, Germany). The intensity profiles are recorded using a CMOS 2-D detector (Pilatus) and the scattering vector q, covers from 0.01 to 0.6$Å^{-1}$. One-dimensional scans of I(q) were extracted from two- dimensional scattering patterns using the analysis package Fit2D.

Angular calibration of the scattered intensities in the small angle regime for the detector is performed using silver stearate (d = 48.68 Å). Sample to detector distance is 271mm and sample exposure to the source is 600 seconds. All experiments are carried out at room temperature, 20°C. Validation experiment is carried out using a reference material, alumina type 150 (Al_2O_3) CRM BAM-PM-104.

2.3. Data analysis

The principle of SAXS technique is large objects scatter to small angle and small objects scatter to large angle. In this aspect, SAXS complement XRD as the minimum of diffraction angle for XRD is typically 10^0 which is easily achievable by SAXS. The relationship between the angle and the scattering vector q is given by [5]:

$$q = \frac{4\pi}{\lambda}\sin\theta \qquad (1)$$

where:

q = scattering vector, $\boldsymbol{\lambda}$ = wavelength of the x-rays, θ = half of the diffraction angle

Scattered data signal in 2 dimensional is reduced to 1 dimensional data profile by radial averaging and typically plotted in scattered intensity, I (arbitrary unit) versus scattering vector q (nm^{-1}). Data in I and q is used to fit three possible model of the scatterer based on the following equation [2] :

$$I(q) = I_0 * e^{-(q^2\frac{Rg^2}{3})} \qquad \text{gives spherical shape} \qquad (2)$$

$$I(q) * q = I_0 * e^{-(q^2\frac{Rc^2}{2})} \qquad \text{gives cylindrical shape} \qquad (3)$$

$$I(q) * q^2 = I_0 * e^{-(q^2\frac{Rt^2}{1})} \qquad \text{gives lamellar shape} \qquad (4)$$

where,

I(q) = scattered intensity, q = scattering vector, I0 = Initial intensity,
Rg = Radius of gyration due to globular, Rc = Radius of gyration due to cross section,
Rt = Radius of gyration due to thickness

The calculation could also be performed on samples which contain nanoporous structure to obtain information on specific surfaces based on the following equation [3] :

$$Si = \frac{\pi * \varphi_1 * \varphi_2 * 10^4 * \frac{k}{Q}}{\rho_A} \qquad (5)$$

where:

Si = specific surfaces, φ1 = volume fraction of solid, φ2 = volume fraction of air,

ρA = apparent density of sample, k/Q = constant in $Angstrong^2$/ $Angstrong^3$

IOP Conf. Series: Materials Science and Engineering **298** (2018) 012026 doi:10.1088/1757-899X/298/1/012026

3. Results and discussion

Figure 4 shows the effect of surface treatment on the nano-filler. The treatment successfully de-agglomerates the raw nano-filler. The success of the treatment is basically stems from the optimum amount of surfactant selected from SEM results. Figure 5-7 clearly show the agglomeration exist when the surfactant is not suffice or in excess. The optimum amount of surfactant was found to be 1.6%wt. Even with the proper selection of surfactant amount, still this does not guarantee that the fillers will not form agglomeration when they are mix with the polymer blend. At this stage SAXS technique plays vital role to probe the fillers condition when they form composite with the polymer blend. In order to have the understanding on the changes of the clustered fillers, some of the treated fillers were subjected to BET test for measuring the specific surface for comparison with SAXS result.

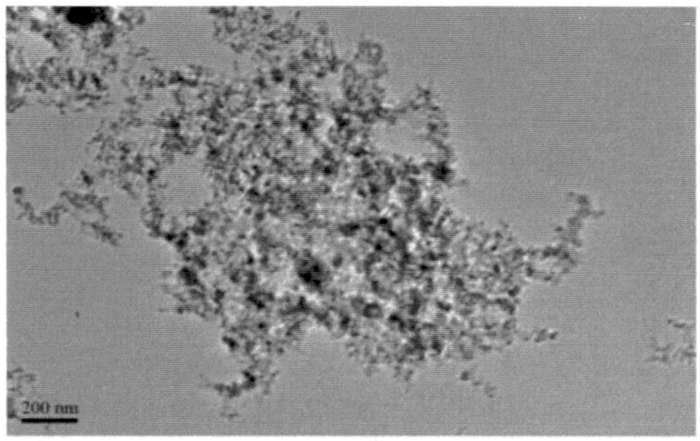

(a)

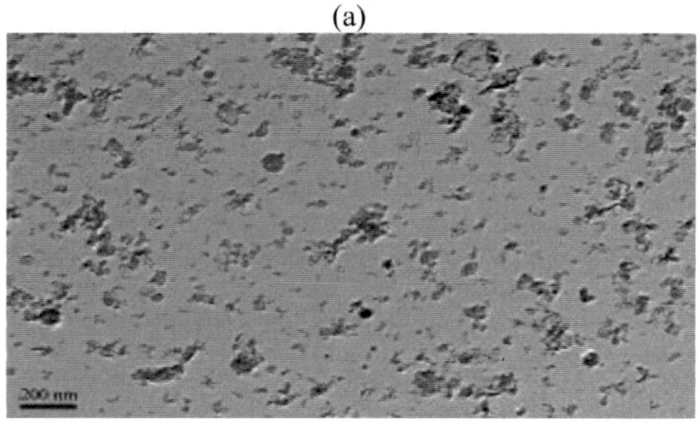

(b)

Figure 4. TEM micrograph of (a) raw nano-filler (x10k) and (b) treated nano-filler (x10k)

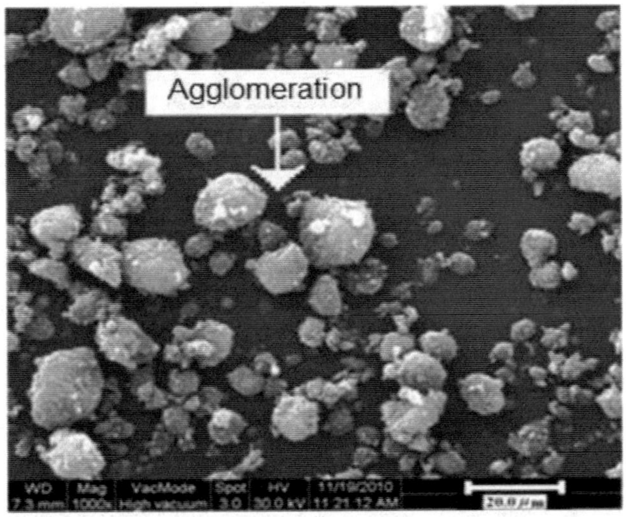

Figure 5. SEM micrograph of treated nano-filler with 0.2%wt surfactant (x10k)

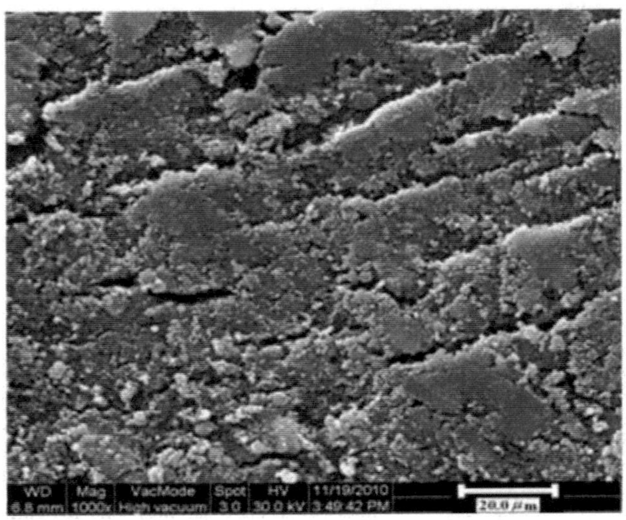

Figure 6. SEM micrograph of treated nano-filler with 1.6%wt surfactant (x10k)

Figure 7. SEM micrograph of treated nano-filler with 2.0%wt surfactant (x10k)

Figure 8 shows an example of comparison between composite with 10%wt nano-filler and polymer blend. Both of them have the same linear scaling in the q range of $0.01 < q_z < 0.25$ Å^{-1}. Exposure time was 600 seconds and the beam centre was at pixel x = 274, y = 99. Fillers or B_4C particles are clearly visible by the excess SAXS signal (A). There is no preferred orientation of the nanostructure within the plane of the samples. This does not exclude the possibilty that a preferred nematic type orientation exists perpendicular to the surface. Rings at larger angles (B) are due to Be window.

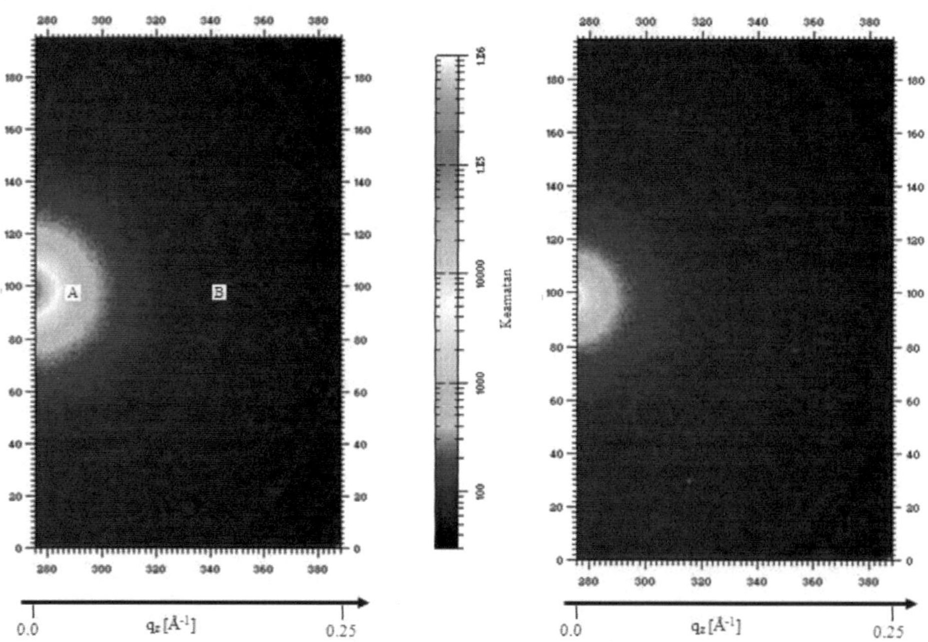

(a) (b)

Figure 8. SAXS 2D scattering pattern (a) composite with 10%wt nanofiller (b) polymer blend

Figure 9 shows 1D scattering profiles of some of the samples. These profiles are subjected to analysis using Guinier plot, Porod plot and Invariant plot [3]. The values obtained from the analysis were substituted in equation 5. The result of this equation were tabulated in Table 2.

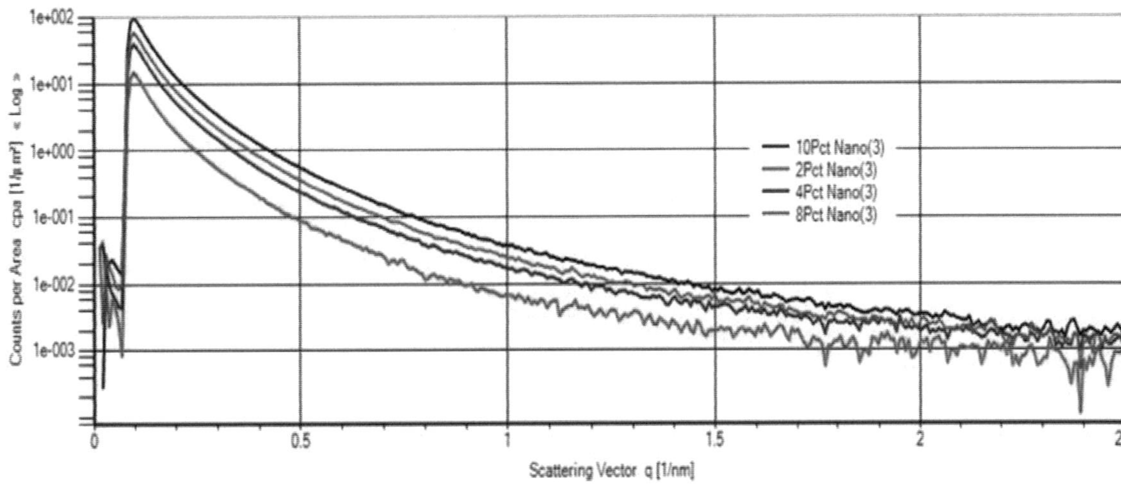

Figure 9. Data reduction from 2D scattering pattern results in 1D scattering profiles ready for data analysis

It is clearly seen from table 1 that as the filler loading is increased, the value of specific surface is reduced. This indicates the fillers experience slight agglomeration. The value of specific surface of the treated filler from BET test was found to be 595 m^2/g. This proves that the result from Table 2 is very reliable as they are close to BET result.

Table 1. Specific surface of fillers inside the polymer blend

Filler loading (%wt)	Treated filler Si/p (m^2/g)
2	554.6
4	509.6
6	481.5
8	460.3
10	457.4

4. Conclusions

The result of this study shows the importance of the advanced characterization technique in implementing the R&D in developing countries. The use of small angle x-ray scattering (SAXS) technique revealed the information on agglomeration based on the value of specific surface (m^2/g). This lab-scale instrument could be used as a platform to further develop knowledge and skills especially to cater the need of using high flux X-ray or synchrotron facility in the near future.

5. References

[1] Hafizal Yazid et. al, December 2012, Journal of Nuclear and Related Technologies, Vol.9, No. 2.
[2] Manfred Kriechbaum, 16 August 2016, International Conference on X-Ray and Related Technique (ICXRI) 2016, Kuala Lumpur, Malaysia.
[3] Heimo Schnablegger and Yashveer Singh, 2011, The SAXS Guide.

IOP Conf. Series: Materials Science and Engineering **298** (2018) 012027 doi:10.1088/1757-899X/298/1/012027

Thermal neutron flux measurement using self-powered neutron detector (SPND) at out-core locations of TRIGA PUSPATI Reactor (RTP)

Nur Syazwani Mohd Ali [1], **Khaidzir Hamzah**[1], **Faridah Mohamad Idris**[2], **Mohamad Hairie Rabir**[2]

[1] Department of Energy Engineering, Faculty of Chemical and Energy Engineering, Universiti Teknologi Malaysia,81310 Johor Bahru, Johor, Malaysia.
[2]Malaysian Nuclear Agency, Bangi, 43000 Kajang, Selangor, Malaysia.

nsyazwanimali68@gmail.com

Abstract. The thermal neutron flux measurement has been conducted at the out-core location using self- powered neutron detectors (SPNDs). This work represents the first attempt to study SPNDs as neutron flux sensor for developing the fault detection system (FDS) focusing on neutron flux parameters. The study was conducted to test the reliability of the SPND's signal by measuring the neutron flux through the interaction between neutrons and emitter materials of the SPNDs. Three SPNDs were used to measure the flux at four different radial locations which located at the fission chamber cylinder, 10cm above graphite reflector, between graphite reflector and tank liner and fuel rack. The measurements were conducted at 750 kW reactor power. The outputs from SPNDs were collected through data acquisition system and were corrected to obtain the actual neutron flux due to delayed responses from SPNDs. The measurements showed that thermal neutron flux between fission chamber location near to the tank liner and fuel rack were between 5.18×10^{11} nv to 8.45×10^{9} nv. The average thermal neutron flux showed a good agreement with those from previous studies that has been made using simulation at the same core configuration at the nearest irradiation facilities with detector locations.

1. Introduction

Neutron flux can be classified as safety parameter used to analyse the neutronic and thermal hydraulic parameters and also convey the neutron behavior during fission reactions. The flux was used to determine the reactor states through neutron production and its evolution as its supplies the information to the reactor protection system and reactor power regulating [1]. This flux can be measured using detectors such as fission chamber (FC), self-powered neutron detectors (SPND), local power range monitors and others. The principal for all neutron detectors applies the activation of the main components materials which produce the electrical signals that are proportionate to the neutron flux at the detector locations.

The benefit of such the neutron detectors is that it can be used as sensors to develop a fault detection system (FDS) for safety improvement, reliability, and availability during reactor operations [2]. The FDS mainly focuses on the neutron flux parameters to detect any abnormal events such as sudden increases in number of neutrons during reactor operations. RTP has implement safety procedure which automatically will trip the reactor when operating limit is exceeded. By implementing the FDS, it will help along diagnose and overcome the sources of failure in real-time mode which prevents the reactor from tripping and allow the reactor to continues is operation in safe manners.to

Generally in research reactors, wide range FC detectors were used as safety instruments to measure the neutron flux and also to monitor the neutron behavior in the reactor core. The measurements carried out were the leakage flux that is proportional to the linear power level calculation, period time calculation and as a reactor trip parameters. These parameters were able to determine the reactor states during operations from start-up to full power level [3]. FC was usually located in fixed positions around the reactor core with leak-tight instruments port so that the detectors were not subjected to moisture [4]. To prevent any build-up of corrosive substances on the detector, an inert gas is provided in this port.

At RTP, the neutron detectors that are currently used are FC which is responsible for detecting the signal changes during reactor operations. However, the prompt response from FC limits the signals output to study the neutronic and thermal hydraulic parameters. The signals measured need to undergo interpretation which requires a precise knowledge of the mass and isotopic composition of the fissile coating materials [5]. FC can only display the fluctuation of the signal from the detectors which difficult to be used as the FDS sensors. The actual flux value at certain power level is unknown. The signal also cannot be manipulated to obtain the flux and power. Therefore, by utilizing the SPNDs, the experiment was conducted with the aims to measure the thermal flux and to estimate the reliability of the output for developing the FDS at RTP core. The FDS method will not discuss further in this paper.

2. Materials and methods

2.1. TRIGA PUSPATI Reactor (RTP)

Malaysia has the one and only research reactor namely TRIGA PUSPATI (RTP) since 1982 where it has been operated safely by Malaysian Nuclear Agency (Nuklear Malaysia) and licensed by Atomic Energy Licensing Board of Malaysia (AELB). The RTP reactor is a TRIGA MARK II, pool type reactor with 1 MW thermal power and has achieved its first criticality on 28 June 1982. The cylindrical core consists of fuel elements from solid Uranium Zirconium-Hydride ($UZrH_{1.6}$) with 8.5 wt. %, 12.0 wt. %, and 20.0 wt. % of uranium fuel (U) and 19.7% of Uranium-235 enrichment. The purpose of this homogenous mixture of the $ZrH_{1.6}$ in the fuel is to moderate neutrons of energy above 0.14eV [6]. There are 4 boron carbide (B_4C) control rods whereby one of them is an air follower type and the rest are fuel follower type which composed of 8.5 wt. % of U and B_4C absorber on top of fuel section. The core were surrounded with graphite as reflector and cooled by natural convection. As for protecting purposes, the core biological shield is made of high density concrete of 2.5m thick wall to attenuate radiation and shield its surrounding from radiological contamination. Recently, the RTP core is on it 15th configuration as shown in figure 1 below.

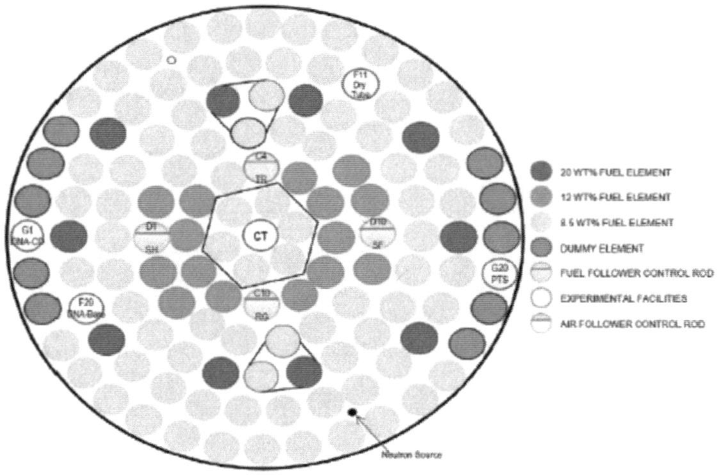

Figure 1. RTP core15th configuration [6].

2.2. *Self-Powered Neutron Detector (SPND)*

SPND has been used widely to measure the flux behavior and as a monitoring devices in core of research and power reactors. The unique characteristics of SPND make it favorable to be used in harsh radiation field associates with high pressure and temperature conditions [7]. It also does not need any external power supply and can be fitted in the in-core locations. The SPNDs consists of three major components and connected to the insulated coaxial cables, centered with emitter surrounded with insulator and collector as shown in figure 2 below. The emitter consists of materials that has high absorption cross-section and capable to react with neutrons through radiative capture like vanadium, rhodium, cobalt and others [8]. Insulators are solid dielectric materials that are capable to retain the high electrical resistivity and high radiation fields. The collector is of made conducting materials like aluminum oxide and magnesium oxide.

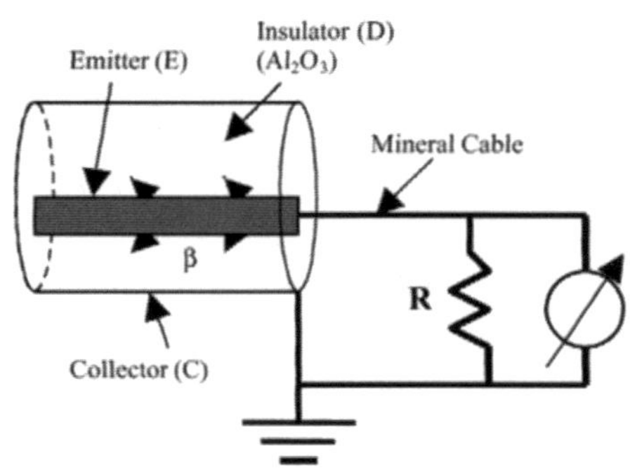

Figure 2. Schematic views of general SPNDs [8]

In this paper, the vanadium SPNDs were used to measure the neutron flux at out-core locations. These SPNDs have been chosen because of the emitter depletion rates were longer but suffer from the delay responses. These SPNDs are desirable to be used because of its high accuracy on detecting the neutrons [9]. The interaction happen when the neutron is being absorbed then emits the energetic electrons after a few minutes due to induction of the beta activity in the emitter [10]. The electrons then travelled and passed through the insulator and were deposited at the collector. This phenomenon will create the collector as negative side and the emitter as positive side thus allow for small currents to flow through the insulated coaxial cable. The observed signal were said to be directly proportional to the rate of neutron absorption in the emitter which represents the neutron flux at the detector locations.

The measurement has been done at the out-core location where the three detectors were coupled and located together at the same locations. Since the vanadium SPNDs suffered from delay responses, each measurement were recorded within 30 minutes at four different locations with full reactor power level at 750 kW. The SPNDs were connected to the data acquisition system to record the signals from the detectors. All the data were then reconstructed to determine the flux using the correction factors in equations 1 and 2 below [11].

$$\varphi_1 = CI \tag{1}$$

$$\varphi_{i+1} = C\left[I_{i+1} - \exp(-\frac{\Delta I_i}{\tau})\right] + \exp(-\frac{\Delta\varphi_i}{M\tau}) \tag{2}$$

φ, C and I are neutron flux, normalisation constant and detector current respectively, whereas τ and Δ are characteristic time decay of V^{52} and sampling time.

The SPNDs position were located at the fission chamber cylinder, at the above graphite reflector, between graphite reflector and tank liner and at the fuel rack as shown in figure 3. The data obtained were used to plot the flux and calculate the responses corrected using correction factors to study the signal's reliability for developing FDS.

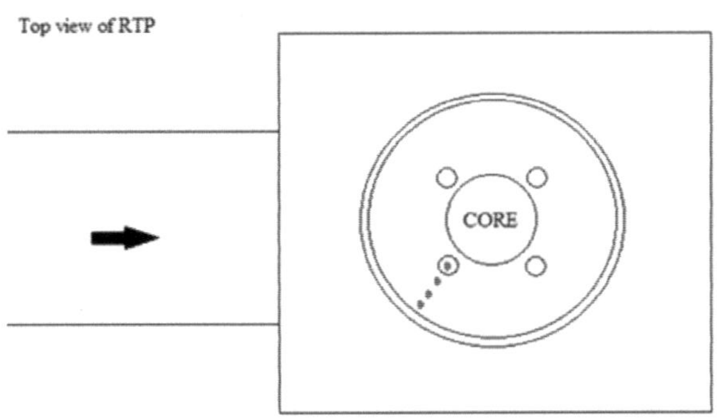

Figure 3. SPNDs location for flux measurements.

3. Data analysis

The outputs from the SPNDs are in current unit which are related proportionately to the flux by using correction factors to calculate the actual flux values. Figure 4 shows the flux and corrected flux calculated using correction factors for each location against the sampling time for all three SPNDs. The flux varied due to the varied distance between the detector and the core. As this measurement were conducted at the out-core locations, the predicted thermal flux was assumed to be less than 3.78 $\times10^{12}$ nv and was obtained from previous simulation at the nearest irradiation facilities (PTS-G20) with fission chamber locations [12]. Based on figure 3 below, the maximum thermal neutron flux recorded in SPND#3 was approximately 4.5×10^{11} nv while the corrected flux calculated was 5.5×10^{11} nv at the location above graphite reflector. The detectors were moving out from fission chamber channel which increases the signals slightly. Besides the detectors movement, the signals may be affected by gamma ray and neutrons which comes from the reflector and also from the core. The minimum flux was located at the fuel rack which less neutrons available in this area.

The measurements were conducted by coupling all three SPNDs together. Figure 4 illustrates the average signals obtained directly from SPNDs and average signal obtained from responses corrected. Based on the graph below in figure 5, the flux started to increase as the power level rises. At 750kW reactor power, the SPNDs were allowed to record the data for 30 minutes at each location to obtain the equilibrium flux. The flux were approximately 3.07×10^{11} nv at fission chamber location, 5.18×10^{11} nv at the above graphite reflector, 4.55×10^{11} nv between the graphite reflector and tank liner and 8.45×10^{09}nv at the fuel rack for both data. By comparing with previous study, the average thermal neutron flux in this work has shown a good agreement, with the assumption which the flux obtained must be less than 3.78×10^{12} nv due to the locations has less neutrons and less gamma radiations to study the reliability of the SPNDs signals.

Based on the result, the signals can be used further to develop a fault detection system in RTP core. However, to ensure the detection system works efficiently, the delayed responses needs to undergo some improvement in real time mode so that the signals produces will promptly transfer to a system for any abnormal events in neutron flux.

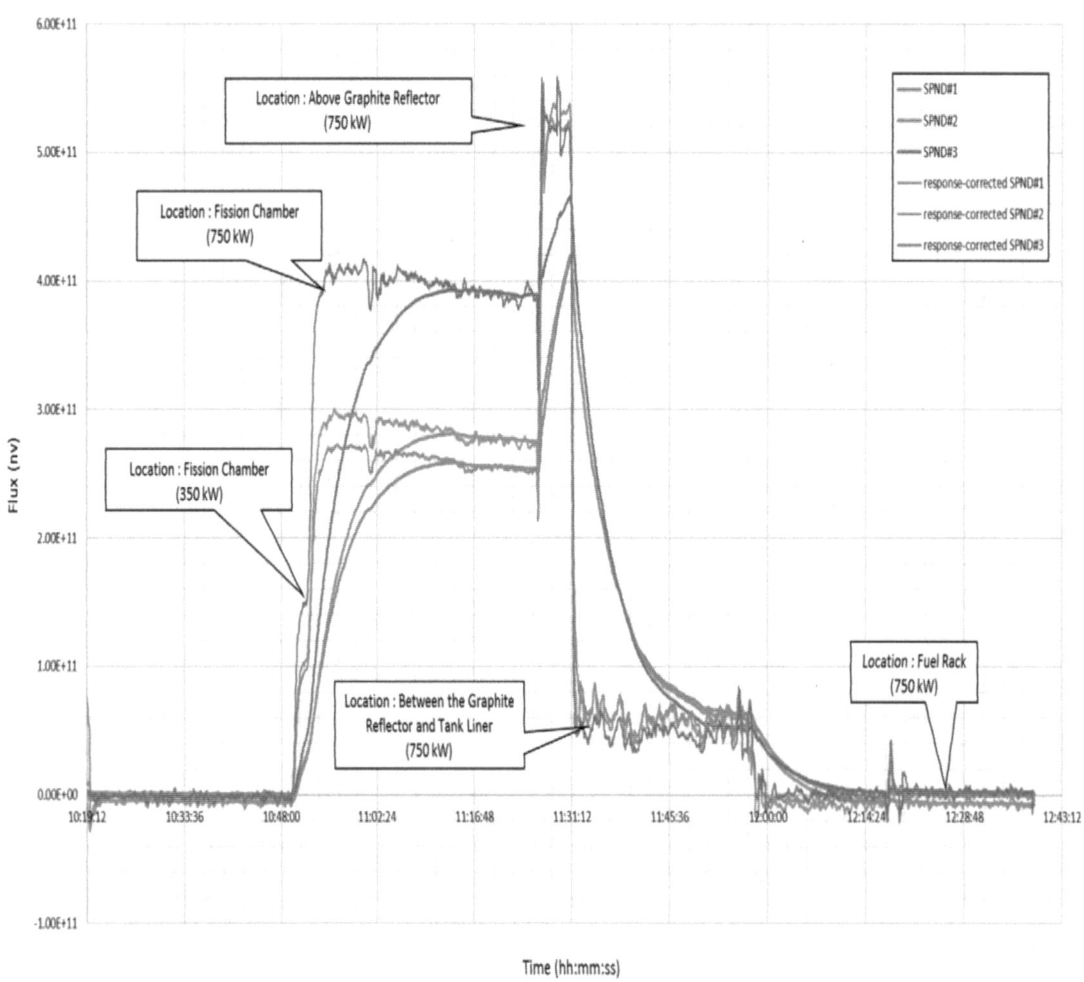

Figure 4. Neutron flux obtained directly from measurements and from responses corrected.

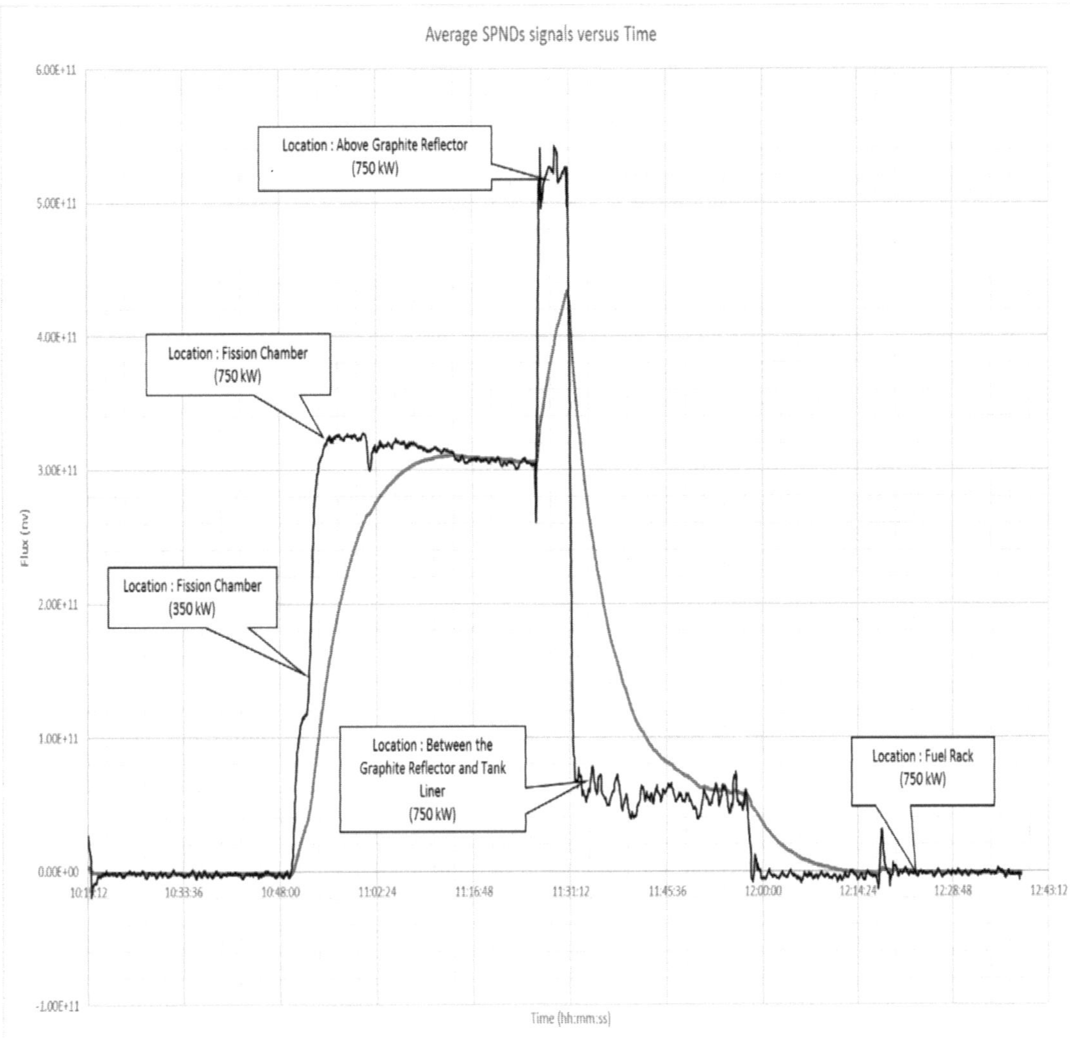

Figure 5. Neutron flux obtained directly from measurements and from responses corrected.

4. Conclusion

Out-core thermal neutron flux measurements has been recorded and compared with previous simulation to study the reliability of the SPNDs signal for preliminary test of FDS development at RTP. The average thermal neutron flux showing a good agreement with those from previous studies that has been made using simulation at the same core configuration at the nearest irradiation facilities with detector locations. The flux obtained from the simulations was 3.78 $x10^{12}$ nv while from this work the average flux measured were in the range of $5.18x10^{11}$ nv to $8.45x10^{09}$ nv. The results has shown that the signals are reliable and capable to be used as a sensor for developing FDS.

5. References

[1] IAEA TRIGA Reactor Main System pp. 17-23.

[2] Jianping Maa and Jin Jiang 2011 Applications of fault detection and diagnosis methods in nuclear power plants: a review *Progress in Nuclear Energy* **53** pp. 255-266.

[3] Regtron K F K I *Wide Range Neutron Flux Monitoring and Safety Assembly* pp. 7-10.

IOP Conf. Series: Materials Science and Engineering **298** (2018) 012027 doi:10.1088/1757-899X/298/1/012027

[4] International Atomic Energy Agency Fission Chamber *Reactor Protection System, Education and Training.*

[5] Fourmentel D, Filliatre P, Barbot L, Villard J F, Lyoussi A, Geslot B, Carcreff HMalo, J Y and Reynard-Carette C 2014*Comparison of Thermal Neutron Flux Measured by* ^{235}*U Fission Chamber and Rhodium Self-Powered Neutron Detector in MTR* (IEEE Transactions On Nuclear Science **61(4)** pp. 2285-2290.

[6] Mohamad Hairie Rabir, Julia Abdul Karim and Mohd Amin Sharifuldin Salleh 2012 Power and neutron flux calculation for The Puspati Triga Reactor using MCNP *Jurnal Sains Nuklear Malaysia* pp. 22-32.

[7] Hilborn J W 1968 U.S. Patent No. 3, 375 (370 Washington DC: U.S. Patent and Trademark Office).

[8] Angelone M, Klix A, Pillon M, Batistoni P, Fischer U, Santagata A 2014 Development of self-powered neutron detectors for neutron flux monitoring in HCLL and HCPB ITER-TBM *Fusion Engineering and Design* **89** pp. 2194–2198.

[9] Rabir M H Online In-Core Thermal Neutron Flux Measurement for The Validation Of Computational Methods.

[10] Mishra A K, Shimjith S R, Bhatt T U and Tiwar, A P 2014 Kalman filter-based dynamic compensator for Vanadium self powered neutron detectors *IEEE Transactions on Nuclear Science* **61(3)** pp. 1360–1368.

[11] Mohamad Hairie Rabir, Muhammad Rawi Md Zin, Mark Dennis Usang, Abi Muttaqin Jalal Bayar, and Na'im Syauqi Bin Hamzah 2016 Neutron flux and power in RTP core-15 *Advancing Nuclear Science and Engineering for Sustainable Nuclear Energy Infrastructure,* (AIP Conference Proceedings **1704** pp. 050018-(1-10).

Adaptive control method for core power control in TRIGA Mark II reactor

Mohd Sabri Minhat[1, a)], Hazlina Selamat[2], Nurul Adilla Mohd Subha[2]

[1]*Reactor Electronic, Instrumentation and Control Section, Reactor Technology Centre, Malaysian Nuclear Agency, 43000 Kajang, Malaysia*
[2]*Control and Mechatronics Engineering Department, Faculty of Electrical Engineering, Universiti Teknologi Malaysia, 81310 Skudai, Malaysia*

[a)]E-mail: sabri@nuclearmalaysia.gov.my

Abstract. The 1MWth Reactor TRIGA PUSPATI (RTP) Mark II type has undergone more than 35 years of operation. The existing core power control uses feedback control algorithm (FCA). It is challenging to keep the core power stable at the desired value within acceptable error bands to meet the safety demand of RTP due to the sensitivity of nuclear research reactor operation. Currently, the system is not satisfied with power tracking performance and can be improved. Therefore, a new design core power control is very important to improve the current performance in tracking and regulate reactor power by control the movement of control rods. In this paper, the adaptive controller and focus on Model Reference Adaptive Control (MRAC) and Self-Tuning Control (STC) were applied to the control of the core power. The model for core power control was based on mathematical models of the reactor core, adaptive controller model, and control rods selection programming. The mathematical models of the reactor core were based on point kinetics model, thermal hydraulic models, and reactivity models. The adaptive control model was presented using Lyapunov method to ensure stable close loop system and STC Generalised Minimum Variance (GMV) Controller was not necessary to know the exact plant transfer function in designing the core power control. The performance between proposed adaptive control and FCA will be compared via computer simulation and analysed the simulation results manifest the effectiveness and the good performance of the proposed control method for core power control.

Keywords: Core Power Control, Adaptive Control, Model Reference Adaptive Control, Self-Tuning Generalised Minimum Variance Controller, Reactor TRIGA PUSPATI

1. Introduction

In this paper, the nuclear Reactor TRIGA PUSPATI (RTP) operating at Malaysian Nuclear Agency is modelled adopting a zero dimensional approach, in order to reproduce the dynamic behaviour of the reactor on the entire operative range. TRIGA Mark II is a nuclear reactor widely used as non-power nuclear reactor in many applications, including production of radioisotopes, neutron radiography, basic research on the properties of material and for education and training.

The RTP has recently been equipped with a new Feedback Control Algorithm (FCA) capable of controlling reactor power [4]. Currently, the system is working not satisfied power tracking performance. In order to improve the current controller, the development of a non-linear dynamic model of the reactor and adaptive control were necessary. The model was validated by comparing the experimental results of the behaviour of the power of the reactor controlled by the FCA.

2. Modelling

2.1. Non-Linear Model
The behaviour of the whole TRIGA reactor can be described by combine the equation of point kinetic, thermal-hydraulic, mass flow rate and reactivity [12].

2.1.1. Point Kinetics Equations:

$$\left.\begin{aligned}\frac{d\psi}{dt} &= \frac{\rho-\beta}{\Lambda}\psi + \sum_{i=1}^{6}\lambda_i\eta_i \\ \frac{d\eta_i}{dt} &= \frac{\beta_i}{\Lambda}\psi - \lambda_i\eta_i \quad i = 1,\dots,6\end{aligned}\right\} \tag{1}$$

2.1.2. Thermal-Hydraulic Equation:

$$\left.\begin{aligned}\frac{dT_m}{dt} &= \frac{P(1-f)}{\tau_m K} + \frac{T_f-T_m}{\tau_m} - \frac{\Gamma C_m}{\tau_m K}(T_m - T_{in}) \\ \frac{dT_f}{dt} &= \frac{P_f}{\tau_f K} - \frac{T_f-T_m}{\tau_f} \\ T_m &= wT_{out} + (1-w)T_{in}\end{aligned}\right\} \tag{2}$$

2.1.3. Mass Flow Rate Equation:

$$\Gamma = \frac{Kw(T_f-T_m)}{C_m(T_m-T_m)} \tag{3}$$

2.1.4. Reactivity Equation:

$$\rho = a_h\Delta h_{cr} + \alpha_m(T_m - T_m^0) + \alpha_f(T_f - T_f^0) \tag{4}$$

2.2. Linear Model
The behaviour of the whole TRIGA reactor can be described by combine the equation of point kinetic, thermal-hydraulic, mass flow rate and reactivity [12].

$$\left.\begin{aligned}\psi(t) &= 1 + \delta\psi(t) \\ \eta_i(t) &= 1 + \delta\eta_i(t) \\ \rho(t) &= \delta\rho(t)\end{aligned}\right\} \tag{5}$$

According perturbations theory, the linearization of the nonlinear equation system is achieved:

2.2.1. Point Kinetics Equations:

$$\left.\begin{aligned}\delta\dot{\psi} &= \frac{-\beta}{\Lambda}\delta\psi + \sum_{i=1}^{6}\frac{\beta_i}{\Lambda}\delta\eta_i + \frac{\alpha_m}{\Lambda}\delta T_m + \frac{\alpha_f}{\Lambda}\delta T_f + \frac{\alpha_h}{\Lambda}\delta h_{cr} \\ \delta\dot{\eta}_i &= \lambda_i\delta\psi - \lambda_i\delta\eta_i \quad i = 1,\dots,6\end{aligned}\right\} \tag{6}$$

2.2.2. Thermal-Hydraulic Equation:

$$\left.\begin{aligned}\delta\dot{T}_m &= \left[\frac{P^0(1-f)}{\tau_m K}\right]\delta\psi - \xi_1\delta T_m + \frac{1}{\tau_m}\delta T_f + \xi_2\delta T_{in} \\ \delta\dot{T}_f &= \left[\frac{P^0 f}{\tau_f K}\right]\delta\psi + \frac{1}{\tau_f}\delta T_m - \frac{1}{\tau_f}\delta T_f\end{aligned}\right\} \tag{7}$$

where,

$$\tau_m = \frac{M_m C_m}{K}, \quad \tau_f = \frac{M_f C_f}{K} \tag{8}$$

$$\xi_1 = \frac{1}{\tau_m} + \frac{C_m \Gamma}{\tau_m K w}, \quad \xi_2 = \frac{C_m \Gamma}{\tau_m K w} \tag{9}$$

2.2.3. *Reactivity Equation:*

$$\delta\rho = \alpha_h \delta h_{cr} + \alpha_m \delta T_m + \alpha_f \delta T_f \tag{10}$$

Since the transient response of a linear system is characterized by the poles of the transfer function, it will be appropriate to formulate the linear stability problem in terms of transfer function and block diagram. Applying the Laplace Transform to the above equations, the system transfer function of neutron density over control rod position, ($\delta h_{cr} = \delta h$) is obtained [12].

$$\frac{\psi(s)}{h(s)} = \frac{\alpha_h}{s\Lambda + \beta - \sum_{i=1}^{6}\left(\frac{\beta_i \lambda_i}{s+\lambda_i}\right) - \frac{\alpha_m[\alpha+\gamma(s\tau_f+1)] + \alpha_f[\gamma+\alpha\tau_m(s+\xi_1)]}{\tau_m(s\tau_f+1)(s+\xi_1)-1}} \tag{11}$$

where,

$$\alpha = \frac{P^0}{K}, \gamma = \frac{P^0(1-f)}{K} \tag{12}$$

In this way, normalized neutron density in the reactor core is possible and equivalent to power density. The reactor power level in percentage can be obtained by multiplied neutron density with 100. In table 1 all the physical quantities and parameters of the model are summarized [16].

Table 1. RTP Model Parameters.

Group	B_i	λ_i
1	0.00023	0.0124
2	0.00153	0.0305
3	0.00137	0.1114
4	0.00277	0.3014
5	0.00081	1.1363
6	0.00029	3.0137

Symbol	*Quantity*	*Value*
T_{in}^0	Initial inlet water temperature	32°C
C_m	Moderator specific heat capacity	4187 J kg^{-1}K^{-1}
C_f	Fuel specific heat capacity	601.95 J kg^{-1}K^{-1}
M_m	Moderator total mass	846 kg
M_f	Fuel total mass	207.45 kg
Γ	Coolant mass flow rate	6.7 kg s^{-1}
K	Global heat transfer coefficient	5000 W ^{0}C^{-1}
P^0	Steady-state power level	1x10^{-6} W
W	Weighting factor for computation of moderator temperature	0.5
F	Fraction of power deposited in the fuel	1
Λ	Mean neutron generation time	36 μs
β	Delayed neutron fraction	700x10^{-5}Δk/k
α_f	Reactivity due to change in temperature fuel	-0.1 m Δk/k/°C
α_m	Reactivity due to change in temperature moderator	1x10^{-8}Δk/k/°C

2.3. Low Order Nominal Plant Model

The control design objective for this research is to obtain a low order controller that can deal with a defined range of power levels. It is important to note that the model representing the RTP is essentially a linearized model of the nonlinear point kinetics with two temperature feedback.

The method of specifying low order nominal plant order for this control design by using an approximation based on the analytical method or 'physics' of the process.

The analytical method approach for choosing $\psi(s)$ uses a one delayed neutron group representing an average response of the delayed neutrons and, uses the singular perturbation techniques to eliminate the fast dynamics represented by the first state. The 2nd order linear model without temperature feedback, based on this approximation of the 'physics', is [18]:

$$\frac{\psi(s)}{h(s)} = \frac{\alpha_h}{s\Lambda+\beta-\frac{\beta\lambda}{s+\lambda}} \tag{13}$$

The dynamics of the system in the frequency domain can be summarised in block diagram form shown in figure 1 [19]:

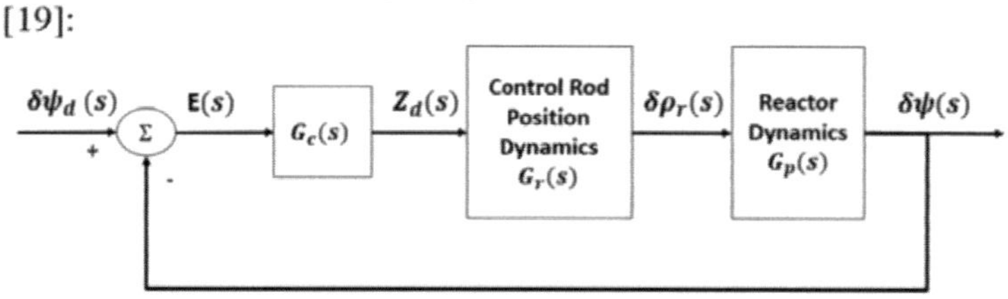

Figure 1. Block diagram of the dynamic system.

The transfer function from reactivity input, $\varrho_r(s)$ to neutron density in reactor core, $\psi(s)$ is [20]:

$$\frac{\psi(s)}{\rho_r(s)} = \frac{\alpha_h}{s\Lambda+\beta-\frac{\beta\lambda}{s+\lambda}} = \frac{\frac{1}{\Lambda}(s+\lambda)}{s^2+\left(\lambda+\frac{\beta}{\Lambda}\right)s} \approx \frac{\frac{1}{\Lambda}(s+\lambda)}{s^2+\left(\frac{\beta}{\Lambda}\right)s} \tag{14}$$

By initially assuming ideal signal conditioning and rod drive dynamics which instantaneously provides a speed that matches the speed demand signal, $Z_d(s)$, control rod position dynamics can be represented as a pure integral function which has a Laplace transform equal to $\frac{1}{s}$. The control rod transfer function $G_r(s)$ can be simply represented as $\frac{W}{s}$ where W is the total reactivity worth of the rod and rod speed, $Z_d(s)$ is computed in units of fraction of core length per second:

$$\delta\dot{\rho}_r = G_r z_d \tag{15}$$

Applying the Laplace Transform to the Eq. (14), the reactivity due to control rod movement is obtained:

$$\rho_r(s) = \left(\frac{W}{s}\right) Z_d(s) \tag{16}$$

The resulting 3rd order linear model, based on this approximation of the 'physics', is:

$$\frac{\psi(s)}{Z_d(s)} = \frac{\frac{W}{\Lambda}(s+\lambda)}{s\left[s^2+(\frac{\beta}{\Lambda})s\right]} = \frac{W(s+\lambda)}{\Lambda s^3+\beta s^2} \quad (17)$$

Table 2 shows parameter which were used in the above dynamic equation.

Table 2. Plant Parameters.

Symbol	***Value***
β	$0.007\ \frac{\Delta k}{k}$
λ	$0.1\ s^{-1}$
W	$0.002274\ cm^{-1}$
Λ	$36\ \mu s$

3. Adaptive controller design

Adaptive control is a type of controller that has the ability to adjust itself to any parameter variations occurring in a control system. It has been applied to many e.g. in Selamat et al. (2008). In classical control strategies, the coefficients of the controller are fixed are pre-specified by the designer. This may not be sufficient to provide satisfactory performance for system there are time-varying, subject to changes in operating conditions or disturbance acting on them.

Adaptive control is needed to overcome difficulties in determining suitable control laws there are satisfactory over a wide range of operating condition [23].

Adaptive control such as MRAC and STC are dealing with complex systems that have unpredictable parameter deviations and uncertainties. The basic objective to maintain consistent performance of a system in the presence of uncertainty and variations in plant parameters. Adaptive control is superior to robust control in dealing with uncertainties in constant or slow-varying parameters [32].

MRAC and STC can be designed using both Direct and Indirect approaches. We consider Direct MRAC design rely on the Lyapunov Stability Theory and Indirect STC design combine with an on-line recursive plant parameter estimator for this research.

3.1. MRAC controller

The structure of MRAC system is shown in figure 2 below [23]. Basically the MRAC consist of four main components: 1) Plant to be controlled, 2) Reference model to generate desire closed loop output response, 3) Controller that is the time-varying and whose coefficients are adjusted by adaptive mechanism, and 4) Adaptive mechanism that uses error to produce controller coefficient.

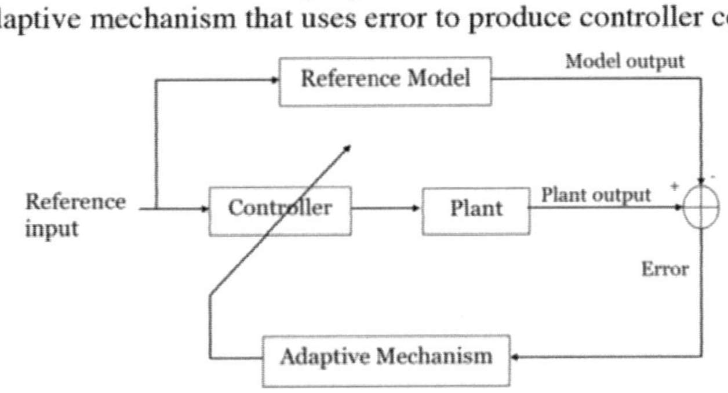

Figure 2. General structure of an MRAC system

Regardless of the actual process parameters, adaptation in MRAC takes the form of adjustment of some or all of the controller coefficients so as to force the response of the resulting closed-loop control system to that of the reference model. Therefore, the actual parameter values of the controlled system do not really matter.

The MRAC can be designed such that the globally asymptotic stability of the equilibrium point of the error different equation is guaranteed. To do this, the Lyapunov Second Method is used. It requires an appropriate Lyapunov function to be chosen, which could be difficult. This approach has stability consideration in mind and is also known as the Lyapunov Method. A closed-loop system with a controller that has the following parameters is considered: 1) $r(t)$ = Reference input signal, 2) $u(t)$ = Control signal, 3) $y(t)$ = Plant output, 4) $y_m(t)$ = Reference model output, and 5) $e(t)$ = Difference between plant and reference model output = $y(t)$-$y_m(t)$.

In this research, a PI-controller with adjustable parameter is parameterized and provides tracking [25]. A PI-controller can be described by the following equation:

$$u(t) = K\left(e(t) + \frac{1}{T_I}\int_0^t e(t)\,dt\right) \tag{18}$$

where T_I is the integration time constant, K is a constant and e is the error. If the controller is tuned to be slow and T_I is large, then the controller first acts like P-controller, but later the as the integration starts to take effects, the steady-state deviation goes slowly to zero. In PI-controller, the steady state deviation will finally go to zero. If the controller is tuned to be fast and T_I is small, then both terms (P and I) will affect the control signal all the way from the beginning. The system becomes faster, but the output signal might oscillate.

The reference model specifies the ideal response $y_m(t)$ to the external command $r(t)$. The use of a reference model for control can be traced to the nuclear reactor system. Often, the situation therein is such that the controls designer are sufficiently familiar with the plant to be controlled and its desired properties. Thus, by choosing the structure and parameters of a reference model suitably, its outputs can be used as the desired plant response. While in principle such a model can be linear or nonlinear, considerations of analytical tractability have made linear reference models more common in practice.

In order to design the controller we have linearized the system and obtain a low order nominal-plant model, and then designed a linear controller for it. RTP model in this work is described by a third order transfer function. For the reactor power to have a smooth tracking, a reference model is used to provide the reactor with appropriate power changes to get to the desired power (Z.S. Alavi et al., 2009). The reference model used is given by Equation (21). The PI-controller has the form given in Equation (22).

3.1.1. Plant:

$$G_p(s) = \frac{W(s+0.1)}{(36x10^{-6})s^3+0.007s^2} \tag{19}$$

3.1.2. Reference Model:

$$G_m(s) = \frac{0.2}{s^2+s+0.2} \tag{20}$$

3.1.3. PI-Controller:

$$u = -ky + \frac{k_i}{s}(r-y) = \theta_1 r - \theta_2 \tag{21}$$

where,

$$\theta_1 = \frac{k_i}{s}, \qquad \theta_2 = \frac{sk + k_i}{s}$$

In designing an MRAC using Lyapunov the following step should be followed:

- Derived a differential equation for error, $e = y - y_m$ $(i.e. \dot{e},\ \ddot{e},\ \dddot{e})$ that contains the adjustable parameter, θ.

$$\left.\begin{aligned} e &= y - y_m \\ \dot{e} &= \dot{y} - \dot{y}_m \\ \ddot{e} &= \ddot{y} - \ddot{y}_m \\ \dddot{e} &= \dddot{y} - \dddot{y}_m \end{aligned}\right\} \quad (22)$$

- Find a suitable Lyapunov function $V(e,\theta)$ – usually in quadratic form to ensure positive definiteness:

$$V(\ddot{e},\dot{e},e,X_1,X_2,X_3,X_4) = \lambda_1\ddot{e}^2 + \lambda_2\dot{e}^2 + \lambda_3\, e^2 + \lambda_4 X_1^2 + \lambda_5 X_2^2 + \lambda_6 X_3^2 + \lambda_7 X_4^2 \quad (23)$$

where $\lambda_1, \lambda_2, \lambda_3, \lambda_4, \lambda_5, \lambda_6 > 0$ positive definite.

The derivative of V become,

$$\dot{V} = 2\lambda_1\ddot{e}\,\dddot{e} + 2\lambda_2\ddot{e}\,\dot{e} + 2\lambda_3 e\dot{e} + 2\,\lambda_4 X_1\dot{X}_1 + 2\,\lambda_5 X_2\dot{X}_2 + 2\,\lambda_6 X_3\dot{X}_3 + 2\,\lambda_7 X_4\dot{X}_4 \quad (24)$$

where for stability $\dot{V} < 0$.

- Derive an adaptation mechanism based on $V(e,\theta)$ such that *e* goes to zero.

$$\dot{X}_1 = \dot{\theta}_2 = \frac{W\lambda_1}{36x10^{-6}\,\lambda_4}\dot{y}\ddot{e} \quad (25)$$

$$\dot{X}_2 = \dot{\theta}_2 = \frac{0.1W\lambda_1}{36x10^{-6}\,\lambda_5}y\ddot{e} \quad (26)$$

$$\dot{X}_3 = \dot{\theta}_1 = \frac{-W\lambda_1}{36x10^{-6}\,\lambda_6}\dot{r}\ddot{e} \quad (27)$$

$$\dot{X}_4 = \dot{\theta}_1 = \frac{-0.1W\lambda_1}{36x10^{-6}\,\lambda_7}r\ddot{e} \quad (28)$$

The summing equation from (26) to (29) obtained:

$$\left.\begin{aligned} \theta_1 &= \frac{k_i}{s} = \frac{-W\lambda_1}{2(36x10^{-6})\lambda_6}\dot{r}\ddot{e} - \frac{0.1W\lambda_1}{2(36x10^{-6})\lambda_7}r\ddot{e} \\ \theta_2 &= \frac{sk+k_i}{s} = \frac{W\lambda_1}{2(36x10^{-6})\lambda_4}\dot{y}\ddot{e} + \frac{0.1W\lambda_1}{2(36x10^{-6})\lambda_5}y\ddot{e} \end{aligned}\right\} \quad (29)$$

The adjustable parameter for MRAC design using Lyapunov method:

$$\left.\begin{aligned} \frac{d\theta_1}{dt} &= \dot{\theta}_1 = -\gamma_1 er \\ \frac{d\theta_2}{dt} &= \dot{\theta}_2 = \gamma_2 ey \end{aligned}\right\} \quad (30)$$

The closed loop system block diagram designed as shown in figure 3 below:

IOP Conf. Series: Materials Science and Engineering **298** (2018) 012028 doi:10.1088/1757-899X/298/1/012028

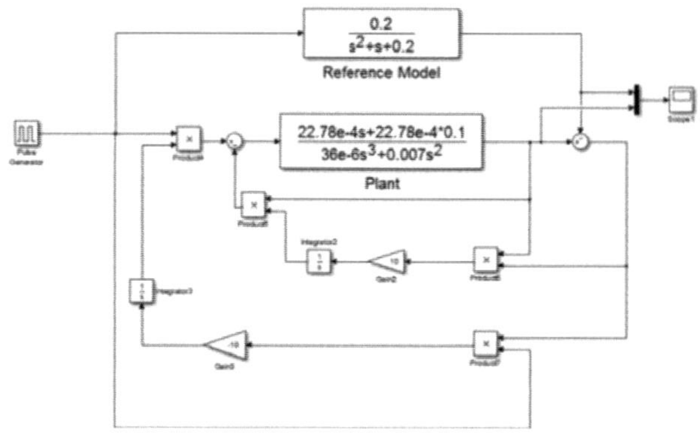

Figure 3. Block diagram for the closed loop adaptive system using the Lyapunov method.

3.2. GMVC controller

In this section we look at how the self-tuning GMVC is used to control the RTP. In self-tuning GMVC it is not necessary to know the exact plant transfer function since the controller parameters are estimated directly.

The discrete-time transfer function can be obtained from converted continuous transfer function in previous 3rd order linear model. The plant can be represented as:

$$G_p(z^{-1}) = \frac{z^{-k}B(z^{-1})}{A(z^{-1})} = \frac{z^{-1}(157.07W-142.78Wz^{-1})}{1-2z^{-1}+2z^{-2}-3.58x10^{-85}z^{-3}} \tag{31}$$

In GMVC controller, a pseudo output as in Eq. (31) is introduced

$$\emptyset(t+k) = Py(t+k) - Rw(t) + Qu(t) \tag{32}$$

where: $y(t+k)$ = Output, $w(t)$ = Set point or Servo signal, $u(t)$ = Input, and P, R, Q value = set by designer.

In GMV controller, the identity is given by equation (33) below:

$$AE + z^{-k}G = PC \tag{33}$$

where

$$\left.\begin{aligned} E &= 1 + e_1 z^{-1} + \cdots e_{n_e} z^{-n_e} \\ G &= g_0 + g_1 z^{-1} \ldots . + g_{n_g} z^{-n_g} \end{aligned}\right\} \tag{34}$$

For a unique solution, the following are ensured:

$$\left.\begin{aligned} n_e &= k - 1 \\ n_g &= max(n_a - 1, n_p + n_c - k) \end{aligned}\right\} \tag{35}$$

The GMV controller law used:

$$u(t) = \frac{H}{F}\, w(t) - \frac{G}{F}\, y(t) \tag{36}$$

where, $F = BE + QC$ and $H = RC$.

Figure 4 illustrates the general structure of a GMV controller block diagram (S.F. Sulaiman et al., 2013).

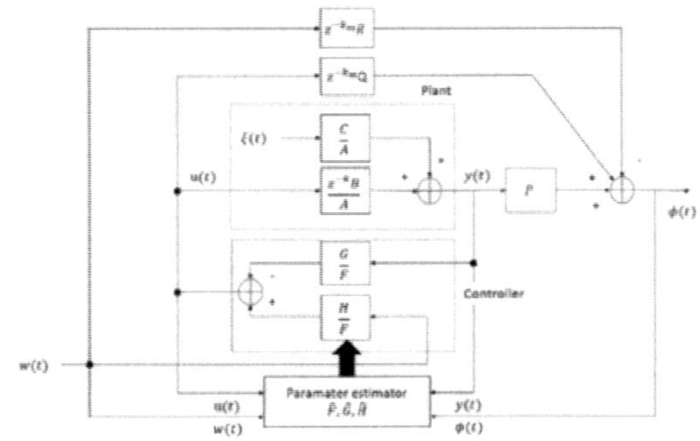

Figure 4. The structure of the Self-Tuning GMVC

The self-tuning GMV controller algorithm used in this study can be summarized as follows, at each controller sample time *t*:

- Step 1

Based on equation (33), equation as below is determined:

$$\emptyset(t) = Py(t) - Rw(t-k) + Qu(t-k) \tag{37}$$

- Step 2

Estimate the controller parameter $\hat{F}, \hat{G}$ and $\hat{H}$ using the recursive least square (RLS) estimation algorithm.

Replacing k=1, equation (37) can be written in regression from as below:

$$\emptyset_{RLS}(t) = [y(t-1)\ y(t-2)\ y(t-3)\ u(t-1)\ u(t-2) - w(t-1)]\begin{bmatrix}\hat{g}_0\\ \hat{g}_1\\ \hat{g}_2\\ \hat{f}_1\\ \hat{f}_2\\ \hat{h}_0\end{bmatrix}$$

- Step 3

Calculate and apply the GMVC control to the system.

$$u(t) = \frac{\hat{H}}{\hat{F}}\ w(t) - \frac{\hat{G}}{\hat{F}}\ y(t) \tag{38}$$

- Step 4

The algorithm is repeated for the next iteration or sampling time.

As mentioned earlier in (32), the value of parameters *P*, *Q* and *R* was set by the designer. Thus, in this work, the value of *P* and *R* is assumed to be 1, while *Q* value to be 0.64.

4. Results and analysis

The MRAC Lyapunov approach is simulated using Matlab SIMULINK with the reference input of a square wave signal with the amplitude of 1 and frequency 0.0025Hz. Figure 5 shows the outputs for *y* and y_m of the system with model reference adaptive controller designed based on the Lyapunov method, with the same adaptation gain value, γ_1 and γ_2, of 10 respectively.

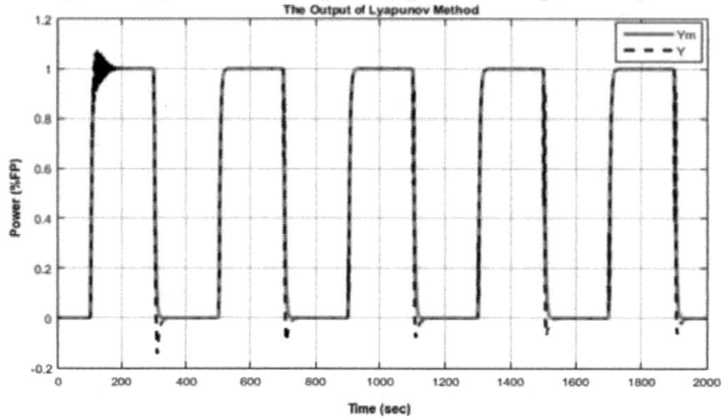

Figure 5. Output signal of the system using the Lyapunov method.

Based on observation from the above output signal, designing an MRAC using the stability approach will ensure a stable closed loop system.

Figure 6 shows the output performance of the GMV controller. From a simulation result obtained, response from Self-Tuning GMV controller gives the best stability result and allowable value of percentage overshoot which is only less than 10% of overshoot.

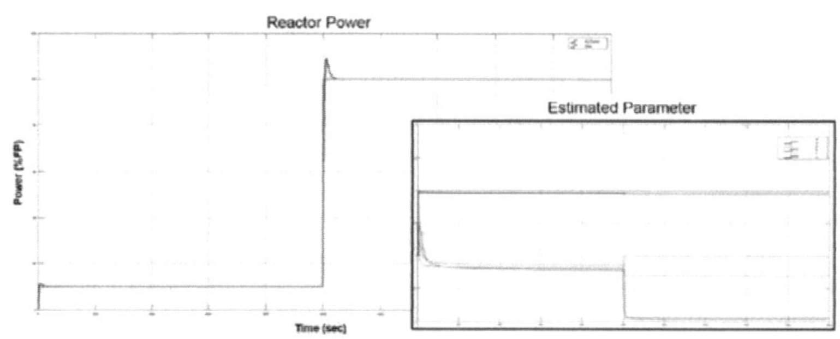

Figure 6. Output response of GMV controller

4.1. *Adaptive Controller Applied to RTP Model*

Dynamic equations of a RTP are nonlinear equations. So some part of complexity of designing a controller and getting answer is because of its nonlinearity. The system has ten states and the ten state equations which increases the complexity.

In our designs the controller is designed in a classic way (a linear controller) and the attempt was to use the reactor model as precise as possible. So we have used all ten states completely and we have used the precise values of coefficients in non-linear RTP model.

The performance of MRAC design based on the reference model shown in fFigure 7. The rising time was very short and rated of power change more than 33.3% per second and overshoot more than 110% FP that causing the reactor to TRIP. However, if the TRIP limit parameters are not considered in this simulation the reactor power is still converges to zero and reached steady-state. The error between the reference model output and the reactor output become relatively large at the moment of sudden change of Power Demand (PDM) at 1000 seconds, but eventually converges to zero quickly. Based on this result shown that modification of MRAC are needed during transient.

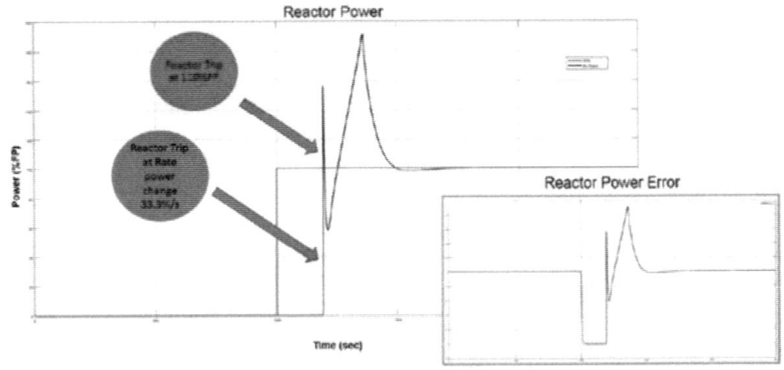

Figure 7. Reactor power performance of MRAC using RTP model

In Self-Tuning GMVC, it is not necessary to know the exact plant transfer function since the controller parameters *F*, *G* and *H* are estimated directly. Figure 8 shows the result of Self-Tuning GMVC that applied to the RTP model. The reactor power increase rapidly which is possible, since there are no limitation on the rate power change consider in the algorithm. The TRIP also happen as the MRAC and modification of GMVC are compulsory.

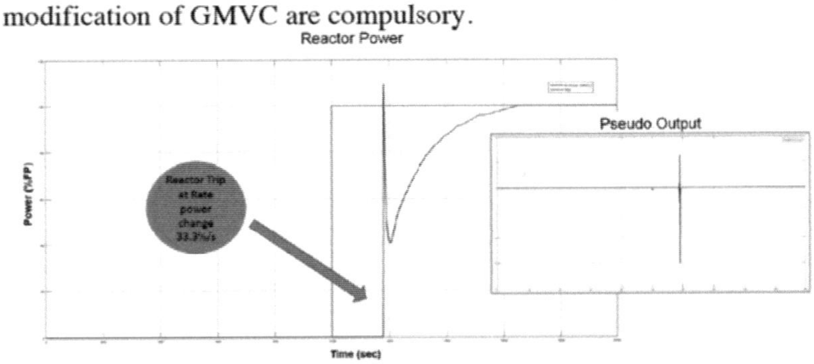

Figure 8. Reactor power performance of GMVC using RTP model

4.2. Modification of Adaptive Controller

The modification of MRAC and GMVC are for safety and operation purpose. Basically the reactor power tracking power during transient is the main issues to be considered. The current controller which is FCA controller was designed based on control theory and calculate the motor control command using error between actual power and PDM. The rate of power change limitation and increment power limitation included during the design of FCA.

The drawback of FCA is slow tracking compare to the adaptive controller. The proposed combination of adaptive and FCA should produce better performance during transient and steady-state. The propose controller for modified MRAC (M-MRAC) shown in figure 9 [26], while for the modified GMVC (M-GMVC) shown in figure 10.

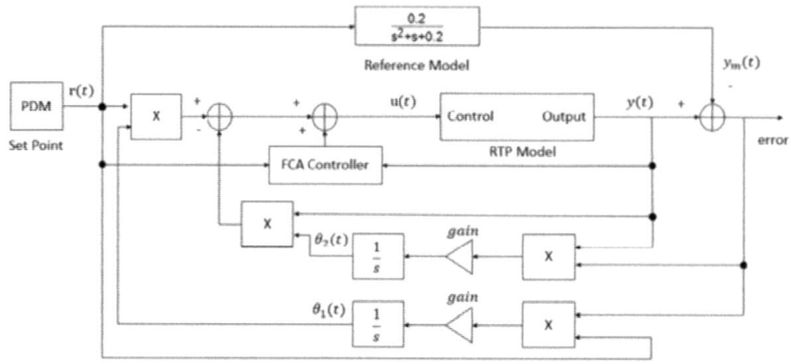

Figure 9. Propose controller M-MRAC

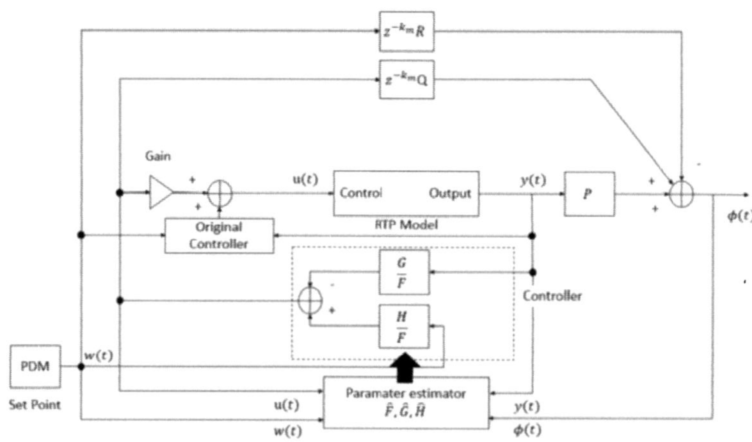

Figure 10. Propose controller M-GMVC

4.2.1. Controller MRAC:

$$u_{MRAC} = -ky + \frac{k_i}{s}(r - y) = \theta_1 r - \theta_2 y \tag{39}$$

4.2.2. Controller M-MRAC:

$$u_{M-MRAC} = u_{MRAC} + u_{FCA} \tag{40}$$

4.2.3. Controller GMVC:

$$u_{GMVC} = gain * \left(\frac{G}{F}y - \frac{H}{F}w\right) \tag{41}$$

4.2.4. Controller M-GMVC:

$$u_{M-GMVC} = u_{GMVC} + u_{FCA} \tag{42}$$

Figure 11 shows the result of estimated self-tuning M-GMVC controller parameter. The sampling time used is 0.2s. The solid lines that indicate in yellow, purple, cyan, red, green and blue colours are shows the estimated values, $\hat{g}_0, \hat{g}_1, \hat{g}_2, \hat{f}_1, \hat{f}_2$ *and* $\hat{h}_0$ over the simulation period. It can be observed that the estimated, whose initial value is $\theta(t) = 0$, slowly converge to the actual values showing that the RLS estimator has effectively estimated the system parameters.

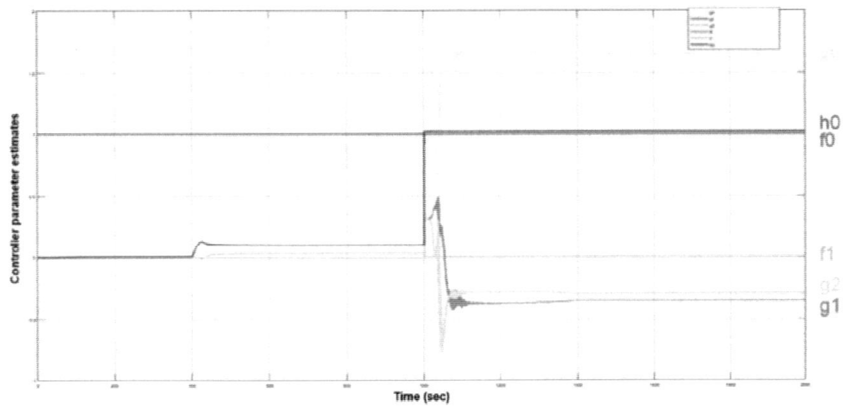

Figure 11. Estimated self-tuning M-GMVC controller parameters.

The result of feedback controller and modified adaptive controller shown in Figure 12. From the graph below, the set point PDM signal which is unit step indicate in the red colour. While the blue represent the FCA controller that need the improvement in this research. The green colour for M-MRAC and black colour for M-GMVC both present the improvement result after modification.

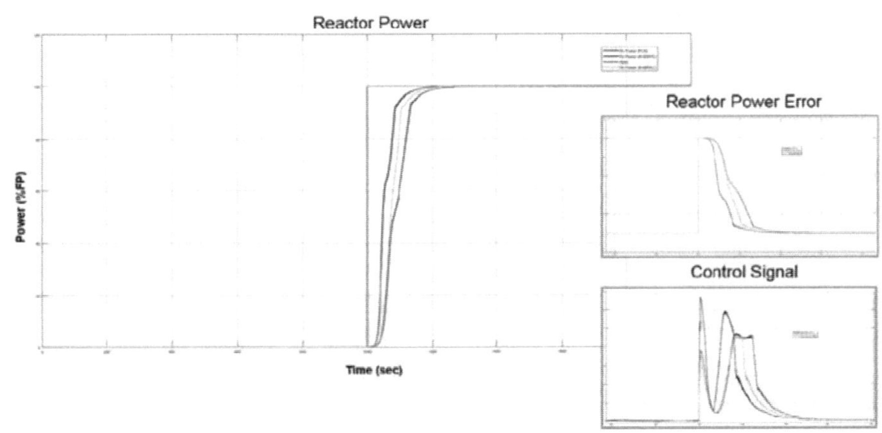

Figure 12. Feedback Controller vs Adaptive Controller.

The summary performance of three controller shown in table 3 quantitatively. Based on table 3, clearly describe the behaviour of each controller. Adaptive controller provide a good result in term of rise time and settling time. However, both type of controller good in eliminating error and make the reactor power output follow the PDM at steady-state.

Table 3. Comparison Transient and Steady State Response of Power Control System at 1MW.

Properties	Feedback Controller	Adaptive Controller	
	Controller 1 (FCA)	**Controller 2 (M-MRAC)**	**Controller 3 (M-GMVC)**
Percent Steady-State Error ($\%e_{ss}$)	0.087%FP	0.235%FP	0.130%FP
Settling Time (T_s)	181.994 s	158.171 s	134.801 s
Percent Overshoot (%)	0.491 %	0.471 %	0.474 %
Rise Time (T_r)	76.740 s	53.902 s	48.124 s

5. Conclusion

A dynamic model of the RTP was developed in order to design FCA and adaptive controllers. Experimental results of the controller in FCA from RTP were compared with the simulation results FCA and adaptive controller (M-MRAC and M-GMVC) showing the goodness of the model. The analysis shows that an M-GMVC offers generally better results than the FCA which reduce settling time up to 25% of the nominal settling time. However further investigations and experimental activities are necessary in order to reach the best optimum of the new adaptive controller for RTP.

6. References

[1] Malaysia Nuclear Agency 2014 Chapter 5: The reactor *safety analysis report of PUSPATI TRIGA reactor* pp 5-1 – 5-34

[2] Malaysia Nuclear Agency 2014 Chapter 8: instrumentation and control systems *safety analysis report of PUSPATI TRIGA reactor* pp 8-1 – 8-24

[3] Reactor TRIGA PUSPATI log book operation (Malaysia Nuclear Agency July 2, 2015)

[4] Minhat M S, Dane D R B, Shick D R S B 2013 Development of a new control algorithm for power controller simulation for 1MWatt research reactor (IAEA)

[5] di Tigliole A Borio, Cammi A., Gadon M A, Magrotti G, Memoli V 2009 Study of a new automatic reactor power control for the TRIGA Mark II reactor at University of Pavia ANIMMA Int. Conf (Marseille, France, June 6-7, 2009).

[6] Rivero-Gutierrez T, Benitez-Read J S, Segovia-De-Los-Rios A, longoria-Gandara L C and Palacios-Hernandez J C 2012 Design and implementation of a fuzzy controller for a TRIGA Mark III reactor *Science and Technology of Nuclear Installations*

[7] Perez-Cruz J H, Chairez I, Poznyak A, De Jesus Rubio J 2011Constrained neural control for the adaptive tracking of power profiles in a TRIGA reactor *Int. Journal of Innovative Computing, Information and Control*

[8] Alang Md Rashid N K 1992 Nuclear reactor control using tunable fuzzy logic controllers (PhD Thesis The University of New Mexico, Albuquerque, New Mexico USA)

[9] Baang D, Yong S S , Young K K 2013Research reactor power control system design by Matlab/Simulink (IGORR)

[10] Alang Md Rashid N K 2015 Modeling Nuclear Process by Simulink (International Islamic University Malaysia AIP Publishing LLC) 978-0-7354-1301-6

[11] MATLAB/SIMULINK User's Guide, Mathwork, 2001

[12] Cammi A, Ponciroli R, di Tigliole A B, Magrotti G, Prata M, Chiesa D and Previtali E 2013A zero dimensional model for simulation of TRIGA Mark II dynamic response

[13] Baang D, Kim S, Noh T, Park J Y, Kim Y K 2012Adaptive Power Control for Nuclear Reactor Korea Atomic Energy Research Institute, Korea, Hongik University Seoul, Korea, The 18th Pacific Basin Nuclear Conf., Busan, Korea, March, 2012)

[14] Baang D, Suh Y and Park C 2015 Power Control Method for Research Reactor, Korea Atomic

Energy Research Institute *Transactions of the Korean Nuclear Society Spring Meeting Jeju*, Korea, May 2015

[15] Alavi Z S, Menhaj M B, Eliansi H 2009 Model Reference Adaptive Control of a Nuclear Reactor University of Iran, *Proc. of the 2009 IEEE*

[16] Malaysia Nuclear Agency 2014 Chapter 17 Operational Limits and Conditions *Safety Analysis Report of PUSPATI TRIGA Reactor* pp 17-8

[17] Ogata K 2002 *Modern Control Engineering* Prentice Hall

[18] Shaffer R A, Edwards R M and Lee K Y 2005 Design and Validation of Robust and Autonomous Control for Nuclear Reactor (The US Nuclear Regulatory Commission and The Pennsylvania State University, USA)

[19] Power M A and Edwards R M Automatic Control of the TRIGA Reactor (Experiment B#6, The Pennsylvania State University)

[20] Schultz M 1950 Automatic Control of Power Reactor (Westinghouse Atomic Power Division, November 6, 1950)

[21] Sulaiman S F, Abdul Rahim H R, Johari S H, Osman K, Zainal Abidin AF and Rahmat M F 2013 Optimizing the Process Parameters of GMV Controller by PSO Tuning Method, *Australian J. of Basic and Applied Science* (Universiti Teknikal Malaysia Melaka) **7(9)** 44-50 ISSN 1991-8178

[22] Malaysia Nuclear Agency, Reactor TRIGA PUSPATI Log Book Operation (April 7, 2016)

[23] Selamat H and Yusof R 2014 Introduction to Adaptive and Self-Tuning Control (Universiti Teknologi Malaysia)

[24] Selamat H, Yusof R and Goodall R M 2008 Self-Tuning Control for Active Steering of a Railway Vechicle aith Solid-Axle Wheelsets *IET Control Theory & Applications J.* **2(5),** 374-383

[25] Sapiee M R, Abdullah F, Noordin A and Jahari A N PI Controller Design Using Model Reference Adaptive Control Approaches for a Chemical Process *Proc. 2008 Student Conference on Research and Development* (SCOReD 2008 Universiti Teknologi Malaysia, November 2008)

[26] Pawar R J and Parvat B J 2015 Design and Implementation of MRAC and Modified MRAC technique for Inverted Pendulum *Int. Conf. on Pervasive Computing (ICPC)* (Pravara Rural Engineering College,Loni, India)

[27] Chen C T 1999Linear System Theory and Design (University Press.)

[28] Slotine J J E and Li W 1991 Chapter 8 Adaptive Control *Applied Nonlinear Control* (Prentice-Hall International Press.)

[29] Chen-Kuo Weng, Robert M. Edwards, Asok Ray, "Robust Wide-Range Control of Nuclear Reactors by Using the Feedforward-Feedback Concept", The Pennsylvania State University, March 4, 1994

[30] Edwards R M, Lee K Y and Ray A 1991 Robust Optimal Control of Nuclear Reactors and Power Plants (The Pennsylvania State University, March 1991)

[31] Mohd Nor A, Ab. Muin M Z and Adom A H 2013 Modelling and Adaptive Control of Palm Oil Biodiesel Engine (PhD Thesis, Universiti Malaysia Perlis)

[32] Lavretsky E Adaptive 2008 Control: Introduction, Overview, and Applications (IEEE Robust and Adaptive Control Workshop June 9, 2008)

7. Acknowledgements

This work was supported by a ScienceFund grant (03-03-01-SF0262) from Ministry of Science, Technology and Innovation (MOSTI), Malaysia. The author would like thank all the individuals who has contribute in this paper especially from my supervisors Assoc. Prof. Ir. Dr Hazlina Binti Selamat and Dr Nurul Adilla Binti Mohd Subha, my parents, my friends and to the Malaysian Nuclear Agency for the support.

Measurement and simulation of thermal neutron flux distribution in the RTP core

Mohamad Hairie B. Rabir[1], Abi Muttaqin B. Jalal Bayar[1], Na'im Syauqi B. Hamzah[1], Muhammad Khairul Ariff B. Mustafa[1], Julia Bt. Abdul Karim[1], Muhammad Rawi B. Mohamed Zin[1], Yahya B. Ismail[1], Mohd Huzair B. Hussain[1], Mat Zin B. Mat Husin[1], Roslan B. Md Dan[1], Ahmad Razali B. Ismail[1], Nurfazila Bt. Husain[1], Zareen Khan B. Abdul Jalil Khan[1], Shaiful Rizaide B. Mohd Yakin[1]Mohamad Fauzi B. Saad[1], Zarina Bt. Masood[1]

[1]Reactor Technology Centre, Technical Support Division, Malaysian Nuclear Agency, Bangi, 43000 Kajang, Selangor, Malaysia

m_hairie@nuclearmalaysia.gov.my

Abstract. The in-core thermal neutron flux distribution was determined using measurement and simulation methods for the Malaysian's PUSPATI TRIGA Reactor (RTP). In this work, online thermal neutron flux measurement using Self Powered Neutron Detector (SPND) has been performed to verify and validate the computational methods for neutron flux calculation in RTP calculations. The experimental results were used as a validation to the calculations performed with Monte Carlo code MCNP. The detail in-core neutron flux distributions were estimated using MCNP mesh tally method. The neutron flux mapping obtained revealed the heterogeneous configuration of the core. Based on the measurement and simulation, the thermal flux profile peaked at the centre of the core and gradually decreased towards the outer side of the core. The results show a good agreement (relatively) between calculation and measurement where both show the same radial thermal flux profile inside the core: MCNP model over estimation with maximum discrepancy around 20% higher compared to SPND measurement. As our model also predicts well the neutron flux distribution in the core it can be used for the characterization of the full core, that is neutron flux and spectra calculation, dose rate calculations, reaction rate calculations, etc.

1. Introduction

The Malaysian 1MW PUSPATI TRIGA Reactor (RTP) was designed to effectively implement the various fields of basic nuclear research and education. It incorporates facilities for advanced neutron and gamma radiation studies as well as for isotope production, sample activation, and student training. RTP has reached its first criticality on 28 Jun 1982 with excess reactivity of $ 0.15. The cross-sectional view of Core-15, the latest core configuration is shown in figure 1. It has annular core surrounded by graphite reflector and cooled by natural convection. Elements are arranged in seven circular rings and the spaces between the fuel rods are filled with water that acts as coolant and moderator.

The core consists of 8.5 wt. %, 12 wt. % and 20 wt. % fuel elements, 4 control rods, some graphite elements and central thimble. RTP uses standard TRIGA $UZrH_{1.6}$ fuel with 20 % of U-235 enrichment. The fuel dimension is shown in figure 2. RTP has 4 control rods which are made up of

IOP Conf. Series: Materials Science and Engineering **298** (2018) 012029 doi:10.1088/1757-899X/298/1/012029

boron carbide. Three of them are from fuel follower type and the other is air follower. The fuel follower control rods (FFCR) made up of 8.5 wt. % $UZrH_{1.6}$ and B_4C absorber on top of the fuel section.

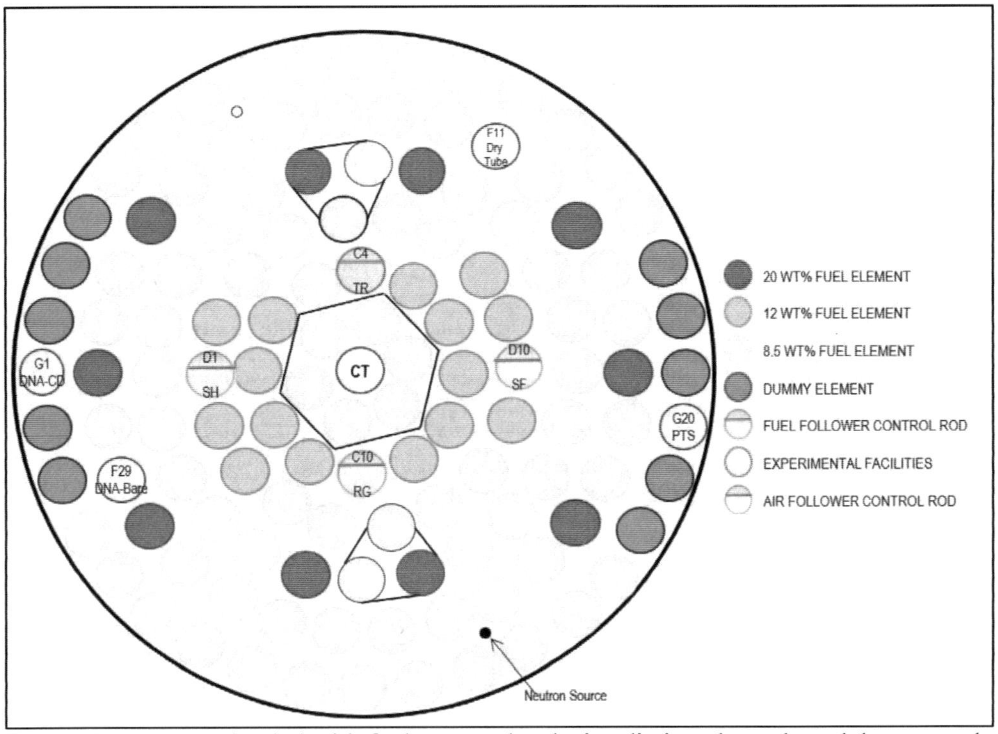

Figure 1. RTP core loaded with fuels, control rods, irradiation channels and dummy rods

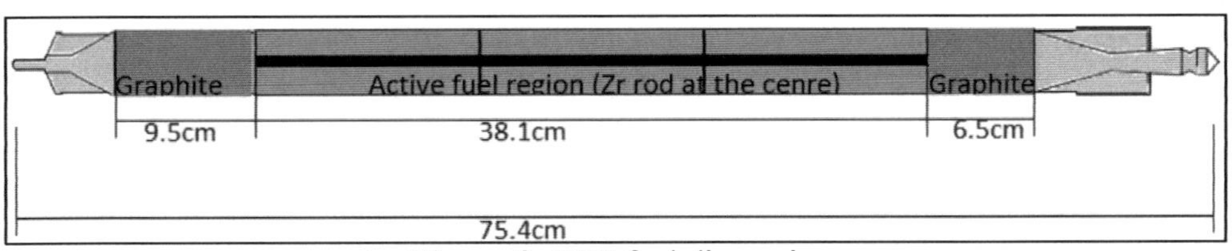

Figure 2. RTP fuel dimension

Most reactors use thermal neutrons to sustain the fission chain reaction. These so called thermal reactors contain neutron moderator that slows down fast neutrons to thermal equilibrium with the atoms ($E < 1$ eV) in the system. Neutron flux (ϕ) is to consider it to be the total path length covered by all neutrons in one cubic centimeter during one second. Mathematically, this is the equation,

$$\Phi = nv\,\phi \tag{1}$$

where,

ϕ: neutron flux (neutrons cm^{-2} s^{-1}),
n: neutron density (neutrons cm^{-3}) and.
V:neutron velocity (cm s^{-1}).

As a very good approximation, the distribution of neutrons inside a nuclear reactor core can be determined using the two group diffusion equations as shown in figure 3. This flux distribution is valid in a homogeneous core. In reality, the distribution of neutrons is greatly affected by different elements in the core and reflector. Thus, to predict the detail neutron flux distribution in real reactor core arrangement, we rely on measurements or the use of Monte Carlo neutron transport method.

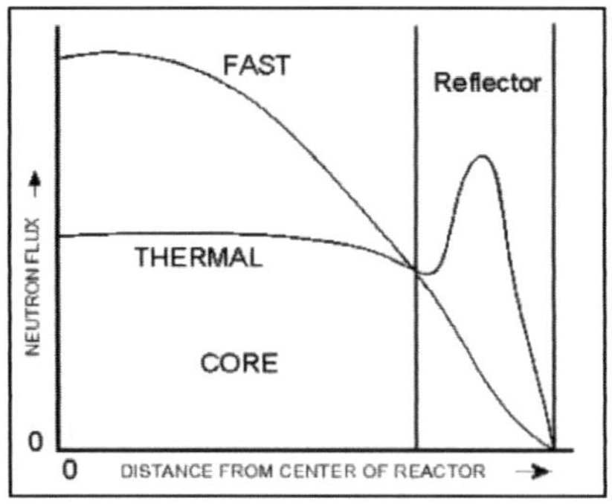

Figure 3. Fast and thermal neutron distribution in core and reflector estimated using the two group diffusion equation.

The data and information on the in-core neutron flux distribution is crucial in research reactors. It is an input variable for many experiments and important for determining the distribution of in-core power. In this work, the radial flux density distribution of the thermal neutrons is measured in RTP core. The used method, i.e. on-line self-powered neutron detector, is one of the most important methods in nuclear technology. The results will then be compared to the Monte Carlo neutron transport code model of the core. This measurement plays a vital role as part of RTP neutronics properties analysis to support the improvement of in-core management strategy. A continuous improvement of core and fuel management strategy is crucial element needed for better utilization and to ensure safe operation of RTP.

2. Method

Self-powered neutron detector (SPND) is a unique type of neutron detector that is widely applied for in-core flux measurement. These devices incorporate a material chosen for its relatively high cross section for neutron capture leading to subsequent beta or gamma decay. In its simplest form, the detector operates on the basis of directly measuring the beta decay current following capture of the neutrons. There is no external bias voltage needed to be applied to the detector, hence the name self-powered. SPND is small in size, low cost, and relatively simple electronics required in conjunction with their use.

The SPND used in RTP consists of 3 mechanical components which is a coaxial cable consisting of an inner electrode (emitter), surrounded by insulator and an outer electrode (collector), as shown in figure 4. The emitter or beta source material absorbs neutrons and emits beta particles. An insulator electrically isolates the beta source material from the collector while collector absorbs the emitted beta particles as illustrated in figure 5. Vital material in this experiment is vanadium (^{51}V) as an emitter, which has a neutron-beta interaction with thermal neutron cross-section of 5 barns (5×10^{-24} cm^2) featuring a $1/v$ characteristic without resonances in the energy range of thermal/epithermal neutrons. Vanadium will react with thermal neutron and produce ^{52}V which is in excited state, based on $^{51}V(n,\gamma)^{52}V$ reaction. The ^{52}V then decay with a half-life of about 3.76 minutes to a stable nuclide which is chromium (^{52}Cr) with the emission of a beta particle. Due to this decay time, the SPND signal needs to be corrected using the following formalism.

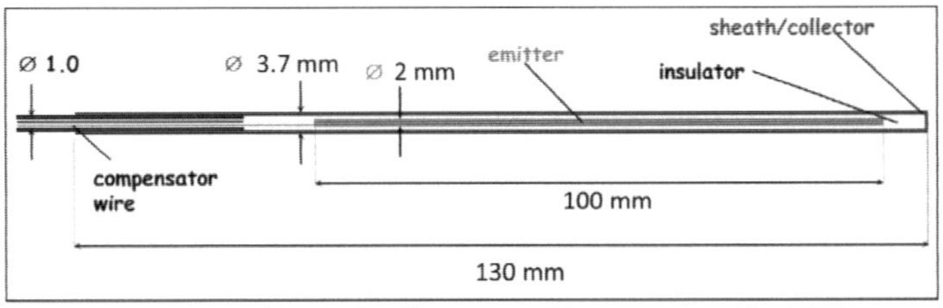

Figure 4. Schematic diagram of SPND used in RTP.

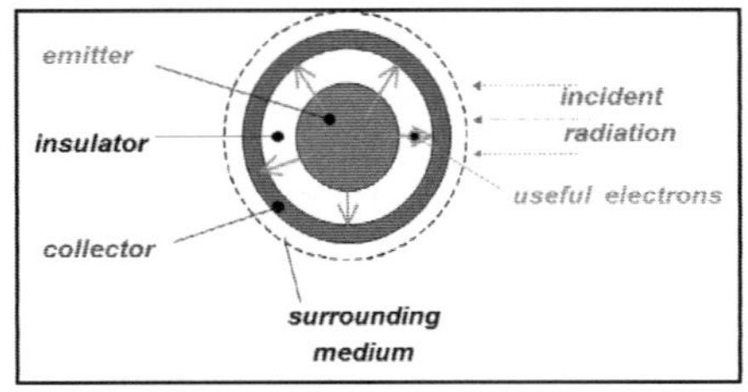

Figure 5. Illustration of emitted electron produced in emitter.

2.1. SPND signal

Due to decay time of emitter, SPND signal needs to be corrected. The basic equation of the population of ^{52}V is the following:

$$\frac{dN(t)}{dt} = \sigma N_0 \varphi(t) - \frac{N(t)}{\tau} \tag{2}$$

with,

φ : neutron flux,
N_0 : atom density of ^{51}V,
σ : neutron absorbtion cross section of ^{51}V to produce ^{52}V (=4.9b),
τ : characteristic time decay of ^{52}V,

and τ , is given by:

$$\begin{aligned} \tau &= \frac{T_{1/2}}{\ln(2)} \\ &= \frac{3.75\,\text{min}}{0.693} = 5.41\,\text{min}. \end{aligned} \tag{3}$$

Signal response correction via digital compensation formalism is given by:

$$\begin{aligned} \varphi_1 &= CI_1 \\ \varphi_{1+1} &= C\left[I_{i+1} - \exp\left(-\frac{\Delta}{\tau}\right)I_i\right] + \exp\left(-\frac{\Delta}{M\tau}\right)\varphi_i \end{aligned} \tag{4}$$

with Δ as the sampling time and C as a normalisation constant and, is given by:

$$\begin{aligned} C &= \frac{1-\exp\left(-\dfrac{\Delta}{M\tau}\right)}{1-\exp\left(-\dfrac{\Delta}{\tau}\right)} \\ M &= \frac{1}{1+\dfrac{K_d}{K_p}} = \frac{K_p}{K_p + K_d} \end{aligned} \tag{5}$$

IOP Conf. Series: Materials Science and Engineering **298** (2018) 012029 doi:10.1088/1757-899X/298/1/012029

Here, K_d is the probability that a ^{52}V decay produces an electron contributing to the current and, K_p as the as the probability that a neutron captured by ^{52}V decay leads instantaneously to an electron contributing to the current.

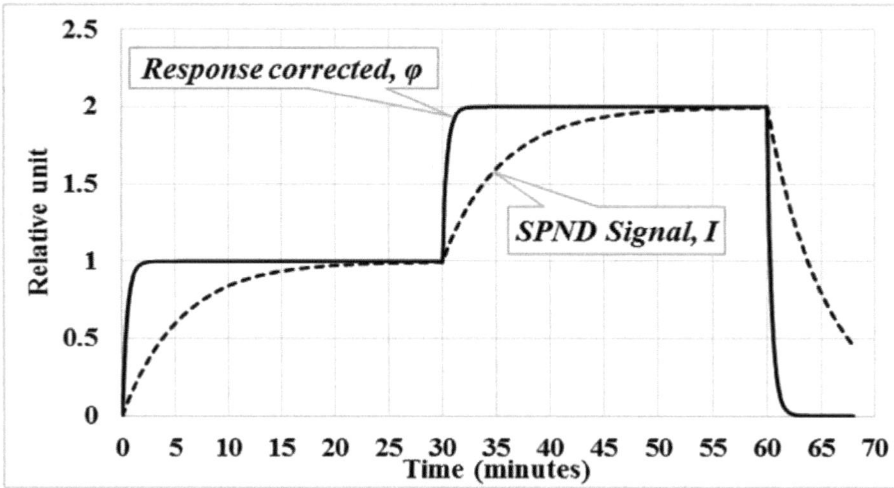

Figure 6. Example of corrected response

2.2. Neutron source

The MCNP tally results are relative, normalised to one source particle. It needs to be scaled to a desired fission source (power) level to get absolute value (flux, reaction rate, fission density, etc). Typical source strength for 1MW power, ***S*** and scaling method is given by:

$$S = \frac{P \times \nu}{Q} \tag{6}$$

where,

P : reactor power (watt),
ν : neutrons emitted per fission for U-235,
Q : energy released per fission (watt.s)

The scaled flux Φ (n.cm^{-2}s^{-1}), is given by:

$$\Phi = \phi_{F4} \times S \times \frac{1}{k_{eff}} \tag{7}$$

where,

ϕ_{F4} : MCNP F4 tally output (cm^{-2})
S : source strength (n.s^{-1}).

Figure 7 shows the measurement set-up where each SPND is attached to a holder made of an Aluminium hollow rod. The holder then can be fixed to a flux hole on the reactor core top grid plate. The location of flux holes used in this measurement is shown in figure 8 and 9 respectively.

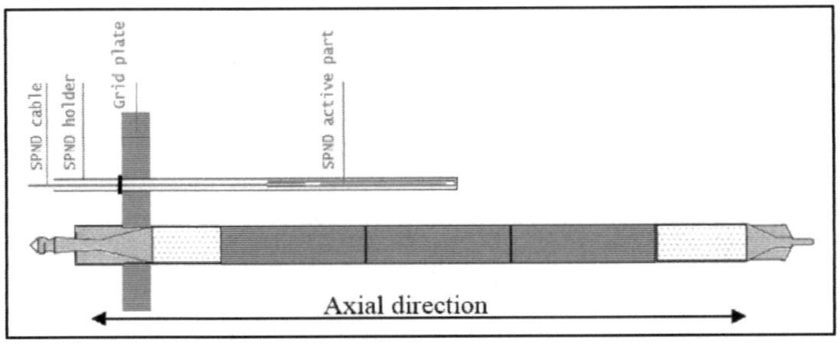

Figure 7. Schematic diagram of RTP fuel with SPND inserted in a holder which left hanging on the top of core grid plate where the axial position can be adjust.

The Monte Carlo N-Paticle Code (MCNP) with ENDF-VII cross section data was used to simulate the thermal neutron flux distribution in the core. There are several works mentioned in reference [1], [2] and [3] on the development of RTP MCNP core calculation. Detail of MCNP calculation of RTP can be found from these references. The RTP core neutronics analysis using MCNP code for current configuration involved burnup and nuclide inventory data. TRIGLAV code produced individual fuel burnup in MWD and then the actinides and fission product build-up calculation performed using MCNPX BURNUP card. The model then was used to re-create thermal neutron flux measurement at the same location of the SPNDs as shown in figure 8 and figure 9.

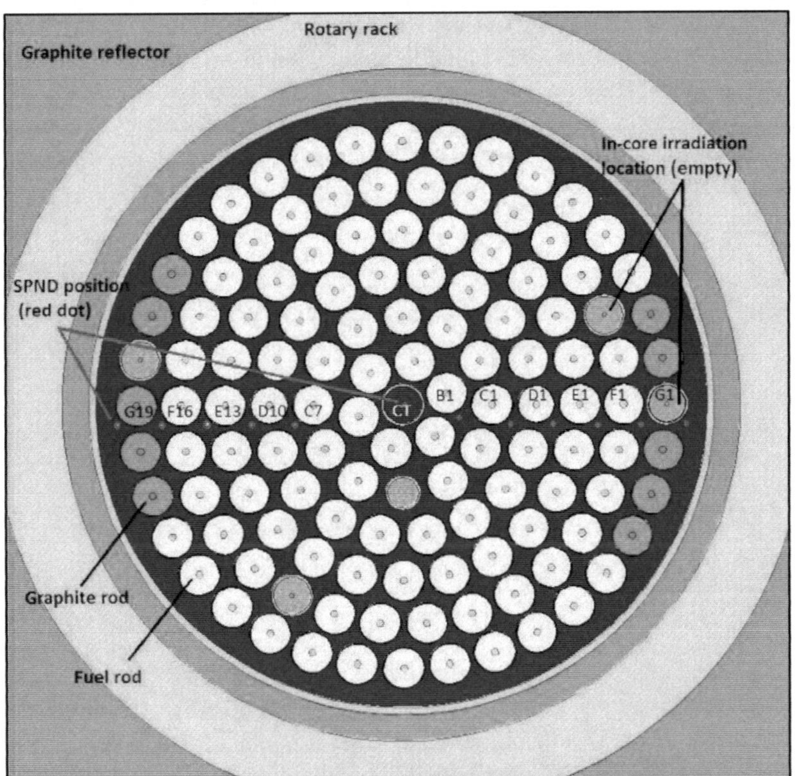

Figure 8: Top view of RTP core showing SPND position.

IOP Conf. Series: Materials Science and Engineering **298** (2018) 012029 doi:10.1088/1757-899X/298/1/012029

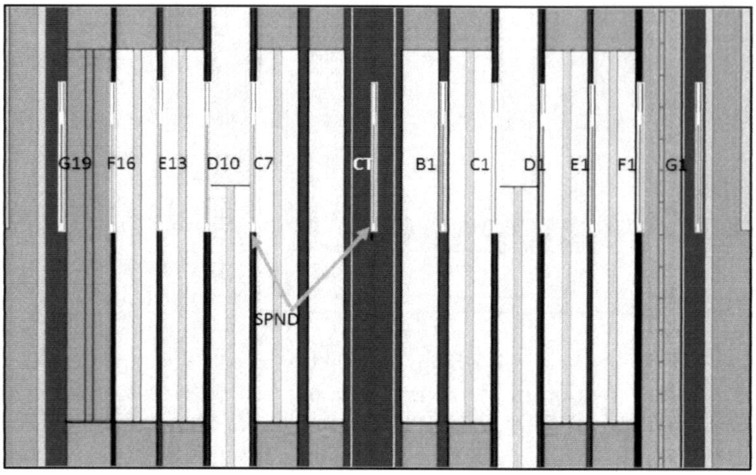

Figure 9: Side view of RTP core model in with actual control rod position during the measurement.

3. Results

Figure 10 shows the measurement results for CT and E1 location at different core power level. Reactor power increased gradually from 15 W to 50 kW, 100 kW, 250 kW, 500 kW, 750 kW and lastly 1 MW. The delayed response with a characteristic half-life value of 3.75 minutes from SPND signal needs to be corrected using special digital compensation method to get actual flux curve as shown in section 2.1. There are several literatures [4][5] on the formulation for the SPND digital compensation method; in this work, we used the provided formulation in reference [6]. The response correction was performed for every SPND signals at every measurement location in the core.

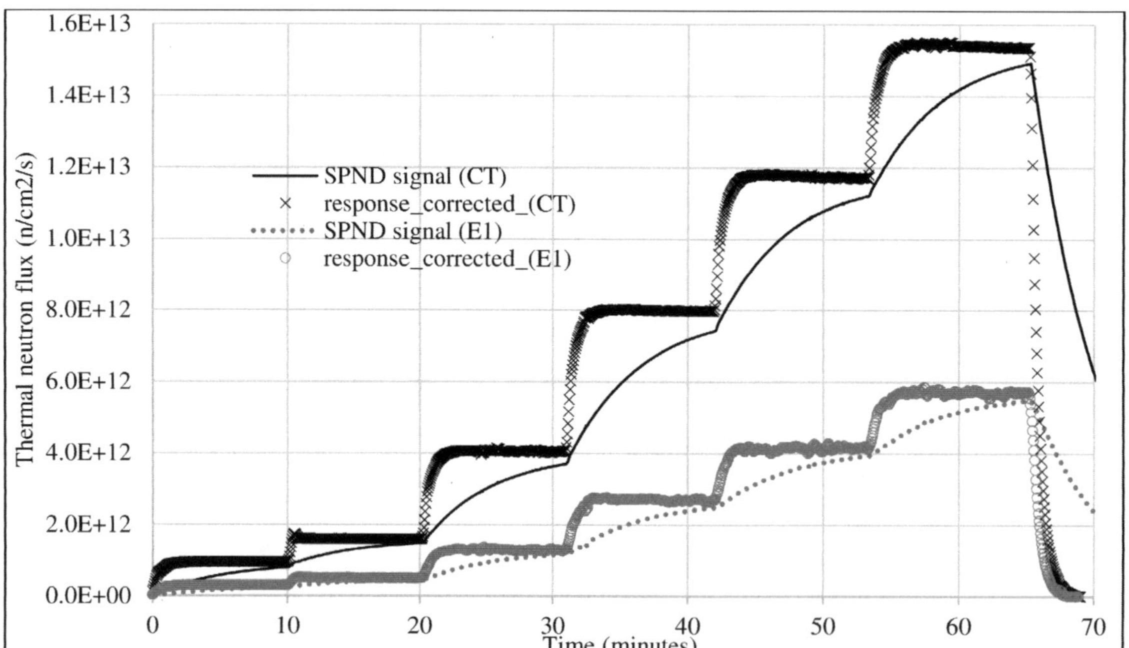

Figure 10. Gradual increase of thermal neutron flux measurement at Central Thimble (CT) and E1 location for gradual increase of core power level from zero power to maximum 1 MW.

Figure 11 shows the spatial thermal neutron flux for the half side of the core. Based on these results, we can see the trend where maximum flux is in the centre and gradually reduced towards the edge due to leakage. The local flux fluctuation is observable heterogeneous core trend. The radial thermal flux profile was found to be non-cosine due to heterogeneity of the core and variation in fuel element types and burnup. The significant peak at the core centre is also because of it has larger water volume compare to other location of measurements in B, C, D, E, F or G ring. Hence more moderation process occur which increased thermal neutron flux at the core centre. The effect of graphite rods on the core flux shape are also noticeable; SPND measurement at the core edge where it was surrounded by graphite elements, cause the flattening of thermal neutron flux.

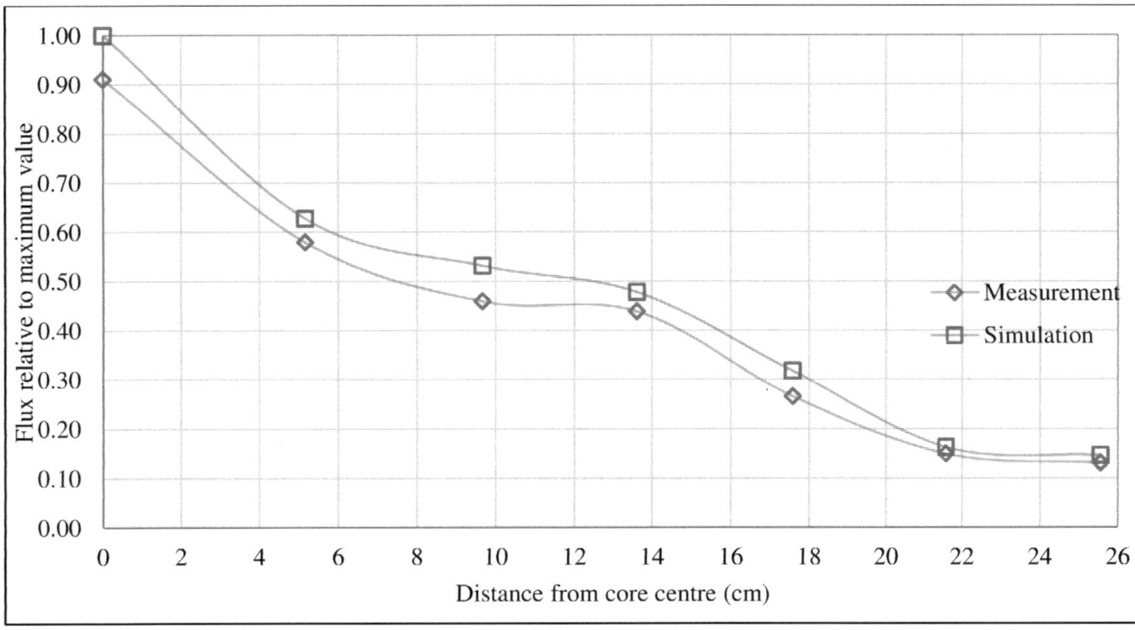

Figure 11. Thermal neutron flux trend normalized to maximum flux value of MCNP simulation and SPND measurement.

Figure 11 also shows the thermal neutron flux radial profile comparison between measured and calculated from MCNP simulation. The thermal neutron flux calculation outputs from MCNP are normalized to reactor power using the method mentioned in section 2.2. MCNP simulation overestimate measured flux at average discrepancy around 10% but the trend for both results are almost the same. There are several factors of this discrepancy and one of them is the linear scaling of power to flux in MCNP. In actual phenomena, neutron flux does not increase linearly with power due to the strong negative temperature feedback as shown in measurement curve in figure 12. Measured fuel temperature at full power level was around 300 °C to 400 °C.

The radial thermal neutron flux distribution (integrated over axial active region) simulation results using RTP MCNP model is shown in figure 13. Notice that large depressions of thermal flux inside the fuel elements, especially the several locations in the F-ring filled with 20 wt. % fuel type. Local thermal flux peaks inside the water gaps between the fuel elements are also recognizable. The calculated excess reactivity is \$5.20 while measured data at the beginning of the 15th core configuration is \$5.18.

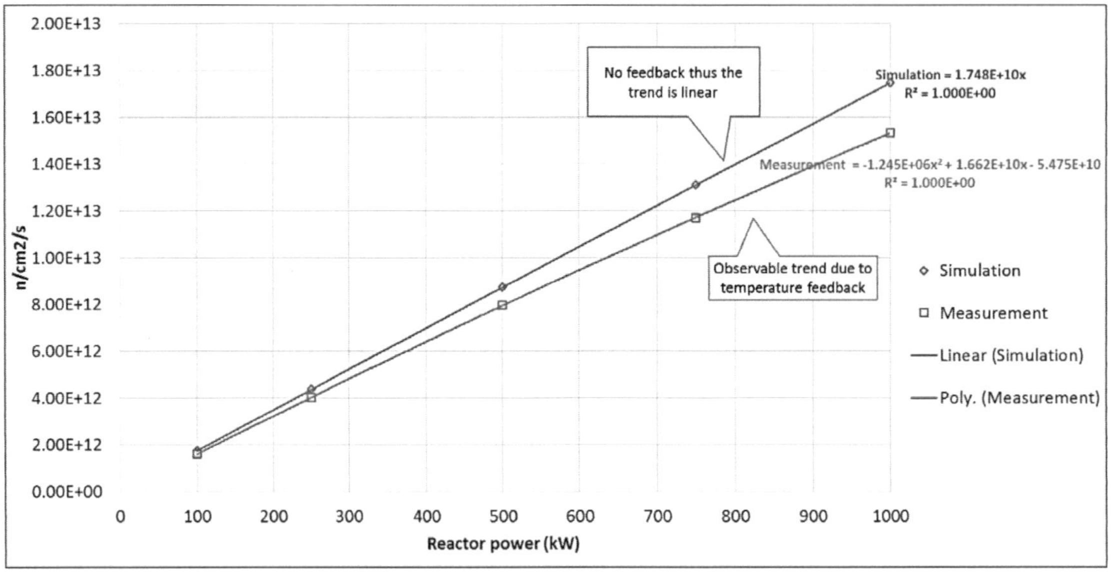

Figure 12. Thermal neutron flux in CT at different reactor power level. Comparison between MCNP simulation linear curve and SPND measured neutron flux with polynomial curve.

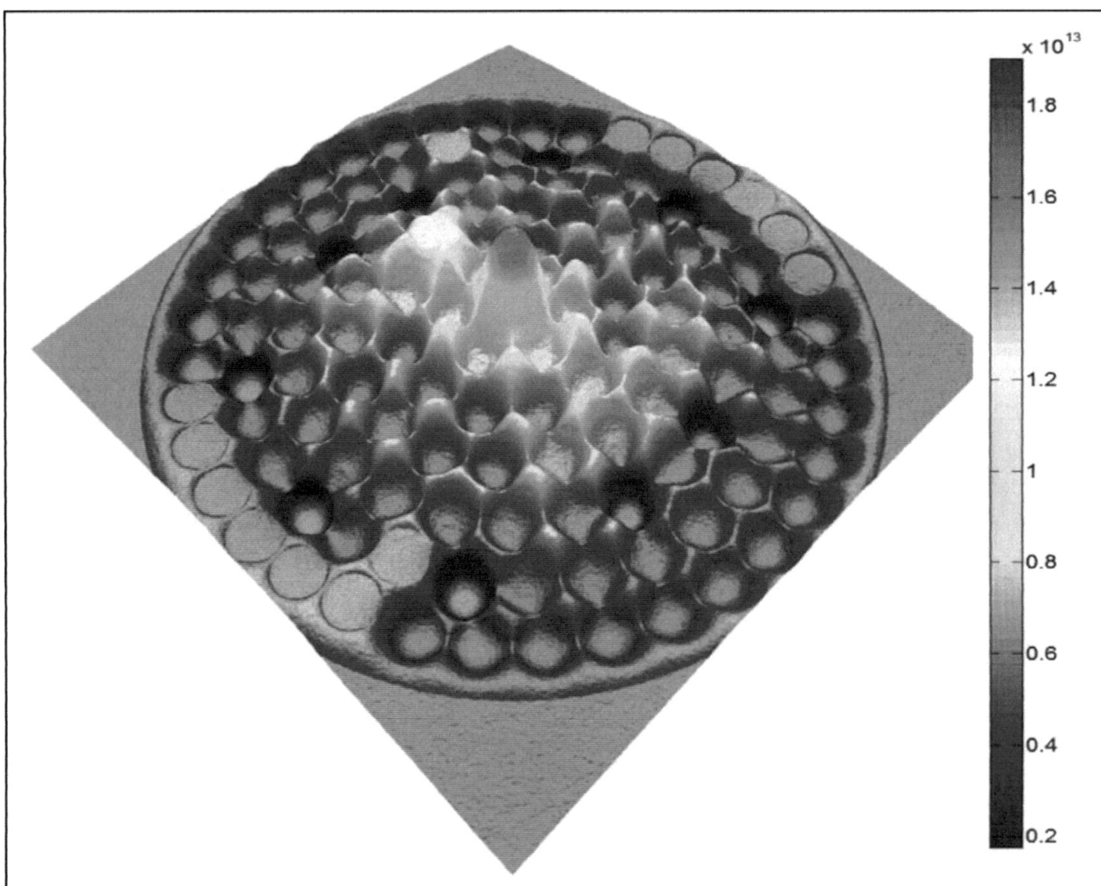

Figure 13. MCNP simulation of thermal neutron flux radial distribution at 1 MW core power level.

IOP Conf. Series: Materials Science and Engineering **298** (2018) 012029 doi:10.1088/1757-899X/298/1/012029

4. Conclusions

Based on the measurement, the thermal neutron flux profile peaked at the centre of the core and gradually decreased towards the outer side of the core. This distribution resulted shows a clear heterogeneous core effect that could not be predicted using neutron diffusion method. Thus a more advanced code; MCNP Monte Carlo simulation was needed due its ability to predict complex core arrangement. Comparison between MCNP code and SPND measurement results showed a good agreement (relatively), where both methods showed the same radial thermal flux trend inside the core. MCNP simulation results over estimation was ~10% higher compare to the SPND measurement. This was due to error in model simplification, cross section data (ENDFVII), feedback phenomena, uncertainty in power and actual measurement setup. As our model also predicts well the neutron flux distribution in the core it can be used for the characterization of the full core, that is neutron flux and spectra calculation, dose rate calculations, reaction rate calculations, etc.

5. References

[1] Mohamad Hairie Rabir, Julia Abdul Karim, Norfarizan Mohd Said, Muhammad KhairulAriff Mustafa, Abi Muttaqin Jalal Bayar, Muhd Husamuddin Abdul Khalil, Muhammad Rawi Md Zin 2016 In-core neutron flux determination *Nuclear Bulletin of Malaysia* (ISSN:13945610, Jun 2016) **18** 26-28

[2] Mohamad Hairie Rabir, Muhammad Rawi Md Zin, Mark Dennis Usang, Abi Muttaqin Jalal Bayar, Na'im Syauqi Bin Hamzah 2016 Neutron Flux and Power in RTP Core-15 *AIP Conf. Proc.* **1704** 050018

[3] Mohamad Hairie Rabir, Mark Dennis Usang, Na'im Syauqi Hamzah, Julia Abdul Karim, Mohd Amin Sharifuldin Salleh 2013 Determination of neutronics parameters for RTP using MCNP code *J. Sains Nuklear Malaysia* (ISSN: 2232-0946) **25(1)** 53-67

[4] Mishra A K and Tiwari A P Delayed Time Response Self-Powered Neutron Detectors for Reactor Control *BARC Newsletter*

[5] Srinivasarengan K., Mutyam L, Belur M N., Bhushan, Tiwari A P, Kelkar M G and Pramanik M 2012 Flux Estimation from Vanadium and Cobalt Self Powered Neutron Detectors (SPNDs): Nonlinear Exact Inversion and Kalman filter approaches (American Control Conf. Fairmont Queen Elizabeth, Montréal, Canada, June 27-June 29, 2012)

[6] Vermeeren L 2015 Self-powered neutron detector modelling and testing (Agensi Nuklear Malaysia, March-April, 2015)

[7] Vermeeren L 2015 Calibration of THERMOCOAX vanadium self-powered neutron detectors for Agensi Nuklear Malaysia via MCNPX modelling (Agensi Nuklear Malaysia, March-April, 2015)

[8] Vermeeren L 2015 Report on the SPND commissioning – phase 2 for Agensi Nuklear Malaysia, (SCK•CEN Laurent Pichon (Thermocoax–THX), Agensi Nuklear Malaysia, 07-09/07/2015)

[9] Agensi Nuklear Malaysia 2014 *Safety Analysis Report (SAR) for RTP*

IOP Conf. Series: Materials Science and Engineering **298** (2018) 012030 doi:10.1088/1757-899X/298/1/012030

Comparative study between single core model and detail core model of CFD modelling on reactor core cooling behaviour

R Darmawan

Malaysian Nuclear Agency, Bangi, 4300 Kajang, Selangor, Malaysia

roslid@nm.gov.my

Abstract. Nuclear power industry is facing uncertainties since the occurrence of the unfortunate accident at Fukushima Daiichi Nuclear Power Plant. The issue of nuclear power plant safety becomes the major hindrance in the planning of nuclear power program for new build countries. Thus, the understanding of the behaviour of reactor system is very important to ensure the continuous development and improvement on reactor safety. Throughout the development of nuclear reactor technology, investigation and analysis on reactor safety have gone through several phases. In the early days, analytical and experimental methods were employed. For the last four decades 1D system level codes were widely used. The continuous development of nuclear reactor technology has brought about more complex system and processes of nuclear reactor operation. More detailed dimensional simulation codes are needed to assess these new reactors. Recently, 2D and 3D system level codes such as CFD are being explored. This paper discusses a comparative study on two different approaches of CFD modelling on reactor core cooling behaviour.

1. Introduction

The design and development of a nuclear reactor has always been focused on its reliability to operate safely throughout its lifetime. The early nuclear reactor designs are mostly dependent upon experimental analysis [1], where complex configuration and various phenomena have to be investigated. This approach would require very specialised experimental facilities which may not be affordable to many researchers or institutions [2][3]. Later, the development of computational method has brought about new approaches on the design and analysis of nuclear reactor safety. This is where the usage of numerical method was employed for most of the nuclear safety assessment activities. The one dimensional simulation codes designed specifically for nuclear reactor safety analysis were then developed by various nuclear engineering institutions specifically to simulate the neutronics and thermal-hydraulic behavior of the reactors [4]. These one dimensional simulation codes have been extensively used since the last 3 decades by the nuclear industry in the nuclear reactor design process [5]. However, the complexity of the recent advanced nuclear reactor design requires three dimensional capabilities to investigate specifically localised phenomenon in the nuclear reactor. Thus, the last decade has shown new interest on using three dimensional simulation codes to analyse nuclear reactor performance and safety designs [3].

For most of the reactor design process and safety analysis, 1D simulation codes are used extensively due to its maturity and abundance usage experience. However, for the last two decades, 3D Computed Fluid Dynamics codes such as FLUENT and CFX have started to be used in addition to

1D simulation codes. The 3D capability is needed in certain cases where 1D code might not be able to predict. Most of the applications are in the Pressurised Water Reactors (PWR) which have more complex design and operating condition compared to Light Water Reactors (LWR). Among the applications are the study on sub cooled boiling phenomena, void fraction distribution, slugging, counter current flow limitation and pressurized thermal shock [6][7]. Although the operation and design of LWR reactors are less complicated, 3D CFD code is still needed in certain cases. Most of the works in LWR reactors, which employing CFD method are related to the emergency analysis of the reactors [3,7,8,9].

CFD study on reactor operational behaviour may involve the modelling of the reactor core as the heat source. 3D CFD modelling would require 3D configuration of the reactor core. Modelling the details of core configuration may incur heavy computational load as well as time consuming. Most CFD studies employ the simplification model to reduce the computational load and simulation time, especially during preliminary analysis [10][11][12][13][14]. Among the typical methods used are the simplified core approach and porous media approach. The simplified core approach would either simplified the core geometric configuration [11][14] or assuming the core as a single element [12]; whereas for porous media approach requires the input on porous media parameters obtained from the empirical study of the actual core [10][13]. In this study, the aspect of model configuration in CFD was investigated, two different model configurations were modelled and the results were compared to study the effect on cooling behaviour of a hypothetical reactor core.

2. Methodology

A typical open pool reactor core was modeled using CFD software FLUENT. The reactor is of cylindrical type with the top exposes to the ambient. The wall was assumed to be adiabatic so as to simulate the typical concrete shielding of an open pool reactor. Two different core configurations were modelled as shown in figure 1. The core of the reactor was modeled as a single cylinder and as an array of rod fuel elements. The two different core models were subjected to the same heat flux, which is expected to undergo cooling process through conduction and convection of the cooling medium. The model is assumed to be in steady state condition to simulate a normal operation of the reactor. The cooling fluid is normal water, operating at pool bulk temperature and at atmospheric pressure. No external force or pressure is imposed on the fluid, thus, it is assumed to be in stationary and in laminar flow condition. The pool wall is considered smooth and no frictional force is exerted on the fluid. The top part of the model is considered as convective wall, exposed to free stream air at room temperature. The convective heat transfer coefficient in free stream air is set to 10W/m^2 K as used by *Rodriguez et. al.* [15] and Gastelurutia *et. al.* [16]. Other fluid properties of the model are shown in figure 2. The validation of the models was conducted on modeling program verification and numerical solution verification aspects as described in [17].

3. Results and discussion

The results of the simulation on both models with adiabatic wall condition are shown in figure 3 and figure 4. Both core models were subject to 3.5W/m^2 heat flux. The detail core model shows that the maximum temperature on the surface of the fuel core is about 127°C, which is a typical maximum surface temperature of low power open pool reactor [18]. However, the result on the single core reactor shows much lower surface temperature at about 63°C. It may resulted from the smaller surface area in the single core model that at the same heat flux per unit area, would produce much lower heat from the core surface. Both core models should be undergoing almost the same rate of heat dissipation process that the model with detail core and higher total heat, ended up with higher core surface temperature.

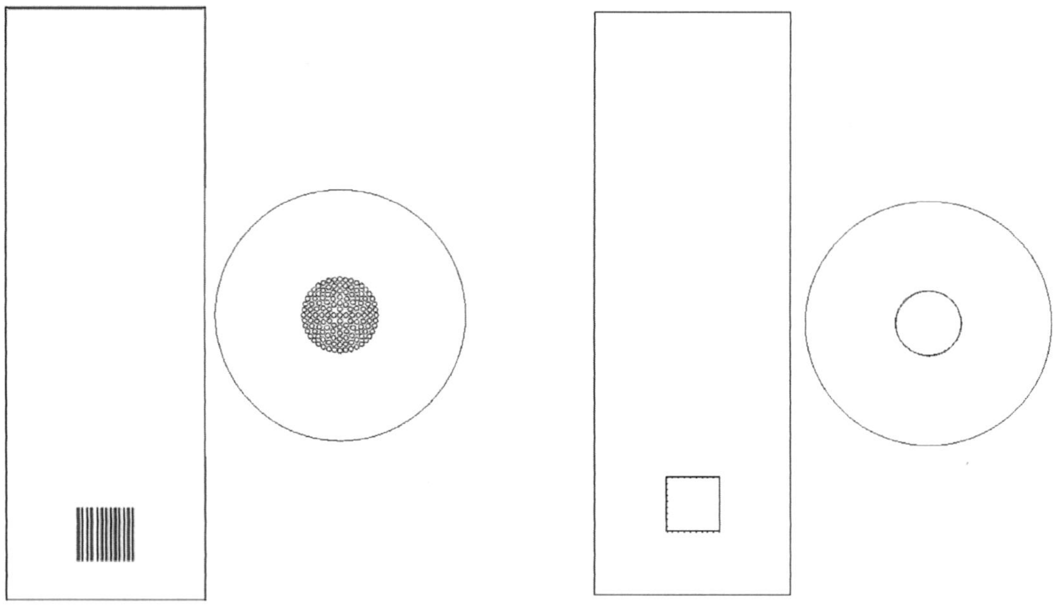

Figure 1. Detail core model (left) and single core model (right) of the reactor pool.

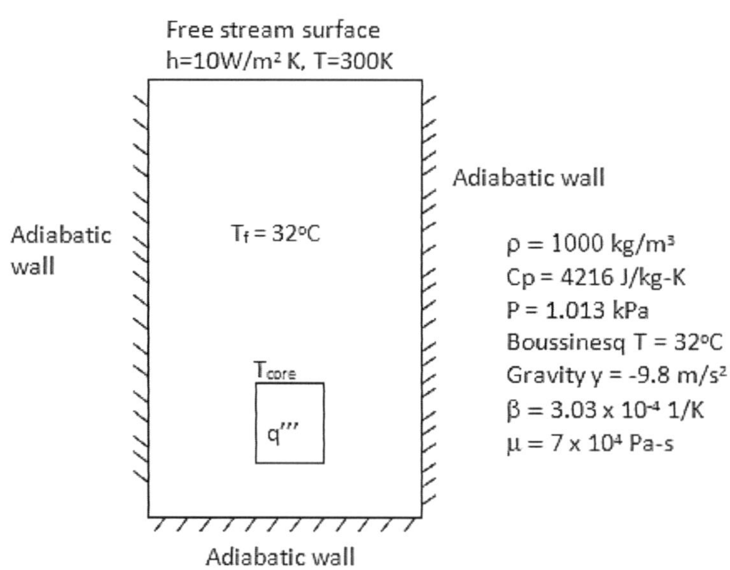

Figure 2. Schematic of the reactor pool model.

IOP Conf. Series: Materials Science and Engineering **298** (2018) 012030 doi:10.1088/1757-899X/298/1/012030

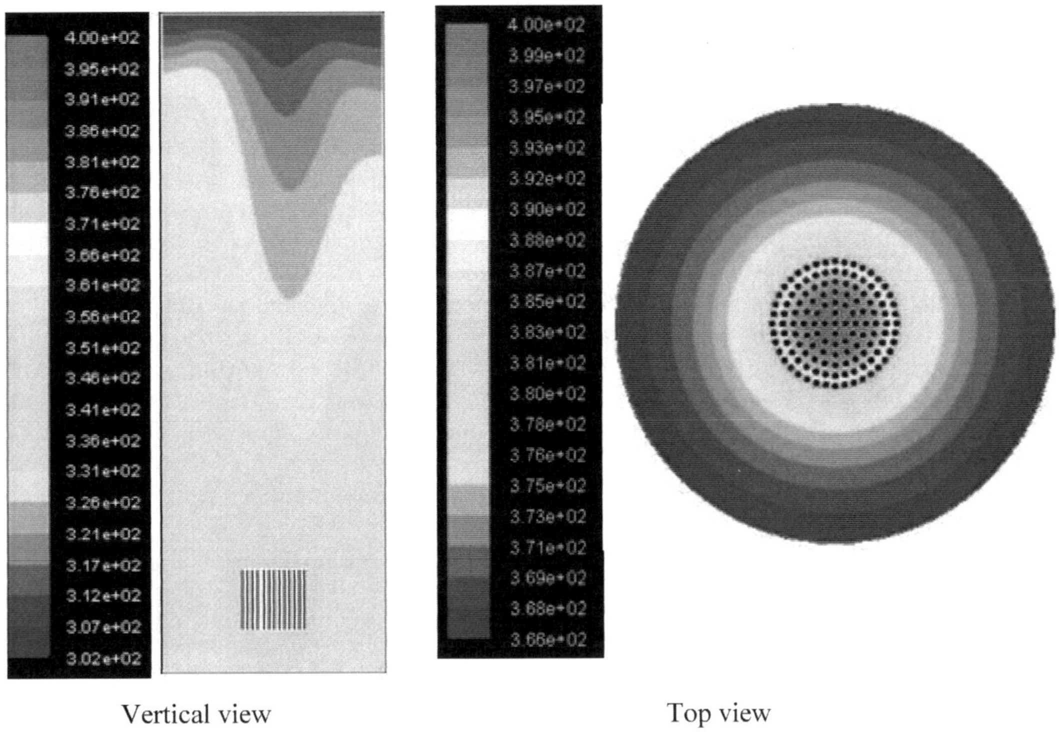

Figure 3. Temperature profiles (in K) of detail core model with 3.5W/m^2 heat flux and adiabatic wall.

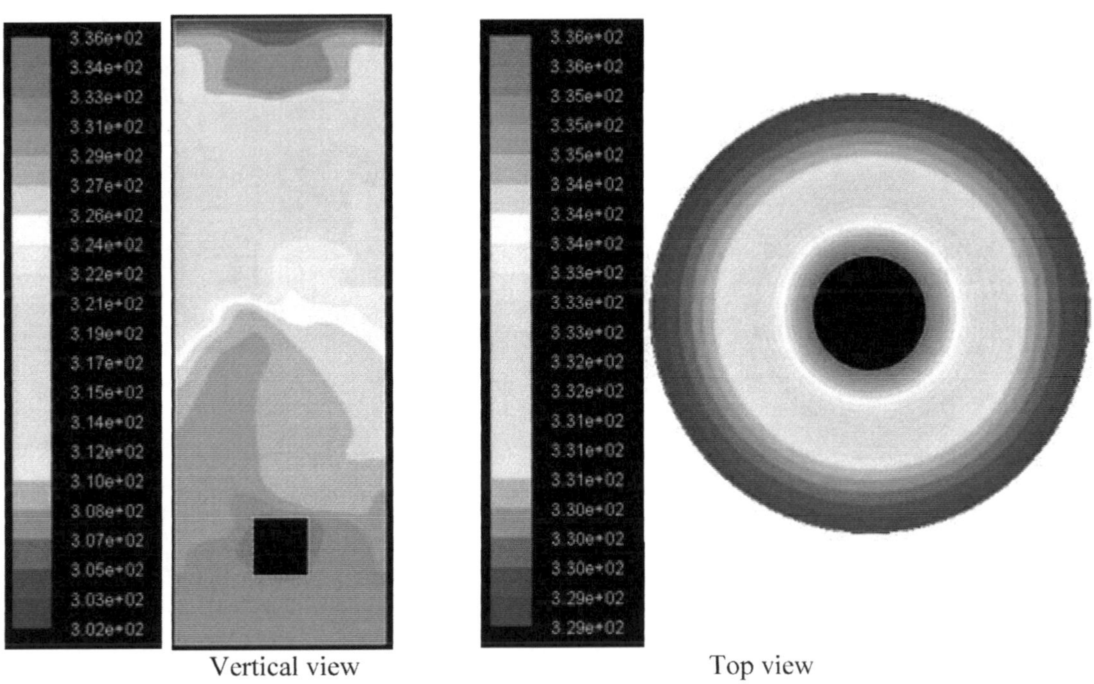

Figure 4. Temperature profiles (in K) of single core model with 3.5W/m2 heat flux and adiabatic wall.

The stream function profiles illustrated in figure 5 shows distinct features, where detail core model has lower stream mass. Most of the recirculation stream occurs around the core with longer recirculation stream above the core and along the cylinder until the top of the cylinder. Two distinct convection modes could be seen that are the rigorous stream around the core and the more stable stream above the core. Whereas, for single core model, the same recirculation mechanism occurs around the core, however the convection streams distributed into three flow behaviours; rigorous around the core, mixed mode at the middle and more stable stream at the top. These features show that the detail core model has higher temperature gradient between the core and the ambient. The maximum mass stream is lower and the convection mechanism occurs in a stable manner up to the top of the pool. In the single core model, the significant temperature gradient may only be up to the middle of the pool where the mixed mode stream occurs. These convection mechanisms substantiated the discussion that the detail model may have higher surface heat due to more surface area, compared to the single model case. From the above results and discussion, the simple model such as single cylindrical core may not be a good representation of the actual core. A more detail modelling such as the detail core model would be a better representation of the actual reactor cooling behaviour as shown by Figure 3, where the results is comparable with the typical maximum core temperature of low power open pool reactor [18].

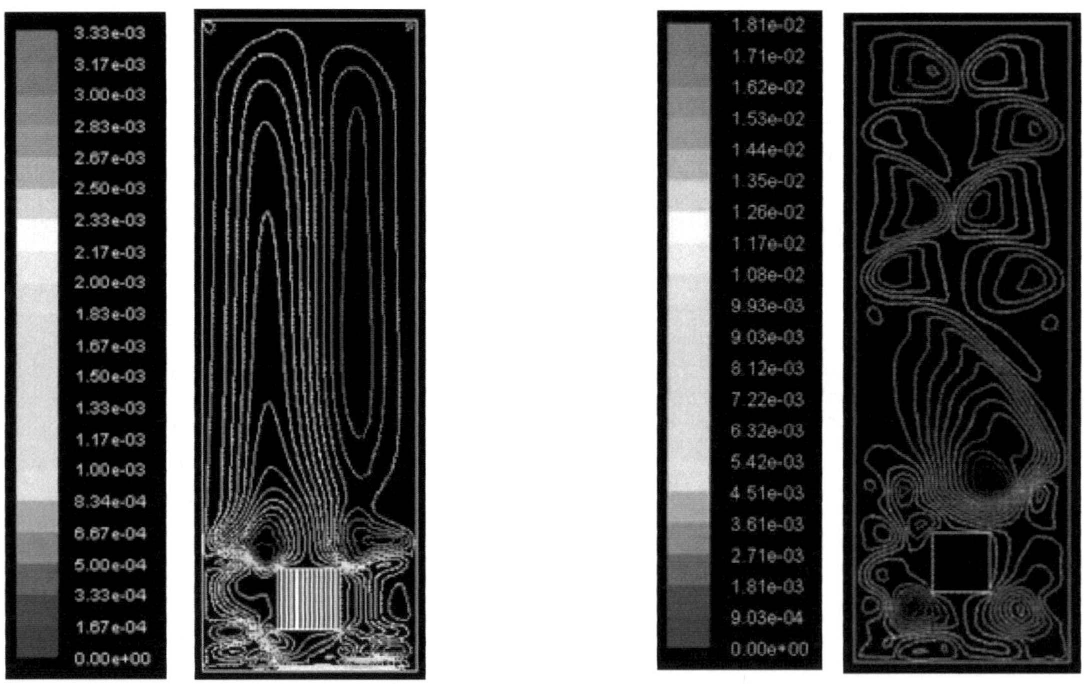

Figure 5. Stream function profiles (in kg/s) for full core model (left) and single core model (right).

Another comparative study was also conducted between the adiabatic wall and fixed wall cases. Both models were subjected to the same heat flux with the pool wall either set as adiabatic or fixed temperature at 300K. The results of the simulations are shown in figure 6 and 7. Both models show the same characteristics where the models with fixed temperature have lower core surface temperature compared to models with adiabatic walls. The models with adiabatic walls have only the top opening as the region of low temperature, where it is set as a free stream with temperature at 300K. This will reduce the heat dissipation capability when compared to the fixed wall temperature. In the fixed wall temperature case, low temperature region is on the surrounding wall as well as the bottom surface of the pool. Thus, heat would be dissipated quickly to the surrounding wall and the floor of the pool.

IOP Conf. Series: Materials Science and Engineering **298** (2018) 012030 doi:10.1088/1757-899X/298/1/012030

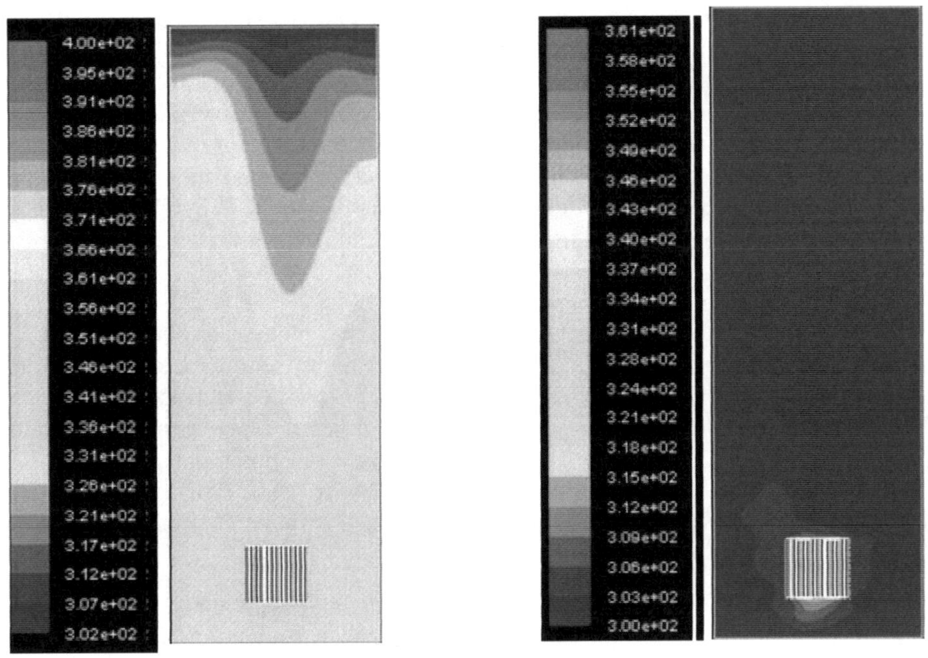

Figure 6. Te… K) of detail core model with … all…

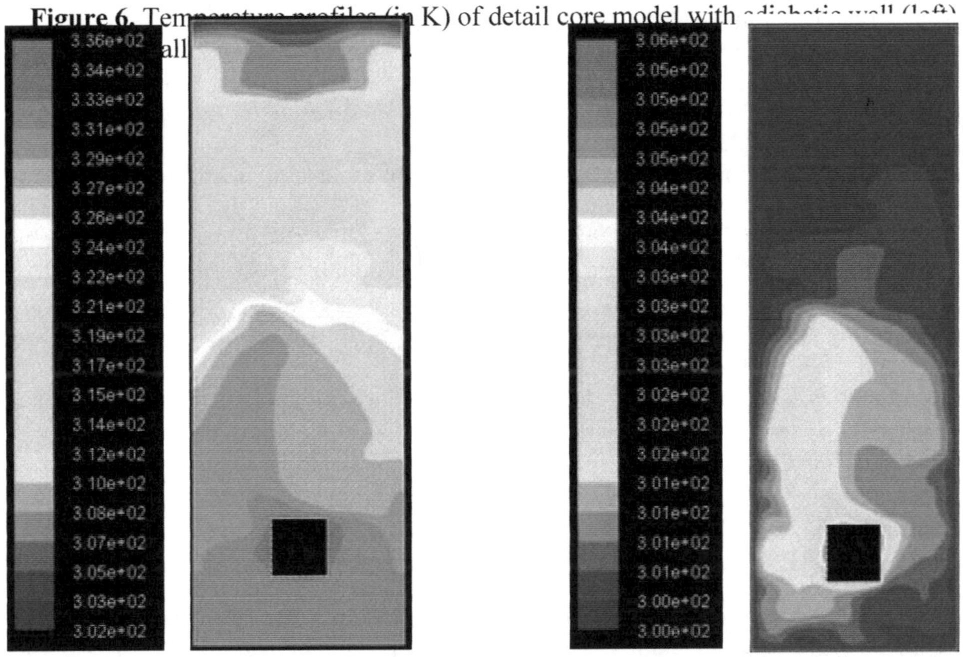

Figure 7. Temperature profiles (in K) of single core model with adiabatic wall (left) and fixed wall temperature (right).

The stream function profiles show similar characteristics for both cases where in the case of detail core model, the stream has two modes and the single core model has three stream modes as discussed earlier. This phenomenon signifies that both the adiabatic wall and fixed wall temperature undergoing the same heat convection mechanism in both models, and the amount of low temperature region will determine the heat dissipation capacity. The results of this comparative study also reveal that the adiabatic wall model is a better representation of the actual reactor pool wall compared to fixed wall temperature model. The adiabatic wall in detail core models resulted in 127°C of maximum core temperature; whereas the fixed wall temperature model has only 88°C maximum core temperature, which is much lower than the typical maximum core temperature for low power open pool reactor.

4. Conclusion

CFD modeling of a reactor core cooling behaviour would be more representative for a detail core model compared to a single core model. The detail core model produced similar characteristics of temperature and flow profiles to the typical low-power open-pool reactor as compared to the single core model. CFD modeling with adiabatic wall model is a better representation to simulate reactor core cooling with concrete shielding wall compared to fixed wall temperature model. The adiabatic wall model produced temperature and flow profiles which have close characteristics with the typical open pool reactor. A comparison with porous media approach may also be pursued to further evaluate the accuracy of each model.

5. References

[1] Bestion D 2004 Recommendation on use of CFD codes for nuclear reactor safety analysis *5th EURATOM Framework Program 1988-2002, Key Action: Nuclear Fission* (European Commission)

[2] Sardain P, Ayrault L, Laffont G, Challet F, Marie L B, Merrill B, Porfiri, M T and Caruso G 2005 The EVITA programme: experimental and numerical simulation of a fluid ingress in the cryostat of a water-cooled fusion reactor *Fusion Engin. and Design* **79** 1265–69

[3] Andreani M, Haller K, Heitsch M, Hemström B, Karppinen, I, Macek, J, Schmid, J, Paillere, H and Toth, I 2008 Benchmarking of CFD codes for application to nuclear reactor safety *Nuc. Engin. and Design* **238**, Issue 3 502-13

[4] Yadigaroglu G 2005 Computational fluid dynamics for nuclear applications: from CFD to multi-scale CMFD *Nuc. Engin. and Design* **235** 153-64

[5] Gao S, Leslie D C and Hewitt G F 2008 Improvements to the modelling of two-phase flow and heat transfer in a transient nuclear reactor analysis code *Appl. Therm. Engin.* **28** 915-22

[6] Koncar B, Krepper E and Egorov Y 2005 CFD modelling of subcooled flow boiling for nuclear engineering application *Inter. Conf. Nuclear Energy for New Europe 2005, Bled, Slovenia, September 5-8, 2005* p140.1-140.14

[7] Deendarlianto H T, Apanasevich P, Lucas D, Vallée C and Beyer M Application of a new drag coefficient model at CFD-simulations on free surface flows relevant for the nuclear reactor safety analysis *Annals of Nuclear Energy* **39(1)** 70-82

[8] Bernard J A and Lin W H 2001 Numerical model of MIT research reactor *IEEE Transc. on Nuc. Science,* Vol **8** p27-3 27-10

[9] Vanderhaegen M, Vierendeels J and Arien B 2001 CFD analysis of the MYRRHA primary cooling system *Nuc. Engin. and Design*, Vol **241** Issue 3 775-84

[10] Yan Y, Rizwan-uddin and Sobh N 2005 CFD modeling on reactor core *Math. and Computation, Supercomputing, Reactor Phys. and Nuc. and Bio. Applications, Palais des Papes, Avignon, France, September 12-15, 2005* (American Nuclear Society, LaGrange Park, IL).

[11] Khanbabaei B, Ghisemizad A and Farajollah H 2008 CFD calculation of fluid flow in VVER-1000 Reactors *Jou. of Appl. Sciences*, Vol **8(5)** p780-87

[12] Prosicky M and Zacha P 2003 Modeling of temperature field in a reactor vessel downcomer during transient *Transaction of the 17th Inter. Conf. on Structural Mechanics in Reactor Tech. Prague, Czech Republic, August 17-22, 2003*
[13] Sipaun S M 2014 Thermal-fluid modeling of the Missouri S&T Reactor *Doctoral Dissertations* **2358** (Missouri University of Science & Technology)
[14] Keshmiri A 2010 Thermal hydraulic analysis of gas cooled reactor flows *Ph.D. thesis* (University of Manchester)
[15] Rodríguez I, Castro J, Pérez-Segarra C D and Oliva A 2009 Unsteady numerical simulation of the cooling process of vertical storage tanks under laminar natural convection *Inter. Jou. of Therm. Sci.* **48** p708–21
[16] Gastelurrutia J, Ramos J C, Larraona G S, Rivas A, Izagirre J and Luis D R 2011 Numerical modeling of natural convection of oil inside distribution transformers *Appl. Therm. Engin.* **31** p493-505
[17] Rosli D, Nor Mariah A, Noraini A A, M Khairol Anuar M A 2017 Investigation on alternative method to enhance the cooling capacity of an open pool research reactor *AIP Conference Proceeding 1799* Vol **10:1** p28-38.
[18] Mesquita A Z 2005 Experimental investigation on temperatures distributions in a nuclear research reactor TRIGA IPR-R1 *Ph.D Thesis* (Universidade Estadual de Campinas, São Paulo, Portuguese)

Neutron measurement at the thermal column of the Malaysian Triga Mark II reactor using gold foil activation method and TLD

Safwan Shalbi[1], Wan Norhayati Wan Salleh[1], Faridah Mohamad Idris[2], Muhammad Aliff Ashraff Rosdi[1], Muhammad Syahir Sarkawi[1], Nur Liyana Jamsari[3],Nur Aishah Mohd Nasir[3]

[1]Nuclear Engineering Department, Faculty of Chemical and Energy Engineering, Universiti Teknologi Malaysia, 81310 Skudai, Johor Bahru, Malaysia
[2]Malaysian Nuclear Agency, 43000 Kajang, Selangor, Malaysia
[3]Universiti Kebangsaan Malaysia, 43600 Bangi, Selangor, Malaysia
[4]Universiti Tenaga Nasional, 43000 Kajang, Selangor, Malaysia
Type the author addresses here

hayati@petroleum.utm.my

Abstract. In order to design facilities for boron neutron capture therapy (BNCT), the neutron measurement must be considered to obtain the optimal design of BNCT facility such as collimator and shielding. The previous feasibility study showed that the thermal column could generate higher thermal neutrons yield for BNCT application at the TRIGA MARK II reactor. Currently, the facility for BNCT are planned to be developed at thermal column. Thus, the main objective was focused on the thermal neutron and epithermal neutron flux measurement at the thermal column. In this measurement, pure gold and cadmium were used as a filter to obtain the thermal and epithermal neutron fluxes from inside and outside of the thermal column door of the 200kW reactor power using a gold foil activation method. The results were compared with neutron fluxes using TLD 600 and TLD 700. The outcome of this work will become the benchmark for the design of BNCT collimator and the shielding

1. Introduction

Cancer is one of the leading causes of death by 42% in the developing country [1]. Even though there were improvements in the common cancer treatments such as surgery, radiotherapy and chemotherapy, the patient of glioblastoma still have a very short overall survival period, which less than 18 months [2]. Thus, boron neutron capture therapy (BNCT) is looked as a promising method was studied around the world because the conventional treatment are not efficient enough to cure cancer [3] like melanoma and glioblastoma multiforme. According to [4], the patient will be irradiated with slow energy of neutron (thermal neutron or epithermal neutron) to reach the ratio of high concentration of boron compound in the tumour cell injected with a Boron-10 compound. The irradiation with slow neutron will caused nucleus of Boron-10 to capture the slow neutron to form

Boron -11 which then decays to Li-7 from the emission of an alpha particle He-4 [5]. BNCT was called binary treatment for cancer based on its nuclear reaction of the slow neutron bombarded with the non-toxic and non-radioactive compound of Boron-10.

In general, most of the BNCT studies use neutron source from the research reactors. This research is aimed at developing a BNCT facility for the cancer treatment studies that is safe and controlled from radiation over exposure and practically uses slow neutrons emitted from the Malaysian TRIGA MARK II reactor. Thus, the first step in order to achieve the long-term goal of this project is to establish the suitability of neutron fluxes produced from the research reactor. Commonly, most of the TRIGA reactor has a thermal column facility, which is comprises of graphite moderator designed to produce epithermal neutrons and thermal neutrons which can be utilised in BNCT facility. Therefore, the thermal column of Malaysian TRIGA MARK II reactor was identified for BNCT study as it has thermal and epithermal neutron source for this research. Previous researches indicated that the thermal column in the Malaysian TRIGA MARK II reactor has a sufficient a quality neutron beam to develop BNCT facility [5]. In order to develop a BNCT facility at the thermal column at this reactor, neutron measurement needs to be carried out comprehensively in the thermal column (graphite), thermal column door and outside of the thermal column to produce benchmark for material to be used for collimator and shielding this research.

2. Methodology

2.1. Experimental Arrangement

In this research, the experimental arrangement for the thermal column was divided into three phases as shown in figure 1.

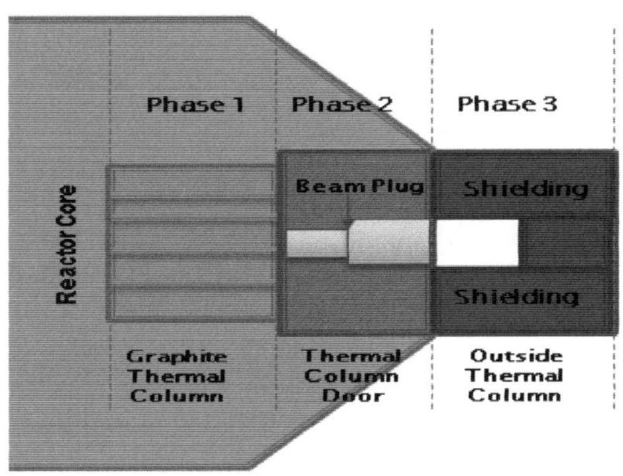

Figure 1. Three phases of thermal column of Malaysia TRIGA MARK II reactor.

Phase 1 consists of two parts; the inner part of the graphite which is right at the peripheral of the graphite reflector, and the second part which is the outer part right behind the door of the thermal column. The outer part is a pile of graphite stringers stacked in crisscross layers along the thermal column. There are 7 graphite stringers which were labelled according to their positions. Most of the graphite stringers have a 127cm length and 10.16 x 10.16 15 cm^2 area, except for the lowermost and the uppermost graphite stringers which have areas about 10.16 x 3.18 cm^2 with the same length. All of the stringers except a central stringer are provided with the six holes with 2 cm diameter and 0.7 cm depth with 24.5cm distance between the holes as a purpose for the sample irradiation in the thermal column. The central stringers were labelled as G7 consists of 13 holes and the 10cm distance between the holes. The central stringers were different compared to others stringers because of the neutron flux passing through these stringers are the highest because the location of central stringers is directly pointing from the reactor core. Figure 2 shows the drawing of removable graphite stringers labelled and their position at the thermal column.

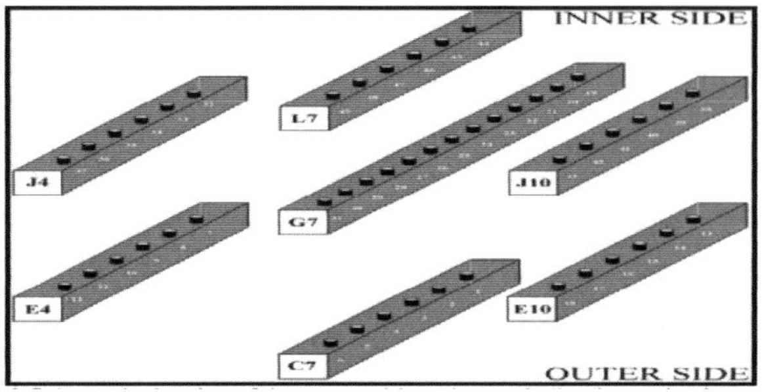

Figure 2. Schematic diagram of removable graphite stringers labeled and their position in the thermal column [6].

Phase 2 consists of thermal column door which have one beam pug across the beam line and G7 stringers. The beam plug divided in two sections which is internal and external beam plug. The internal beam plug has small diameter compare to external beam plug. Currently, both beam plugs are made of fully solid stainless steel and function as a beam stopper. During the experiment, both beam plug will be pulled out and replace with 8 cm diameter of temporary collimator made from polyethylene and filled with paraffin. The beam plug is potentially being developed for BNCT collimator.

Phase 3 are referring to the outside of the thermal column, which is covered about 100 cm from the exit of the thermal column door (Phase 2). The measurement of neutron in this region is important to estimate the neutron and gamma flux across the beam line and the surrounding of the thermal column for the safety and precautions prepared for the future BNCT facility.

2.2. Gold foil activation method

The neutron flux measurement is determined by using gold foil activation method that utilizes ^{197}Au reaction with neutrons to produce ^{198}Au. In this experiment, there are two types of samples being prepared which is pure 99.99% gold foil and Cadmium (Cd). The gold foil is used in this experiment because the gold foil will decay and emitting single gamma of well-known energy 412keV and the half-life of 2.81 days. The cadmium is well known as the thermal neutron filter that will be used to cover gold foil for to separate the thermal and epithermal neutron fluxes. There are 20 samples of bare gold foils with less than1 mm in length and 20 gold foil covered cadmium samples with less than 1.3mm length were prepared for irradiation in Phase 1. All the samples were put in the irradiation vials which were labelled and that make the samples easier to handle and reduce contamination from irradiation of materials and for safety precautions. The positions of irradiated samples loaded into the stringers are shown as table 1 below:

The irradiation starts and continues for exactly 6 hours before getting the reactor shutdown. The time of reactor shutdown is also recorded for the purpose of activity calculations. The thermal column is opened again in the following day for unloading samples. In the Phase 2 and Phase 3, the position of the samples is shown as table 2 and figure 3:

IOP Conf. Series: Materials Science and Engineering **298** (2018) 012031 doi:10.1088/1757-899X/298/1/012031

Table 1. Stringers and holes use in experiments

Stringers	Holes
E4	E07
	E09
	E12
E10	E13
	E16
	E18
G7	G19
	G24
	G26
	G28
	G31
J4	J32
	J35
	J37
J10	J38
	J40
	J43
L7	L44
	L46
	L47

Table 2. The sample of bare gold foil and cadmium-covered gold foil for the Phase 2 and Phase 3 with the distance from G7 stringer

Phase	Sample	Distance from Graphite Thermal Column (cm)
2	A1	142
	A2	182.7
	A3	208
	A4	233.3
3	A5	258.6
	A6	283.9

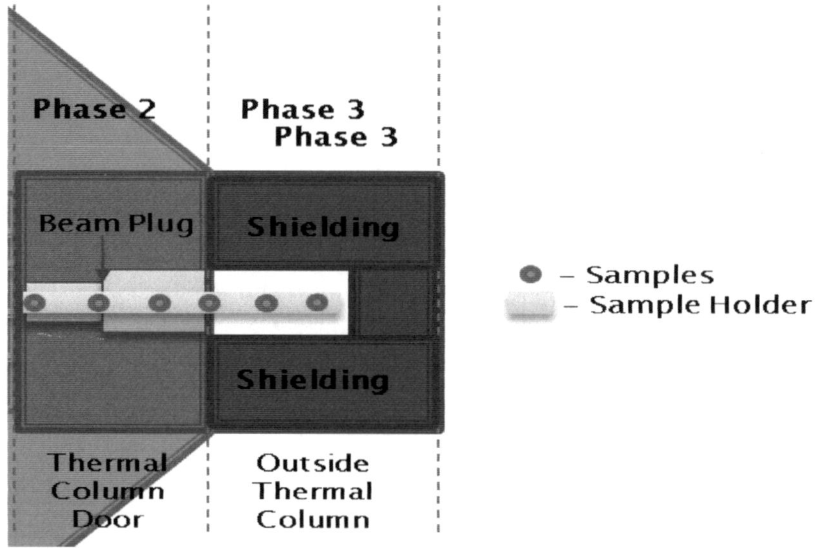

Figure 3. Illustration of sample position for the Phase 2 and Phase 3.

After irradiation, all of the samples in Phase 1, Phase 2 and Phase 3 were counted by a detector. The detector uses a germanium-detector gamma ray spectrum acquisition and analysis with 4.59 x 10^{-3} of detector efficiency. The count rate of radionuclide took more than two weeks to fully complete as the sample must have less than 25% of dead time or to repeat and, each sample took more than 30 minutes to be counted. The data then analysed by using the specific equation for neutron activation analysis to obtain thermal neutron flux and epithermal neutron flux [7]. At first, the count rate was converted into absolute foil activity by incorporating the detector efficiency by using the relation in equation (2.1):

$$A_{abs}=\frac{C_r}{\eta_d} \tag{2.1}$$

where A_{abs} absolute foil activity (counts/s), C_r= observed count rate (count/s), η_d= detector relative efficiency

By finding the activity of bare gold foil and the activity of cadmium covered gold foil, the thermal neutron flux was obtained by using equation (2.2):

$$\emptyset_{th} = \frac{\rho(A_{abs} - A_{cd})}{m\Sigma_a(1 - e^{-\lambda_{ti}})e^{-\lambda_{tc}}} \tag{2.2}$$

where A_{cd} = activity of cadmium covered gold foil, ρ = density of the foil material (gm/cm^3), m = weight of the foil (gm), Σ_a = macroscopic thermal absorption cross-section (cm^{-1}), λ = decay constant of radioactive isotope formed during radiation (s^{-1}), t_c= cooling time (s), t_i= irradiation time (s)

For gold, the response region that can absorb thermal neutron and epithermal neutron is in between 0.0015eV to 5.8eV and the cut-off energy of cadmium of 0.55eV was used to mark both thermal and epithermal neutron region. Therefore, the difference between the activity of bare gold foil and the cadmium covered gold foil, if both were irradiated in the same neutron flux under the same circumstances, is the activity caused by thermal neutron flux. The ratio of thermal to epithermal fluxes was determined by cadmium ratio, which is defined by the equation (2.3):

$$Cadmium\ ratio = \frac{Ativity\ of\ bare\ gold\ foil}{Activity\ of\ Cd\ covered\ gold\ foil} \quad (2.3)$$

By using equations 2.2 and 2.3, the epithermal neutron flux can be calculated by the equations given below.

$$R_{cd} - 1 = \frac{\varnothing_{th} \cdot \sigma_{th}}{\theta 1526 barn} \quad (2.4)$$

and,

$$\varnothing_{epi} = \theta\ Log(\frac{E_2}{E_1}) \quad (2.5)$$

where $\varnothing_{th}$ = thermal neutron cross-section (barns), θ = intermediate neutron flux density per unit lethargy, $\varnothing_{epi}$ = epithermal neutron flux (neutrons $cm^{-2}s^{-1}$), E_1 = 0.5 eV, E_2 = 2 MeV, I_r = resonance integral (1526 barns).

2.3 Thermoluminescent dosimeter

In this experiment, two types of TLD was used which is TLD-600 and TLD-700 for measuring gamma and neutron dose rate. The advantage of using TLD in this research is that a TLD has a long-term storage of information data and have high sensitivity towards radiation. TLD-600 could detect and is sensitive to both neutron and photon such as gamma while TLD-700 is sensitive to detect and measure photon only [8]. On the other hands, the use of TLD detector is convenient because the atomic density of LiF is definitely close to the human body. TLD detector is also cheap and affordable compare to another detector use in the recent world market. The experiment carried out using TLD-600 and TLD-700 only for the Phase 3 (A5 and A6), because of the time limitation with only 60 seconds of irradiation time and the safety issues as the Phase 1 and Phase 2 are highly exposed to radiation of gamma and neutron. Figure 3 and table 2 above illustrates the position of the TLD- 600 and TLD-700 for sample A5 and A6.

3. Results and discussion

3.1. Gold foil activation measurement

Based on experimental results shown in table 3, the thermal neutron flux from the phase 1 proved that stringer G7 provide highest thermal neutron flux compared to other stringers. Thermal neutron is defined as slow neutrons that are approximately in thermal equilibrium with a moderator. The highest thermal neutron flux came from hole G19 from stringer G7 with 7.42 x $10^{11} n.cm^{-2}s^{-1}$ while the lowest is 1.35E+09 $n.cm^{-2}s^{-1}$ from hole E18 of stringer E10. Most of the samples loaded in stringer G7 have higher thermal neutron flux due to the position of the stringer that located closest to the reactor core. As the distance of the stringer is close to the reactor core, more neutron fluxes would be produced. Meanwhile, the position of stringer E10 (which is located above the reactor core) gave the lowest thermal neutron flux due to the high volume of shielding material there. Besides that, the epithermal neutron flux graph indicated that the highest result of epithermal neutron come from hole G19 with 1.16 x $10^{11} n.cm^{-2}s^{-1}$ as compared to the lowest result from stringers L7. Epithermal neutron is defined as neutron having energy in the range immediately above the thermal range, roughly between 0.02 and 100 eV. This result also occurs due to the position of the stringer with stringer G7 located near to the reactor core. Meanwhile, the position of stringers L7 is above of G7 and located near to the moderator. The interaction of the neutrons in irradiated samples and coolant from thermalisation effect has produced less epithermal neutron flux. This indicates that both thermal and epithermal neutron flux in G7 at Phase 1 are applicable for BNCT purpose.

Table 3. Thermal and epithermal neutron flux measured using the gold foil activation method for Phase 1

Stringers	Holes	Thermal Neutron Flux $(\mathrm{neutron.cm^{-2}s^{-1}})$	Epithermal Neutron Flux $(\mathrm{neutron.cm^{-2}s^{-1}})$
E4	E07	7.05 x 10^{10}	1.11 x 10^{10}
	E09	2.34 x 10^{10}	3.10 x 10^{8}
	E12	2.24 x 10^{10}	2.18 x 10^{9}
E10	E13	5.75 x 10^{10}	2.64 x 10^{9}
	E16	1.04 x 10^{10}	1.33 x 10^{9}
	E18	1.35 x 10^{9}	6.61 x 10^{7}
G7	G19	7.42 x 10^{11}	1.16 x 10^{11}
	G24	2.81 x 10^{11}	2.30 x 10^{9}
	G26	2.41 x 10^{10}	1.32 x 10^{8}
	G28	1.47 x 10^{10}	2.03 x 10^{8}
	G31	1.85 x 10^{10}	1.73 x 10^{8}
J4	J32	5.39 x 10^{10}	2.94E+09
	J35	8.31 x 10^{10}	2.12 x 10^{8}
	J37	7.31 x 10^{10}	1.81 x 10^{8}
J10	J38	6.78 x 10^{10}	2.50 x 10^{9}
	J40	1.63 x 10^{10}	2.07 x 10^{8}
	J43	9.79 x 10^{9}	1.13 x 10^{8}
L7	L44	2.10 x 10^{10}	1.24 x 10^{9}
	L46	1.48 x 10^{10}	6.65 x 10^{7}
	L47	1.20 x 10^{10}	3.98 x 10^{7}

Table 4 shows the result of thermal and epithermal neutron flux measurement at Phase 2 and Phase 3. From table 4, the highest thermal neutron flux is 3.62 x $10^{7}\mathrm{n.cm^{-2}s^{-1}}$ at the distance of 142 cm from the stringer G7. The thermal neutron flux decreased linearly with the distance at Phase 2. At the Phase 3, the thermal neutron flux at position A5 decreased about 25% from the 3.53 x 10^{6} $\mathrm{n.cm^{-2}s^{-1}}$ and increase back to 128 x 10^{6} $\mathrm{n.cm^{-2}s^{-1}}$ when the distance measured at 283.9 cm from the thermal column. The surrounding factor become the main factor of this fluctuate of thermal neutron from the Phase 2 to the Phase 3 as the Phase 3 is outside of the thermal column and exposed to the air.

The epithermal at the Phase 2 and Phase 3 proved that the neutron flux decrease linearly towards the distance from the graphite thermal column. The sample A1 at the distance 142 cm was the highest epithermal neutron flux with 5.91 x 10^{6} $\mathrm{n.cm^{-2}s^{-1}}$ while the sample A6 at the 283.9 cm was measured the lowest with 3.89 x $10^{4}\mathrm{n.cm^{-2}s^{-1}}$. Technically, the epithermal neutron flux reduced to about 99.34% from Phase 2 to the Phase 3.

3.2. Thermal and epithermal neutron flux across the beam line

The result of neutron measurement from the Phase 1, Phase 2 and Phase 3 was combined to show the thermal and epithermal neutron flux across the beam line with 283.9 cm from the reactor core of TRIGA MARK II. The measurement start with the stringers G7 as a beam line for the Phase 1 and continued by Phase 2 and Phase 3 as shown in the Table 5 and Figure 4.

IOP Conf. Series: Materials Science and Engineering **298** (2018) 012031 doi:10.1088/1757-899X/298/1/012031

Table 4. Thermal neutron and epithermal neutron measured using the gold foil activation method for Phase 2 and Phase 3.

Phase	Sample	Distance from Graphite Thermal Column (cm)	Thermal Neutron Flux ($neutron.cm^{-2}s^{-1}$)	Epithermal Neutron Flux ($neutron.cm^{-2}s^{-1}$)
2	A1	142	3.62 x 10^7	5.91 x 10^6
	A2	182.7	5.51 x 10^6	3.98 x 10^6
3	A3	208	5.27 x 10^6	1.90 x 10^6
	A4	233.3	3.53 x 10^6	8.90 x 10^5
	A5	258.6	8.89 x 10^5	2.78 x 10^5
	A6	283.9	1.28 x 10^6	3.89 x 10^4

According to the table 5 and figure 4, both neutron thermal and epithermal neutron flux decreased linearly with the distance from 142 cm to 283.9cm from graphite thermal column as the flux generally is exponentially decreasing with distance[9]. From the Phase 1 to the Phase 2, the thermal and epithermal neutron flux are significantly reduced to about 99.8% for thermal neutron flux and 96.5% for epithermal neutron flux. The use of graphite as a neutron moderator for the research reactor in Phase 1 become the main factor for the tremendously decreases of flux from the Phase 1 to the Phase 2. The property of graphite that have higher thermal conductivity result in the increase the measured thermal and epithermal neutron fluxes at Phase 1. Across the beam line, the thermal neutron flux was higher than epithermal neutron flux with 1.00 x 10^2 $n.cm^{-2}s^{-1}$ difference. The aim of this measurement to clearly see the thermal and epithermal neutron across the beam line for the optimization purpose for collimator. The result also shows that the thermal and the epithermal neutron flux at the thermal column of the Malaysian Triga Mark II reactor can be used for BNCT research purpose according to the IAEA standard[10].

Table 5. Thermal neutron and epithermal neutron measured using the gold foil activation method across the beam line for Phase 1,Phase 2 and Phase 3

Distance (cm)	Thermal Neutron Flux ($neutron.cm^{-2}s^{-1}$)	Epithermal Neutron Flux ($neutron.cm^{-2}s^{-1}$)
0	7.42 x 10^{11}	1.16 x 10^{11}
28.4	2.81 x 10^{11}	2.30 x 10^9
56.8	2.41 x 10^{10}	1.32 x 10^8
85.2	1.47 x 10^{10}	2.03 x 10^8
113.6	1.85 x 10^{10}	1.73 x 10^8
142	3.62 x 10^7	5.91 x 10^6
182.7	5.51 x 10^6	3.98 x 10^6
208	5.27 x 10^6	1.90 x 10^6
233.3	3.53 x 10^6	8.90 x 10^5
258.6	8.89 x 10^5	2.78 x 10^5
283.9	1.28 x 10^6	3.89 x 10^4

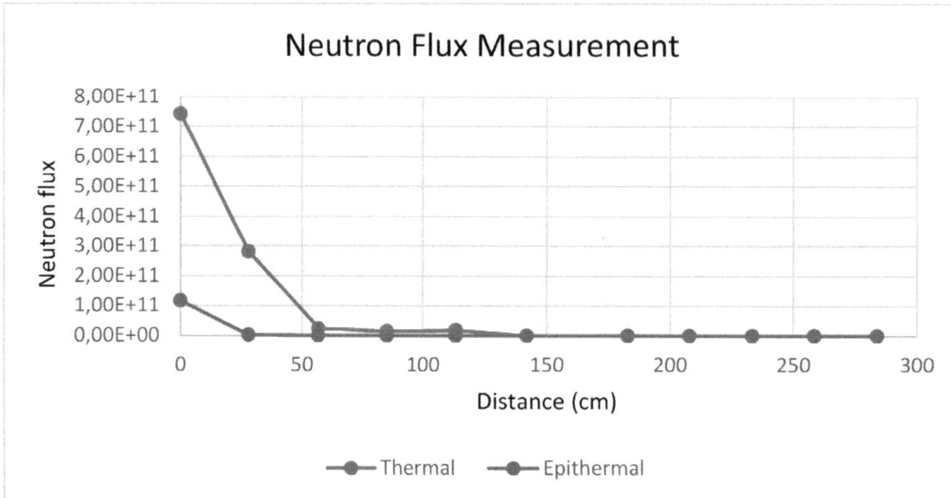

Figure 4. The neutron flux measurement across the beam line for the Phase1, Phase 2 and Phase 3.

3.3. TLD-600 and TLD-700 measurements

Table 6 shows the result of neutron and gamma at the Phase 3 obtained from TLD-600 and TLD-700. Only Phases 3 was available to use TLD as the TLD required shorten time to measure dose and to prevent any safety issues regarding to over exposure of radiation. The previous results revealed that the outside of the thermal column contained higher neutron than gamma. Therefore, at 200 kW experiment, the outside part of the thermal column is safe with low gamma radiation[11]. By comparing both gold foil activation and TLD method, the neutron (mSv) at sample A5 is higher with TLD. As can be showed on table 6, the gold foil activation measurement are significantly not in sync with TLD experiment measurement due to different count rate. The TLD measurement was counted right after the experiment been conducted while for the gold foil activation analysis, the sample was counted when the activity of the sample are low. The gap of the time of sample counted for both experiments effected the measurements [11].

Table 6. The comparison between the measurement of neutron and gamma using TLD and gold foil activation method

Sample	**Gold foil activation measurement**		**TLD-600 and TLD-700 measurement**	
	Neutron (mSv)	**Gamma (mSv)**	**Neutron (mSv)**	**Gamma (mSv)**
A5	3.00	-	4.20	0.25
A6	4.20	-	2.38	0.27

4. Conclusion

The thermal and epithermal neutron fluxes at thermal column of Malaysian TRIGA MARK II reactor were successfully measured using two methods; gold foil activation and TLD. The purpose of this measurement is to explore the feasibility of the neutron source from the thermal column to be used for BNCT application. From the experiments conducted, the thermal column produced significantly higher thermal and epithermal flux at the stringers G7. Besides that, the data also proved that the thermal and epithermal fluxes decreased linearly with distance as the experiment are set up across the beam line on the Phase 1, Phase 2 and Phase 3. This experimental data would be further used in designing collimator and shielding for BNCT facility.

5. References

[1] Chiragkumar J G and Malleshappa N N 2015 Selective cancer treatment by Boron Neutron Capture Therapy (BNCT)–a review *Int. J.l of Pharmaceutical Chemistry and Analysis* **2(3)** 136-138.

[2] Hopewell JW, Gorlia T, Pellettieri L, Giusti V, H-Stenstam B and Sköld K 2011 Boron neutron capture therapy for newly diagnosed glioblastoma multiforme: an assessment of clinical potential. *Appl. Radiation and Isotopes* **69(12)** 1737-1740.

[3] Auterinen, Iiro, Kotiluoto, Petri, Hippeläinen E, Kortesniemi Mika, Seppälä, Tiina, Seren, Tom, Collan, Juhani 2004 Design and construction of shoulder recesses into the beam aperture shields for improved patient positioning at the FiR 1 BNCT facility *Appl. radiation and isotopes* **61(5)** 799-803.

[4] Durisi E, Zanini A, Manfredotti C, Palamara F, Sarotto M, Visca L and Nastasi U 2007. Design of an epithermal column for BNCT based on D–D fusion neutron facility *Nuc. Instr. and Meth. in Phys. Research Section A: Accelerators, Spectrometers, Detectors and Associated Equipment* **574(2)** 363-369.

[5] Faião-Flores F, Coelho P R P, Muniz R O R, Souza G S, Arruda-Neto J and Durvanei M A. 2011 Antitumor potential induction and free radicals production in melanoma cells by Boron Neutron Capture Therapy *Appl. Radiation and Isotopes* **69(12)** 1748-1751.

[6] Munem, Eid Mahmoud Eid Abdel. (2007). Neutron flux measurements with Monte Carlo verification at the thermal column of a Triga mark II reactor: Feasibility study for a BNCT facility

[7] Ali L, Ansari S A and Iqbal M. 1994 *Neutron flux measurements in the full power LEU core of PARR-1* (PINSTECH--138 Pakistan)

[8] Kortov V S, Milman I I, Nikiforov S V and Gorelova E A 2000 The use of thermoluminescent detectors for radiation monitoring on territories of atomic power plants *J. of Int. Research Publication* **1** 2000/01 ISSN, pp.1311-8978

[9] Abdel_Munem E A and Tajuddin A A 2005b Beam parameters and dosimetric characteristic for neutrons from the MINT TRIGA reactor thermal column as part of a design and feasibility study for BNCT *Posiding dan bengkel Reactor Interest Group (RIG) ' Research reactor for sustainable development* (24-25 June Cyberjaya. Malaysia, Malaysia Institute for Nuclear Technology research (MINT)) p 393-403.

[10] IAEA, International Atomic Energy Agency (2001) Current status of neutron capture therapy (IAEA-TECDOC-1223)

[11] Solleh M R M, Tajudin A A, Mohamed A A, Munem E M EA, Rabir M H, Yoshiaki K 2011 Collimator And Shielding Design For BNCT Research At TRIGA MARK II Reactor *Proc. of Nuclear Science, Technology and Engineering Conference (NUSTEC2011)* (Malaysian Nuclear Society)

IOP Conf. Series: Materials Science and Engineering **298** (2018) 012032 doi:10.1088/1757-899X/298/1/012032

Analysis of Water Volume Changes and Temperature Measurement Location Effect to the Accuracy of RTP Power Calibration

T.Lanyau[1, a)], N.S.Hamzah[1, b)], A.M.Jalal Bayar[1, c)], J.Abdul Karim[1], P.K.Phongsakorn[1], K.Mohammad Suhaimi[1], Z.Hashim[1], H.Md Razi[1], Z.Mohd Fazli[1], A.S.Ligam[1], M.K.A.Mustafa[1], M.Maskin[1]

[1]Malaysian Nuclear Agency, Bangi, 43000 Kajang, Selangor, Malaysia.

[a)]tonny@nuclearmalaysia.gov.my
[b)]naimsyauqi@nuclearmalaysia.gov.my
[c)]abi@nuclearmalaysia.gov.my

Abstract. Power calibration is one of the important aspect for safe operation of the reactor. In RTP, the calorimetric method has been applied in reactor power calibration. This method involves measurement of water temperature in the RTP tank. Water volume and location of the temperature measurement may play an important role to the accuracy of the measurement. In this study, the analysis of water volume changes and thermocouple location effect to the power calibration accuracy has been done. The changes of the water volume are controlled by the variation of water level in reactor tank. The water level is measured by the ultrasonic measurement device. Temperature measurement has been done by thermocouple placed at three different locations. The accuracy of the temperature trend from various condition of measurement has been determined and discussed in this paper.

1. Introduction

Reaktor TRIGA PUSPATI (RTP) is the only nuclear reactor in Malaysia, designed for research, training and radioisotope production. It comes to more than 35 years of operation since its first criticality in June 1982. The RTP is a pool type reactor, using demineralized light water as coolant. The operational power up to 1MWth with mixed core configuration. The core is cooled by natural convection. Heat removal is provided by the primary and secondary cooling system, whereby the reactor coolant water is circulated by the primary loop through the heat exchangers. The heat carried by the coolant water is transferred to the secondary loop for rejection to the ambient through cooling tower.

In nuclear research reactor, power monitoring is done by detection of neutron flux. The main reason to monitor neutron flux in a reactor is that it is proportional to the power density [1].In RTP, the reactor power is measured by fission chambers installed inside the reactor pool. Three fission chambers are used for power monitoring. The output from the fission chamber is translated into electrical power through Reactor Digital Instrumentation and Control System (ReDICS). In order to maintain functionality and accuracy of the power measurement, calibration have to be conducted prior to major changes in the core including maintenance activities. The main reason to conduct the power calibration is to ensure that the power measured by the fission chamber agrees with the thermal power released by the reactor in the form of temperature rise. In the existence of deviation between indicated power by the fission chamber with the measured thermal power through power calibration, the position of the fission chambers need to be adjusted physically or alternatively by tuning the electronic gain. The common practice of TRIGA reactor worldwide is to use calorimetric method for power calibration. This method is widely used in research reactor through measurement of thermal power released from the reactor. The heat produced during the operation of the reactor will contribute to temperature rise in the coolant water. With knowing the heat capacity constant K, the thermal power can be obtained by [2]–[8] :

$$P = K\frac{\Delta T}{\Delta t} \tag{1}$$

where P is the reactor power [kW], $\frac{\Delta T}{\Delta t}$ is temperature-rise rate [°C/h] and K is experimentally determined heat capacity constant [kWh/°C] given by

$$K = V_w \rho C_p \tag{2}$$

Where V_w is coolant volume [m^3], ρ is coolant density [kg/m^3] and C_p is the coolant specific heat capacity [kJ/kg°C]. The reactor pool heat capacity constant can be calculated as well with assumption that reactor pool temperature is constant throughout the pool and heat losses are neglected hence the reactor pool is treated as insulated when water temperature is close to ambient temperature[2], [3], [6]. For TRIGA system, the mass is mainly the water in the tank because of its large heat capacity[2], [3], [6]. Therefore, water volume in the reactor tank is an important parameter that may influence the accuracy of the power calibration.

For RTP, power calibration is done using the power calibration constant also called tank constant furnished by General Atomic at its early operation. The tank constant of the RTP is 9.56°C/200kW-hr [9]. The reactor tank is filled with demineralized light water about 17.41m^3[9]. Considering no major modification of the original design of the reactor tank that possibly contribute to significant changes in the water volume, therefore the tank constant still being used in the power calibration.

This paper presents the study conducted through the experiments to evaluate the effect of water volume changes and measurement location to the power calibration result. With the assumption that the total water volume is maintained, so the tank constant can be used in this experiment. Therefore, the water volume changes are introduced by changing the water level. The result obtained from this study is very important to improve the effectiveness of temperature measurement in order to reduce error of the power calibration.

2. Experimental set-up and procedure

The experiments that have been conducted are based on the procedure of power calibration conducted at RTP. The set-up of the experiment is illustrated in Figure 1. During the normal operation, the water level is control within 15.24cm to 19.30cm from the pool top. To introduce the water volume changes, the level is set to 15.24cm (high), 17.53cm (intermediate) and 19.30cm (low) from the pool top. The water level is measured by the ultrasonic measurement device and monitored through the display panel located at the top of the reactor pool. Three multi-meters attached with two thermocouples type K each are installed at various location in the pool for temperature measurement. Combination of multi-meter and thermocouples are shown in Table 1. These combinations are used in each experiment condition as mentioned in Table 2.The bulk temperature detector also used for temperature measurement. The temperature of bulk is measure by the Resistance Temperature Detector (RTD) installed under the bridge. The arrangement of the thermocouples and location of bulk temperature detector is illustrated in Figure 1. Two thermocouples are installed for each combination, such as TC1A and TC1B in order to observe the effect of particular location to the temperature measurement at same depth of the thermocouple.

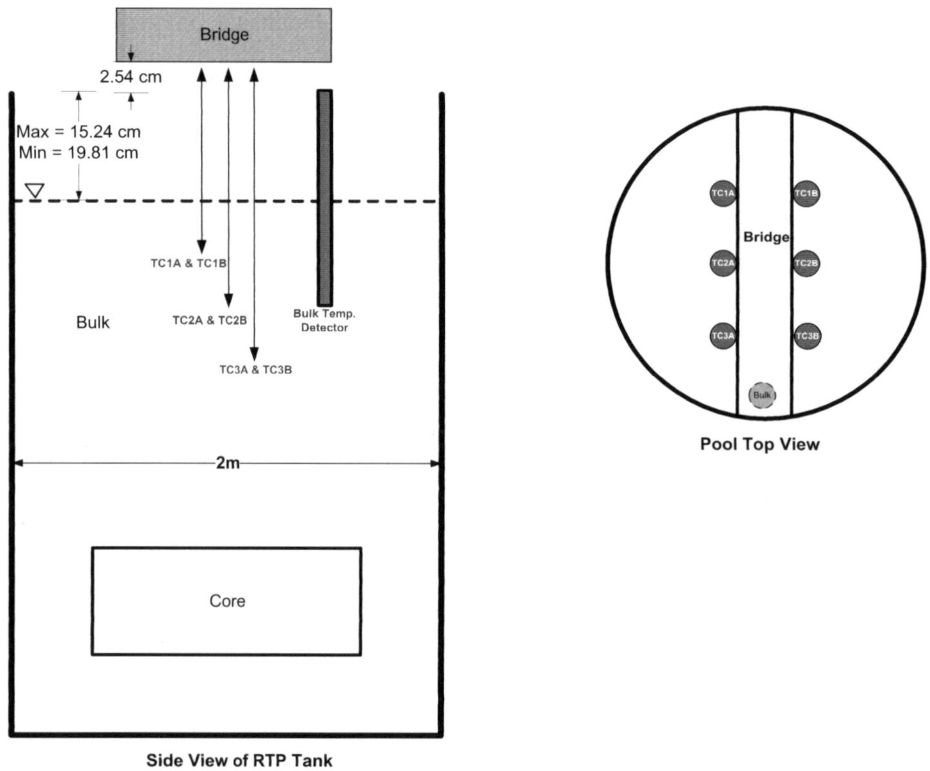

Figure 1. Set-up of experiment

Table 1. Devices Set-up

Multi-meter ID	Thermocouple ID	Thermocouples Depth from bottom of bridge [cm]
M1	TC1A & TC1B	50.8
M2	TC2A & TC2B	63.5
M3	TC3A & TC3B	76.2

Table 2. Experiment Condition

Experiment Set	Water Level [cm]
1.1	15.24
1.2	15.24
2.1	17.53
2.2	17.53
3.1	19.30
3.2	19.30

The experiment was conducted by the following procedure.

1. Install the multi-meters and thermocouples based on the combination in Table 1.
2. Set pool bulk water temperature close to ambient temperature (25°C ~ 30 °C) by running the cooling system. After the temperature achieve an equilibrium state, then switch-off the primary pump and demineralizer pump.
3. Switch-off cooling tower and secondary pump.
4. Isolate the reactor tank by closing the incoming and outgoing valve (PV1, PV2, PV7, and PV8).
5. Set the water level to specified level. Water topping and draining can be done through the demineralizer system if necessary.
6. Operate the reactor at power constant power 200kW in auto mode for 75 minutes. In the event of temperature of the RTD (bulk temperature detector) reach near 45°C, the reactor will be shutdown manually.
7. Record reading from all thermocouples and bulk temperature every 5 minutes time interval.
8. Shutdown the reactor after completing 75 minutes.
9. Run primary and secondary cooling system to cool down the water until ambient temperature.

3. Result and discussion

The results of temperature rise at different water level are presented in Figure 2, Figure 3 and Figure 4.These results show the agreement of linear relationship between temperature rise and time period of the reactor operation at constant power 200kW. The slope of the graphs which represent the temperature rise rate show about the same value of 0.15°C/min. This taught shows that the variation of water level which indicate the changes in the water volume within the defined range of water level 15.24cm to 19.30cm has minimum effect to the accuracy of the temperature measurement. The calculated average power based on the slope of each experiment using Equation (1) are presented in Table 3.

IOP Conf. Series: Materials Science and Engineering **298** (2018) 012032 doi:10.1088/1757-899X/298/1/012032

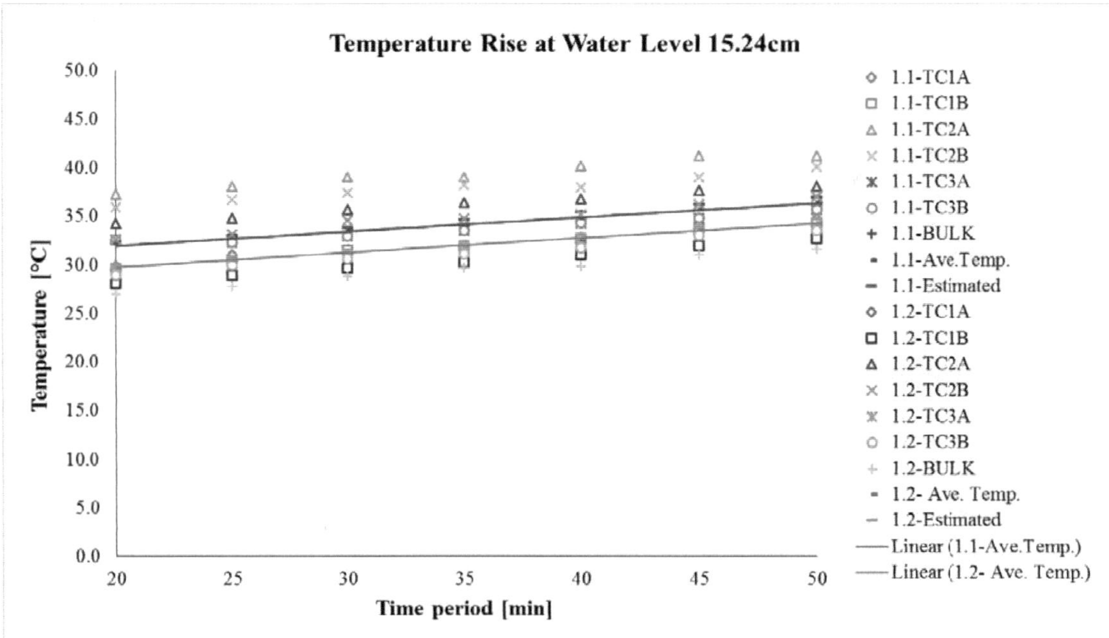

Figure 2. Temperature rise at water level 15.24cm

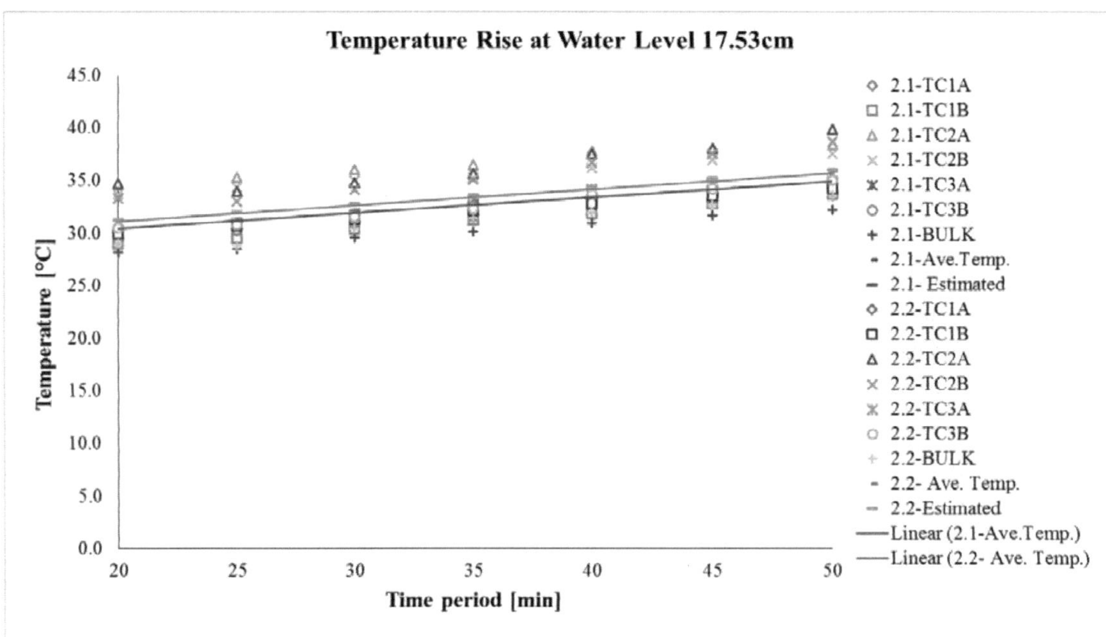

Figure 3. Temperature rise at water level 17.53cm

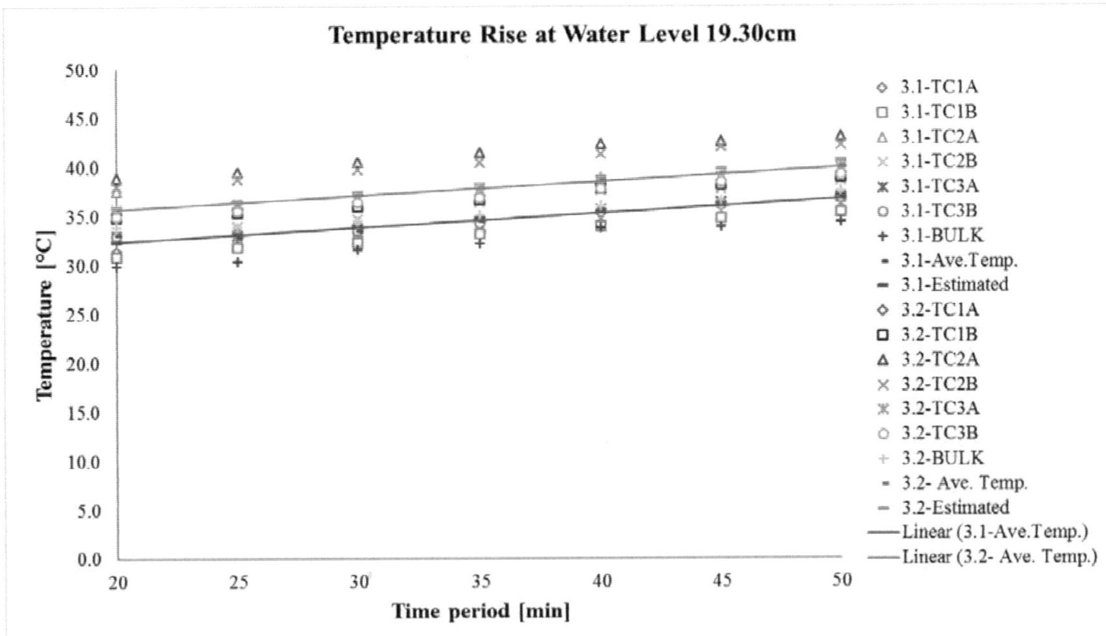

Figure 4. Temperature rise at water level 19.30cm

Table 3. Evaluation of power measurement accuracy at different water level

Water Level [cm]	Experiment ID	Slope [°C/min]	Average Power [kW]	Standard Deviation	Error from indicated power 200kW [%]
15.24	1.1	0.147	186.59	2.66	6.70
	1.2	0.150			
17.53	2.1	0.150	189.85	3.11	5.07
	2.2	0.153			
19.30	3.1	0.148	183.0	3.54	8.40
	3.2	0.144			

The analysis of the temperature rise at different measurement location shows that different depth of the thermocouples has small influence to the result of tempereature measurement. The trend of the temperature rise shows the linear relationship with the slope is about the same. The average temperature rise at different measurement point are plotted in Figure 5, Figure 6 and Figure 7. Table 4 shows the evaluation of the accuracy of measurent at different thermocouple depth. The deviation of average power at different depth of thermocouples is very small.

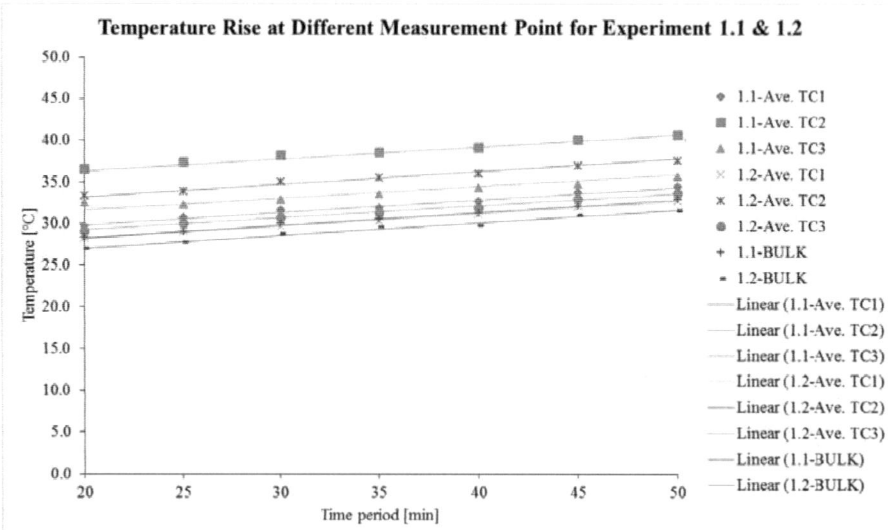

Figure 5. Temperature rise at different measurement point for experiment 1.1 and 1.2

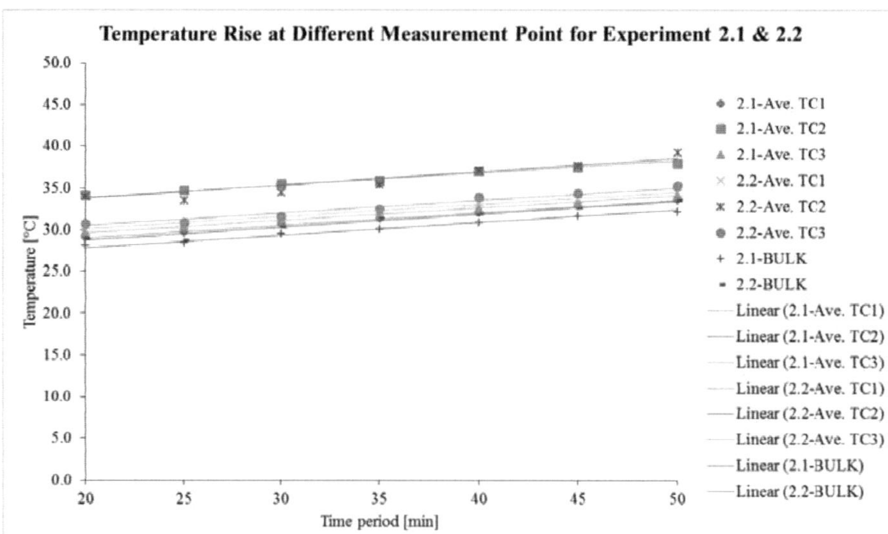

Figure 6. Temperature rise at different measurement point for experiment 2.1 and 2.1

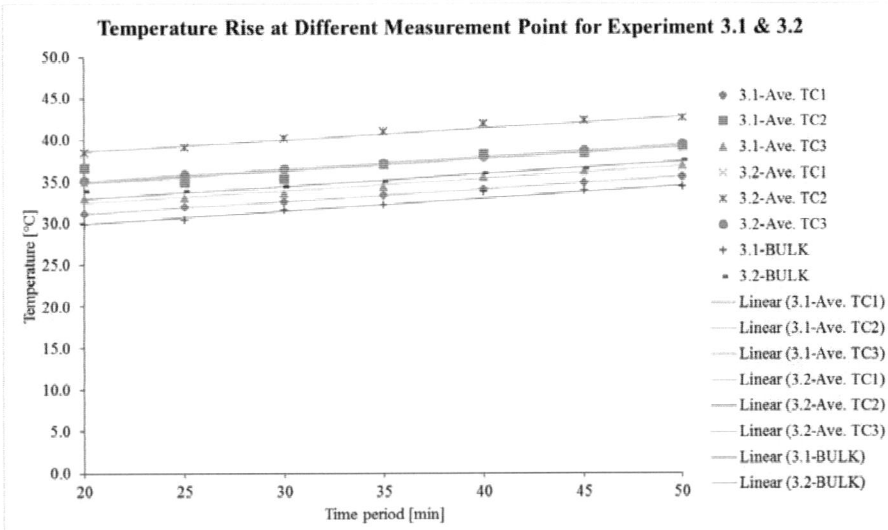

Figure 7. Temperature rise at different measurement point for experiment 3.1 and 3.2

Table 4. Evaluation of power measurement accuracy at different measurement point

Depth of thermocouple [cm]	Experiment ID	Average Slope [°C/min]	Average Power [kW]	Standard Deviation	Error from indicated power 200kW [%]
50.8	1.1	0.151	186.61	3.41	6.7
	1.2	0.150			
	2.1	0.150			
	2.2	0.146			
	3.1	0.150			
	3.2	0.145			
63.5	1.1	0.144	185.77	9.85	7.12
	1.2	0.153			
	2.1	0.147			
	2.2	0.161			
	3.1	0.144			
	3.2	0.139			
76.2	1.1	0.143	184.10	3.52	7.95
	1.2	0.146			
	2.1	0.150			
	2.2	0.150			
	3.1	0.146			
	3.2	0.145			

4. Conclusion
This paper conclude that the water volume changes introduced by the variation of the water level at range 15.24cm to 19.30cm has small influence to the temperature measurement. The measurement position also does not affect much the temperature measurement result. From the thermal power calculation based on the slope of the graph shows some error from the indicated power by the fission chamber. Therefore, parallel to the objective of power calibration, some adjustment on the position of fission chambers need to be done to tune the indicated power level.

Acknowledgement
The authors express their sincere thanks to those involved directly or indirectly in the experiments especially to the staffs in Reactor Technology Centre, Malaysian Nuclear Agency for their kind help, comments and recommendations all the way to complete this paper. Their contributions are highly appreciated.

References
[1] IAEA, "Triga Reactor," *Nucl. Install. Saf. Div. IAEA*, 2005.
[2] A. Z. Mesquita, H. C. Rezende, and A. Augusto, "Development of Methods for Monitoring and," vol. 13, no. 1, pp. 24–27, 2012.
[3] A. Z. Mesquita, H. C. Rezende, and R. M. Gomes Do Prado Souza, "Thermal power calibrations of the IPR-R1 TRIGA reactor by the calorimetric and the heat balance methods," *Prog. Nucl. Energy*, vol. 53, no. 8, pp. 1197–1203, 2011.
[4] A. P. Tomaz Azgar, Matjaz Ravnik, "Analysis of TRIGA Reactor Thermal Power Calibration Method," 1999.
[5] G. Zerovnik, L. Snoj, A. Trkov, L. Barbot, D. Fourmentel, and J. F. Villard, "Measurements of thermal power at the TRIGA Mark II reactor in Ljubljana using multiple detectors," *IEEE Trans. Nucl. Sci.*, vol. 61, no. 5, pp. 2527–2531, 2014.
[6] M. A. Salam, A. Haque, M. M. Uddin, M. B. Shohag, and M. A. Malek Soner, "Thermal Power Calibration and Error Minimization of 3MW TRIGA Mark-II Research Reactor," *Int. J. Sci. Eng. Res.*, vol. 7, no. 5, pp. 710–716, 2016.
[7] S. A. Agbo, Y. A. Ahmed, I. O. B. Ewa, and Y. Jibrin, "Analysis of Nigeria Research Reactor-1 Thermal Power Calibration Methods," *Nucl. Eng. Technol.*, vol. 48, no. 3, pp. 673–683, 2016.
[8] Ž. Štancar and L. Snoj, "Thermal Power Calibration of the TRIGA Mark II reactor," *Proc. 23rd Int. Conf. Nucl. Energy New Eur.*, pp. 1–8, 2014.
[9] Malaysian Nuclear Agency, "Safety Analysis Report for PUSPATI TRIGA Reactor 2017," 2017.

Neutron and gamma ray fluences measurement at radial Beam Port 1 of TRIGA MARK II PUSPATI research reactor

Muhammad Aliff Ashraff Rosdi[1], Pei Sean Goh[1], Faridah Idris[2], Safwan Shalbi[1], Muhammad Syahir Sarkawi[1], Nur Syazwani Mohd Ali[1], Nur Liyana Jamsari[3], Ahmad Syafiq Ramli[4], Azraf Azman

[1]Nuclear Engineering Department, Faculty of Chemical and Energy Engineering, Universiti Teknologi Malaysia, 81310 Skudai, Johor Bahru, Malaysia
[2]Malaysian Nuclear Agency, 43000 Kajang, Selangor, Malaysia
[3]Universiti Kebangsaan Malaysia, 43600 Bangi, Selangor, Malaysia
[4]Universiti Tenaga Nasional, 43000 Kajang, Selangor, Malaysia

peisean@petroleum.utm.my

Abstract. Neutron and gamma ray fluences are the important criteria to be looked at in a neutron diffractometer system facility design. The fluence is defined as the number of neutrons and gamma rays travel through a unit area. Currently, the facility for neutron diffractometer system are planned to be developed at radial beam port 1 of TRIGA MARK II PUSPATI research reactor (RTP). The aim of this research is to determine the value of neutron and gamma ray fluences produced at the end of radial beam port 1 and to identify the shielding materials suitable for neutron and gamma ray. In order to achieve this aim, an experiment has been designed to obtain the neutron and gamma dose rates by using TLD-600 and TLD-700. The results from this experiments are converted into neutron and gamma ray fluences and are then compared with the results from simulation. The comparison shows that both results meet an agreement on the feasibility of shielding material for neutron diffractometer system. Our research results may be of help in the design of shielding material for neutron diffractometer facility at RTP.

1. Introduction

Radiation shielding is a physical barrier that is able to lower radiation dose rate by reflecting or absorbing radiation particles when dealing with ionizing radiation such as neutrons, gamma, and other charge particles. Shielding plays a vital role in all radiation practices especially in nuclear applications. In neutron diffractometer system, shielding is one of the essential instruments that needs to be emphasized. Aside from being used as biological shielding for human, the shielding is also important to protect sensitive instruments from radiation [1]. There are several considerations need to be taken when choosing the suitable materials for shielding such as strength of shielding, radiation particle attenuation factor, damage resistance, mechanical properties and cost. Generally, high density materials are more practical to shield gamma radiations compare to their low density counterparts. The thickness of the shielding materials can also influence the efficiency of the shielding materials. As for neutron radiation, materials with low atomic number such as water and hydrogen based materials are known to be more effective. However, one significant drawback of low atomic number of materials is the formation of gamma radiation upon their interactions with neutron. The common materials which have been used as radiation shielding materials are lead, boron, paraffin, cadmium, polyethylene and concrete [2][3][4].

In this paper, the performance of the shielding materials has been tested in order to identify most suitable material to shield ionizing radiations and to determine the fluences obtained at radial beam port 1 of RTP. An experimental investigation has been performed for neutron and gamma radiations emitted from radial beam port to identify the shielding properties of each materials. The neutron and gamma dose rate was obtained by using Thermoluminescence Detector (TLD). Monte Carlo N-Particle (MCNP) simulation also has been conducted for the same shielding arrangement from source of ^{241}Am/Be. Both results from experiment and simulation were compared and thus obtained the best suitable shielding materials to be used for neutron diffractometer system applications.

2. Methodology

2.1. Materials

Total of eight materials have been chosen such as boron block, lead sheet, pure polyethylene block, ordinary concrete block, aluminium plate, cadmium sheet, steel slab, and pure paraffin block, to be irradiated. The thickness of all samples are set to be constant. Table 1 shows the details of shielding materials used in this experiment.

Table 1. Characteristics of shielding materials.

No	Material	Thickness, cm	Molecular weight, g/mol	Density, g/cm^3
1	Boron	5	10.81	2.50
2	Lead	5	207.2	11.4
3	Polyethylene	5	84.16	0.94
4	Concrete (Ordinary)	5	13.92	2.30
5	Aluminium	5	26.98	2.70
6	Cadmium	5	112.41	8.70
7	Steel	5	55.84	8.05
8	Paraffin	5	226.44	0.90

2.2. Experimental arrangement

The experiment was performed at the TRIGA MARK II PUSPATI research reactor (RTP), Malaysian Nuclear Agency (MNA), Kajang, Malaysia. Among four beam ports available, radial beam port 1 was chosen because the neutron diffractometer system was proposed to be developed there [6]. The experimental arrangement consist of neutron and gamma source directly from reactor core with the total flux intensity of $5.9\text{x}10^{10}\text{cm}^{-2}\text{s}^{-1}$ at maximum power of 750 kW [7], two type of detectors (TLD-600 and TLD-700), and shielding materials. According to simulation by Farhi in 2015, $2.0\text{x}10^{8}\text{cm}^{-2}\text{s}^{-1}$ flux yielded from the end of the beam port 1 [6]. Figure 1 shows the setup of the experiment. The sample material was placed inside of a shielding box at a distance of 30cm away from the end of beam tube. Then, a detector was located in front of the box at distance of 5cm away from sample. The TLD-600 was used to measure dose rate of neutron and gamma ray while TLD-700 to measure gamma particles only. The detectors were found to be the most suitable method to determine the neutron and gamma ray dose rate [8].

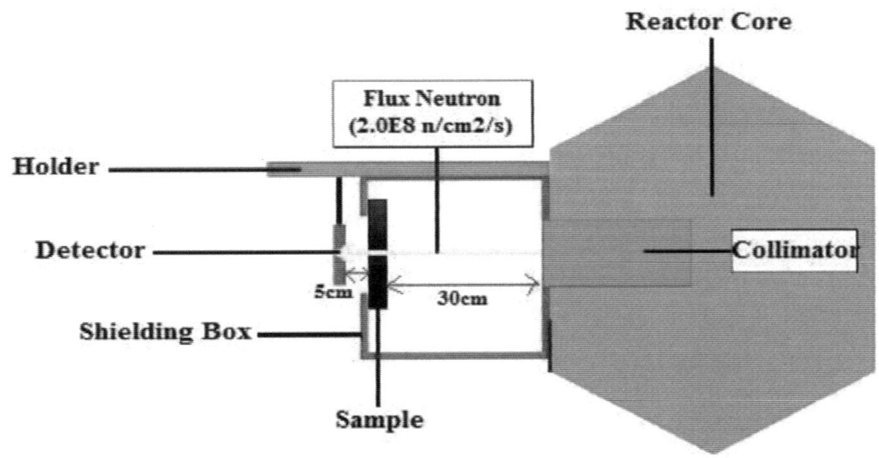

Figure 1. Experimental setup

Initially, the power of reactor is set to 15 kW. The reason of choosing low thermal power is because of the safety issue where insufficient shielding blocks were available around the experimental site. Then, TLD detectors from both types are placed in front of the shielding box without any sample for a minute in order to measure background radiation dose emitted from radial beam port 1 of RTP. Next, first sample of experiment is put inside the shielding box. The reactor need to be scrammed before the following step as for radiation precaution. The steps was followed by placing the TLDs 5cm away from sample for a minute to measure radiation dose rate produced after the sample. The experimental procedures were repeated for the other 7 samples.

Radiation dose rate of neutron and gamma measured from the TLD-600 and TLD-700 was analyzed at Secondary Standard Dosimetry Laboratory (SSDL). The neutron and gamma radiation dose was calculated using the following equation [10]:

$$\mathrm{H_p(10)_\eta = \big(Q4(nC) - Q3(nC)\big)\ x\ CF_\eta(\tfrac{mSv}{nC})} \tag{2.1}$$

$$\mathrm{H_p(10)_\gamma = \big(Q2(nC)\big)\ x\ CF_\gamma(\tfrac{mSv}{nC})} \tag{2.2}$$

where equation 2.1 is used to calculate neutron dose rate and equation 2.2 to calculate gamma dose rate respectively. The neutron and gamma dose rate obtained are then converted into fluence using the following formula [5]:

$$P = \frac{H}{\phi} \tag{2.3}$$

where P is dose equivalent per fluence (9.1pSvcm^2), H is radiation quantity dose equivalent in Sv, and ϕ is neutron fluence in cm^{-2}.

2.3. Monte Carlo simulation code MCNP

The results from the experiment were compared with the simulation results of the general purpose MCNP code MCNPX. The MCNPX code is a Fortran90 radiation transport code developed in 1994 that can transport particles with various energy ranges. Basically, the interaction of neutrons and photons with matter can be described through statistical means which best simulated using Monte Carlo method. This method is a computational algorithm that can provide approximate solutions to a variety of nuclear by the simulation of random quantities [9].

IOP Conf. Series: Materials Science and Engineering **298** (2018) 012033 doi:10.1088/1757-899X/298/1/012033

3. Results and discussion

3.1. Neutron and gamma dose rate measurement.

The result from the calculations was then interpolated to the power of 750 kW and shown in table 2.

Table 2. Neutron and gamma dose rate.

Material	Thickness, cm	Hp(10)η, mSv	Hp(10)γ, mSv
Background	5	$19.3x10^{-1}$	$6.88x10^{-1}$
Boron	5	$3.36x10^{-5}$	$14.2x10^{-1}$
Lead	5	$9.46x10^{-1}$	$3.43x10^{-1}$
Polyethylene	5	$12.8x10^{-1}$	$9.83x10^{-1}$
Concrete	5	$15.4x10^{-1}$	$10.4x10^{-1}$
Aluminium	5	$18.9x10^{-1}$	$11.8x10^{-1}$
Cadmium	5	$4.83x10^{-5}$	$22.8x10^{-1}$
Steel	5	$2.70x10^{-2}$	$11.9x10^{-1}$
Paraffin	5	$15.3x10^{-1}$	$8.40x10^{-1}$

3.2. Neutron and gamma fluence measurement

Fluence is defined as the number of radiation particles that intersect at a unit area of material. The unit of fluence is cm^{-2}. Based on the neutron and gamma dose rate obtained from TLDs, fluence can be calculated by using the equation 2.3 as stated before. The fluence of neutron and gamma measured in this experiment are tabulated and plotted in table 3 and figure 2.

Table 3. Neutron and gamma fluence

Material	Thickness, cm	Absorption Cross Section, σ_a	Fluence, cm^{-2}	
			Neutron	Gamma ray
Background	5	-	$2.12x10^{8}$	$7.56x10^{7}$
Boron	5	767	$3.69x10^{3}$	$1.56x10^{8}$
Lead	5	0.171	$1.04x10^{8}$	$3.77x10^{7}$
Polyethylene	5	0.336	$1.41x10^{8}$	$1.08x10^{8}$
Concrete	5	6.4	$1.69x10^{8}$	$1.14x10^{8}$
Aluminium	5	0.231	$2.08x10^{8}$	$1.30x10^{8}$
Cadmium	5	2520	$5.30x10^{3}$	$2.51x10^{8}$
Steel	5	2.56	$2.91x10^{6}$	$1.31x10^{8}$
Paraffin	5	0.337	$1.68x10^{8}$	$9.23x10^{7}$

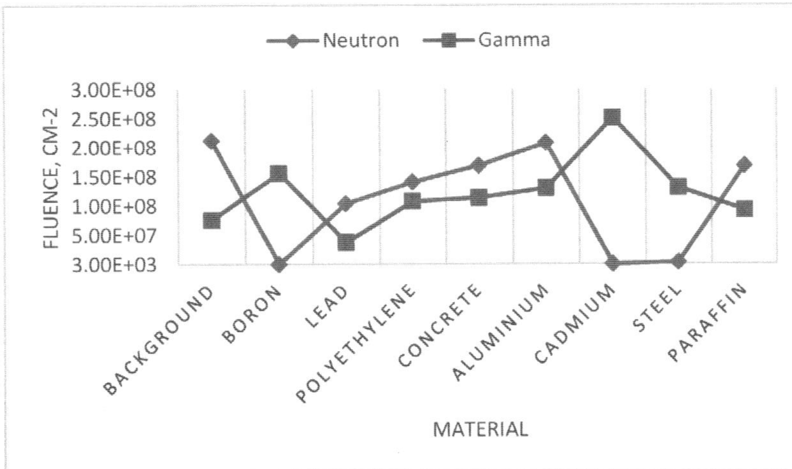

Figure 2. Neutron and gamma fluence

Based on the experimental results, average fluence measured at the end of the beam port 1 at 750 kW power is $1.44\text{x}10^{10}\text{cm}^{-2}$ which is in the acceptable range as reported in the previous study [6]. Specifically, neutron fluence and gamma fluence measured from the experiment at the end window of radial beam port 1 was $2.12\text{x}10^{8}\text{ncm}^{-2}$ and $7.56\text{x}10^{7}\gamma\text{cm}^{-2}$ respectively. The neutron fluence decreased after a shielding sample was placed in front of the beam port. Based on figure 2, the lowest neutron flux was obtained with boron sample block which is $3.69\text{x}10^{3}\text{ncm}^{-2}$ followed by cadmium sheets, steel slabs, lead sheets, pure polyethylene block, paraffin block, ordinary concrete block and aluminium plates. Boron has significantly reduced the neutron fluence due to the absorption of neutron particle as boron is known to be effective in capturing neutron, especially thermal neutron as it has high neutron absorption cross section [11]. Boron (^{10}B) based element is one of the well-known materials to be used in nuclear applications because of its strong neutron absorption properties. Similarly, cadmium also has high neutron absorption cross section which enables it to absorb thermal and epithermal neutrons [11]. As for steel slabs, iron based materials are known to exhibit high neutron scattering cross section. The reduced neutron fluence was mainly attributed to the scattering of neutron particles by the materials. However, iron based materials is not commonly used to shield neutron as its can be easily activated upon interaction with ionizing radiations [12]. In case of pure polyethylene, paraffin and ordinary concrete, the neutron fluence was reduced owing to the presence of light hydrogen element in these materials which are very effective to slow down and stop the neutron particles [13]. Aluminium was found to be neutron transparent as the neutron fluence was only slightly reduced. The low neutron activation properties of aluminium has prompted the utilization of aluminium based materials for instrumentation inside the reactor core [14].

On the other hand, the increase in gamma fluence was observed when the beam passed through all the shielding samples except when lead was used, in which the fluence was reduced to $3.77\text{x}10^{7}\gamma\text{cm}^{-2}$. The gamma fluence measured was in the following increasing order: lead sheets, paraffin block, pure polyethylene block, ordinary concrete block, aluminium plate, steel slabs, boron block and cadmium sheets. The increase in gamma fluence was mainly due to the interaction of these materials with neutron emitted secondary gamma radiation, particularly for boron and cadmium. Lead was a very common shielding material used to shield gamma radiation as it can reduce gamma particle. High gamma absorption cross section and high atomic number properties of lead enable it to be very effective in shielding gamma and x-ray radiations [15].

Results from experiment was compared with the results obtained from MCNPX code simulation as shows in table 4. Figure 3 shows the comparison of the neutron fluence measured and figure 4 compare the gamma fluence obtained. Most of the experiment results was measured a bit higher than the simulation. This was affected by the different incident fluence yielding from the reactor core and the source used in simulation code. However, the results obtained agreed well as both shows the identical trend. The peak produced in Figure 4 at cadmium sample indicates that the cadmium emitted high energy of secondary gamma radiation compare to secondary gamma radiation emitted by boron. In addition, higher measurement of gamma fluence at cadmium during experiment compared to simulation was recorded because of the strong emission of secondary gamma radiations that been detected by detectors.

Table 4. Comparison between simulation and experimental results.

Material	Thickness, cm	Neutron			Gamma		
		Simulation ncm^{-2}	Experimental ncm^{-2}	Error %	Simulation γcm^{-2}	Experimental γcm^{-2}	Error %
Background	5	$1.97x10^8$	$2.12x10^8$	7	$6.06x10^7$	$7.56x10^7$	20
Boron	5	$9.35x10^2$	$3.69x10^3$	75	$1.60x10^8$	$1.56x10^8$	3
Lead	5	$8.13x10^7$	$1.04x10^8$	22	$3.36x10^7$	$3.77x10^7$	11
Polyethylene	5	$1.13x10^8$	$1.41x10^8$	20	$7.48x10^7$	$1.08x10^8$	31
Ordinary Concrete	5	$1.28x10^8$	$1.69x10^8$	24	$9.49x10^7$	$1.14x10^8$	17
Aluminium	5	$1.39x10^8$	$2.08x10^8$	33	$1.12x10^8$	$1.30x10^8$	14
Cadmium	5	$1.54x10^3$	$5.30x10^3$	71	$1.38x10^8$	$2.51x10^8$	45
Steel	5	$1.26x10^6$	$2.91x10^6$	57	$1.28x10^8$	$1.31x10^8$	2
Paraffin	5	$1.14x10^8$	$1.68x10^8$	32	$7.51x10^7$	$9.23x10^7$	19

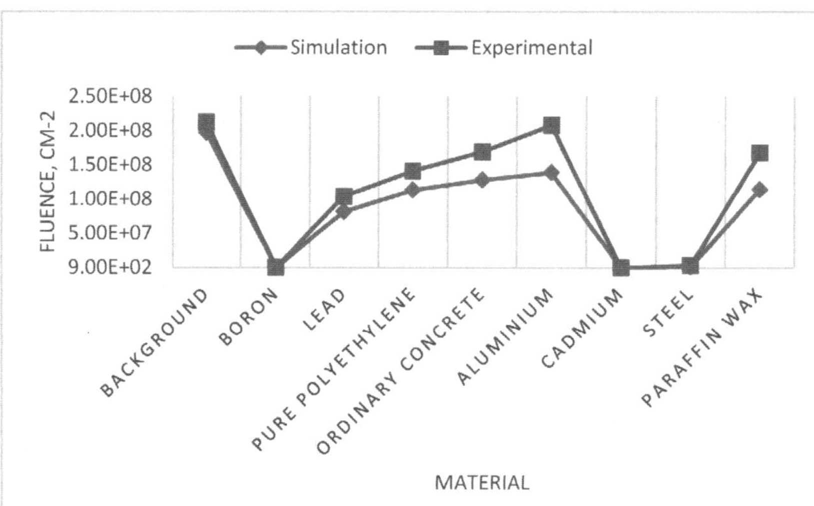

Figure 3. Comparison of neutron fluence between simulation and experiment.

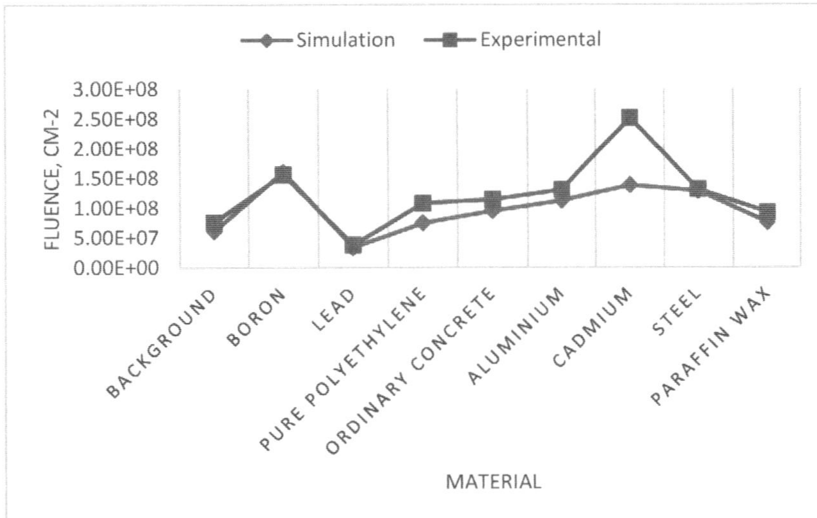

Figure 4. Comparison of gamma fluence between simulation and experiment.

4. Conclusion

Comparisons between simulation and experimental parameters were carried out in this paper. As result, the average fluence produced from the radial beam port 1 at maximum power of 750 kW is $\sim 10^8 cm^{-2}$. Besides, the appropriate biological shielding materials for neutrons and gamma rays also has been identified. Boron shielding sample has been found to be the best material to shield neutron particles while lead shielding material is good for gamma radiation particles. Both results from simulation and experimental agreed well in term of trend with average error of 38% and 18% for neutron and gamma respectively.

5. References

[1] Paranjpe S 2005 Measurement of residual stress in materials using neutrons (Proc. of a Technical Meeting Held in Vienna, 13–17 October 2003, International Atomic Energy Agency).

[2] Mergia K, Salevris A and Messoloras S 2004 The new neutron powder diffractometer at "Demokritos" Research Reactor *Physica B: Condensed Matter* **350(1)** 162-165.

[3] Abdo A, Ali E.-S, and Ismail M 2003 Influence of magnetite and boron carbide on radiation attenuation of cement–fiber/composite *Ann. of Nuc. Energy* **30(4)** 391-403.

[4] Abdullah Y, Yusof M R, Muhamad A, Samsu Z and Abdullah N E 2010 Cement-boron carbide concrete as radiation shielding material *J. of Nuclear and Related Technologies* **7(2)** 1-6.

[5] Cossairt J D and Vaziri K 2008 Neutron dose per fluence and weighting factors for use at high energy accelerators (FERMILAB-PUB-08-244-ESH-REV).

[6] Farhi E, Mohamad Idris F, Mohamed A A, Yazid H B, Jamro R, Harun Al Rashid M and Zin M 2015 Neutron scattering instrumentation at low power reactors for science, engineering and education, *J. of Neutron Research* **18(2-3)** 61-77.

[7] Rabir M H 2017 Power and neutron flux calculation for the puspati TRIGA reactor using MCNP *Jurnal Sains Nuklear Malaysia* **24(1)** 22-32.

[8] Kim J L, Lee J I, Chang I, Pradhan A S, Kim S I and Kim B H 2013 TL Response of pairs of ^{6}LiF: Mg,Cu,Si/^{7}LiF:Mg,Cu,Si and TLD-600/TLD-700 to 0.1-12 MeV neutrons *Rad. Measurements.* **56(1)** 223-227.

[9] Jamro R, Kardjilov N, Rabir M H, Mohamed Zain M R M, Ali A A, Idris F M, Ahmad M H A R M, Yazid K, Yazid H and Azman A 2017 Monte Carlo simulation for designing collimator of the neutron radiography facility in Malaysia *Physics Procedia* **88** 361-368.

[10] Agensi Nuklear Malaysia 1990 Pengiraan dos menggunakan TLD (NM: SSDL.CLN.1990).
[11] Massachusetts Institute of Technology 2005 Properties for materials shielding.
[12] Cossairt J D 2014 Radiation physics for personnel and environmental protection (Fermilab Report Tm-1834 Rev. 13 March 2014).
[13] Madsen D 2014 Borated polyethylene - fire properties and other issues (Report 5443. Lund University, Sweden).
[14] Joshua AR 2009 Design, construction, and characterization of a prompt gamma activation analysis facility at the Oregon State University TRIGA® Reactor. (MSc thesis Oregon State University).
[15] Erdem M, Baykara O, Doğru M, and Kuluöztürk F 2010 A novel shielding material prepared from solid waste containing lead for gamma ray *Rad. Phys. and Chem.* **79(9)** 917-922.

Utilization of thorium and U-ZrH$_{1.6}$ fuels in various heterogeneous cores for TRIGA PUSPATI Reactor (RTP)

Abdul Hannan Bin Damahuri[1, a)], Hassan Mohamed[1, b)], Abdul Aziz Mohamed[1, c)], Faridah Idris[2, d)]

[1]College of Engineering, Universiti Tenaga Nasional, Jalan IKRAM-UNITEN, 43000 Kajang, Selangor, Malaysia
[2]Reactor Technology Centre, Malaysian Nuclear Agency, Bangi, 43000 Kajang, Selangor, Malaysia.

[a)]ahannan@uniten.edu.my
[b)] mhassan@uniten.edu.my
[c)]azizm@uniten.edu.my
[d)]faridah_idris@nuclearmalaysia.gov.my

Abstract. The use of thorium as nuclear fuel has been an appealing prospect for many years and will be great significance to nuclear power generation. There is an increasing need for more research on thorium as Malaysian government is currently active in the national Thorium Flagship Project, which was launched in 2014. The thorium project, which is still in phase 1, focuses on the research and development of the thorium extraction from mineral processing ore. Thus, the aim of the study is to investigate other alternative TRIGA PUSPATI Reactor (RTP) core designs that can fully utilize thorium. Currently, the RTP reactor has an average neutron flux of 2.797 x 10^{12} cm^{-2}/s^{-1} and an effective multiplication factor, k_{eff}, of 1.001. The RTP core has a circular array core configuration with six circular rings. Each ring consists of 6, 12, 18, 24, 30 or 36 U-ZrH$_{1.6}$ fuel rods. There are three main type of uranium weight, namely 8.5, 12 and 20 wt.%. For this research, uranium zirconium hydride (U-ZrH$_{1.6}$) fuel rods in the RTP core were replaced by thorium (ThO_2) fuel rods. Seven core configurations with different thorium fuel rods placements were modelled in a 2D structure and simulated using Monte Carlo n-particle (MCNPX) code. Results show that the highest initial criticality obtained is around 1.35101. Additionally there is a significant discrepancy between results from previous study and the work because of the large estimated leakage probability of approximately 21.7% and 2D model simplification.

1. Introduction

Malaysian government is interested to utilize thorium as an economic, clean and sustainable nuclear fuel under the thorium flagship project that was launched in 2014, which focuses on thorium research and development in Malaysia [1]. One of the objectives of the national project is to produce nuclear-grade fuel from local source for Malaysia's research reactor, the TRIGA PUSPATI Reactor (RTP). Meanwhile, for other countries, thorium research is typically carried out for light water reactors and molten salt reactors, such as a liquid fluoride thorium reactor [2].

Thorium offers a lot of benefits for future energy solution and should be explored for potential usage. It is a fertile material that cannot cause fission reaction on its own. Thorium needs to be transmuted from ^{232}Th into fissile material, namely ^{233}U by neutron bombardment [3].When the thorium bombarded with neutron, the ^{232}Th will absorb the neutron and become ^{233}Th. Next, ^{233}Th will undergo decay process and become ^{233}Pa, which will then decay and become ^{233}U. From here, ^{233}U will undergo fission and the energy will be released with neutrons.

TRIGA PUSPATI Reactor (RTP) is the first and only research reactor in Malaysia. It is located in Bangi and managed by Malaysian Nuclear Agency (MNA). The reactor provides various applications on nuclear technologies, including Neutron Activation Analysis (NAA), Small Angle Neutron Scattering (SANS), Neutron Radiography (NR) and educational training for schools and universities. RTP is an open pool type reactor that uses uranium zirconium hydride ($U\text{-}ZrH_{1.6}$) as the main fuels, which have 8.5, 12 and 20 wt.% of uranium enrichment. RTP uses light water as its coolant and moderator while graphite as a reflector.

The reactor has been operated since 1982 and managed to reach its first criticality on the 28th of June 1982 by only using 66 rods of $U\text{-}ZrH_{16}$ that contain about 2.5 kg of ^{235}U. As for the operational system, the reactor has 80 rods of $U\text{-}ZrH_{1.6}$, comprising of 2.9 kg of ^{235}U. The reactor is operating at nominal power of 750 kW for 1 to 6 hours a day per session. Although the maximum operational power of the reactor is 750 kW, it is licensed to operate up to 1000 kW of power. The reactor only runs normaly from Monday to Thursday in a single week. By excluding weekends and 4 weeks of maintenance period, the reactor works for 175 day yearly. Figure 1 shows the operational hours of RTP for the past 10 years, starting from 2001 to 2010.

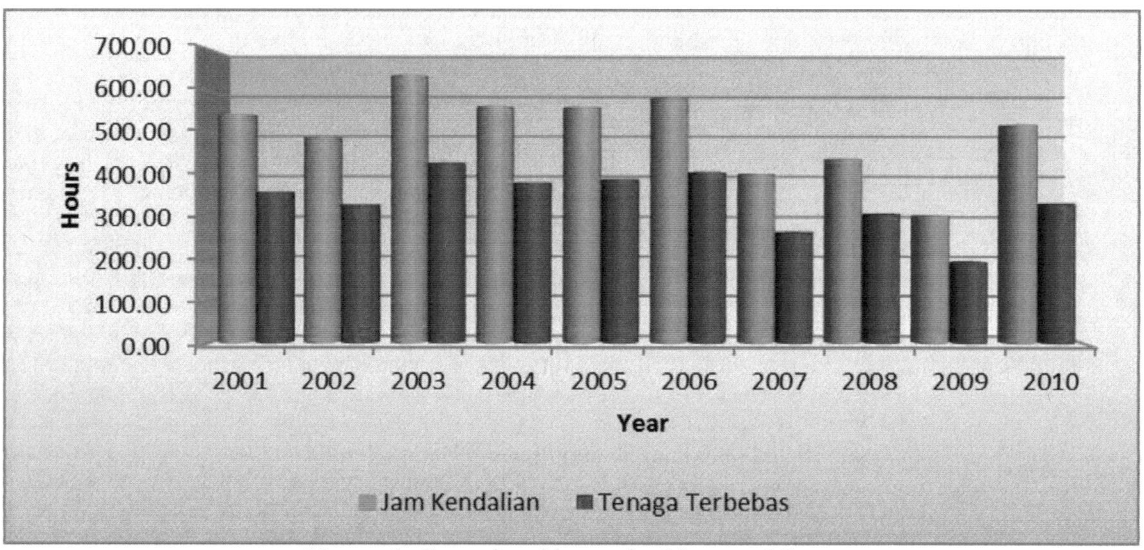

Figure 1. Operational hours for 10 years [4].

To support the national initiative in developing thorium nuclear fuels for RTP, the main objective of the work is to investigate the neutronic feasibility of thorium fuels in RTP. This study is particularly interested to calculate the initial multiplication factor or criticality, *k*, of the RTP core with thorium fuels. Various core configurations of RTP with thorium fuels are modeled and analyzed using Monte Carlo n-Particle.

2. Methodology

The RTP core consists of 6 fuel rings that encircle the central timber, which are labeled as rings B, C, D, E, F and G. The innermost ring is labeled as ring B and the outermost ring is ring G. Ring B contains 6 fuel rods and increasing to 12 rods for ring C. whereas ring D, E, F and G consist of 18, 24, 30 and 36 rods respectively. In each ring, there are several holes used to place fuel or control rods. Figure 2 shows the RTP core that consists of $U\text{-}ZrH_{1.6}$ fuel rods with 8.5, 12 and 20 wt.% of ^{235}U. In the analysis, U-$ZrH_{1.6}$ fuel rods in each ring are gradually replaced with thorium fuels (thorium oxide, ThO_2) beginning from ring B until ring G.

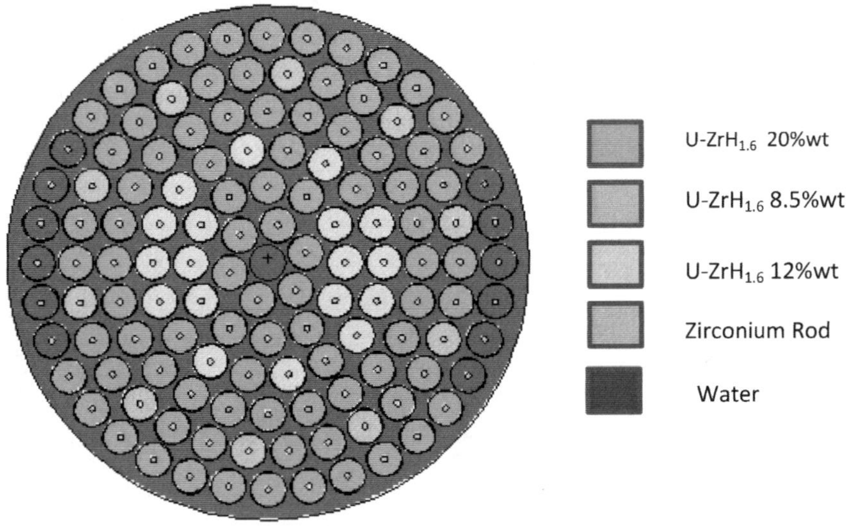

Figure 2. Top view of the RTP core.

Monte Carlo method has gained more interest for neutronic computation because of its capability in accurately modeling 3D geometries [5]. As for this study, all cores were designed in an infinite geometry (infinitely long reactor with 2D design configuration) using MCNPX with ENDF data libraries. The initial criticality, *k*, calculation was carried out using Kcode with 50 million neutron histories. Criticality is defined as the ratio of the neutron production to neutron loss in a reactor. In a supercritical condition, the number of neutrons produced by fission reaction is larger than that of neutrons loss due to absorption or leakage (*k* is larger than 1.0). The criticality, *k*, under 1.0 is subcritical Most supercritical cores are designed for fast breeding reactors because the reactor needs neutron to sustain its system. There were seven configurations that were modeled and analyzed as shown in Figure 3. Results were then compared to the previous study carried out by Abdul Rahman [6].

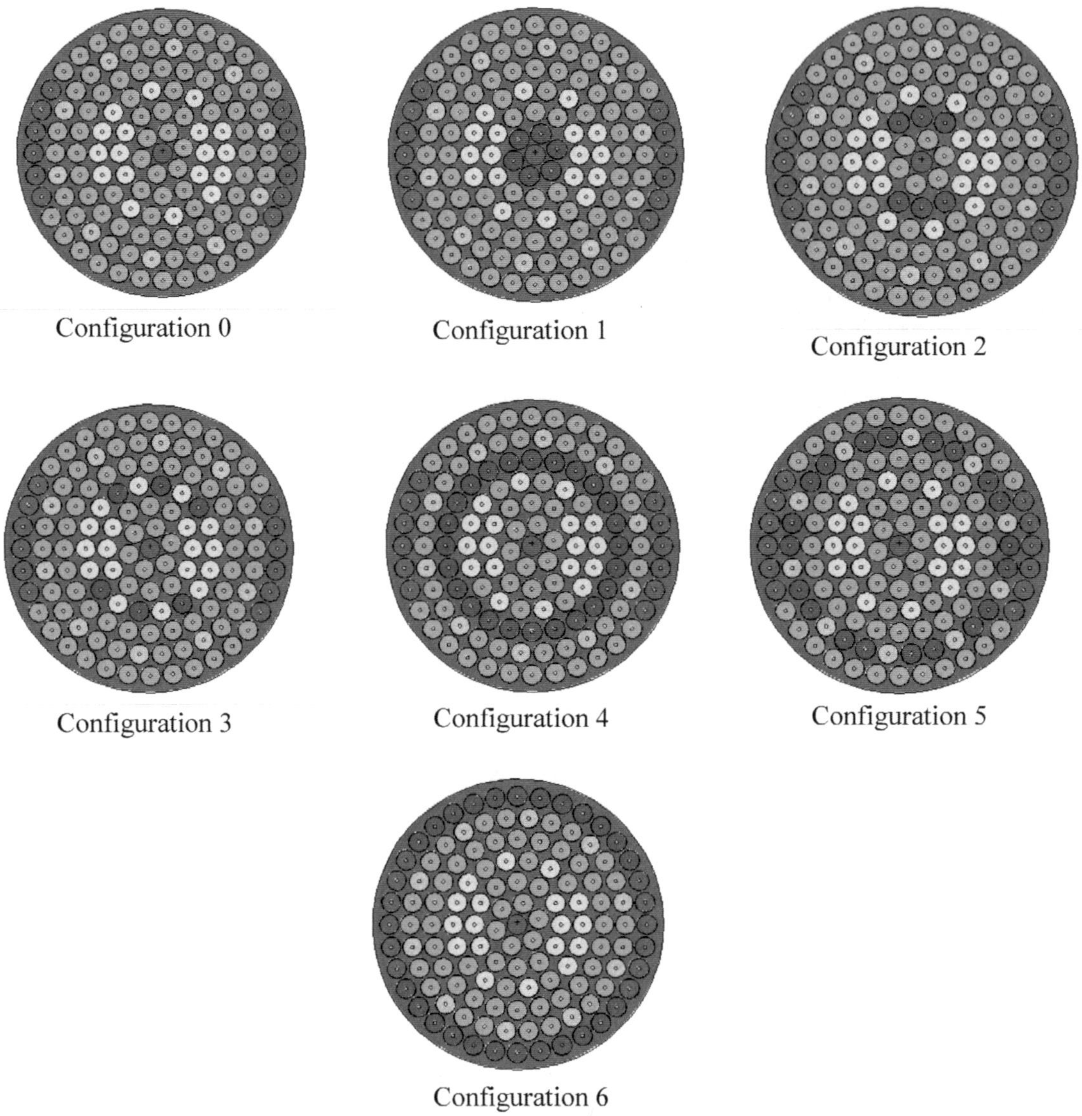

Figure 3. Seven configurations with different thorium fuels placements, including Configuration 0 (reference core without thorium).

Figure 3 shows seven configurations of cores that were simulated in MCNPX. Configuration 0 acts as the main reference of the study. It is the original RTP core without thorium fuels. Next is configuration 1 that has thorium oxide fuel rods in Ring B, which are marked with purple color. All six U-$ZrH_{1.6}$ fuel rods with 8.5 wt.% uranium are being replaced in ring B. For configuration 2, only six U-$ZrH_{1.6}$ (8.5 wt.%) are replaced in ring C. Similarly, in configuration 3, thorium fuels replace six U-$ZrH_{1.6}$ (8.5 wt.%) in ring D. Configurations 2 and 3 are almost similar with each other because thorium fuels are not fully surrounding the core due to the presence of U-$ZrH_{1.6}$ (12 wt.%) fuel rods inside rings C and D. In configuration 4, thorium fuel rods are placed in ring E, substituting all U-$ZrH_{1.6}$ (8.5% wt.%) fuel rods. Thorium fuels are located in ring F for configuration 5, replacing all original uranium fuels except the U-$ZrH_{1.6}$ fuel rods with 20 wt.% uranium. Lastly, configuration 6 the outermost ring G is completely filled with thorium fuel rods.

IOP Conf. Series: Materials Science and Engineering **298** (2018) 012034 doi:10.1088/1757-899X/298/1/012034

In the previous study [6], there are four core configurations that were studied, namely Configurations 0, 1, 2, and 3 (Figure 3). From the results of the study, the most supercritical configuration is Configuration 2, which has thorium oxide fuels in ring F with the initial criticality of 1.16371.

Table 1. Results from previous study of thorium analysis in RTP by Abdul Rahman [6].

Configuration	Position Thorium Oxide	Initial criticality, k	Reactivity difference compared to Configuration 0 (PCM)
0	No thorium fuel	1.0288	0
1	Ring B	1.13454	-9060
2	Ring D	0.90509	13290
3	Ring F	1.16371	-11270

3. Result and discussion

Results in Table 5 shows that all configurations can achieve supercritical. The nearest value to critical state, k = 1.0, is configuration 4 that has thorium located in ring E. The highest initial criticality is 1.35101, which is obtained from configuration 1

Table 2. Initial criticality for all studied configurations.

Configuration	Position Thorium fuels in the core	Criticality, k	Reactivity difference compared to Configuration 0 (PCM)
0	None	1.388	0
1	Ring B	1.35101	1973
2	Ring C	1.34989	2034
3	Ring D	1.3503	2012
4	Ring E	1.25713	7500
5	Ring F	1.32281	3551
6	Ring G	1.30649	4495

Table 3. Comparing results with previous study

Configuration	Previous criticality, *k* [6]	Criticality, *k*	Reactivity Difference (PCM)
0	1.0288	1.388	25154
1	1.13454	1.35101	14122
2	0.90509	1.3503	36428
3	1.16371	1.32281	10335

Comparing the results for configurations 0, 1, 2, and 3 from the previous study, the values for initial k of the previous study are lower. This is mainly due to the configuration of the core design, in which the previous core design is in 3D finite structure. Whereas, our configurations are modelled as a 2D infinite structure (infinitely long reactor). The criticality, k, obtained for a 2D infinite configuration will be larger due to neglected leakage in the axial direction. By comparing the 3D and 2D simulation results, the average leakage probability is estimated to be around 21.7%.

4. Conclusion

All studied configurations for this study can achieve supercritical condition. Configuration 4 has the lowest value of criticality 1.2573. On the other hand, configuration 1 has the highest initial criticality, which is around 1.35101. There is a large discrepancy between previous study and our results due to the large leakage probability, which is estimated around 21.7%. This is mainly due to oversimplification of the core from a 3D core to a 2D structure that was made during the simulation. This study is still at a preliminary stage and hence studying initial criticality alone is insufficient as thorium fuels inside the core need more time to be self-sustainable by having more fission reactions. Additionally, longer period of time is required for ^{232}Th to be transmuted into more fissile isotope of ^{233}U in order to fully utilize the benefits of thorium. Future work that will be carried out includes fuel depletion analysis that will estimate the life cycle of the fuel and fission products produced.

5. References

[1] Agency, M.N. Thorium Flagship Project [26 October 2017] Available from: http://www.nuclearmalaysia.gov.my/new/RnD/energy/fuel/thorium.php
[2] Jaradat S Q 2017 Studies on the liquid fluoride thorium reactor: Comparative neutronics analysis of MCNP6 code with SRAC95 reactor analysis code based on FUJI-U3-(0) *Nuclear Engineering and Design* **314** pp 251-255
[3] Sipaun S 2017 Thorium fueled reactor *AIP Conference Proceedings*: *iNusTEC2016* **1799** 050012
[4] Karim J A, Abu M P, Masood Z, Rabir M H, Salleh M A S, Zakaria M F and Iorgulis C 2012 PUSPATI TRIGA Reactor upgrading: towards the safe operation and feasibility of neutronic approach *International Atomic Energy Agency (IAEA)*
[5] Rabir M H, Hamzah N S, Abu M P, Karim J A and Sharifuldin Salleh M A 2012 Modeling the PUSPATI TRIGA Reactor using MCNP code in *R and D Seminar 2012: Research and Development Seminar*
[6] Mohammed A A and et al. 2016 Simulation on reactor TRIGA Puspati core kinetics fueled with thorium (Th) based fuel element *AIP Conference Proceedings*: *iNusTEC2015* **1704** 020004

Acknowledgements

Special thanks to Malaysian Nuclear Agency for supporting this work and Universiti Tenaga Nasional for sponsoring the research project under the UNITEN Start-up grant Bold 2025 project code RJO10289176/B/1/2017/20

IOP Conf. Series: Materials Science and Engineering **298** (2018) 012035 doi:10.1088/1757-899X/298/1/012035

An investigation into the feasibility of thorium fuels utilization in seed-blanket configurations for TRIGA PUSPATI Reactor (RTP)

Abdul Hannan Bin Damahuri[1, a)], **Hassan Mohamed**[1, b)], **Abdul Aziz Mohamed**[1, c)], **Faridah Idris**[2, d)]

[1]College of Engineering, Universiti Tenaga Nasional, Jalan IKRAM-UNITEN, 43000 Kajang, Selangor, Malaysia
[2]Reactor Technology Centre, Malaysian Nuclear Agency, Bangi, 43000 Kajang, Selangor, Malaysia.

a)ahannan@uniten.edu.my
b) MHassan@uniten.edu.my
c)AzizM@uniten.edu.my
d)faridah_idris@nuclearmalaysia.gov.my

Abstract. Thorium is one of the elements that needs to be explored for nuclear fuel research and development. One of the popular core configurations of thorium fuel is seed-blanket configuration or also known as Radkowsky Thorium Fuel concept. The seed will act as a supplier of neutrons, which will be placed inside of the core. The blanket, on the other hand, is the consumer of neutrons that is located at outermost of the core. In this work, a neutronic analysis of seed-blanket configuration for the TRIGA PUSPATI Reactor (RTP) is carried out using Monte Carlo method. The reactor, which has been operated since 1982 use uranium zirconium hydride ($U\text{-}ZrH_{1.6}$) as the fuel and have multiple uranium weight which are 8.5, 12 and 20 wt.%. The pool type reactor is one and only research reactor that located in Malaysia. The design of core included the Uranium Zirconium Hydride located at the centre of the core that will act as the seed to supply neutron. The thorium oxide that will act as blanket situated outside of seed region will receive neutron to transmute ^{232}Th to ^{233}U. The neutron multiplication factor or criticality of each configuration is estimated. Results show that the highest initial criticality achieved is 1.30153.

1. Introduction

Thorium is an element that occurs naturally in the world, mostly in Canada, Kazakhstan and Australia. Thorium offers several potential advantages, which include a higher abundance in nature compared to uranium, improved proliferation resistance and lower waste radiotoxicity [1]. Thorium-based fuels have excellent neutronic properties during irradiation process [2]. Nonetheless, one of the disadvantages of thorium is in the bred ^{233}U, which is a high-energy gamma emitter that requires well-shielded facilities.

The main thorium isotope is ^{232}Th, which occurs about 99.98% in natural. Unlike ^{235}U, thorium is a fertile material that can be transmuted to the fissile uranium isotope, ^{233}U. A thorium fuel cycle consists of four stages, in which the first stage is the production of ^{233}Th from ^{232}Th by a neutron capture. The second stage is the change from ^{233}Th to ^{233}Pa thru beta decay and within a long period of time, ^{233}Pa eventually becomes ^{233}U. More fission reactions will then take place when ^{233}U exists in a reactor.

Studies on thorium fuel cycle have been limited to few reactors such as molten salt and light water reactors. For light water reactors, thorium fuels are mixed with either uranium or plutonium. As for molten salt reactor, there are two types, in which the first type is thorium in a solid form, surrounded by molten salt [3]. The second type is when thorium is homogenously mixed with molten salt in a liquid form, increasing the heat transfer capability.

One of the commonly studied configurations for thorium reactors is the seed-blanket configuration, which is also known as Radkowsky Thorium Fuel concept. The concept was propose by Professor Radkowski and it is one of the Thorium Seed-Blanket design. Radkowski thorium reactor (RTR) is the design of core reactor based on the arrangement of the core of pressurized water reactor. Design of RTR involves the process of seed-blanket configuration. The element that are situated in the seed are usually metal element or oxide element. As for RTR, the seed element is UO2. For the blanket, the element is THO2. The designs are focusing more on heterogeneous configuration seed-blanket unit (SBU) fuel assembly. The blanket part is separate with the seed part to allow fuel independent with the seed while the seed part are supplying neutrons for the blanket.

While in this study the configuration contains thorium and a fissile material, either uranium or plutonium, in a single core. The seed is usually represented by uranium that provides neutrons to a blanket of thorium. The thorium blanket acts as a neutron receiver from the seed. Figure 1 below shows the seed-blanket configuration in a reactor core. As displayed in the figure, fuels that are situated inside and outside of the rectangle are typically filled with uranium and thorium, respectively.

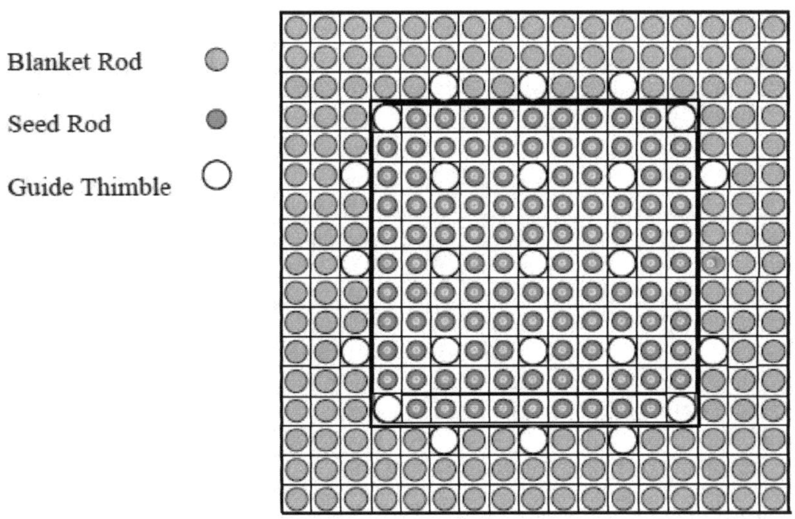

Figure 1. An example of seed-blanket configuration. Reproduced from Ref. [5].

One known advantage of the configuration is that the quantity of 'weapon quality' plutonium produced in a pressurized water reactor can be potentially reduced by a factor of 3 to 5 [4]. As for this research, the seed-blanket configuration is implemented in the TRIGA PUSPATI Reactor (RTP), a research reactor that has been operated by Malaysian Nuclear Agency since 1982. The reactor is a pool-type reactor that uses uranium zirconium hydride, U-$ZrH_{1.6}$, as the main fuel with 8.5, 12 and 20 wt.% of total U [6]. The production of U-$ZrH_{1.6}$ fuel for this reactor, however, has been discontinued. Nonetheless, there is an ongoing initiative to replace the U-$ZrH_{1.6}$ with thorium. Hence, this project investigates the neutronic behavior of thorium fuel as a potential substitute for the RTP, by comparing the initial multiplication factors or criticality, *k*, of the core with and without thorium. The core that contains thorium is modeled based on the seed-blanket concept.

Rector TRIGA Puspati (RTP) is the only research reactor in Malaysia. It has been operated for almost 35 years of ages. The reactor achieved it criticality in 28 June 1982. The reactor was constructed to fulfill various nuclear applications such as for research, education and services for companies. There are several facilities that are available in the reactor such as neutron activation analysis (NAA), small angles neutron scattering (SANs) and neutron radiography (NR). Besides, there is also prompt gamma neutron activation analysis (PGNAA) and boron neutron capture therapy (BNCT).

The model of reactor TRIGA Puspati is TRIGA Mark II Pool-Type Reactor with its maximum power is 1MW thermal. The fuel of the reactor is Uranium Zirconium Hydrate U-$ZrH_{1.6}$ with three different weight percentages which are 8.5, 12 and 20 wt.% of ^{235}U. The reflector of the reactor is graphite which is to reflect the neutron in the reactor to prevent it escaping from the core. The flux RTP is at the rotary rack and the control material is boron carbide and there are 4 types of control rods. Besides, the coolant for the reactor is light water that had been undergoes demineralized process.

2. Methodology

The RTP core consists of 6 rings that encircle the central timber beginning from ring B, which is the nearest ring to the central timber, to ring G that is located at the outermost ring of the core. The number of fuel rods of each ring is not similar as the size of the ring becomes larger when going from the innermost ring B (6 fuel rods) to ring G (36 fuels rods). The core with a seed-blanket configuration is modeled in a 2D structure using MCNPX as shown in Figure 2.

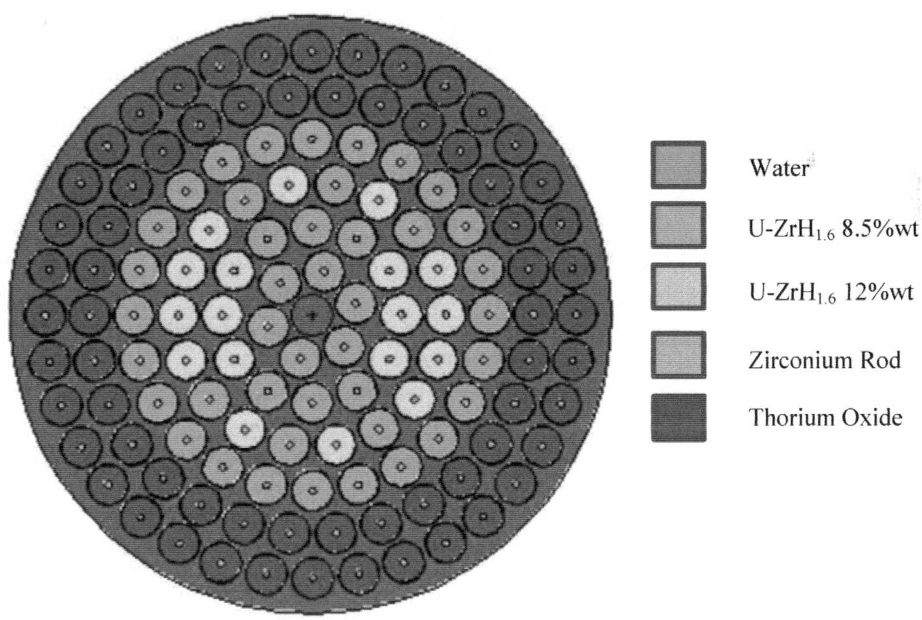

Figure 2. Seed-Blanket Configuration for RTP.

By using KCODE with 10000 neutron per second (NPS) and 5000 active cycles, the initial criticality, *k*, of the core was determined. The criticality of the core, *k*, is 1.0 when the number of neutron produced is the same as the neutron loss. If the criticality is lower than 1.0, the reactor system is considered subcritical due to the number of neutron loss is higher than the number of neutron gain. To achieve supercritical state, the number of neutron produced needs to be higher than that of neutron loss. For a fast breeder reactor, the supercritical condition is essential because thorium in the reactor needs more neutrons to transmute it to ^{233}U for fission to occur.

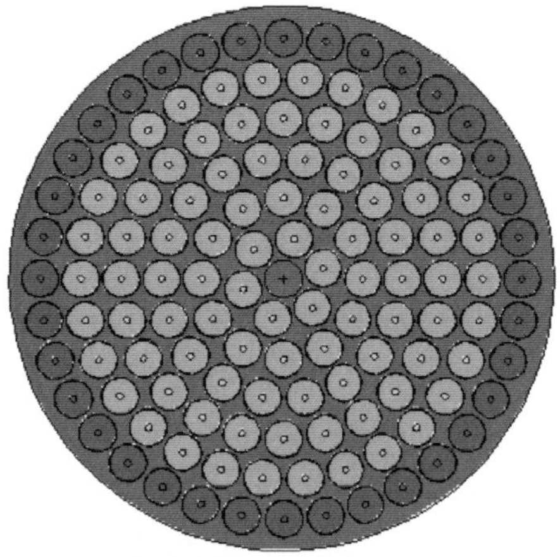

Figure 3. Configuration 1

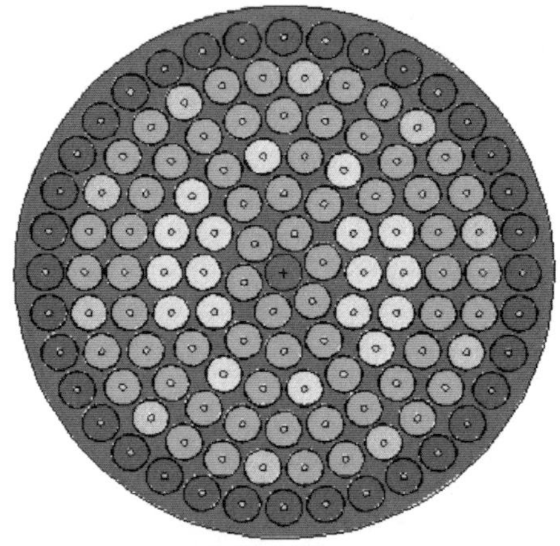

Figure 4. Configuration 2

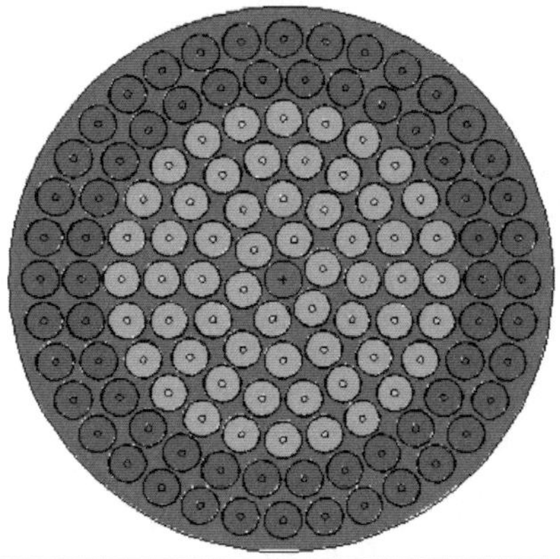

Figure 5. Configuration 3

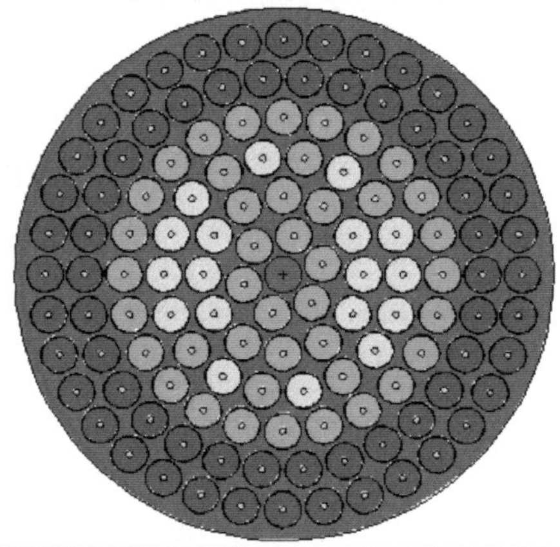

Figure 6. Configuration 4

Four seed-blanket configurations with various thorium and uranium compositions have been modeled and analyzed in this work, as displayed in Figure 3 to Figure 6. For Configurations 1 and 3, both cores contain thorium oxide (ThO_2) and U-$ZrH_{1.6}$ fuels with 8.5 wt.% only. Whereas, Configuration 2 has ThO_2 and U-$ZrH_{1.6}$ fuels at 8.5, 12 and 20 wt.%. Lastly, Configuration 4 comprises of ThO_2 and U-$ZrH_{1.6}$ fuels with 8.5 and 12 wt.%. The initial k of each analyzed seed-blanket configuration was obtained using MCNPX. Additionally, the reactivity difference between each configuration and the original core without thorium was calculated using Equation 1 below,

$$\Delta\rho = \frac{k_1 - k_2}{k_1 \times k_2} \times 10^5 \text{ pcm}, \tag{1}$$

where k_1 is the criticality of the seed-blanket configuration and k_2 is that of the core without thorium, which is approximately 1.388 as reported by Ref. [6].

3. Result and discussion

There are four seed-blanket configurations that have been analyzed for this study and the results are shown in Table 1.

Table 1. Results from MCNPX

Configuration	Criticality, *k*	Reactivity difference (pcm) from configuration without thorium fuel
1	1.26483	-7015
2	1.30153	-4787
3	1.16882	-13510
4	1.19247	-11813

Based from the result above, criticality *k* of the core that contains more thorium oxide seems to be smaller than that of the core that holds less thorium oxide, as observed from Configurations 1 and 2 criticalities that are higher than Configurations 3 and 4. This happens because thorium fuels in the core might have captured neutrons that being released from uranium during the initial state, before having more neutrons to be generated by breeding ^{233}U.

In addition, results show that the criticality of the core that contains all uranium enrichments is higher than that of the core that contains U-$ZrH_{1.6}$ with a single 8.5 wt.%. It is also observed that the configuration that has the smallest *k* value is Configuration 3 with 1.16882 compared to all other variations.

As for reactivity difference, the smallest difference that was estimated is obtained from Configuration 2 (-4787 pcm). This finding seems reasonable since Configuration 2 has the closest look and fuel composition as the original core without thorium. On the other hand, Configurations 3 and 4 have quite large differences, namely -13510 and 11813 pcm, due to their very distinct fuel compositions compared to the original U-$ZrH_{1.6}$ only core.

The study is only at a preliminary stage, the criticality that have been archive was only in early stage. The thorium fuel need more time to be self-sustainable for fission reaction. This is because the thorium fuels need to be transmute first before it begin the process of fission reaction. Besides, the configurations are in 2D infinite configuration. There might be some small leakage probability occur. The reactor are design with slightly different from the original reactor design, the outer part of the reactor such as the concrete and reflector are not included in the design. Thus, it might have different result from the original 3D design.

4. Conclusion

From the study, it shows that all seed-blanket configurations can achieve supercritical condition. The core that contains only uranium zirconium hydride with 8.5 wt.% has a small criticality compared to the core that contains all uranium total U weight values, namely 8.5, 12 and 20 wt.%. Criticality, *k*, for the core that contains less thorium is higher than the core that has more thorium because of the absorption of neutrons in thorium oxide fuel, before uranium can be bred.

IOP Conf. Series: Materials Science and Engineering **298** (2018) 012035 doi:10.1088/1757-899X/298/1/012035

5. References

[1] Mohamed H, Kotlyar D, Parks G T, Shwageraus E 2015 An investigation into the feasibility of utilizing thorium-based fuel cycles in a small fluoride-salt-cooled high-temperature reactor *Proc. In 23rd International Conference on Nuclear Engineering, ICONE23-1659*

[2] OECD Introduction of thorium in the nuclear fuel cycle *OECD Publishing*

[3] Jaradat S Q 2017 Studies on the liquid fluoride thorium reactor: Comparative neutronics analysis of MCNP6 code with SRAC95 reactor analysis code based on FUJI-U3-(0) *Nuclear Engineering and Design* **314** pp 251-255

[4] International Atomic Energy Agency Nuclear Fuel Cycle and Materials Section 2005 Thorium fuel cycle - Potential benefits and challenges *IAEA-TECDOC-1450*

[5] Galperin A and Radkowsky A 1997 Thorium fuel for light water reactors--reducing proliferation potential of nuclear power fuel cycle *Science & Global Security* **6**(3) pp 265-290

[6] Rabir M H, Hamzah N S, Karim J A and Sharifuldin Salleh M A 2012 Modeling the PUSPATI TRIGA Reactor using MCNP code in *R and D Seminar 2012: Research and Development Seminar*

Acknowledgements

Special thanks to Malaysian Nuclear Agency for supporting this work and Universiti Tenaga Nasional for sponsoring the research project under the UNITEN Start-up grant Bold 2025 project code RJO10289176/B/1/2017/20

IOP Conf. Series: Materials Science and Engineering **298** (2018) 012036 doi:10.1088/1757-899X/298/1/012036

Recent advances in Optical Computed Tomography (OCT) imaging system for three dimensional (3D) radiotherapy dosimetry

Ahmad Taufek Abdul Rahman[1,2,3], Nurul Farah Rosli[4], Shafirah Mohd Zain[4] and Hafiz M. Zin[4]

[1]Faculty of Applied Sciences, Universiti Teknologi MARA, 40450 Shah Alam, Selangor, Malaysia
[2]Institute of Science, Universiti Teknologi MARA, 40450 Shah Alam, Selangor, Malaysia
[3]Departmet of Physics, University of Surrey, GU2 7XH, Guildford, Surrey, United Kingdom
[4]Advanced Medical and Dental Institute, Universiti Sains Malaysia, 13200 Bertam, Pulau Pinang, Malaysia

ahmadtaufek@salam.uitm.edu.my

Abstract. Radiotherapy delivery techniques for cancer treatment are becoming more complex and highly focused, to enable accurate radiation dose delivery to the cancerous tissue and minimum dose to the healthy tissue adjacent to tumour. Instrument to verify the complex dose delivery in radiotherapy such as optical computed tomography (OCT) measures the dose from a three-dimensional (3D) radiochromic dosimeter to ensure the accuracy of the radiotherapy beam delivery to the patient. OCT measures the optical density in radiochromic material that changes predictably upon exposure to radiotherapy beams. OCT systems have been developed using a photodiode and charged coupled device (CCD) as the detector. The existing OCT imaging systems have limitation in terms of the accuracy and the speed of the measurement. Advances in on-pixel intelligence CMOS image sensor (CIS) will be exploited in this work to replace current detector in OCT imaging systems. CIS is capable of on-pixel signal processing at a very fast imaging speed (over several hundred images per second) that will allow improvement in the 3D measurement of the optical density. The paper will review 3D radiochromic dosimeters and OCT systems developed and discuss how CMOS based OCT imaging will provide accurate and fast optical density measurements in 3D. The paper will also discuss the configuration of the CMOS based OCT developed in this work and how it may improve the existing OCT system.

1. Introduction

Approximately one-third of the population will develop cancer during their lifetime and about half of these are curable with advances in cancer treatment techniques. Patients are treated using surgery, radiotherapy or chemotherapy or in combination of several treatment techniques. The latest National Cancer Registry reported that the number of patients diagnosed with cancer in Malaysia is increasing;

about 18000 new cancer cases were diagnosed and registered in 2007 [1]. The number is expected to rise steeply by half-fold by 2030 according to a World Health Organization report in 2012 [2].

Modern radiotherapy is one of the main treatment technique to cure cancer. The technique delivers high radiation dose to the tumour whilst minimising dose to the healthy tissues. Development of advanced radiotherapy techniques such as Intensity Modulated Radiation Therapy (IMRT) has improved the ability to better target the radiation by delivering higher and more conformal radiation dose to the target, further improving the use of radiotherapy as a cancer treatment technique and increasing cancer cure rates [3]. However, the complexity of the advanced radiotherapy delivery can severely compromise the dosimetric accuracy of the beam delivery, hence, verification of the treatment is important to ensure treatment success. Radiotherapy treatment needs highly sensitive dosimetry systems with high spatial resolution to accurately map doses to tissues. According to the International Commission of Radiation Units (ICRU) recommendations, the accuracy of the dose delivered to the target should be within ± 5% [4]. To produce at least 95% accurate results, the standard deviation of the dosimeter factors should be less than 2% [5].

IMRT delivers the focused radiation beam by collimating the radiation with a multi leaf collimator (MLC) to match the shape of tumour region. This is to ensure radiation is delivered mostly to the target and minimising exposure to surrounding healthy tissue and other organs at risk. The field shape and dose rate also changes dynamically around the patient. These advanced treatments are prone to delivery errors due to their complex nature. The rapid advances in radiotherapy delivery techniques has outpaced the dosimetry system to accurately monitor the delivery. To achieve precise and accurate measurements of the associated absorbed dose distribution of the treatment, the medical physicist utilises dosimetric devices such as ionisation chambers, film, thermoluminescence detectors and diodes. Each of these devices is either a 1D or 2D measurement device, while conversely the patient and the delivery of radiotherapy is a 3D problem. These complex radiotherapy techniques require verification of the 3D dose distribution. The Monte Carlo simulation is well known to be capable of providing dose deposited in a medium; however, the end result is only as good as the model assumptions. Therefore, an experimental 3D verification is of significant importance in the delivery of dose in radiotherapy treatment. In this respect, there is a need for an advanced dosimetric system that can quantify and verify the accurate dose distribution as well as being capable in recording the integrated absorbed dose in 3D for radiotherapy dosimetry application.

An optical computed tomography (OCT) is a 3D imaging system consists of a light source and an image sensor to detect the light and mapped the optical density of a dosimeter in 3D. The dosimeter is a radiochromic polymer material whose chemical properties change predictably upon exposure to x-rays. Light transmitted through the object is refracted and recorded by the image sensor as optical density. Current commercial OCT system uses a charged coupled device (CCD) as the image sensor [6-9]. CCD has matured and can provide good image quality, but the chip has limitation in providing on-pixel image processing that is useful in an OCT imaging system. The conventional CCD based OCT's scanning process of the radiosensitive polymer is also time consuming, thus, limit the usefulness of this instrument as a rapid system for complex radiotherapy dose verification.

This paper will review development of radiochromic dosimetry system and OCT imaging system for radiotherapy dosimetry. The work investigates application of complementary metal oxide semiconductor (CMOS) image sensor to improve the optical computed tomography (OCT) system. The aim is to improve the imaging speed of the CCD-based OCT.

2. Review of 3D Radiochromic Dosimeter

Three dimensional (3D) radiochromic dosimeter is based on chemical reactions in the dosimeter material which are induced by ionising radiation, that enables the dosimeter to record the radiation dose distribution in 3D. Radiation sensitive dosimeter were first suggested by Day and Stein in 1950 when radiation produced colour changes in gels containing dyes such as methylene blue and phenol-indo-2:6-di-chlorol. Several studies have been reported using gel dosimetry for 3D verification in radiotherapy. The development of 3D dosimetry started from use of Fricke Gels [10], Polymer Gels [11] and recently with the 3D radiochromic plastic material of PRESAGE™ [12].

The Fricke Gel dosimeter was introduced by Gore et al, in 1984, based on chemical dosimeters with a gelling agent to stabilise or fix the chemical reaction as a result of absorbed radiation [10]. The dosimeter is based on oxidation of ferrous ions (Fe+2) to ferric ions (Fe+3) as an aerated acid solution of ferrous sulphate is irradiated. The absorbed dose can then be estimated by measuring the concentration of ferric ions before and after irradiation. Gore et al (1984) proposed that nuclear magnetic resonance (NMR) relaxation measurements of a standard Fricke solution could be used to determine the absorbed dose of ionising radiation [13]. Ferrous sulphate solutions were subsequently incorporated into a gel matrix in order to stabilise the irradiated magnetic resonance imaging (MRI) absorbed dose signature spatially. Longitudinal and transverse relaxation rate images could then be used to map the spatial distribution of absorbed radiation dose [14]. Several studies have concerned the application of Fricke gel dosimeters in radiotherapy. This combination of chemical dosimeter, gelling agent and evaluation technique was the first truly 3D dosimeter to be used to investigate radiotherapy dose distributions. However the diffusion of ferric ions (Fe+3) throughout the gel give rise to significant blurring of the observed dose distribution [15]. The dosimetric properties of the Fricke gel dosimeters are strongly dependent on gel preparation. Since the dosimeter is based on chemical dosimeter principles, thus the chemical component will affect the radiation chemical yield.

The use of polymer gels for radiation dosimetry based on radiation polymerisation has also been proposed. Early study of polymer gel dosimetry was made in 1954 by Alexander et al [16] and Mesrobian et al [17], describing the effects of polymerisation in acrylamide following irradiation. The polymer gel dosimeters are made-up from radiation sensitive chemical interaction with ionising radiation. Thus, a number of reactions are initiated that lead to the conversion of monomers to polymers, resulting in the formation of an insoluble polymer network. Polymerisation leads to a change in the optical properties of the dosimeter, and it therefore becomes opaque [11]. The rate of polymerisation is presumed to depend on the rate of formation of the initiating fragments, the intrinsic reactivity of the monomer and the concentration of radicals and monomers present in the solution [18]. Therefore, the process of polymerisation can be controlled by the type of monomer used, monomer concentration and dose rate [19]. Further investigation included studies by Hoecker in 1958, examining the dosimetric characteristics of radiation-induced polymerisation in liquids [20]. The application of polyacrylamide for gamma dosimetry was proposed by Boni in 1961 [11]. Several studies have reported changes in NMR transverse and longitudinal relaxation rate in irradiated polyethylene oxide and irradiated aqueous solution of N,N'-methylene-bis-acrylamide and agarose, showing these rates to increase with absorbed dose [11, 21].

In 1992 a new gel dosimetry formulation was proposed, which was based on the polymerisation of acrylamide and N,N'-methylene-bis-acrylamide (bis) monomers infused in an aqueous agarose matrix [22]. This system was named BANANA (bis, acrylamide, nitrous oxide and agarose) [23]. This formulation did not suffer the same drawback that Fricke gels did with continual post-irradiation diffusion [11] and showed stability to prevail several months post-irradiation. The polymerisation reaction occurs by cross-linking of the monomers induced by free radical products of water radiolysis. This gradually increases the rate of water proton spin relaxation, up to doses of 15 Gy [23].

In 1994 Maryanski et al [24] refined the BANANA formulation by replacing agarose with gelatine and the first of the BANG gels was born, now consisting of bis (3%), acrylamide (3%), nitrogen and gelatine (5%). In 1994 this formulation was patented by MGS Research, Inc., Madison as a registered trademark and became commercially available and known as BANG® [24]. The BANG gel, readout using MRI facilities resulted in having a lower background for the non-irradiated gel, and a more transparent medium in which the irradiated region is clearly visible [25]. In 1996, a new BANG type gel formulation, which contained bis, acrylic acid, sodium hydroxide, water and gelatine was described by Maryanski et al [26]. The gel differs from BANG mainly in the substitution of acrylic acid for acrylamide and was named BANG-2. This also became commercially available under the

MGS Research Inc. Madison, later BANG - 3 and - 4 were developed and also became commercially available.

A manufacturing procedure for BANG gel was later described by Baldock et. al., [21], and is now most commonly referred to in the literature as PAG, which is an acronym for polyacrylamide gel. Further revision of the characteristics of polymer gel dosimeters were suggested with alternative monomers such as, N-vinylpyrrolidone argon (VIPAR) gels [27], sodium methacrylate [28], 2-hydroxyethyl methacrylate (HEMA) and 2-hydroxyethyl acrylate (HEA) [29]. However, the PAG formulation consisting of co-monomers N, N'- methylene-bis-acrylamide and acrylamide has been the most characterised from in the literature, being studied by numerous research groups, and is known to have desirable qualities for radiation dosimetry, such as reliability, stability and reproducibility [30].

Polymer gel dosimetry systems can also suffer major limitations. Oxygen inhibits the polymerisation reaction as it scavenges free radical initiators (such as OH* and H*) that are produced during water radiolysis [31] and therefore, the gels should only be manufactured under a oxygen free atmosphere (typically nitrogen rich), otherwise referred to as an hypoxic environment. While several groups have performed characterisation of the gels, polymer gel dosimeters have not yet been widely accepted in routine clinical radiotherapy environments. One reason for this is the laborious and complex manufacturing process involved, due in large part to the procedure for eliminating oxygen from the gel system.

Further development in polymer gel dosimetry occurred when Fong et al [32] described a method for developing normoxic gels. This was a significant advancement for polymer gel dosimetry as normoxic gels can be made under normal atmospheric conditions, which does away with the time-consuming process of oxygen elimination during the manufacture process of polymer gels. This meant that polymer gels could be manufactured on the bench top. Perhaps the most widely known normoxic formulation is MAGIC gel [32]. The principle behind the MAGIC gel is in the ascorbic acid oxygen scavenger, commonly known as vitamin C. Ascorbic acid binds free oxygen contained within the aqueous gelatine matrix into metallo-organic complexes and this process is initiated by copper sulphate. The normoxic gel has been reported to be tissue equivalent and has been shown to offer superior dose sensitivity and dynamic range using MRI evaluation when compared with other polymer gel formulations.

De Deene et al [33] have investigated the role of the different chemical constituents of the MAGIC gel. The sensitivity of MAGIC gel to ionising radiation was optimised by maximising the amount of oxygen scavenger and monomer required, while minimising the amount of overall ingredients. Further studies were performed on various oxygen scavengers and it was found that tetrakis (hydroxymethyl) phosphonium chloride (known hence forth just as tetrakis) was effective at scavenging oxygen. The rate of effective oxygen scavenging was found to be dependent upon its concentration, so that a greater concentration of tetrakis corresponded with a decrease in sensitivity to ionising radiation. However, due the increased effectiveness of the tetrakis a reduced concentration was required compared to other oxygen scavengers. Tetrakis was used to replace ascorbic acid and copper sulphate in the MAGIC formulation and a new formulation consisting of methacrylic acid, gelatine and tetrakis, named MAGAT was proposed [33]. Other normoxic gel formulations were also developed such as MAGAS (methacrylic acid gelatine and ascorbic acid), and PAGAS (polyacrylamide gel and ascorbic acid) gels [34]. This work was critical in further developing the potential of the normoxic dosimeter and due to its considerable advantages over current gel dosimeters it was proposed that this required further development. Following the development of normoxic polymer gels, a number of studies were performed on their characterisation. Studies included the ultrasonic properties of MAGIC gel [35], manufacturing methodology studies for MAGAS gel [36] and depth dose studies for MAGAS and MAGAT gels. The radiological water equivalence of MAGIC, MAGAS and MAGAT gels was investigated and studies were also performed on the R2 Dose response optimisation and stability of MAGAT gel. Further radiological attenuation properties of MAGAT and PAGAT gels were also investigated. From these studies, it became apparent that problems exist with the stability of some of the normoxic gels when using oxygen scavengers. It was also clear that normoxic polymer gels responded to diagnostic doses in the same way as radiotherapy doses.

A relatively new class of polymer dosimeter has been proposed to overcome the radiochromic diffusion problem, the dosimeter being known as PRESAGETM (Heuris Pharma, Skillman, NJ, USA) [6, 12, 37]. PRESAGETM is a solid dosimeter which is based on clear polyurethane doped with leuco-dye leucomalachite green. The components of the dosimeter include an alkyl diisocyanate prepolymer, a hydroxyl reactive polyol along with a catalyst, which polymerises into optically clear polyurethane. The leuco dyes have a maximum absorbance at a wavelength of 633 nm and therefore is compatible with a He-Ne laser-based optical scanning system. It has been suggested that PRESAGETM has a number of potential advantages over both conventional polymer and Fricke gels. PRESAGE™ is a robust solid that does not require a container and therefore makes optical matching much easier. Irradiated regions exhibit negligible diffusion of the coloured medium and provided the sample is kept away from light while scanning, it appears highly stable. Initial studies of PRESAGETM have been reported to show very promising results. Further development and optimization techniques have since been published for radiotherapy, brachytherapy and microbeam radiotherapy [38-42]. It is clear that PRESAGETM has advantages over normoxic polymer gels in that they are better suited for evaluation optically and has demonstrated stability. PRESAGE™ dosimeters has a nominal elemental composition of C (63.27%), H (9.1%), N (4.91%), O (21.24%), Cl (2.7%) and Br (0.92%) with an empirical formula of C421H722N28O106Cl6Br giving an electron density of Z_{eff} = 9.33 and a mass density of 1.08 gcm-3 measured by manufacturer using gravimetric analysis [12]. The polymeric matrix is formed in two steps, the first step is prepolymer processes and the second step is integration with the leuco dye, a free radical initiator and catalyst. The PRESAGE™ dosimeter is in the form of cylinders with various sizes i.e. (i) diameter 60 mm and height 60 mm, (ii) 22 mm diameter and 60 mm height, (iii) 18 mm diameter and 60 mm height and (iv) 9.7 mm diameter and 60 mm height (Figure 1) [40, 43].

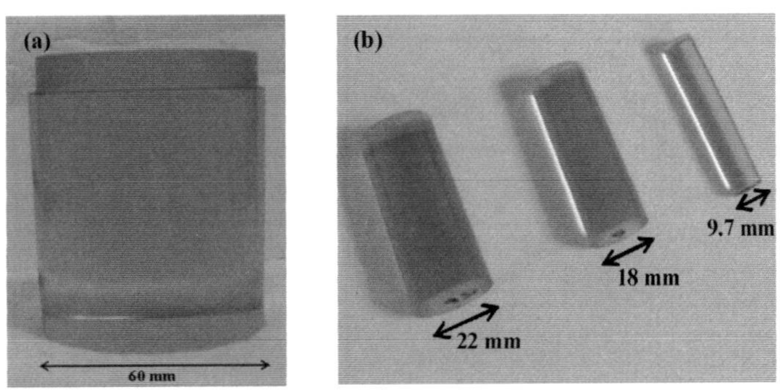

Figure 1. PRESAGETM of different diameter (a) 60 mm, (b) 22 mm, 18 mm and 9.7 mm [40, 43].

3. Review of 3D Optical Computed Tomography Imaging System

Optical Computed Tomography (OCT) is a bench top 3D imaging system for translucent object. The system is more accessible compared to a magnetic resonance imaging (MRI) system. This technique reconstructed 3D dose distributions in a radiochromic dosimeters such as polymer gel or PRESAGE^{TM} [44] using back projection of the 2D images. The optical properties of the polymer gel and PRESAGE^{TM} changes when irradiated. It becomes opaque and develops colour changes. This is due to the increasing in optical density in response to the absorbed dose delivered to the PRESAGE^{TM} dosimeter.

Gore *et al.*, described the first-generation OCT system in 1996. In this setup, a laser beam was stepped across a sample by mechanically moving two synchronous mirrors and the resulting beam is measured using and a photodiode detector. The first-generation OCT suffered from slow scanning speed due to the mechanical movement of the mirrors and only one path through the sample is measured at a time. A single slice with 128 x

128 pixels were acquired in about 12 minutes. The rapid movement of the laser beam requires more complex optics that could lead to artefacts or distorted images, which needs more post-processing. In terms of image quality, the laser scanners are provided signal-to-noise ratio is superior due to the high intensity of the laser beam. This system shown to be insensitive to geometrical distortion although refractive and scattering effects have been observed. The OCT was eventually developed into the first commercially available optical CT system, the Octopus scanner (MGS Research Inc., Madison, CT, USA), with a reported 30-minute scan time.

The second-generation OCT uses an incoherent broad light source and an image sensor such as Coupled Charged Detector (CCD) (Mark Oldham, 2006). The projection data for the optical CT technique acquired by passing a broad light beam through sample and collected by an image sensor. The pixelated detectors are based on CCD to obtain a whole 2D projection instantly at each rotation step, which render a significant speed advantage. The speed and resolution of this type of scanner is superior, but the sensitivity to refractive index inhomogeneities and stray photons from the dosimeter and the walls surrounding it can be a problem when the absorbed dose is reconstructed into a 3D map. The reason for this is that the entire projection is formed from one single exposure of the entire dosimeter [7].

CCD-based optical CT as a viable alternative to MRI for readout of 3D radiation dosimeters. Development of clinically applicable optical CT scanners has been moving in two different sub-classes with different geometries; the cone-beam and the parallel beam [7]. A commercially available cone-beam OCT scanner is the Vista (Modus Medical Devices Inc., London, ON, Canada) with a reported scan time of 5 minutes. Due to the geometry of the cone-beam CT, various scatter correction methods have been applied to counter the scatter issues in pixelated area detectors [45]. Cone beam OCT has requires correction from stray light artefacts.

At the University of Surrey, a group of researcher has developed an OCT based on the parallel beam geometry using tele-centric optics to create beams that travel parallel through the sample and are focused on the detector [7]. The parallel beam OCT has greatly reduced stray light artefacts due to rejection of light ray that deviate. Several studies have been performed using this system with PRESAGE^{TM} dosimetry [7, 8, 38-40, 42, 43, 46-50]. Figure 2 shows some of reconstruction images of 9.7 mm diameter size of PRESAGE^{TM} using CCD based OCT for characterisation study using Microbeam Radiation Therapy (MRT) facilities at ESRF, Grenoble.A subsubsection. *The paragraph text follows on from the subsubsection heading but should not be in italic.*

4. Application of CMOS Image Sensor Technology In Optical Computed Tomography (OCT) Imaging System

The CMOS image sensor (CIS) is a solid state imager manufactured using the standard CMOS process. The CIS is relatively new technology compared to the CCD. Chronologically, CMOS imager was first introduced by Fossum in 1993 [51], more than 20 years after Smith and Boyle's invention of the CCD , which over the past, has long matured, produces superior image quality and is widely used in both commercial and scientific application. The CIS has evolved since its first introduction as a Passive Pixel Sensor (PPS) into a competitive Active Pixel Sensor (APS) imaging technology due to technology scaling of CMOS and the feasibility of high numbers of pixels integrated on-chip [52]. The major breakthrough for the CMOS APS is the ability to integrate sensing with signal processing at pixel level. CISs allow row and column addressing for random pixel access and on-chip fast multiple region-of-interest read-out, which allows the user to trade resolution, or array size, with read-out speed. Some CMOS sensors are integrated with on-pixel intelligence such as in-pixel analogue-to-digital converters. These features were recognised as advantages over the mature CCD technology despite the read noise in CMOS being higher than CCD [53].

As the technology has developed, APS has progressed in quality and can now achieve superior performance comparable to CCDs with lower noise and higher spatial resolution. Furthermore, the fast frame rate achievable with the CIS gives an extra edge to the sensor compared to the slow frame rate of CCD. CIS has now gained increasing popularity as a competitive imaging technology to CCD and is now been adapted in high end consumer products (e.g. mobile imaging devices, i.e. mobile phones and digital cameras) as well as scientific applications.

IOP Conf. Series: Materials Science and Engineering **298** (2018) 012036 doi:10.1088/1757-899X/298/1/012036

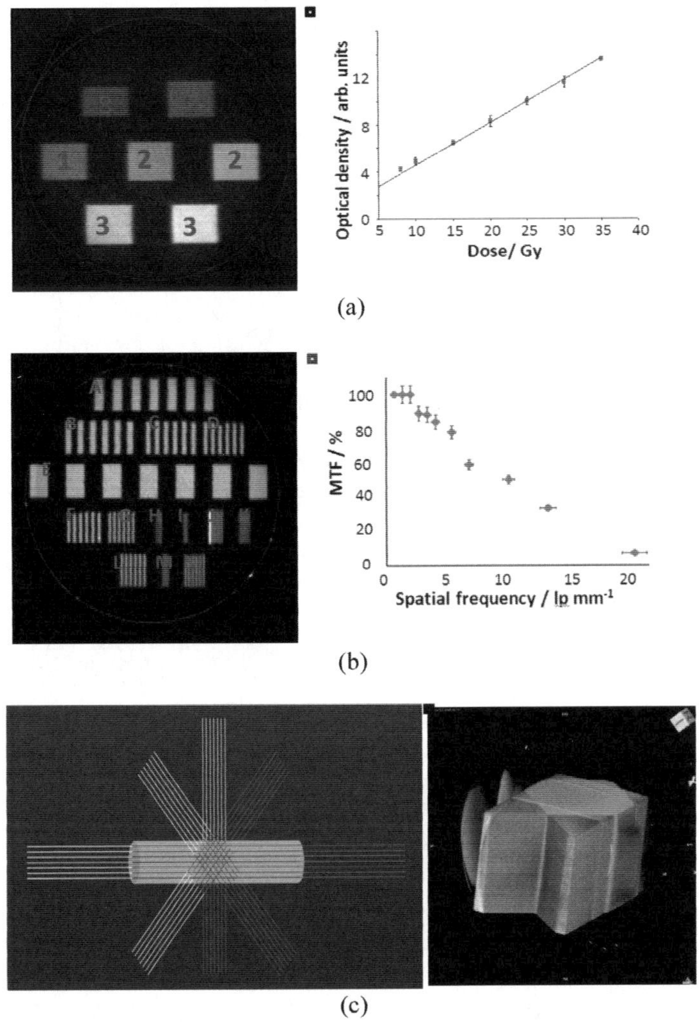

Figure 2. Reconstruction image of 9.7 mm diameter size of PRESAGE™ using CCD-based OCT. (a) Characterise the dose linear response, (b) characterise the spatial resolution of the OCT system and (c) visualise 3D dose deposition using microbeam radiation therapy (MRT) treatment [40, 43].

5. Description Of CMOS-Based Optical Computed Tomography (Oct) Imaging System Developed

In this study, a CMOS based OCT system was developed for imaging radiochromic dosimeter in 3D. A preliminary characterisation of the system was performed to study the performance of the image sensor and other components to acquire projection images of the 3D object, such as the field of view, the sensor frame rate and the speed of the stepper motor. Figure 3 displays the schematic diagram for the optical CT that is placed in a dark room.

The imaging system consists of a 10 Megapixels CMOS image sensor coupled with a lens, model M3520-MPV (Computar, NC, USA). The CMOS image sensor was housed inside a Basler ace camera enclosure, model acA3800-14um (Basler, Ahrensburg, Germany). The CMOS image sensor is a monochrome sensor with a resolution of 3840 by 2748 pixels (ON Semiconductor, Phoenix, AZ). The size of the sensor array is 6.4 mm × 6.4 mm. Each pixel is 1.67 µm × 1.67 µm and is readout using rolling shutter mechanism. The temperature of the camera housing is monitored by thermometer and was kept under 50 C. The sensor is illuminated with a 10 cm × 10 cm large array LED of 633 nm wavelength. The LED light is controlled by light intensity controller (LC-15-4CH-A1, TMS Lite, Penang, Malaysia) and connected to a 12V DC supply.

IOP Conf. Series: Materials Science and Engineering **298** (2018) 012036 doi:10.1088/1757-899X/298/1/012036

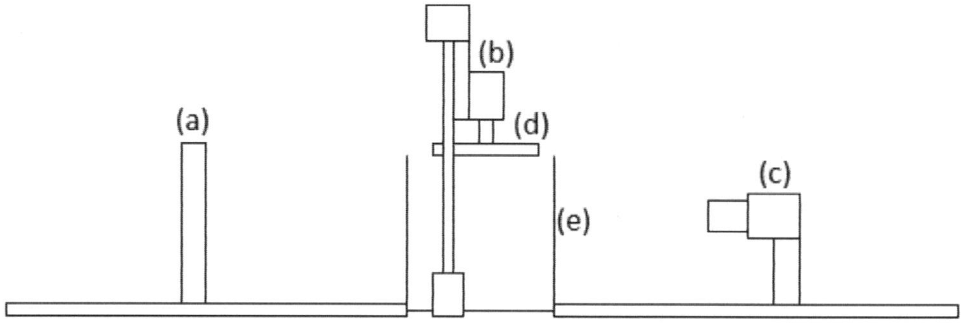

Figure 3. The schematic diagram of the in-house developed OCT (a) large array LED, (b) stepper motor, (c) CMOS image sensor, (d) dosimeter holder of 60 mm diameter, and (e) matching liquid tank.

A GUI was also developed using LabVIEW to capture and display the projection images of the dosimeter image. Using the GUI, the region of interest (ROI) and the frame rate can be controlled. The ROI of the sensor can be determined by adjusting the height and width of the pixels and x-, y-coordinates. The maximum ROI is 3840 x 2748 pixels and the minimum ROI that the sensor can changed is 64 x 64 pixels. With full resolution, the frame rate of the sensor is 8.6 fps and started to increase until 28.5 fps by decreasing the ROI. As shown is Figure 4, the complete GUI for displaying, capturing and saving the image were developed and the image is saved as RAW file and can be analysed by MATLAB. The ROI can be changed and the frame rate can be monitored as listed in Table 1.

Table 1. The relationship between the region of interest and the frame rate. The coordinates are chosen from the centre of the sensor.

Frame rate (fps)	Width (x)	Height (y)	Offset (x)	Offset (y)
8.6	3840	2748	0	0
10.8	2000	2000	920	374
13.0	1500	1500	1170	624
16.4	1000	1000	1420	874
18.3	800	800	1520	974
22.1	500	500	1670	1124
25.6	300	300	1770	1224
28.5	100	100	1870	1324
28.5	80	80	1880	1334

A rotary stage for the dosimeter is positioned between the camera and the LED. The rotary stage for the dosimeter consists of a stepper motor and a stepper motor controller (National Instrument, Austin, TX, USA). The stepper motor controller was connected to a 12 V power supply and a motion interface and was controlled from the PC via peripheral component interconnect (PCI) card. The diameter of dosimeter PRESAGE^{TM} is 60 mm (fixed diameter for this holder) and was placed inside a water tank (130 x 120 x 95) mm, made of acrylic. The stepper motor can hold a maximum weight of 4.4 kg which is sufficient for a 3D dosimeter. The motor rotates the dosimeter while the CMOS sensor captures the image for every step angle specified in the GUI. The minimum step angle of the motor is 1.8^{o}, which means 200 steps per revolution for 360° rotation. The components of the rotary stage are integrated with the imaging system and controlled using LabVIEW (National Instrument, Austin, TX, USA). The step angle, time delay, velocity and acceleration of the motor can be controlled in the GUI as shown in Figure 4(b).

IOP Conf. Series: Materials Science and Engineering **298** (2018) 012036 doi:10.1088/1757-899X/298/1/012036

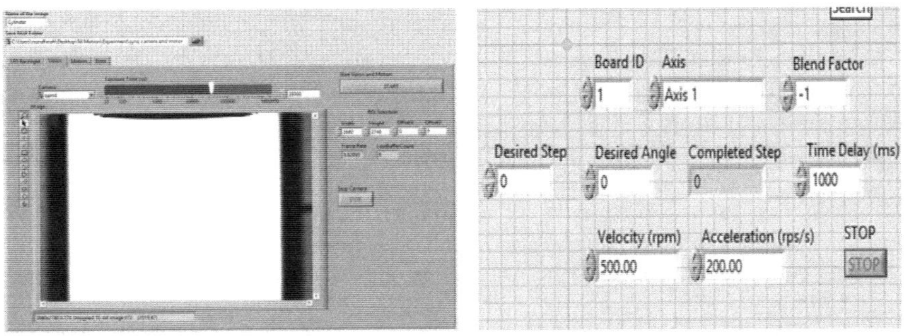

(a) (b)

Figure 4. a) GUI for the CMOS image sensor, b) GUI for the stepper motor

Working Distance (mm)	Height of CMOS (mm)	Horizontal View (mm)	Vertical View (mm)	Distance from the Ground (mm)
600	35	107	77	20
650	35	115	83	12
700	35	125	90	16
750	50	135	97	39
800	50	145	103	30
850	50	153	110	37
900	50	160	115	36
950	50	171	122	30
1000	50	181	128	27
1050	58	191	135	11

Table 2. The FOV is measured with different working distance and height of the CMOS sensor.

The sensor captures the images at every 1.8^o of step angle, and requires 200 images for complete one revolution, 360°. For every rotation of the dosimeter, the sensor will capture the image and save it as RAW File. Later the image will be reconstructed and analysed using MATLAB. To obtain the right working distance, several distance and height of the sensor were tested. The FOV of the sensor at certain working distance was measured by using FOV chart and is shown in Table 2. The optimum distance between the imaging system and the rotary stage is 700 mm, where the field of view (FOV) can covers the whole dosimeter. The development of algorithm for image reconstruction and combination between sensor and motor GUI are of interest for further studies.

6. Conclusion

In this paper, an in-house CMOS base OCT has been developed and the setup has been described. The system can capture image at higher frame rate using ROI compared to existing CCD based OCT. Further experiments are needed to characterise the image quality of the system and application of the system for dosimeter readout. Future advances in CMOS image sensor would enable even faster scanning system with on-pixel image processing.

7. References

[1] Omar, Z.A. and N.S.I. Tamin, *National Cancer Registry Report, Malaysia Cancer Statistics – Data and Figure 2007*. 1, ed. 12011, Putrajaya: Ministry of Health, Malaysia.

[2] WHO, *World health statistics 2014*2014, Geneva, Switzerland: World Health Organization.

[3] D'Agostino, E., et al., *Peripheral doses in radiotherapy: A comparison between IMRT, VMAT and Tomotherapy*. Vol. 57. 2013. 62-67.

[4] ICRU, *Determination of absorbed dose in a patient irradiated by beams of X- or gamma rays in radiotherapy procedures*, in *ICRU Report 24*1976, ICRU: Bethesda, MD.

[5] ICRU, *Radiation dosimetry: Electron beams with energies between 1 and 50 MeV*, in *ICRU Report 35*1984, ICRU Bethesda MD.

[6] Adamovics, J. and M.J. Maryanski, *New 3D radiochromic solid polymer dosimeter from leuco dyes and a transparent polymeric matrix*. Med. Phys. 30, 2003: p. 1349.

[7] Doran, S.J., *The history and principles of chemical dosimetry for 3-D radiation fields: Gels, polymers and plastics*. Applied Radiation and Isotopes, 2009. **67**(3): p. 393-398.

[8] Doran, S.J., et al., *A CCD-based optical CT scanner for high-resolution 3D imaging of radiation dose distributions: equipment specifications, optical simulations and preliminary results*. Physics in Medicine and Biology, 2001. **46**(12): p. 3191.

[9] Oldham, M., Guo, Pengyi, Gluckman, Gary and Adamovics, John, *IMRT verification using a radiochromic/optical-CT dosimetry system*. Journal of Physics: Conference Series, 2006. **56**: p. 221-224.

[10] Schreiner, L.J., *Review of Fricke gel dosimeters*. Journal of Physics: Conference Series, 2004. **3**(1): p. 9.

[11] Baldock, C., *Historical overview of the development of gel dosimetry: a personal perspective*. Journal of Physics: Conference Series, 2006. **56**: p. 14-22.

[12] Adamovics, J. and M.J. Maryanski, *Characterisation of PRESAGETM: A new 3-D radiochromic solid polymer dosemeter for ionising radiation*. Radiat Prot Dosimetry, 2006. **120**(1-4): p. 107-112.

[13] Gore, J.C. and Y.S. Kang, *Measurement of radiation dose distributions by nuclear magnetic resonance (NMR) imaging*. Physics in Medicine and Biology, 1984. **29**(10): p. 1189.

[14] Gore, J.C., Y.S. Kang, and R.J. Schulz, *Measurement of radiation dose distributions by NMR imaging*. Magnetic Resonance Imaging, 1984. **2**(3): p. 244-244.

[15] Baldock, C., et al., *Experimental determination of the diffusion coefficient in two-dimensions in ferrous sulphate gels using the finite element method*. Australasian Physical & Engineering Science in Medicine, 2001. **24**(1): p. 19-30.

[16] Alexander, P., A. Charlesby, and M. Ross, *The Degradation of Solid Polymethylmethacrylate by Ionizing Radiation*. Proceedings of the Royal Society of London. Series A, Mathematical and Physical Sciences, 1954. **223**(1154): p. 392-404.

[17] Usanmaz, A., *Crystal Structure Effect in Radiation-Induced Solid-State Polymerization of Acrylamide*. Turkey J. of Chemistry, 1997. **21**: p. 304-312.

[18] Lepage, M., et al., *The relationship between radiation-induced chemical processes and transverse relaxation times in polymer gel dosimeters*. Physics in Medicine and Biology, 2001. **46**(4): p. 1061-1074.

[19] Lepage, M., et al., *Modelling of post-irradiation events in polymer gel dosimeters*. Physics in Medicine and Biology, 2001. **46**(11): p. 2827-2839.

[20] Hoecker, F.E. and I.W. Watkins, *Radiation polymerization dosimetry*. The International Journal of Applied Radiation and Isotopes, 1958. **3**(1): p. 31-32, IN13-IN14, 33-35.

[21] Baldock, C., et al., *Experimental procedure for the manufacture and calibration of polyacrylamide gel (PAG) for magnetic resonance imaging (MRI) radiation dosimetry*. Physics in Medicine and Biology, 1998. **43**(3): p. 695-702.

[22] Maryanski, M.J., Gore J C and Schulz R J. *3-D radiation dosimetry by MRI: solvent proton relaxation enhancement by radiation-controlled polymerisation and cross-linking in gels*. in *Proc. Intl. Soc. Mag. Reson. Med.* . 1992 New York.

[23] Maryanski, M., et al., *NMR relaxation enhancement in gels polymerized and cross-linked by ionizing radiation: A new approach to 3D dosimetry by MRI*. Magnetic Resonance Imaging, 1993. **11**(2): p. 253-258.

[24] Maryanski, M.J., Gore J C and Schulz R J, *Three-dimensional detection, dosimetry and imaging of an energy field by formation of a polymer in a gel* 1994 Patent 5,321,357: United State of America (USA).

[25] Maryanski, M.J., et al., *Radiation therapy dosimetry using magnetic resonance imaging of polymer gels*. Medical Physics, 1996. **23**(5): p. 699-705.

[26] Maryanski, M., Y.Z. Zastavker, and J.C. Gore, *Radiation dose distributions in three dimensions from*

tomographic optical density scanning of polymer gels: II. Optical properties of the BANG polymer gel. Physics in Medicine and Biology, 1996. **41**(12): p. 2705-2717.
[27] Pappas, E., et al., *A new polymer gel for magnetic resonance imaging (MRI) radiation dosimetry.* Physics in Medicine and Biology, 1999. **44**(10): p. 2677-2684.
[28] Murphy, P.S., et al., *A modified polymer gel for radiotherapy dosimetry: assessment by MRI and MRS.* Physics in Medicine and Biology, 2000. **45**: p. 3213-3223.
[29] Lepage, M., et al., *Dose resolution optimization of polymer gel dosimeters using different monomers.* Physics in Medicine and Biology, 2001. **46**(10): p. 2665-2680.
[30] De Deene, Y., et al., *An investigation of the chemical stability of a monomer/polymer gel dosimeter.* Physics in Medicine and Biology, 2000. **45**(4): p. 859-878.
[31] Hepworth, S.J., M.O. Leach, and S.J. Doran, *Dynamics of polymerization in polyacrylamide gel (PAG) dosimeters: (II) modelling oxygen diffusion.* Physics in Medicine and Biology, 1999. **44**(8): p. 1875-1884.
[32] Fong, P.M., et al., *Polymer gels for magnetic resonance imaging of radiation dose distributions at normal room atmosphere.* Physics in Medicine and Biology, 2001. **46**(12): p. 3105-3113.
[33] De Deene, Y., et al., *A basic study of some normoxic polymer gel dosimeters.* Physics in Medicine and Biology, 2002. **47**(19): p. 3441-3463.
[34] De Deene, Y., Venning,A, Hurley,C, Healy,B. J and Baldock,C, *Dose - Response stability and integrity of the dose distribution of various polymer gel dosimeters.* Physics in Medicine and Biology, 2002. **47**(14): p. 2459-2470.
[35] Mather, M.L., A.K. Whittaker, and C. Baldock, *Ultrasound evaluation of polymer gel dosimeters.* Physics in Medicine and Biology, 2002. **47**(9): p. 1449-1458.
[36] Venning, A.J., et al., *Investigation of the MAGAS normoxic polymer gel dosimeter with Pyrex glass walls for clinical radiotherapy dosimetry.* Nuclear Instruments and Methods in Physics Research Section A: Accelerators, Spectrometers, Detectors and Associated Equipment, 2005. **555**(1-2): p. 396-402.
[37] Adamovics, J. and M.J. Maryanski, *A new approach to radiochromic three-dimensional dosimetry-polyurethane.* Journal of Physics: Conference Series, 2004. **3**: p. 172-175.
[38] Abdul Rahman, A.T. and et al., *Verification of synchrotron microbeam radiation therapy using a purpose-built optical CT microscope.* Journal of Physics: Conference Series, 2010. **250**(1): p. 012083.
[39] Abdul Rahman, A.T. and et al., *Creation of sophisticated test objects for quality assurance of optical computed tomography scanners.* Journal of Physics: Conference Series, 2010. **250**(1): p. 012046.
[40] Abdul Rahman, A.T. and et al., *Sophisticated test objects for the quality assurance of optical computed tomography scanners.* Physics in Medicine and Biology, 2011. **56**(14): p. 4177.
[41] Doran, S., et al., *True-3D scans using PRESAGETM and Optical-CT: A case study in proton therapy.* Journal of Physics: Conference Series, 2006. **56**: p. 231-234.
[42] Doran, S.J. and et al., *An investigation of the potential of optical computed tomography for imaging of synchrotron-generated x-rays at high spatial resolution.* Physics in Medicine and Biology, 2010. **55**(5): p. 1531.
[43] Simon, J.D., et al., *Establishing the suitability of quantitative optical CT microscopy of PRESAGE® radiochromic dosimeters for the verification of synchrotron microbeam therapy.* Physics in Medicine & Biology, 2013. **58**(18): p. 6279.
[44] Gore, J.C., et al., *Radiation dose distributions in three dimensions from tomographic optical density scanning of polymer gels: I. Development of an optical scanner.* Physics in Medicine and Biology, 1996. **41**(12): p. 2695-2704.
[45] Jordan, K. and J. Battista, *A stable black-refractive-index-matching liquid for optical CT scanning of hydrogels.* Journal of Physics: Conference Series, 2009. **164**: p. 012045.
[46] Al-Nowais, S., et al., *A preliminary analysis of LET effects in the dosimetry of proton beams using PRESAGE(TM) and optical CT.* Applied Radiation and Isotopes, 2009. **67**(3): p. 415-418.
[47] Krstajic, N. and S.J. Doran, *Focusing optics of a parallel beam CCD optical tomography apparatus for 3D radiation gel dosimetry.* Physics in Medicine and Biology, 2006. **51**(8): p. 2055-2075.
[48] Krstajic, N. and S.J. Doran, *Fast laser scanning optical-CT apparatus for 3D radiation dosimetry.* Physics in Medicine and Biology, 2007. **52**(11): p. N257-N263.
[49] Krstajic, N. and S.J. Doran, *Characterization of a parallel-beam CCD optical-CT apparatus for 3D radiation dosimetry.* Physics in Medicine and Biology, 2007. **52**(13): p. 3693-3713.
[50] Wai, P., Adamovics, J., Krstajic, N., Ismail, A., Nisbet, A. and Doran, S., *Dosimetry of the microSelectron-HDR Ir-192 source using PRESAGE(TM) and optical CT.* Applied Radiation and Isotopes, 2009. **67**(3): p. 419-422.

[51] Fossum, E.R., *Active pixel sensors: are CCDs dinosaurs?* Proc. SPIE 1900, Charge-Coupled Devices and Solid State Optical Sensors III, 1993.
[52] Boyle, W.S. and G.E. Smith, *Charge coupled semiconductor devices*. The Bell System Technical Journal, 1970. **49**(4): p. 587-593.
[53] Janesick, J., *Dueling detectors: CCD or CMOS?*, in *Spie OE-Magazine*2002. p. 30-33.

Acknowledgments
We would like to acknowledge University of Surrey for their CCD based OCT facility. We would like to thank Universiti Sains Malaysia for their help to expand our study for the new development of CMOS based OCT (Short Term Grant 304.CIPPT.6312148). Thanks to Universiti Teknologi MARA to provide the financial support under BESTARI Grant 600-IRMI/MyRA 5/3/BESTARI (030/2017).

IOP Conf. Series: Materials Science and Engineering **298** (2018) 012037 doi:10.1088/1757-899X/298/1/012037

Radiation shielding properties of ferro-boron concrete

Muhammad Syahir Sarkawi [1, a)], Muhammad Rawi Mohamed Zain [2], Mohamad Hairie Rabir [2], Faridah Mohamad Idris [2], Jasman Zainal [1]

[1] Department of Energy Engineering, Faculty of Chemical and Energy Engineering, Universiti Teknologi Malaysia, 81310 Johor Bahru, Malaysia
[2] Malaysian Nuclear Agency, 43600 Kajang, Selangor, Malaysia

syahirsarkawi@gmail.com

Abstract. The problem of shielding against gamma and neutron radiation in nuclear facility have always attracted a great deal of attention. Typically, the best-known materials for shielding both gamma-ray and neutrons are concrete. However, due to low neutron absorption cross section in ordinary concrete, it can only weakly absorb thermal neutrons. In order to increase the neutron capture cross section, additional compound was mixed into the original concrete. In this paper, we have used ferro-boron compound to enhance the radiation shielding properties of concrete. Ferro-boron is an alloy, which is formed by combining iron with boron compound content between 10% to 17%. In this work, Monte Carlo simulation codes of MCNPX was used to simulate the radiation shielding properties of ferro-boron concrete. According to the simulation results, it is clearly shows that adding ferro-boron compound into concrete mixture can significantly enhance the radiation shielding properties of the concrete.

1. Introduction

Protection of biological entities from the harmful effects of radiation exposure is fundamental requirement in the application of nuclear technology [1]. Nowadays, many fields use nuclear technology in their application such as agriculture, cancer treatment centre, nuclear research reactors, nuclear power plant, industry inspections and research facilities has set people and environment around the facilities in serious exposure from ionizing radiation. Hence, an adequate protection is needed to ensure people and environment around the nuclear facilities is safe from any harmful effects of ionizing radiation. Radiation exposure can be avoided by methods involving time, distance and shielding [2]. Shielding is the most important method as compared to the others [3]. Typically, concrete is widely used as radiation shielding in many nuclear facilities [4].

Concrete is mainly made up from cement, aggregates and water. The properties of the materials used in the concrete give significant affect to the radiation shielding capability of the concrete [5]. In this study, we have used ferro-boron compound as aggregates for the concrete. Ferro-boron is an alloy which is formed by combining iron with boron compound between 10% to 17% [6]. Boron has large absorption cross section (700 -760 b) and ^{10}B which 19.9% of nature boron, has huge thermal neutron absorption cross-sections (3840 b) [2]. Thus, it is important to include boron compounds in concrete mixture to enhance its capability as a biological shielding material.

Several studies have been carried out on the radiation shielding properties using different materials. D. Venakata [7] has compared neutron attenuation properties of ferro-boron slabs with other high density materials including ferro-tungsten, boron carbide and mild steel plates. S.Özavci [3] and S. J. Stankovic [5] have measured and simulated linear attenuation coefficients for concrete containing different type of materials.

In this paper, we have calculated the minimum shielding thickness wall of concretes containing ferro-boron against the thermal neutrons (0.025 eV) and gamma radiation (1173 keV and 1332 keV) by using Monte Carlo simulation codes of MCNPX.

2. Methodology

In this work, Monte Carlo simulation code of MCNPX version 2.6 was used to simulate the transmitted thermal neutron and gamma-ray from ferro-boron concrete with different density. The elemental compositions and densities of the materials used in this work shown in table 1. Table 2 shows five different samples that have been simulated in this study.

For simulations, input files have been prepared. In this input files, irradiation geometry, material definition, physical settings and tally options have been represented in sequential order. A cube of 100 cm dimensions was used in this code as a shielding concrete. A beam of thermal neutron (0.025 eV) and gamma-rays (1173 keV and 1332 keV) were directed toward the shielding concrete. Cubical ferro-boron are added to the concretes. The code was run for 100 million particles history. F2 and F4 tally have been used to simulate the average surface flux and average flux in the cell for every 10 cm shielding thickness. Simulation results have been read from MCNPX output files.

Table 1. The elemental compositions and density of the materials [6][8].

Material	Element	Fraction (%)	Density (g/cm^3)
Standard Concrete	C	23.00	2.34
	O	40.00	
	Si	12.00	
	Ca	12.00	
	H	10.00	
	Mg	2.00	
Ferro-Boron	C	0.50	4.00
	Fe	79.40	
	B	17.00	
	Al	1.00	
	Si	2.00	
	P	0.05	
	S	0.05	

Table 2. Ferro-boron contents and density of simulated samples.

Samples	Ferro-Boron Content (%)	Density (g/cm^3)
Standard Concrete	0	2.30
Concrete + FeB 1	30	2.82
Concrete + FeB 2	50	3.21
Concrete + FeB 3	70	3.53
Concrete + FeB 4	90	3.79
Ferro-Boron	100	4.00

3. Results and discussion

Radiation shielding properties of concretes containing ferro-boron compounds were investigated using Monte Carlo method. MCNPX code was used to calculate the minimum shielding thickness of concrete containing the additional material which is ferro-boron.

The concentration of ferro-boron in concrete is changed from 0 % to 100 %. The density of standard concrete is smaller than ferro-boron and concrete density can be increase depending on the additional materials. The densities versus ferro-boron content in concrete were plotted in figure 1. From figure 1, the densities of the concrete significantly increase as the ferro-boron content increases.

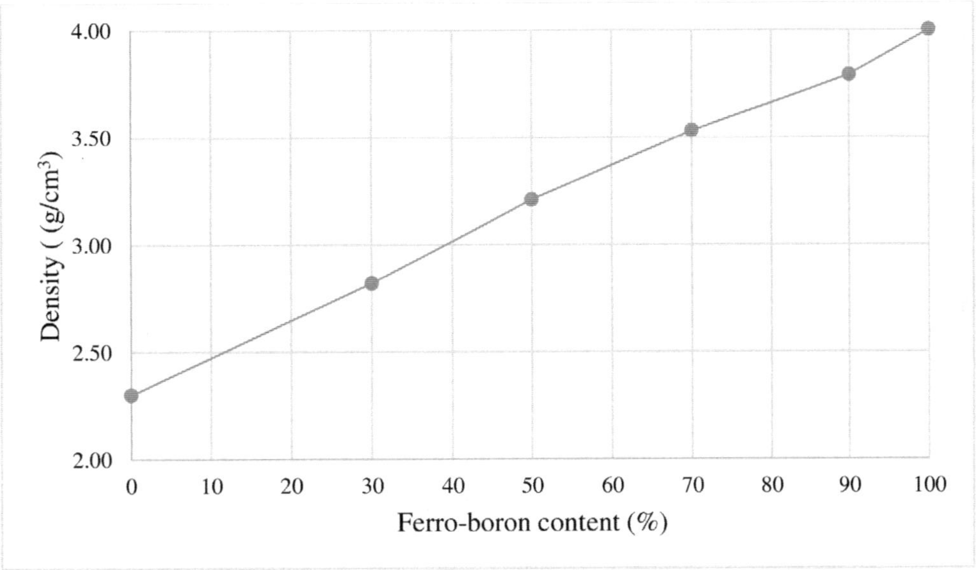

Figure 1. The densities of concrete samples with different ferro-boron content.

Simulated gamma and thermal neutron transmission versus shielding thickness were given in figure 2 and figure 3 respectively. From previous investigations, it is known that high density concrete increased its radiation shielding capability and decreased the shielding thickness depending on the properties of additional materials [9]. From figure 2 and figure 3, it is seen that concrete containing ferro-boron capable to block almost all the gamma radiation and thermal neutrons at thickness of 40 cm and 30 cm respectively, while standard concrete needs shielding thickness around 60 cm for shielding gamma and thermal neutron radiation.

Standard concrete curve from figure 3 shows higher transmitted thermal neutron at 2nd layer of MCNPX model flux tallied, this is due to multiple scattering within the tallied area. From figure 2 and figure 3, it also seen that concrete containing ferro-boron has better shielding properties for thermal neutron as compared to gamma radiation. This is due to ferro-boron that contain boron compound which has high neutron absorption cross section. Hence, it shows that ferro-boron is capable of enhancing shielding properties of concrete.

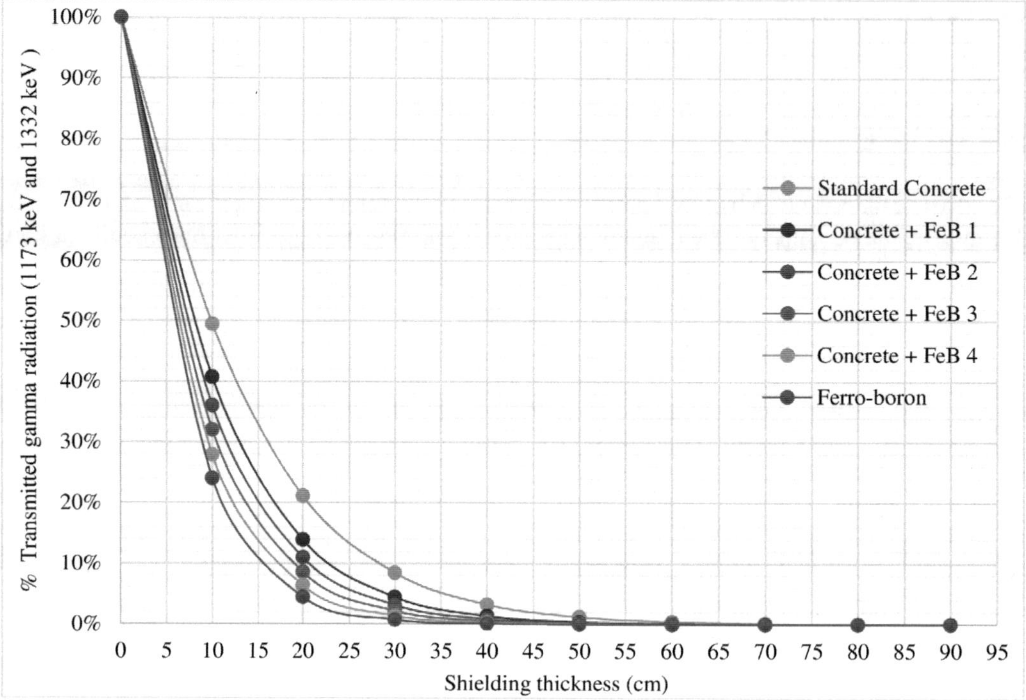

Figure 2. Transmitted gamma radiation versus shielding thickness.

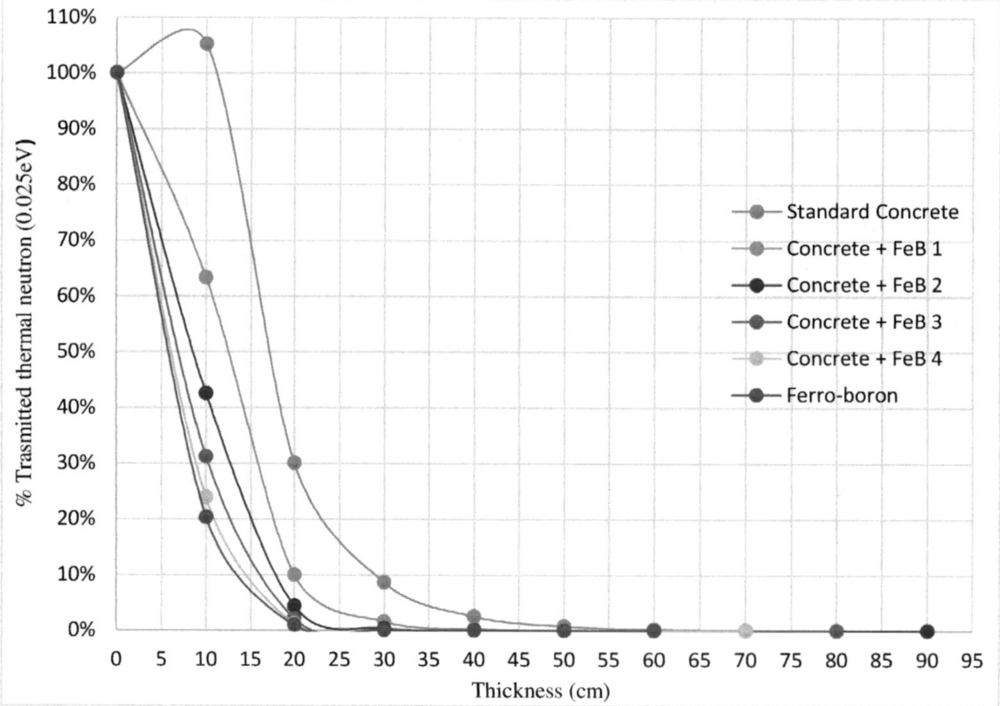

Figure 3. Transmitted thermal neutron versus shielding thickness.

4. Conclusion

In this study, the shielding thickness of concrete containing ferro-boron at different ratios was determined using Monte Carlo simulation code of MCNPX. According to the simulation results, it can be considered that ferro-boron is a good shielding material for enhancing shielding properties of concrete.

It is known that boron is a great thermal neutron absorber material while iron is a crucial shielding material to moderate fast neutrons below 1MeV. Therefore, we should consider adding ferro-boron into concrete mixture as a promising shielding material for neutron application facilities.

5. References

[1] Murray, R.L. and K.E. Holbert, *Chapter 11 - Radiation protection**, in *Nuclear Energy (Seventh Edition)*. 2015, Butterworth-Heinemann: Boston. p. 153-176.

[2] Martin, J.E., *Radiation Shielding*, in *Physics for Radiation Protection*. 2008, Wiley-VCH Verlag GmbH. p. 367-423.

[3] Özavcı, S. and B. Çetin, *Radiation shielding properties of concretes including quiclime (CaO)*. AIP Conference Proceedings, 2017. **1815**(1): p. 130009.

[4] El-Khayatt, A., *Radiation shielding of concretes containing different lime/silica ratios*. Annals of Nuclear Energy, 2010. **37**(7): p. 991-995.

[5] Stanković, S., et al., *Gamma Radiation Absorption Characteristics of Concrete with Components of Different Type Materials*. Acta Physica Polonica A, 2010. **117**(5): p. 812-816.

[6] Resources, W. *Properties of Ferro Boron*. ; Available from: http://www.wbrl.co.uk/ferro-boron.html. [cited 2017 9/5/2017]

[7] Subramanian, D.V., et al., *Comparison of neutron attenuation properties of ferro boron slabs containing 5% natural boron with other high density materials*. Radiation Protection and Environment, 2015. **38**(3): p. 109.

[8] Demir, F., et al., *Determination of radiation attenuation coefficients of heavyweight-and normal-weight concretes containing colemanite and barite for 0.663 MeV γ-rays*. Annals of Nuclear Energy, 2011. **38**(6): p. 1274-1278.

[9] Sarıyer, D., R. Küçer, and N. Küçer, *Neutron Shielding Properties of Concretes Containing Boron Carbide and Ferro – Boron*. Procedia - Social and Behavioral Sciences, 2015. **195**: p. 1752-1756.

Acknowledgements

The simulations in this paper were performed at Malaysian Nuclear Agency, Bangi, Malaysia. The author would like to acknowledge all the contributors in completing this paper.

The design and dosimetric evaluation of tannin-based *Rhizophora* spp. particleboards as phantoms for high energy photons and electrons

M F Mohd Yusof[1,3], R Abdullah[3], A A Tajuddin[1], R Hashim[2], S Bauk[1], P N K Abd Hamid[1]

[1] School of Physics, Universiti Sains Malaysia, 11800 Penang, Malaysia
[2] School of Industrial Technology, Universiti Sains Malaysia, 11800 Penang, Malaysia
[3] School of Health Sciences, Universiti Sains Malaysia, 16150 Kelantan, Malaysia

mfahmi@usm.my

Abstract. A set of phantom with an external dimension of 30 cm x 30 cm was constructed from tannin-based *Rhizophora* spp. particleboards similar to the solid water phantoms. The dosimetric characteristics of the particleboard phantoms were evaluated at high energy photons and electrons by measuring the beam output at 6 MV photons and 6 MeV electrons based on the IAEA TRS 398:2000 protocol. The tissue-phantom ratio ($TPR_{20,10}$) was measured at 6 and 10 MV photons. The beam output calibration of the particleboards was in good agreement to water and solid water phantoms at 6 MV photons with percentage difference of 1.7 and 6.2% respectively. The beam output calibration of the tannin-based *Rhizophora* spp. particleboards at 6 MeV electrons on the other hand were in excellent agreement to water with percentage difference of 0.3. The percentage depth dose of tannin-based *Rhizophora* spp. particleboards were in agreement to water and solid water within 4.5% when measured using ionization chamber and EBT2 film. The electron beam parameters of R_{50}, R_{80} and R_{90} at 6 MeV electrons also were in good agreement to water and solid water phantoms. The overall results had indicated the suitability of tannin-based *Rhizophora* spp. particleboards as water substitute phantom materials for high energy photons and electrons.

1. Introduction

Phantom materials are the most important tools in quality control (QC) and dosimetry works in radiotherapy. The phantom materials shall be made water equivalent so that the dosimetric evaluation made in the phantom can be directly related to the human soft tissues [1]. Solid water phantoms is the most commonly used water substitute phantom materials for dosimetry works and quality assurance in radiotherapy. The solid water phantoms are epoxy resin-based materials fabricated with mass density of 1.0 g/cm^3 to simulate the attenuation properties human soft tissue towards ionizing radiations.

Earlier studies suggested *Rhizophora* spp. as potential phantom material for dosimetry works and quality control for photons. *Rhizophora* spp. is the largest species of mangrove tree that grow in the saline coastal area of subtropical regions including Malaysia. Earlier study found that *Rhizophora* spp. wood had mass density and attenuation coefficient close to water [2]. *Rhizophora* spp. wood also showed similar absorption and scattering properties to water towards low energy photons [3]. A study on percentage depth dose at high energy photons and electrons also showed similarities within 4% to water and Perspex [4]. The use of *Rhizophora* spp. solid woods as phantom materials however has several limitations including limited size of the wood to construct full size phantoms. The *Rhizophora* spp. wood also have tendency to crack over period of times besides its density inhomogeneity across the trunk. The binderless particleboards made of *Rhizophora spp.* was introduced with advantages including better density uniformity and able to be fabricated at various size and shapes without compromising its attenuation properties [5,6]. The fabrication of binderless *Rhizophora* spp. particleboards however suffers significant reduction of physical and mechanical strength to endure the rigid and heavy workload as phantom.

The use of synthetic-based adhesives such as urea-formaldehyde (UF) commonly used in industrial particleboards manufacturing failed to retain the attenuation properties of *Rhizophora spp.* particleboards in comparison to water [7,8]. The use of biological-based adhesive materials was identified to be the suitable adhesive materials for *Rhizophora* spp. particleboards in compared to the commonly used synthetic-based adhesives. Several studies had indicated that the use of biological-based adhesives had increased the physical properties of *Rhizophora* spp. particleboards and retained their water equivalent and attenuation properties towards X-ray energies in the diagnostic imaging ranges [9-11].

Tannin has been used as alternative binder for particleboards and plywood to replace commonly used formaldehyde-based adhesives [12-14]. Tannin is a biological-based adhesive that can be extracted mainly from barks including *Rhizophora spp.* [15]. Previous study had suggested the suitability of tannin as biological-based adhesive for *Rhizophora spp.* particleboards as phantom at low energy photons [16]. The use of tannin-based particleboards at low treatment level of tannin (5%) however still failed to satisfy the industrial standards for physical and mechanical properties. Previous studies had suggested that the suitable percentage of biological-based adhesives for *Rhizophora spp.* is within 10% without changing their attenuation properties [9,10]. This study focused on the fabrication and characterisations of *Rhizophora spp* particleboards with 10% percentage of tannin adhesive as phantom materials in application of photon and electron beams.

2. Methodology

2.1. Fabrication of the Rhizophora spp. Particleboard Phantoms

The *Rhizophora* spp. particleboards were constructed based on the previous study by Mohd Yusof et al., [11]. The wood particles with particle size ranges of ≤ 104 μm were used to fabricate the particleboards. Amounts of 10% tannin based on the dry weight of wood particles were added in to the particleboards as binder to increase the physical and mechanical properties of the particleboard. The particleboards were fabricated with target density of 1.0 g/cm^3 based on the calculation of the amount of wood particles required to construct the particleboards at the desired size and density. The physical and mechanical properties of the particleboards were evaluated based on the internal bond (IB) strength and modulus of rupture (MOR) according to the Japanese Industrial Standards, JIS A 5908 [17]. A total of 32 particleboards with 30 cm x 30 cm size and thicknesses of 0.5 and 1.0 cm were constructed based on the external dimensions of typical solid water phantom as shown in figure 1. Two units of the particleboards were fabricated with insertion of plane-parallel ionization chamber inserts similar to that in solid water phantoms. The density of the particleboards were measured using gravimetric method based on the external dimension given by the equation

$$\rho = \frac{length \times width \times thickness}{mass} \tag{1}$$

Figure 1. The fabricated *Rhizophora* spp. particleboard phantoms

2.2. Elemental Composition and Effective Atomic Number

The attenuation properties of medium are related to the effective atomic number, therefore the effective atomic number value is used to determine the water equivalent and attenuation properties of the particleboard in comparison to water. The percentages of elemental composition of the particleboards were determined using energy dispersive x-ray analysis (EDXA). The effective atomic number, Z_{eff} was calculated based on the measured percentages of elemental composition of the particleboards. The effective atomic number is calculated according to the study by Duvauchelle et al., [18] using the equation:

$$Z_{eff} = \left[\sum_{i=1}^{n} (\alpha_i \, z_i^m)\right]^{\left(\frac{1}{m}\right)} \quad (2)$$

with a_i and z_i are electron fraction and atomic number of ith element in the sample respectively. m is the experimental coefficient for biological materials and water with value of 3.4. The electron fraction of the ith element can be calculated by the equation:

$$\alpha_i = \frac{wi\left(\frac{z_i}{A_i}\right)}{\Sigma w_i\left(\frac{z_i}{A_i}\right)} \quad (3)$$

with w_i and A_i are fractional weight and atomic mass of the ith element respectively.

2.3. Determination of Tissue-Phantom Ratio ($TPR_{20,10}$) and Output Calibration of the Particleboards

The measurement of tissue-phantom ratio ($TPR_{20,10}$) is used to determine quality of the clinical photon beams in the medium. The $TPR_{20,10}$ of the tannin-based *Rhizophora* spp. particleboards were evaluated at 6 and 10 MV photons based shown the IAEA TRS 398:2000 protocol. The average charge at 10 and 20 cm depths were collected at respective photon energies and $TPR_{20,10}$ values were calculated using the equation:

$$TPR_{20,10} = \frac{charge\ collected\ at\ 20\ cm\ depth\ (nC)}{charge\ collected\ at\ 10\ cm\ depth\ (nC)} \quad (4)$$

The value of $TPR_{20,10}$ from the tannin-based *Rhizophora* spp. particleboards were compared to that in water and solid water phantoms.

The beam output calibration was performed on the tannin-based *Rhizophora* spp. particleboards at 6 MV photons and 6 MeV electrons based on the IAEA TRS 398:2000 protocol. The absorbed dose at the depth of maximum dose, d_{max} was measured and compared to that in water and solid water phantoms. The calibration of Gafchromic EBT2 film dosimeter on *Rhizophora* spp. particleboards were conducted at 6 MV photons and 6 MeV electrons. The EBT2 film was placed at the depth of maximum dose, D_{max} and exposed to 6 MV photons and 6 MeV electrons with dose between 50 and 600 cGy. The optical density of the exposed film is calculated using the equation:

$$OD = -Log_{10}\frac{PV}{65535} \quad (5)$$

with *PV* is the pixel value or the scanner value. The exposed films were scanned using a flatbed film dosimetry scanner and the calibration curve between the OD and radiation dose is plotted and compared to that in solid water phantoms.

2.4. Measurement of Percentage Depth Dose at High Energy Photons and Electrons

The percentage depth dose (PDD) of the tannin-based *Rhizophora* spp, particleboards at 6MV photons and 6 MeV electrons were measured using Markus plane-parallel ionization chamber and EBT2 film. The measurements using ionization chamber were conducted at calibration condition (10 cm x 10 cm field size, 100 cm source to surface distance and 100 monitor units). The PDD of the particleboards were measured using the equation

$$PDD = \frac{D}{D_{max}} \times 100\% \quad (6)$$

with *D* is the dose measured at any depth and D_{max} is the maximum dose across the central axis of the beam. The PDD curve of 6 MV photons and 6 MeV electrons were plotted and compared to that of water and solid water phantoms. The percentage of discrepancies of PDD between the tannin-based *Rhizophora* spp. particleboards to water and solid water phantoms were calculated at each measured depths. The depths of electron beam parameters including R_{50}, R_{80} and R_{90} were determined form the PDD curve of electrons and compared to that in water and solid water phantoms.

3. Results and Discussions

3.1. Evaluation on Physical and Mechanical Properties of the Particleboards

The average IB strength and of the particleboards is illustrated in figure 2 and 3 respectively. The results showed that the addition of tannin adhesive had significantly increased the average IB strength and MOR of the particleboards in comparison to the binderless *Rhizophora* spp. particleboards [6] and *Rhizophora* spp. particleboards with lower percentage of tannin (Safian, 2012). The average IB strength of the tannin-based *Rhizophora* spp. particleboards in this study satisfied the Type 18 (3.0 N/mm^2) for the JIS A 5908. Similar results were seen in the average MOR of the particleboards. The tannin-based *Rhizophora* spp. in this study satisfied the Type 8 for the JIS A 5908 with the increase of 18% from the binderless *Rhizophora* spp. particleboards.

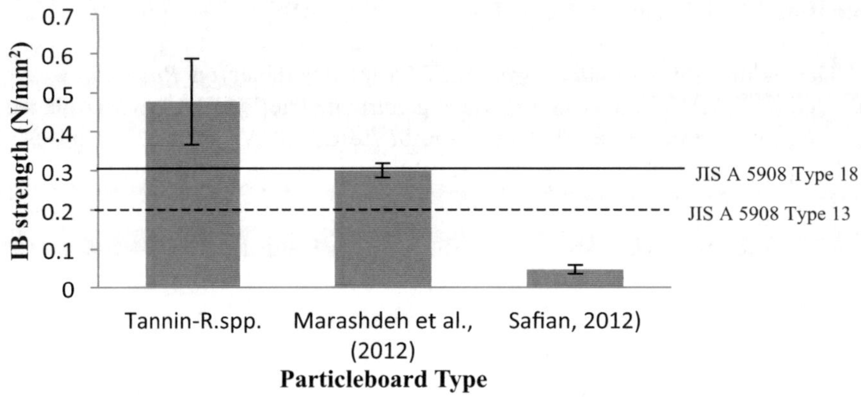

Figure 2. The average internal bond (IB) strength of the *Rhizophora* spp. particleboards

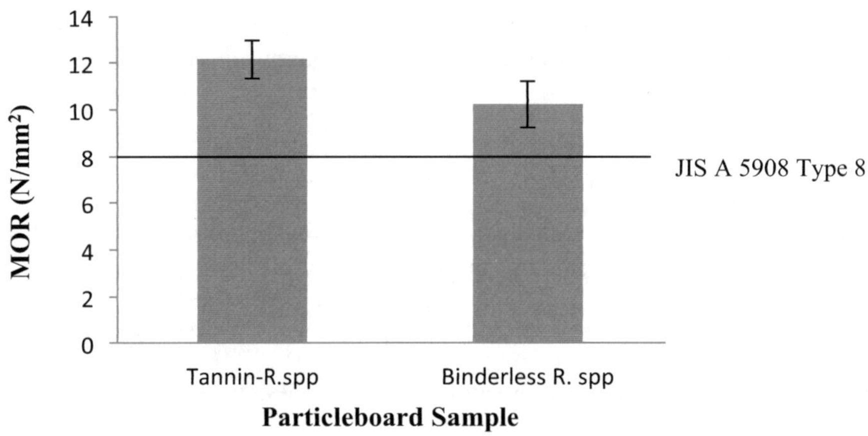

Figure 3. The average MOR of the *Rhizophora* spp. particleboards

The measured density, percentage of elemental compositions and calculated effective atomic number of the tannin-based *Rhizophora* spp. particleboards is presented in table 1. The average density of the particleboards was close to the value of water. The result is in good agreement to the study on the solid raw *Rhizophora* spp. wood by Bradley *et al.*, [2] and Banjade *et al.*, [4]. The results is also in good agreement to the study on the fabrication of *Rhizophora* spp. particleboards as phantom by Marashdeh et al., [6], Abuaraa *et al.*, [9] and Tousi *et al.*, [10]. The elemental composition of the tannin-based *Rhizophora* spp. particleboard was measured using the energy dispersive X-ray analysis (EDXA). The results showed high percentage of carbon and oxygen which is similar to the human soft tissue. It is postulated that the particleboards having similar attenuation properties to human soft tissue based on the similarity of elemental compositions. The effective atomic number of the particleboards was measured using the elemental compositions of the particleboards based on the study by Duvauchelle et al., [18]. The results showed that the effective atomic number of the tannin-based *Rhizophora* spp. particleboards was close to the value of water [19]. The attenuation properties of a material are proportional to the effective atomic number of the material [20]. Therefore, it is postulated that the attenuation properties of tannin-based *Rhizophora* spp. particleboards are similar to that of water. The average CT number of the particleboards was measured at 120 kVp CT X-ray energy and 250 mAs tube current setting based on the abdominal imaging protocol. The average CT number of tannin-based *Rhizophora* spp. particleboards was close to the value of water indicating close attenuation properties to water towards diagnostic X-ray energy. The results was in good agreement to the previous studies on the CT numbers of *Rhizophora* spp. particleboards by Marashdeh et al., [6], Abuarra *et al*., [9] and Tousi *et al*., [10].

Table 1. The physical properties of tannin-based *Rhizophora* spp. particleboard phantoms.

Physical property	Description
Average density ± standard deviation	1.003 ± 0.014
Elemental composition	Carbon (51.25%), Oxygen (43.11%), Fluorine (5.64%)
Effective atomic number, Z_{eff}	7.22 [a] (water = 7.42 [a])
Average CT number	-13.69 (water = -2.5, solid water = -11.61)

[a]AAPM-21, [19]

3.2. Evaluation of Tissue-Phantom Ratio ($TPR_{20,10}$) and Output Calibration of the Particleboards

The tissue-phantom ratio ($TPR_{20,10}$) of tannin-based *Rhizophora* spp. particleboards at 6 and 10 MV photons is presented in table 2. The value of $TPR_{20,10}$ in tannin-based *Rhizophora spp*. particleboards was compared to the value measured in water and solid water phantoms commonly used in the dosimetry works involving high energy photons and electrons. The results showed good agreement of tissue-phantom ratio of tannin-based *Rhizophora spp*. particleboards in comparison to water and solid water phantoms with percentage of variation within 2% and in agreement the study by Banjade *et al.*, [4]. The value of $TPR_{20,10}$ remains as the most appropriate method to determine the beam quality of a clinical photon beams Therefore, a material with close value of $TPR_{20,10}$ to the water is postulated to have similar attenuation properties to water and soft tissues.

Table 2. The tissue-phantom ratio of tannin-based *Rhizophora* spp. particleboards in comparison to water and solid water at 6 and 10 MV photons.

Photon energy (MV)	$TPR_{20,10}$		
	Water (W)	Solid water (S)	Tannin-based *Rhizophora* spp. (R)
6	0.678	0.675	0.689
10	0.746	0.743	0.754

The beam output calibration of tannin-based *Rhizophora* spp. particleboards at 6 MV photons and 6 MeV electrons is presented in table 3. The absorbed dose at d_{max} in the particleboards showed an excellent agreement to the value in water at 6 MV photons with 1.69% percentage of discrepancy. The result was consistent with the earlier study by Banjade *et al.*, (2001) that measured the absorbed dose at d_{max} in solid raw *Rhizophora* spp. wood (4%). A comparison to solid water phantoms at 6 MV photons also showed good agreement within 6% percentage of discrepancy and in agreement to the study by White and Constantinou [1] and Reft [21]. The output calibration of tannin-based *Rhizophora* spp. particleboards at 6 MeV electrons also showed good agreement within 0.3 and 2.4% percentage of discrepancy to that in water and solid water phantom respectively. The EBT2 film calibration curve of tannin-based *Rhizophora* spp. particleboards at 6 MV photons and 6 MeV electrons is shown in figure 4(a) and 4(b) respectively. The calibration curve showed good dose-dependence of EBT2 film on *Rhizophora* spp. particleboards at between 50 and 700 cGy. The results were also in good agreement to the solid water phantoms indicating similar attenuation properties of the particleboards to solid water phantoms at 6 MV photons and 6 MeV electrons.

IOP Conf. Series: Materials Science and Engineering **298** (2018) 012038 doi:10.1088/1757-899X/298/1/012038

Table 3. The beam output calibration of tannin-based *Rhizophora* spp. particleboards at 6 MV photons and 6 MeV electrons based on IAEA TRS 398:2000.

Energy	Output (cGy/MU)		
	Water	Solid water	Tannin-based *Rhizophora spp.* particleboards
6 MV	1.001	1.083	1.018
6 MeV	1.004	0.977	1.001

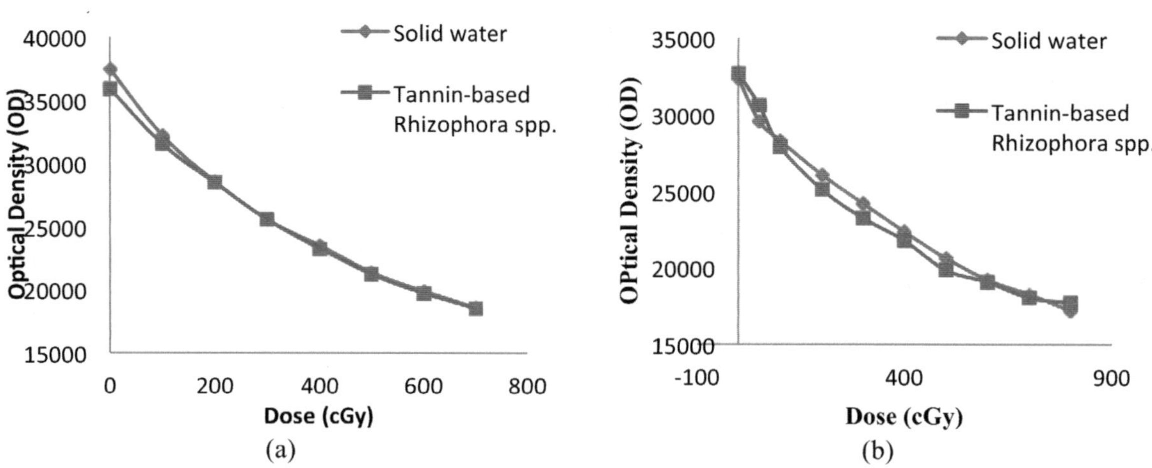

Figure 4. The EBT2 film calibration curve at (a) 6 MV photons and (b) 6 MeV electrons.

3.3. Evaluation on Percentage Depth Dose in High Energy Photons and Electrons

The PDD curve of tannin-based *Rhizophora* spp. particleboards at 6 MV photons measured using ionization chamber and EBT2 film is illustrated in figure 5(a) and 5(b) respectively. The PDD curve of the particleboards was in good agreement to water within 4.39% percentage of discrepancies at all measured depths when measured using ionization chamber. The PDD curve of the particleboards was also in good agreement with solid water phantoms within 4.05% percentage of discrepancies at all measured depths. The results are in good agreement to the study on the PDD curve of solid raw *Rhizophora* spp. wood by Banjade at al., [4]. The PDD measurement using EBT2 film showed that the PDD curve of tannin-based *Rhizophora* spp. particleboards was in good agreement to solid water phantoms within 4% at all measured depths. The results had indicated the similarity of attenuation properties of tannin-based *Rhizophora* spp. to that in water and solid water phantoms and can be potentially used as phantom for high energy photons.

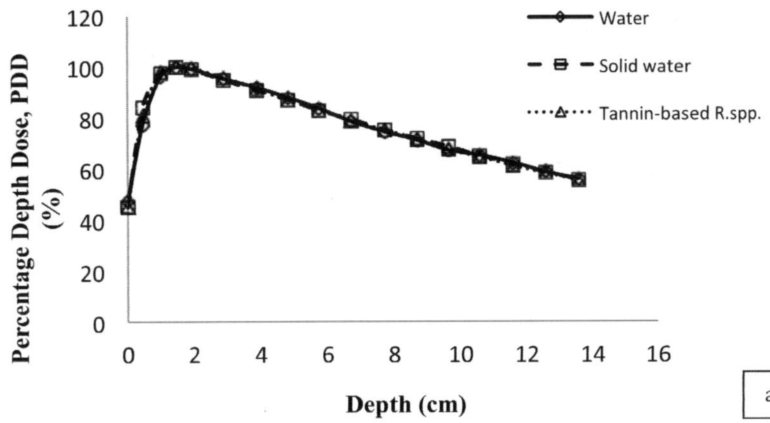

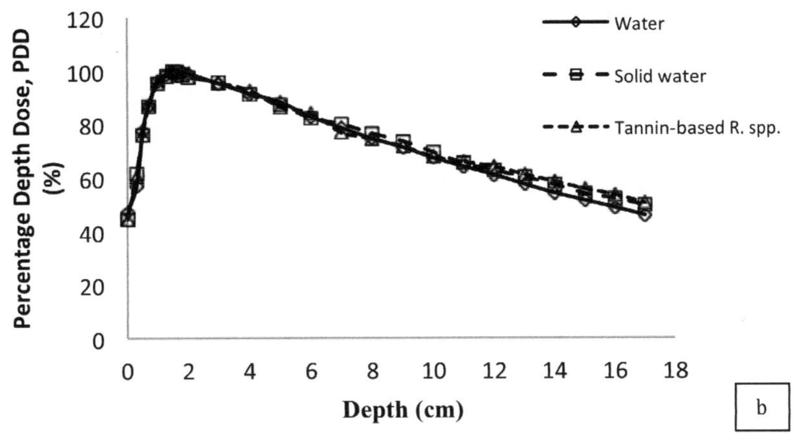

Figure 5. The percentage depth dose (PDD) curve of tannin-based *Rhizophora* spp. particleboards in comparison to water and solid water phantoms at 6 MV photons measured using (a) ionization chamber and (b) EBT2 film.

The PDD curve of tannin-based *Rhizophora* spp. particleboards at 6 MeV electrons measured using ionization chamber and EBT2 film is illustrated in figure 6(a) and 6(b) respectively. The PDD curve of the particleboards was in good agreement within 5.12% to solid water phantoms at all measured depths. A comparison to water also showed good agreement within 5.24% at all measured depths. The results were in good agreement to the study by Banjade et al., [4]. The measurement of PDD using EBT2 film showed an excellent agreement to solid water phantoms within 2% percentage of discrepancies at all measured depths.

The electron beam parameters of *Rhizophora* spp. particleboards at 6 MeV electrons in comparison to water and solid water phantom are presented in Table 4. The results showed that all electron beam parameters of R_{50}, R_{80} and R_{90} of the particleboards were close to that in water and solid water phantoms. The R_{80} and R_{90} has been used to determine the effectiveness of phantom material as it indicated the therapeutic range of electrons where the target volume received 80 and 90% of the prescribed dose respectively [22]. The overall results showed that the attenuation properties and dosimetric profiles of tannin-based *Rhizophora* spp. particleboards was similar to water and solid water phantoms as standard phantom materials for dosimetry works and quality assurance in high energy photons and electrons. The results also showed the compatibility of the particleboards when measured using ionization chamber and EBT2 film as the most common dosimeters used in dosimetry works and quality assurance of high energy photons and electrons.

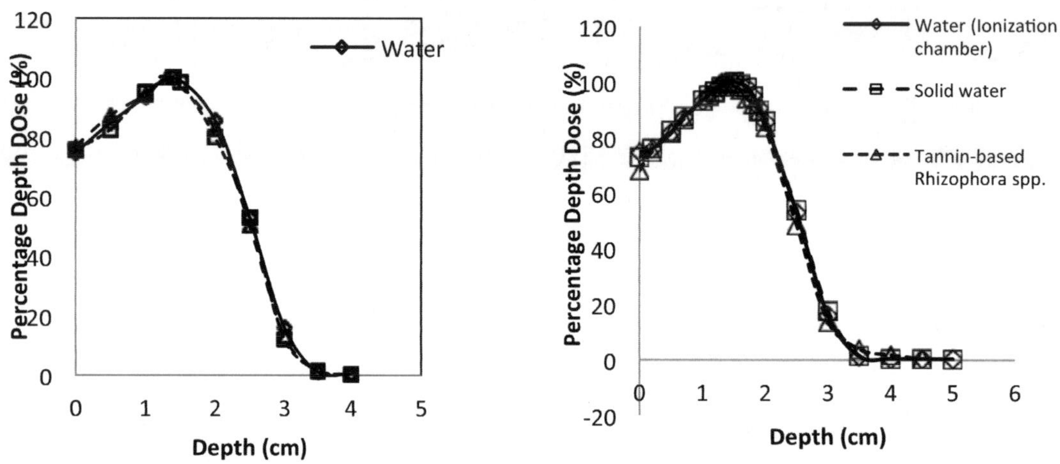

(a) (b)

Figure 6. The percentage depth dose (PDD) curve of tannin-based *Rhizophora* spp. particleboards in comparison to water and solid water phantoms at 6 MeV electrons measured using (a) ionization chamber and (b) EBT2 film.

Table 4. The electron beam parameters of 6 MeV electrons in tannin-based *Rhizophora* spp. particleboards in comparison to water and solid water phantoms.

Parameter	Depth (cm)		
	Water	**Solid water**	**Tannin-based *Rhizophora* spp.**
R_{50}	2.36	2.32	2.18
R_{80}	1.86	1.86	1.85
R_{100}	1.50	1.40	1.50

4. Conclusions

The addition of tannin had increased the physical and mechanical properties of the *Rhizophora* spp. particleboards shown by the average IB strength and MOR. The fabricated tannin-based *Rhizophora* spp. particleboards showed similar physical properties to water indicated by the mass density, effective atomic number and average CT number. The $TPR_{20,10}$ values of the tannin-based *Rhizophora* spp. particleboards was close to the value of water and solid water phantoms measured at 6 and 10 MV photons. The beam output calibration at 6 MV photons and 6 MeV electrons showed that the absorbed dose in the tannin-based *Rhizophora* spp. particleboards was close to that in water and solid water phantoms. The calibration curve of tannin-based *Rhizophora* spp. particleboards showed good linearity between 50 and 700 cGy radiation doses of 6 MV photons and 6 MeV electrons. The PDD curves of tannin-based *Rhizophora* spp. particleboards showed an excellent agreement within 5% percentage of discrepancies when measured using ionization chamber and EBT2 film at 6 MV photons and 6 MeV electrons. The overall results had indicated the potential use of tannin-based *Rhizophora* spp. particleboards as phantom for high energy photons and electrons.

5. References

[1] White D R and Constantinou C 1982 Anthropomorphic phantom materials *PMRP* **1** 133-193

[2] Bradley D A, Tajuddin A A, Che Wan Sudin C W A, and Bauk, S 1991 Photon attenuation studies on tropical hardwoods. *International Journal of Radiation Applications and Instrumentation. Part A. Appl. Radiat. Isotopes* **42** 771-773

[3] Tajuddin A A, Che Wan Sudin C W A, and Bradley D A 1996 Radiographic and scattering investigation on the suitability of *Rhizophora spp.* as tissue equivalent medium for dosimetric study *Radiat. Phys. Chem.* **47** 739-740

[4] Banjade D P, Tajuddin A A, and Shukri A 2000 A study of *Rhizophora spp.* wood phantom for dosimetric purposes using high-energy photon and electron beams *Appl. Radiat. Iso.***55** 297-302

[5] Shakhreet B Z, Bauk S Tajuddin A A, and Shukri A 2009 Mass attenuation coefficnet on natural *Rhizophora spp.* woods for X-rays in the 15.77 to 25.27 keV range *Radiat. Prot. Dosim* **135** 4-53

[6] Marashdeh M W, Hashim R, Tajuddin A A, Bauk S, and Sulaiman O 2011 Effect of particle size on the characterization of binderless particleboard made from *Rhizophora spp.* Mangrove wood for use as phantom material *Bioresource Technol.* 6 4028-4044

[7] Surani B T 2008 The suitability of PF, UF and PRF resins in term of structure and attenuation properties to be used in *Rhizophora* spp. particleboard phantom. *M.Sc. Thesis*, Universiti Sains Malaysia

[8] Ngu K T, Bauk S, Hashim R, Tajuddin AA, and Shukri A 2015 Fabrication of formaldehyde-based particleboards and their mass attenuation coefficients at 15.77, 17.48, 21.18 and 25.27 keV photon energies. *JPS* **26(1)**, 27-33

[9] Abuarra A, Hashim R, Bauk S, Kandaiya S, and Tousi ET 2014 Fabrication and characterization of gum Arabic bonded *Rhizophora spp.* particleboards, *Mater. Design* **60** 108-115

[10] Tousi E T, Bauk S, Hashim R, Jaafar M S, Abuarra A 2014 Measurement of mass attenuation coefficients of *Eremurus-Rhizophora spp.* particleboards for X-ray in the 16.63-25.30 keV energy range. *Radiat. Phys. and Chem.* **103** 119-125

[11] Mohd Yusof M F, Hashim R, Tajuddin A A, Bauk S and Sulaiman O 2017 Characterization of tannin-added *Rhizophora* spp. as phantom materials for photon beams *Ind. Crop. Prod.* **95** 467-474

[12] Pizzi A and Merlin M 1981 A new class of tannins adhesives for exterior particleboard. *Int. J. Ahhes. Adhes.* **1(5)** 261–264

[13] Pizzi A and Scharfetter H 1989 Adhesives and techniques open new possibilities for the wood processing industry. Part1: Experience with tannin based adhesives *Eur. J. Wood Wood Prod.* **39** 85–89

[14] Oo C, Pizzi A, Pasch H, and Kassim M 2008 Study on the structure of mangrove polyflavonoid tannins with MALDI – TOF mass spectrometry. *J. Appl. Polym. Sci* **109** 963-967

[15] Mohd Yusoff M N, Chew L T, Mohd Ali A R, Abdul Kadir A 1988 Mangrove tannin as an adhesive for wood-based panels *JTFS.* **1** 97-102

[16] Safian A 2012 Tannin-based *Rhizophora spp.* particleboard breast phantom at mammographic energy *M.Sc. Thesis*, Universiti Sains Malaysia

[17] Japanese Industrial Standard Particleboards (JIS A 5908) 2003 Standardization Promotion Department, Japanese Standard Association, Tokyo

[18] Duvauchelle P, Peix G, and Babot D 1999 Effective atomic number in the Rayleigh to Compton scattering ratio *Nucl. Instrum. Methods Section B: Beam Interactions with Materials and Atoms,* **155** 221-228

[19] AAPM-21, 1983 Protocol for the determination of absorbed dose from high energy photon and electron beams; Task Group 21 *Med. Phys.* **10** 741-771

[20] Kurudirek M 2014 Effective atomic numbers and electron densities of some human tissues and dosimetric materials for mean energies of various radiation sources relevant to radiotherapy and medical applications *Radiation Phys. Chem.* **102** 139-146

[21] Reft C S 1989 Output calibration in solid water for high energy photon beams. *Med. Phys.* **16** **(2)** 299-301

[22] Khan M F 2010 The physics of radiation therapy, Lippincott William & Wilkins, Philadelphia

Acknowledgements

The authors also would like to acknowledge the Oncology, Radiotherapy and Nuclear Medicine Department, Universiti Sains Malaysia for the use of the linear accelerator in this study.

IOP Conf. Series: Materials Science and Engineering **298** (2018) 012039 doi:10.1088/1757-899X/298/1/012039

Automatic sample changer control software for automation of neutron activation analysis process in Malaysian Nuclear Agency

N. Yussup[1,a)], M. M. Ibrahim[1], N. A. A. Rahman[1], M. Mokhtar[1], N. A. A. Salim[2], S. C. Soh@Shaari[1], A. Azman[1], L. Lombigit[1], A. Azman[2] and S. A. Omar[2]

[1]Technical Support Division, [2]Waste and Environmental Technology Division, Malaysian Nuclear Agency, Bangi, 43000 Kajang, Malaysia.

[a)]nolida@nm.gov.my

Abstract. Most of the procedures in neutron activation analysis (NAA) process that has been established in Malaysian Nuclear Agency (Nuclear Malaysia) since 1980s were performed manually. These manual procedures carried out by the NAA laboratory personnel are time consuming and inefficient especially for sample counting and measurement process. The sample needs to be changed and the measurement software needs to be setup for every one hour counting time. Both of these procedures are performed manually for every sample. Hence, an automatic sample changer system (ASC) that consists of hardware and software is developed to automate sample counting process for up to 30 samples consecutively. This paper describes the ASC control software for NAA process which is designed and developed to control the ASC hardware and call GammaVision software for sample measurement. The software is developed by using National Instrument LabVIEW development package.

1. Introduction

Neutron activation analysis (NAA) activity has been started in Malaysian Nuclear Agency (Nuclear Malaysia) since 1983 with the establishment of the NAA laboratory, not long after the installation of the only nuclear research reactor in Malaysia, Nuclear Malaysia TRIGA Mark II reactor in 1982 which provides the neutron source for the NAA work [1]. It was set up as part of the reactor utilisation programme to provide elemental analysis needs for the research and training activities in this research centre as well as for other potential users in the country. Since then, the NAA has become a major analytical tool and has been used as a reference analytical technique for elemental analysis, owing to its better accuracy compared to other analytical techniques. In Nuclear Malaysia, the NAA activity is performed by personnel of the Analytical Chemistry Application (ACA) group under the Division of Waste and Environmental Technology.

Most of the NAA processes established in Nuclear Malaysia from sample receiving and preparation to reporting of result as shown in figure 1 are performed manually. These manual procedures carried out by the NAA lab personnel are tedious, time consuming and inefficient especially in sample counting and measurement process. Currently, the NAA Counting Laboratory had only one detector equipped with a commercial automatic sample changer (ASC) which can perform round the clock measurement for only up to 20 samples. However, at present the commercial ASC is broke down and cannot be operated due to motor problem. Meanwhile, the other seven units of detector system are utilised manually with limited counting hours per day. More radiation is also exposed to NAA personnel during sample handling and manual counting process.

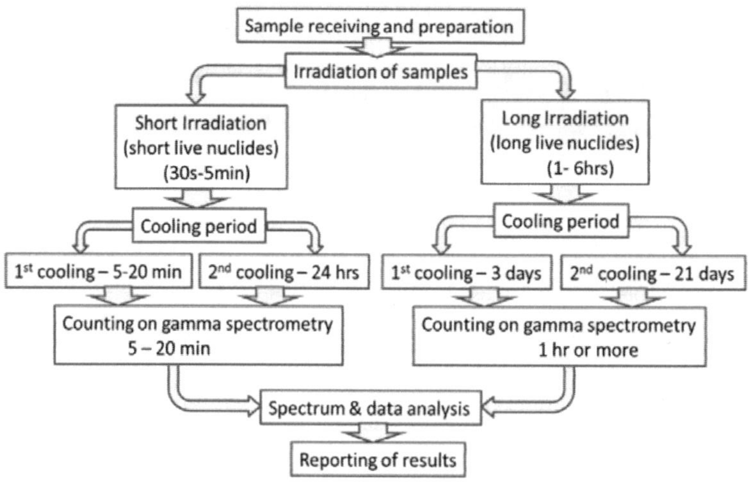

Figure 1. Established NAA procedure.

The purpose of NAA automation system as a whole is to analyse multiple samples efficiently by reducing human intervention during the process from sample preparation stage to computation of elemental concentrations. There are three stages in the system which are sample registration, ASC and sample analysis. The in-house developed ASC consists of hardware as shown in figure 2 and software development. The system itself contains three separate software developments including ASC control software. The other two software programs are Sample Registration Software [2] and Sample Analysis Software. On the ASC part, the control software must communicate via RS-232 communication interfacing with the peripheral interface controller (PIC) microcontroller implemented on the ASC hardware control board as illustrated in figure 3. The main goal of this project is to develop a user-friendly interface in order to automate sample counting process and reduce the human intervention in the NAA laboratory process.

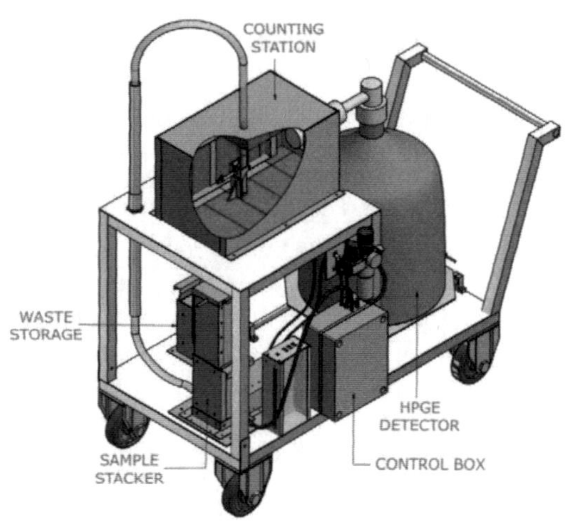

Figure 2. ASC hardware.

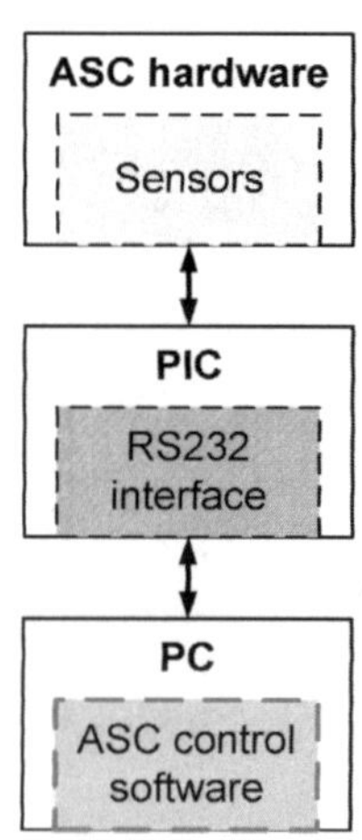

Figure 3. ASC block diagram.

IOP Conf. Series: Materials Science and Engineering **298** (2018) 012039 doi:10.1088/1757-899X/298/1/012039

2. Software Development

The ASC control software is developed to control the sample transfer of ASC hardware [3] and initiates the data acquisition process in gamma spectroscopy software. The graphical user interface (GUI) is developed by using National Instruments LabVIEW 2014 development package. VISA functions are used to communicate via serial interfacing of the ASC hardware. File I/O functions are used to create, open, read from and write to files. The controls and functions like String, Array and Structures offered by LabVIEW are utilized in order to develop user-friendly control software.

For data acquisition and sample measurement, the selected gamma spectroscopy software is Ortec GammaVision version 7 since most of the nuclear counting systems at the NAA laboratory are from Ortec and GammaVision version 7 which has several additions and improvements in JOB functions in order to optimize automation process compare to the previous version. Two files are required to run GammaVision software automatically which are .SDF file to setup GammaVision acquisition and analysis setting; and .JOB file to automate the functions. Nuclides library file (.LIB) and calibration file (.CLB) are optional based on user requirement for sample measurement and analysis during the sample counting process.

Prior to the control software development, three steps are identified in the software execution in order to run accordingly as desired. The three steps and the respective main task are as below:

- Initialization: Set acquisition and analysis setting
- Data acquisition settings: Set sample counting sequence and parameters
- Data acquisition status: Run sample counting process

Figure 4 shows the flowchart of the ASC control software execution. The more detailed tasks for each step are shown in figure 5. The software interface and program are designed and developed based on these steps and the tasks.

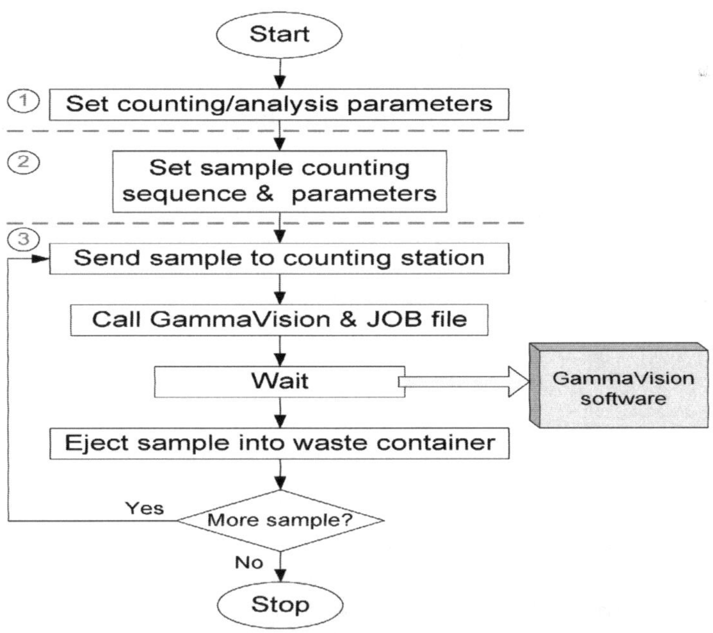

Figure 4. ASC control software flowchart.

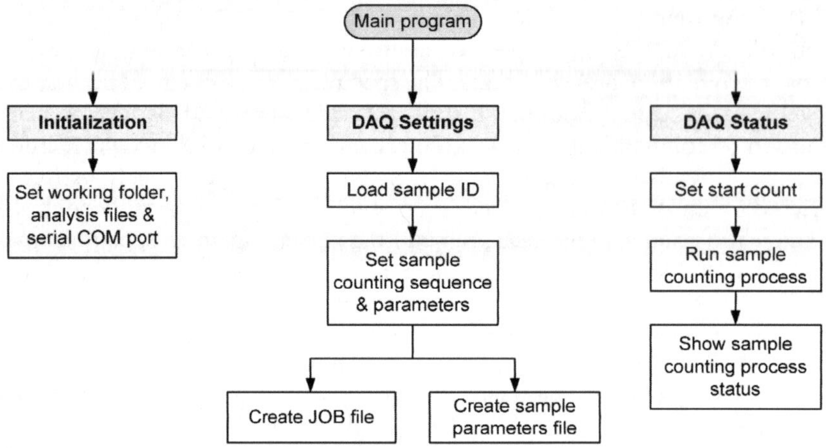

Figure 5. The hierarchy chart of ASC control software development.

The most important task of ASC control software is the sample counting process. This process is performed in the final step of the software execution which is Data Acquisition Status as shown in figure 4 and figure 5. In this process, the ASC hardware mechanism is controlled to change sample one by one and GammaVision spectroscopy software is called for data acquisition of sample count. The data acquisition process is performed by GammaVision by using JOB Control Service where .JOB file for each sample is created by the ASC control software in text file format based on sample parameters entered by user [4]. The .JOB file is called upon by the control software following the sample counting sequence set up by the user. The flowchart of sample counting process is shown in figure 6.

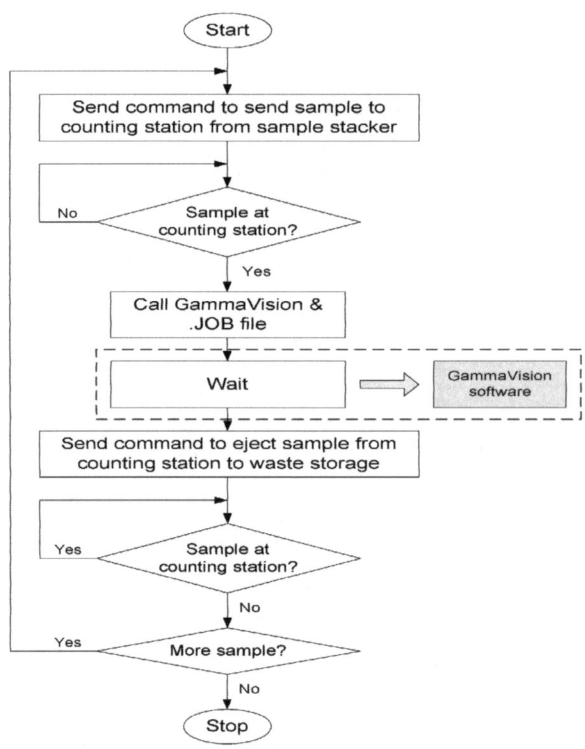

Figure 6. Sample counting process flowchart.

There are six states involved in the design of the sample counting process program including Finish state. The state machine of the program is illustrated in figure 7. During the states of Send Sample and Eject Sample, character 's' and 'e' is transmitted respectively via serial communication to the control box where PIC microcontroller is implemented in order to send sample to counting station and to eject sample from counting station to waste storage of ASC hardware. GammaVision software and .JOB file for the current sample is called by ASC control software during the Sample Count state. Timers with certain delay period are set up for every state transition. The transitions for Sample Ready and Sample Out state are based on a reading of infrared (IR) sensor located at the counting station of ASC hardware. Both states will move to the next state only when they read the desired condition of IR sensor.

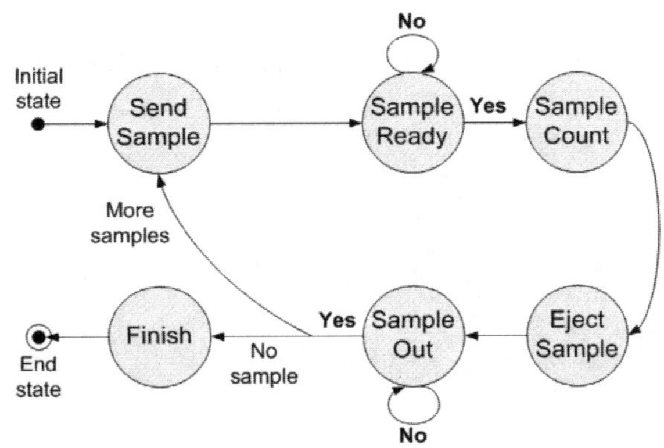

Figure 7. Sample counting process state machine.

The .JOB file is created by ASC control software based on user requirements and standard gamma spectroscopy data acquisition setting performs using GammaVision software during manual handling. The analysis setting is saved in .SDF file which is called by the .JOB file created for each sample during the automation of sample changing and counting process. The flowchart of .JOB file created in the program to be run by GammaVision software is shown in figure 8. The LabVIEW program for calling GammaVision and .JOB file for each sample is shown in figure 9. As mentioned previously, the GammaVision software is called during Sample Count state of sample counting process.

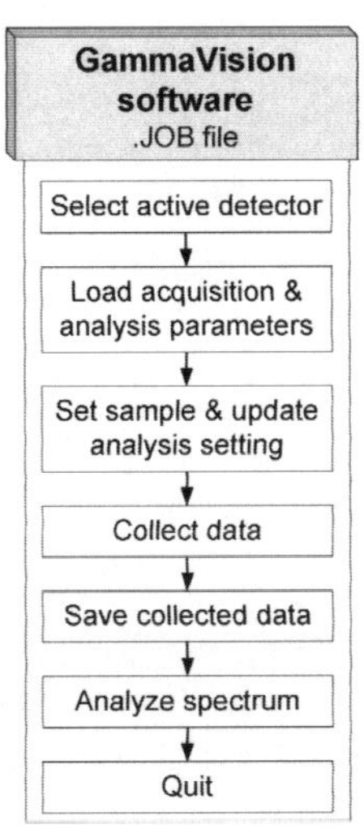

Figure 8. .JOB file function automation flowchart.

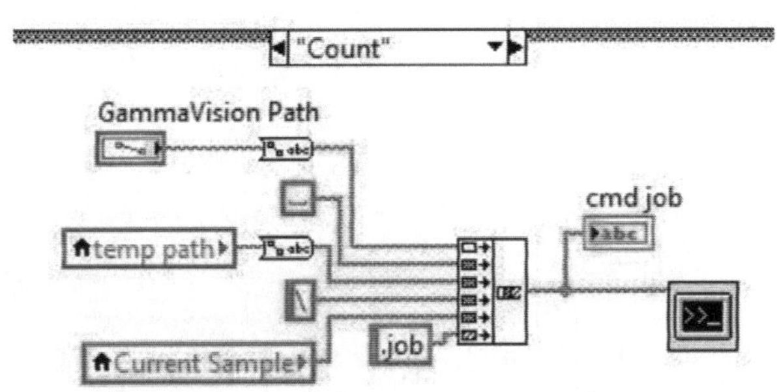

Figure 9. LabVIEW program for calling GammaVision and .JOB file.

IOP Conf. Series: Materials Science and Engineering **298** (2018) 012039 doi:10.1088/1757-899X/298/1/012039

3. Results and Discussion

The ASC control software is developed not only to control ASC hardware and automate the sample changing and counting process. It is designed to also capable to load sample ID and the respective weight automatically based on the samples text file created by Sample Registration Software. In counting process, the software ensures output report (.RPT) text file is generated by GammaVision to be read by Sample Analysis Software for sample analysis and reporting purposes. Besides .RPT file, GammaVision software also generates ORTEC spectrum (.AN1) file, spectrum (.SPE) file which is in ASCII format and analysis result (.UFO) file.

The ASC control software is capable to automate sample counting process for up to 30 samples per batch. This limitation is due to the ASC hardware design where only 30 samples are fit into the sample stacker. Each state of sample counting process requires certain amount of delay period which resulted in total delay required for each sample count is 24 seconds. This delay is necessary to write command to control the ASC, read the sensor at counting station and analyse the data. Therefore, the developed ASC can automate sample changing and counting process for up to 23 samples per day for one hour counting live time.

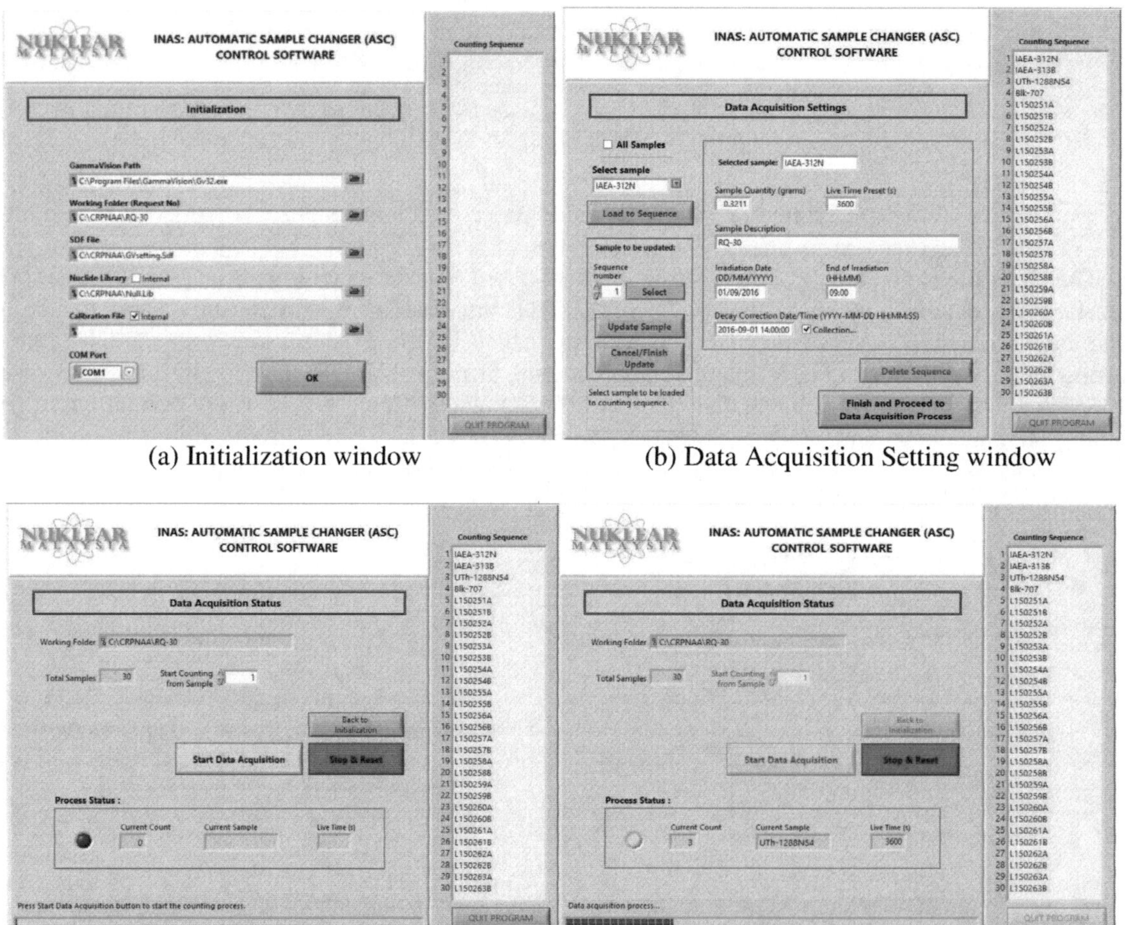

(a) Initialization window (b) Data Acquisition Setting window

(c) Data Acquistion Status window before starting data acquisition process. (d) Data Acquistion Status window during data acquisition process.

Figure 10. ASC control software GUI.

Each window of the ASC control software shown in figure 10 represents the three steps of the ASC control software execution as described in previous section. On Initialisation window, user has to determine the GammaVision software path, .SDF file that will be used by GammaVision to set the data acquisition parameters, nuclide library (.LIB) and calibration file (.CLB) that can be leaved as default. These parameters will ensure GammaVision software can be called by the ASC control software correctly with the required analysis settings for data acquisition and measurement. Besides that, user is also required to determine the working folder path where the samples text file produced by Sample Registration Software is saved.

On Data Acquistion Settings window, as shown in figure 10(b), the user is required to set up the sample counting sequence along with the parameter for each sample in the sequence. The .JOB file is created and saved in a temporary folder located in the working folder. The final window will show the status of data acquisition process as shown in figure 10(d). During this step, the ASC control software will communicate with the ASC hardware via the serial interface circuit to change sample accordingly and call GammaVision software with the respective .JOB file for each sample to execute data acquisition process. The .RPT files and other measurement files generated by GammaVision are saved directly in the working folder path.

4. Conclusions

The main responsibilities of the ASC control software are setting up sample counting sequence, automate the sample changing process and calling GammaVision software for sample counting process. The software also acts like an interface between Sample Registration Software and Sample Analysis Software to produce comprehensive NAA automation software. Computer serial connection needs to be stable to ensure the reliability of the automatic sample counting process. The software is open for modification and improvement based on the user requirement for sample changing and counting process.

5. References

[1] Malaysian Nuclear Agency 2007 Safety Analysis Report for PUSPATI TRIGA Reactor (Malaysian Nuclear Agency: MINT/L/2006/42 (s))

[2] Rahman N A A, Yussup N *et. al.* 2015 Sample registration software for process automation in the neutron activation analysis (NAA) facility in Malaysia Nuclear Agency *AIP Conf. Proc.* **1659** 030011

[3] Azman A, Hamzah M A, Atan M N, Yussup N *et. al.* 2014 Design and development of automatic sample changer for neutron activation analysis (NAA) counting station (Nuclear Science and Tec.Conf. 2014)

[4] Yussup N, Rahman N A A, Ibrahim M M, Mokhtar M, Salim N A A, Soh@Shaari S C and Azman A 2017 Development of automation software for neutron activation analysis process in Malaysian Nuclear Agency *AIP Conf. Proc.* **1799**, 050009

IOP Conf. Series: Materials Science and Engineering **298** (2018) 012040 doi:10.1088/1757-899X/298/1/012040

GSM module for wireless radiation monitoring system via SMS

Nur Aira Abd Rahman[1,3], Noor Hisyam Ibrahim[2], Lojius Lombigit[1], Azraf Azman[1], Zainudin Jaafar[1], Nor Arymaswati Abdullah[1], and Glam Hadzir Patai Mohamad[1]

[1]Malaysian Nuclear Agency, 43000, Bangi, Selangor, Malaysia
[2]Rocksolid Simplified Solution PLT, 43000, Kajang, Selangor, Malaysia
[3]nur_aira@nm.gov.my

[3]nur_aira@nm.gov.my

Abstract. A customised Global System for Mobile communication (GSM) module is designed for wireless radiation monitoring through Short Messaging Service (SMS). This module is able to receive serial data from radiation monitoring devices such as survey meter or area monitor and transmit the data as text SMS to a host server. It provides two-way communication for data transmission, status query, and configuration setup. The module hardware consists of GSM module, voltage level shifter, SIM circuit and Atmega328P microcontroller. Microcontroller provides control for sending, receiving and AT command processing to GSM module. The firmware is responsible to handle task related to communication between device and host server. It process all incoming SMS, extract, and store new configuration from Host, transmits alert/notification SMS when the radiation data reach/exceed threshold value, and transmits SMS data at every fixed interval according to configuration. Integration of this module with radiation survey/monitoring device will create mobile and wireless radiation monitoring system with prompt emergency alert at high-level radiation.

1. Introduction

There are various types of nuclear instruments that are used to detect, measure, and identify radiation. These instruments are typically intended for safety purposes; to monitor the radiation level and to ensure the safety of radiation worker, public, and environment. An essential feature that could be added to enhance the capability of these instruments is wireless communication. Integration of the instruments with wireless communication will enable continuous radiation monitoring especially for remote locations, early radiation emergency warning, as well as wireless device setup, change, or query that satisfy the needs specific application.

Wireless technology has evolved from 1G (First Generation) to 4G LTE (Fourth Generation Long Term Evolution); each generation has brought improvement in terms of communication speed, bandwidth, security, and reliability. However, detail consideration must be done in selecting the best wireless technology that will suit the application of the system. Factors that should be taken into consideration are the strength and quality of cellular signal especially in rural monitoring sites and offshores, power supply types and power consumption, size of data transfer, speed of data transfer or bandwidth, as well as design and production costs.

This paper reports the development of GSM module for radiation safety instruments such as survey meter and area monitor. The module is intended for wireless monitoring and early emergency warning. Data transmission involves radiation measurement data, instrument status, and configuration parameters that did not require big bandwidth or high transfer speed. On the other hand, reliability of the data to be sent on time, no data loss, low power consumption and low telecommunication signal are crucial. Hence the Short Messaging Service or SMS fits perfectly as it enable the device to reliably send data at low signal, and no data loss.

2. System design

Hardware and firmware development of the GSM module were done in parallel. Initially, the firmware was developed on Arduino platform; Arduino Uno [1] and GSM Shield for Arduino. The Serial Monitor in Arduino IDE is a very useful tool to echo communication between the microcontroller and GSM shield. This tool enables programmer to monitor and verify firmware operation in GSM shield responds and SMS data processing. Hence, the proof-of-concept prototype is successfully developed on Arduino platform. However, the Arduino GSM library was not utilized; instead, the microcontroller communicates with GSM shield by using AT Command to avoid dependency on the library as well as to maintain the flexibility of the firmware.

2.1. Hardware Design

A customized PCB is designed to create a single board that housed all the necessary design blocks as shown in figure 1. The voltage regulator circuit provides voltage supply to all the components on board based on their respective voltage specifications. Microcontroller Atmega328P controls and synchronizes the operation of the module; it controls the operation of GSM module, and handles data and commands from the external device. Telit GL865-DUAL/QUAD V3 module (Telit) is a compact GSM/GPRS module that suits portable and battery operated device. This IC acts as GSM modem that transmits and receives all the SMS for the GSM module.

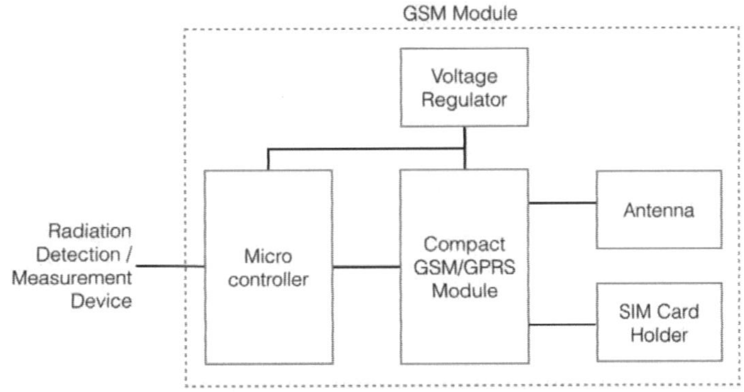

Figure 1. Block diagram of GSM PCB Module.

The PCB is designed based on Telit design recommendation. Voltage regulators are 3.8 Volts for GSM Module and 3.3 Volts for Atmega328P microcontroller. The resulting PCB is shown in figure 2. The compact size enables the PCB to be easily incorporated into the same enclosure of survey meter/area monitor or small compact casing for external connection.

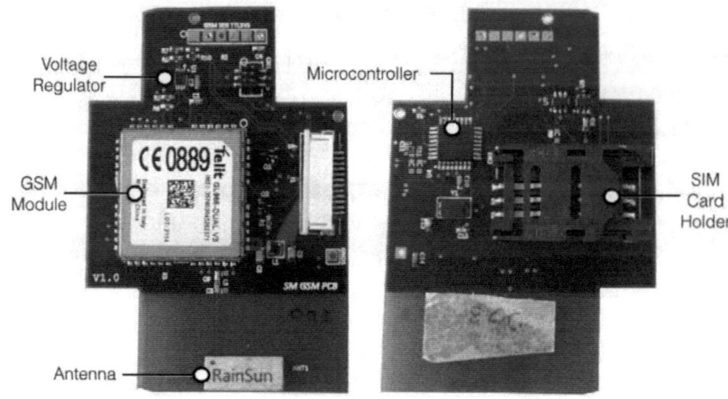

Figure 2. GSM PCB module and the main components placements for front view (left) and back view (right).

2.2. Firmware Design

Communication between GSM module and microcontroller is done using AT commands by using standard serial connection [2]. The GSM module is designed to enable wireless communication for radiation monitoring instrument intended for continuous data monitoring and emergency alert. There are three configuration parameters that are essential to complete the task; the Host number, time interval for data transmission, and threshold level for alert SMS. These parameters are stored in EEPROM of microcontroller. User will be able to change and update the configuration parameters via SMS. Firmware of GSM microcontroller is implemented as a finite state machine as shown as the state diagram in figure 3. The firmware is responsible to handle task related to GSM/SMS communication with the Host server.

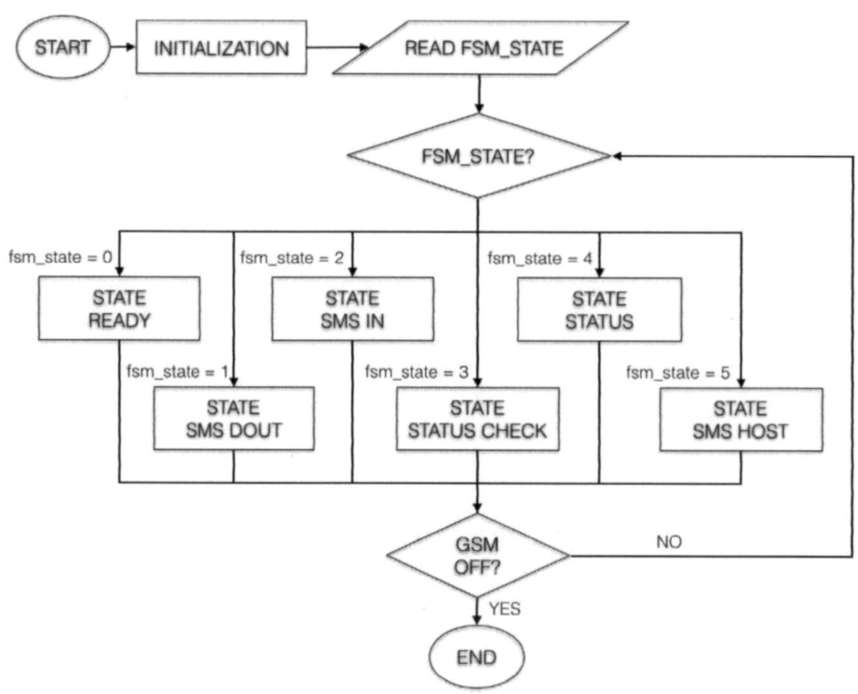

Figure 3. Finite State Machine (FSM) for the GSM microcontroller firmware.

Each state in the FSM handles specific tasks of the GSM module as listed below:

- State Ready (St_Ready) - the default state that determines the other states priorities and initiate the state events execution. Each state priority is shown in St_Ready flowchart in figure 4.
- State SMS Data Out (St_SMS_Dout) - send SMS at every fixed time interval according to configuration parameter.
- State SMS In (St_SMS_In) - process all incoming SMS. There are there possibilities for the received SMS; SMS from host to change parameter configuration (either host number, data transmission time interval, or alert threshold value), SMS from any senders to query instrument status, and unrelated SMS. St_SMS_In compute the SMS type, and execute further process accordingly. All the SMS will be deleted at the end of the process to avoid full SMS inbox in the SIM card.
- State Status Check (St_Status) - handles in-situ status query from the connected instrument. It will returns all three configuration parameters to the instrument. This is useful if the user on-site wants to view the information. Optionally the information can be displayed on the instrument display. This state will also update new configuration status upon configuration change via SMS.
- State SMS Alert (St_SMS_Alert) - handles alert notification SMS to Host whenever the radiation data reach or exceed threshold value parameter.

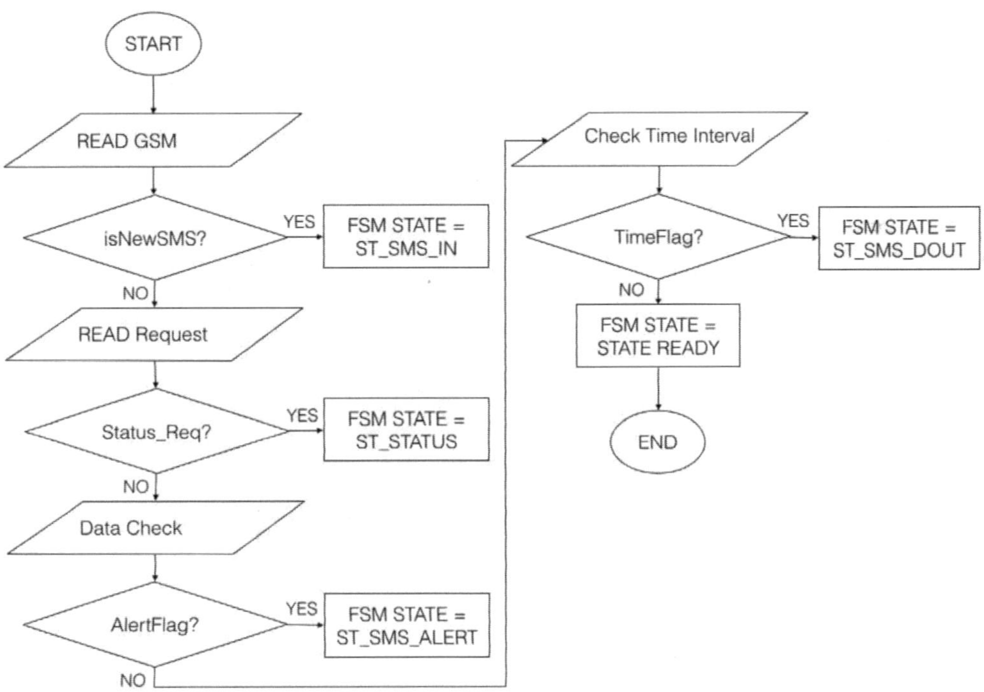

Figure 4. St_Ready (default state) flowchart and the states priorities.

2.3. AT Command and SMS PDU Processing

Communication between microcontroller and Telit is done by using AT Command via serial connection. AT commands are list of standard instructions that is used to control a modem. For example, AT+CMGD is a command used to instruct Telit to delete SMS. In order to simplify the SMS processing, AT+CMGF=1 is used to instruct Telit for SMS in text mode. In default state (St_Ready), microcontroller frequently checks its serial receive buffer in case there is new notification or status from Telit GSM (ReadGSM). All incoming SMS is temporarily copied and stored in the firmware array buffer. It is crucial for the microcontroller to identify and process SMS intended for the system. In this case, the module is expected to receive three types of SMS:

- SMS from Host or any sender to check the status
- SMS from Host to change configuration parameter (Host number, data time interval, threshold value for alert)
- Other unrelated SMS

An example of a received SMS is as follows:

+CMGL: 5, "REC UNREAD", "+60123030224","","2016/06/09 11:27:35+32"
@HOST SCFG H+60162324045 I5 T3

The first line consists of information on location index of the SMS in the message storage area, status of the SMS message ("received unread", "received read", etcetera), originator phone number, and timestamp at which the SMS message arrived at the SMSC. The second line is the body of SMS that might contain request instruction or configuration parameter from the host. Hence, the SMS body is designed to contain markers or keywords that could be identified by the microcontroller. The keyword '@HOST' is used to identify that the SMS is a host SMS and 'SCFG' is a marker for configuration SMS. Subsequently, the microcontroller will be able extract the information from the SMS and execute the next function respectively.

3. Results and Discussions

The first variant of this GSM module is designed to be integrated with a survey meter prototype. The PCB is connected to survey meter PCB by a customised flexible PCB cable and fitted inside the same enclosure as shown in figure 5.

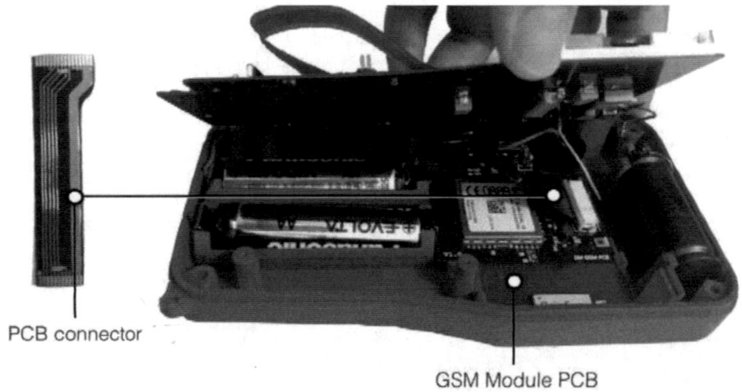

Figure 5. GSM module for survey meter.

The GSM module has been tested to verify its functionality. These involved transmitting a series of SMS to the GSM module to update the configuration parameters i.e. the time interval, host number, and threshold values. Results had shown that the GSM module was able to identify and compute incoming SMS and responded accordingly. The result for time interval configuration update for 3, 5, and 10 minutes is shown by the timestamp in figure 6.

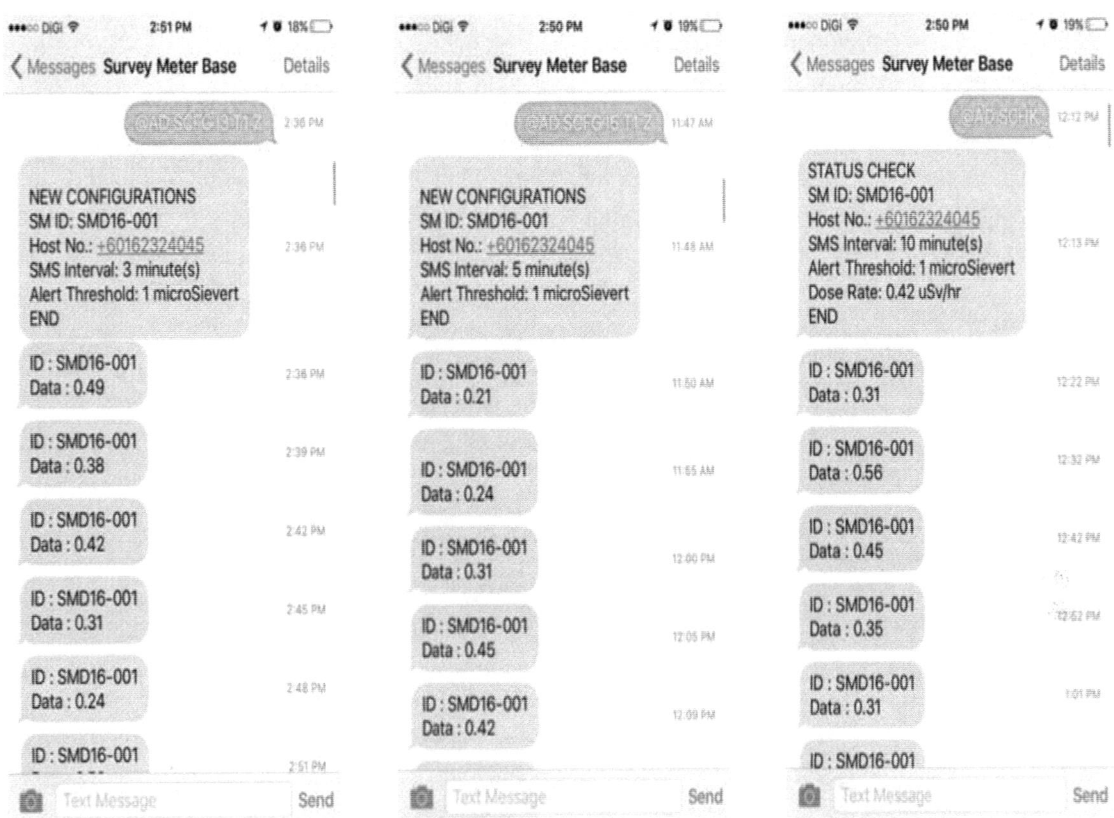

Figure 6. Data SMS at interval of 3, 5 and 10 minutes.

The GSM module also enables user to query the status of device and current data at anytime. The microcontroller firmware is programmed to identify keywords "@AD SCHK" for status query and returns status SMS to the sender as shown in figure 7. On the other hand, Host can change the configuration parameter via SMS with keywords "@AD SCFG". The microcontroller will extract the configuration data and update the EEPROM respectively. This new configuration will be retained until the next configuration SMS. Upon successful update, the GSM module will return the configuration SMS to Host as shown in figure 7.

Finally, the GSM module is tested to send alert SMS whenever the radiation measured exceed the threshold value. This feature is beneficial for prompt or early emergency alert in radiation monitoring and safety. The GSM module should respond by transmitting alert SMS immediately, regardless of the data time interval. In this case, the GSM module is tested with a check source and the corresponding alert SMS is shown in figure 8.

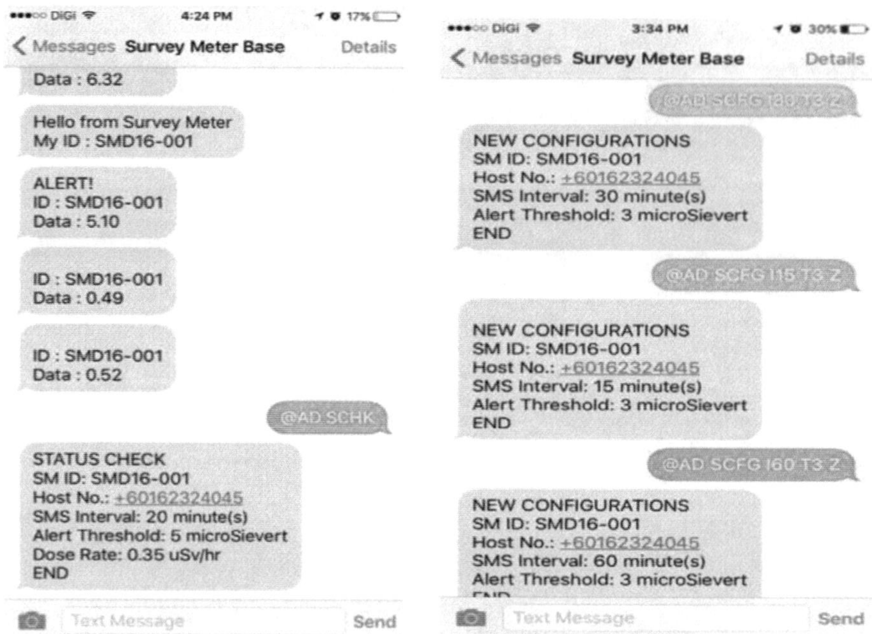

Figure 7. Status query SMS (left) new configuration update SMS (right).

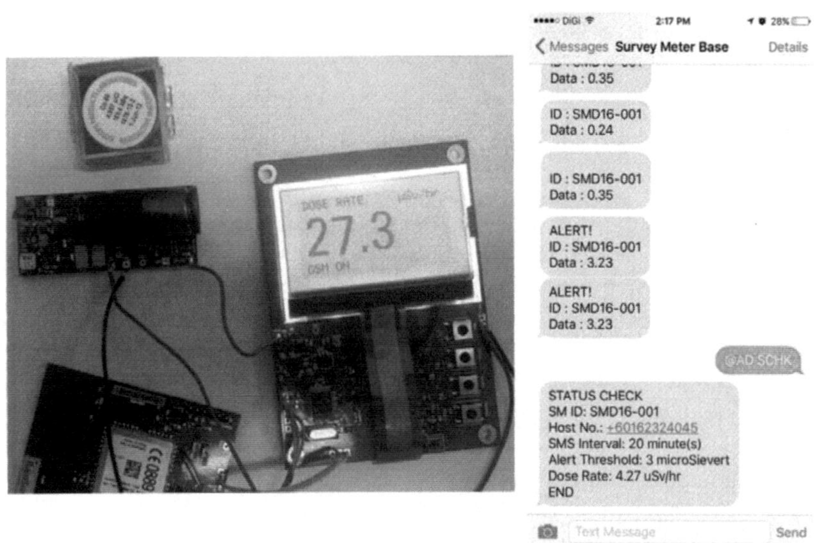

Figure 8. Alert SMS for radiation level higher than threshold value.

4. Conclusion and Future Works

A small and compact GSM module has been successfully designed, fabricated, and tested. The GSM module was designed to be integrated with radiation detection and measurement instrument such as survey meter and area monitor to enable wireless radiation monitoring and early warning system. The main components of the module were Atmega 328P microcontroller (Arduino Uno) and Telit compact GSM/GPRS module. A series of tests has proven that the module was functioning according to designed specifications. Future works includes power saving implementation on the firmware to prolong the battery life.

5. References

[1] Projects M 2015 The Evil Goat (Arduino Project) 102–103
[2] Zhang L, Wang X, Wang C, and Gu X 2011 The application of stolen radioactive source tracking system based on internet of things technology *Proc. 3rd Int. Conf. on Measuring Tech. and Mechatronics Automation ICMTMA 2011* **3** 696–698.

Acknowledgments

This module is part of the project funded by Ministry Science Technology and Innovation (MOSTI), Malaysia under the Technofund Grant.

IOP Conf. Series: Materials Science and Engineering **298** (2018) 012041 doi:10.1088/1757-899X/298/1/012041

Design and evaluation of corn starch-bonded *Rhizophora* spp. particleboard phantoms for SPECT/CT imaging

Puteri Nor Khatijah Abd Hamid[1], Mohd Fahmi Mohd Yusof[2], Abd Aziz Tajuddin[1], Rokiah Hashim[3] and Rafidah Zainon[4]

[1]School of Physics, Universiti Sains Malaysia, 18000 Penang, Malaysia
[2]School of Health Sciences, Universiti Sains Malaysia, Kota Bharu, Kelantan, Malaysia
[3]School of Industrial Technology, Universiti Sains Malaysia, 11800 Penang, Malaysia
[4]Advanced Medical and Dental Institute, Universiti Sains Malaysia, 13200 Bertam, Penang Malaysia

pnkah14_phy063@student.usm.my

Abstract. The aim of this study was to design and evaluate of corn starch-bonded *Rhizophora* spp. particleboards as phantom for SPECT/CT imaging. The phantom was designed according to the Jaszczak phantom commonly used in SPECT imaging with dimension of 22 cm diameter and 18 cm length. Six inserts with different diameter were made for insertion of vials filled with 1.6 µCi/ml of ^{99m}Tc unsealed source. The particleboard phantom was scanned using SPECT/CT imaging protocol. The contrast of each vial for particleboards phantom were calculated based on the ratio of counts in radionuclide volume and phantom background and compared to Perspex® and water phantom. The results showed that contrast values for each vial in particleboard phantomis near to 1.0 and in good agreement with Perspex® and water phantoms as common phantom materials for SPECT/CT. The paired sample *t*-test result showed no significant difference of contrast values between images in particleboard phantoms and that in water. The overall results showed the potential of corn starch-bonded *Rhizophora* spp. as phantom for quality control and dosimetry works in SPECT/CT imaging.

1. Introduction

The potential use of *Rhizophora* spp. wood as phantoms for dosimetry and diagnostic imaging had been studied [1-4]. The fabrication of particleboards made of *Rhizophora* spp. wood have the advantages over its solid raw wood including enable to be fabricated at various sizes and shapes and density homogeneity across the particleboards. The potential use *Rhizophora* spp. particleboards as phantom had been studied in various applications using external beam of X-rays including mammography, diagnostic radiography and computed tomography (CT) imaging [4-8]. To date, there are limited information regarding the potential use of *Rhizophora* spp. particleboards for in-vivo dosimetry and imaging. A previous study by Bradley et al. [2] suggested that the solid raw wood of *Rhizophora* spp. having similar dose characteristics to solid water phantoms when measured using ^{192}Ir brachytherapy sources. The present study focused on the in-vivo detection and imaging using phantoms made of *Rhizophora* spp. particleboards based on the single photon emission computed tomography (SPECT) imaging.

SPECT is an imaging technique that provides three dimensional maps of in-vivo radiopharmaceuticals distribution [9]. CT is an imaging of cross-sectional transmission acquired by using a high output X-ray tube. The integrated SPECT/CT systems provide precise localization of physiologic data from nuclear medicine studies. SPECT imaging is superior to planar imaging with regard to abnormal localization by improving the object contrast [10]. Image contrast refers to different intensity in image corresponding to different levels of radioactive uptake in patient. Contrast is the ratio of signal change of an object of interest, such as lesion, relative to the signal level of surrounding parts of the image. Tomographic contrast is important to determine the detectability of small lesions.

2. Materials and methods

2.1. Fabrication of Rhizophora spp. Particleboards

The *Rhizophora* spp. wood trunks were obtained from charcoal factory in Kuala Sepetang, Perak, Malaysia. The wood trunks were cut and chipped into wood chips using planner machine. The wood chips were ground into wood particles using a grinder machine. The wood particles were filtered using horizontal sieving machine to obtain only wood particle with ≤ 74 μm sizes range. The commercial corn starch was used as bio-adhesive with 10% treatment level. The dry mass of wood particles and corn starch were calculated based on target density of particleboard (1.0 g/cm^3). The wood particles and corn starch were mixed well and distributed uniformly on the mould. The mixture was pressed without temperature applied for 5 minutes to removes air trapped. Then, the mixture was compressed at 190°C for 20 minutes using hot press machine into particleboard. The elemental compositions of *Rhizophora* spp. particleboard were determined using energy dispersive X-ray analysis (EDXA). The effective atomic number, Z_{eff} of the particleboard was calculated based on the study by Duvauchelle et al. [11] shown by the equation:

$$Z_{eff} = \sum_{i=1}^{N}(\alpha_i Z_i^m)^{\frac{1}{m}} \tag{1}$$

with α_i and Z_i are the electronic fraction and atomic number of the ith element in the compound respectively. The experimental coefficient (m) for biological material such as water, wood or human organs has been evaluated to be equal to 3.4 [11]. The electronic fraction of the ith element was calculated using the equation:

$$\alpha_i = \frac{w_i\left(\frac{Z_i}{A_i}\right)}{\sum w_i\left(\frac{Z_i}{A_i}\right)} \tag{2}$$

with $w_i$and A_i are the fractional weight and atomic mass of the ith element respectively.

2.2. Design of Rhizophora spp. Particleboard Phantoms for SPECT/CT

The *Rhizophora* spp. particleboards were fabricated based on the study by Abd Hamid *et al.* [12]. An amount of 10% of corn starch was added into the particleboards as adhesive material based on the dried weight of the wood particles. The particleboards were fabricated at 1.0 g/cm^3 target density. A number of 18 particleboards were fabricated with thickness of 1.0 cm. The particleboards were then trimmed circularly to obtain particleboards with diameter of 22 cm similar to the diameter of Jaszczak phantom used for SPECT imaging. A number of 6 particleboards were prepared with six inserts with diameters between 1.6 cm and 4.1 cm for insertion of vials as shown in figure 1. The particleboards were stacked together using 4 long rods to minimize the air spaces between the particleboards. The Perspex® phantom was constructed with the similar design and dimension to particleboards phantom. The hollow tank of Jaszczak phantom filled with water was used as reference. A number of 6 vials with diameter between 1.9 and 4.1 cm diameter were constructed using Perspex® for containing ^{99m}Tc radionuclide. The details of the dimension of the Perspex® vials are presented in Table 1. The vials were constructed with rubber cap to prevent spillage of the radionuclide.

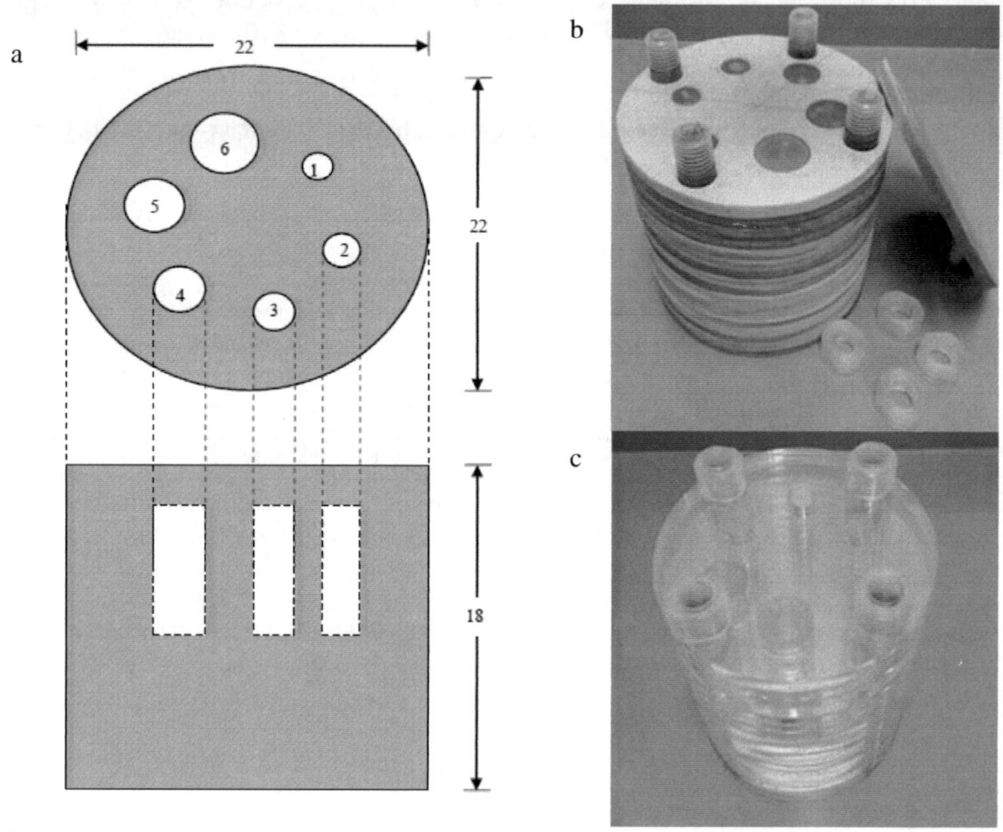

Figure 1. (a) The layout design of SPECT/CT phantom made of corn starch-bonded *Rhizophora* spp. particleboards and Perspex®, (b) the constructed *Rhizophora* spp. particleboards phantom and, (c) Perspex® phantom.

Table 1. The external dimensions of vials used to contain the ^{99m}Tc radionuclide for *Rhizophora* spp. particleboard phantoms.

Vial no.	Diameter (cm)		Length (cm)	Volume (cm^3)
	Internal	External		
1	1.3	1.9	6.0	6.0
2	1.6	2.2	6.0	8.6
3	1.9	2.5	6.0	11.6
4	2.5	3.1	6.0	21.5
5	3.0	3.6	6.0	30.0
6	3.5	4.1	6.0	41.5

2.3. Tomographic Contrast Study on SPECT/CT Images

The vials of different volumes and diameters were filled with ^{99m}Tc radionuclide with activity of 1.6 μCi/ml calibrated using a dose calibrator. The vials were inserted into the *Rhizophora* spp. particleboards. The phantom was positioned on the couch of gamma camera model Discovery NM/CT/670. A SPECT/CT imaging was performed and the SPECT images were reconstructed into transaxial view using an ordered-subsets expectation maximization (OSEM) algorithm with Butterworth filter at 0.48 filtration frequency. The details of the SPECT/CT imaging parameters are presented in Table 2. These same acquisition and reconstruction parameters were used for Perspex® and Jaszczak phantoms imaging.

Table 2. The SPECT/CT imaging parameters on the *Rhizophora* spp. particleboards, Perspex® and water phantoms.

Parameter	Description
Matrix size	128 × 128
View angle per projection	3°
No. of view	120
CT scan exposure factors	120 kVp, 20 mA

Regions of interest (ROI) were drawn on the SPECT/CT images of radionuclide within each vial and at background region. The contrast of the image from each vial was measured using the equation:

$$Contrast = \frac{C_{sph} - C_{bgd}}{C_{bgd}} \quad (3)$$

with C_{sph} and C_{bgd} are the sphere and background count respectively. The contrast values from the particleboards phantom were compared to that in Perspex® and water phantoms.

3. Results and discussion

3.1. Analysis of Physical and Attenuation Properties of Rhizophora spp. Particleboards

The calculated Z_{eff} of the corn starch bonded *Rhizophora* spp. particleboards in comparison to the previous works on *Rhizophora* spp. particleboard phantoms is presented in Table 3. The result showed that the Z_{eff} of corn starch bonded *Rhizophora* spp. particleboards was close to the value of water [13]. Two materials are said to have similar attenuation properties towards ionizing radiations when the effective atomic numbers are equal [11]. Therefore, it is postulated that the attenuation properties of the corn starch bonded *Rhizophora* spp. particleboards are close to water as indicated by the effective atomic number.

Table 3. The calculated effective number (Z_{eff}) of *Rhizophora* spp. particleboards in comparison to water and Perspex®.

Sample	**Percentage of elemental composition (%)**				Z_{eff}
	Carbon (C)	**Hydrogen (H)**	**Oxygen (O)**	**Nitrogen (N)**	
Water [a]	-	11.11	88.89	-	7.480
Perspex®[a]	59.98	31.97	8.05	-	5.210
Binderless *Rhizophora* spp.[c]	32.93	-	38.98	28.08	7.186
Corn Starch[b]	46.82	6.94	46.24	-	6.877
Corn starch bonded *Rhizophora*spp.[b]	27.20	-	72.55	0.25	7.587
Rhizophora spp. raw wood[d]	40.16	3.78	47.90	3.78	-

[a]AAPM-21 [13], [b]Current study, [c]Mohd Yusof et al. [8], [d]Bradley et al. [2]

3.2. Analysis of SPECT/CT Images using Rhizophora spp. Particleboard Phantoms

The SPECT/CT images of the corn starch bonded *Rhizophora* spp. particleboards in comparison to Perspex and water (Jaszczak) phantoms are illustrated in figure 2. The contrast values of vials of different diameter at different time of projections are presented in figure 3. The results showed similarity of SPECT/CT image by *Rhizophora* spp. particleboard in comparison to water and Perspex® phantoms. The images of radionuclide in the vials were also consistent with their CT images. The measured contrast of different diameter of vials showed good agreement between *Rhizophora* spp. particleboard to water phantom. The contrast values were increase when the vials diameter larger at all time of projections.

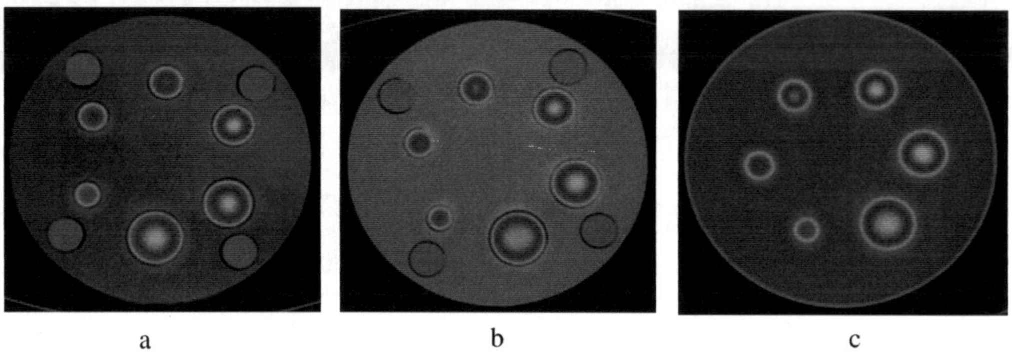

Figure 2. The SPECT/CT images of (a) *Rhizophora* spp. particleboard (b) Perspex® and (c) water (Jaszczak) phantoms.

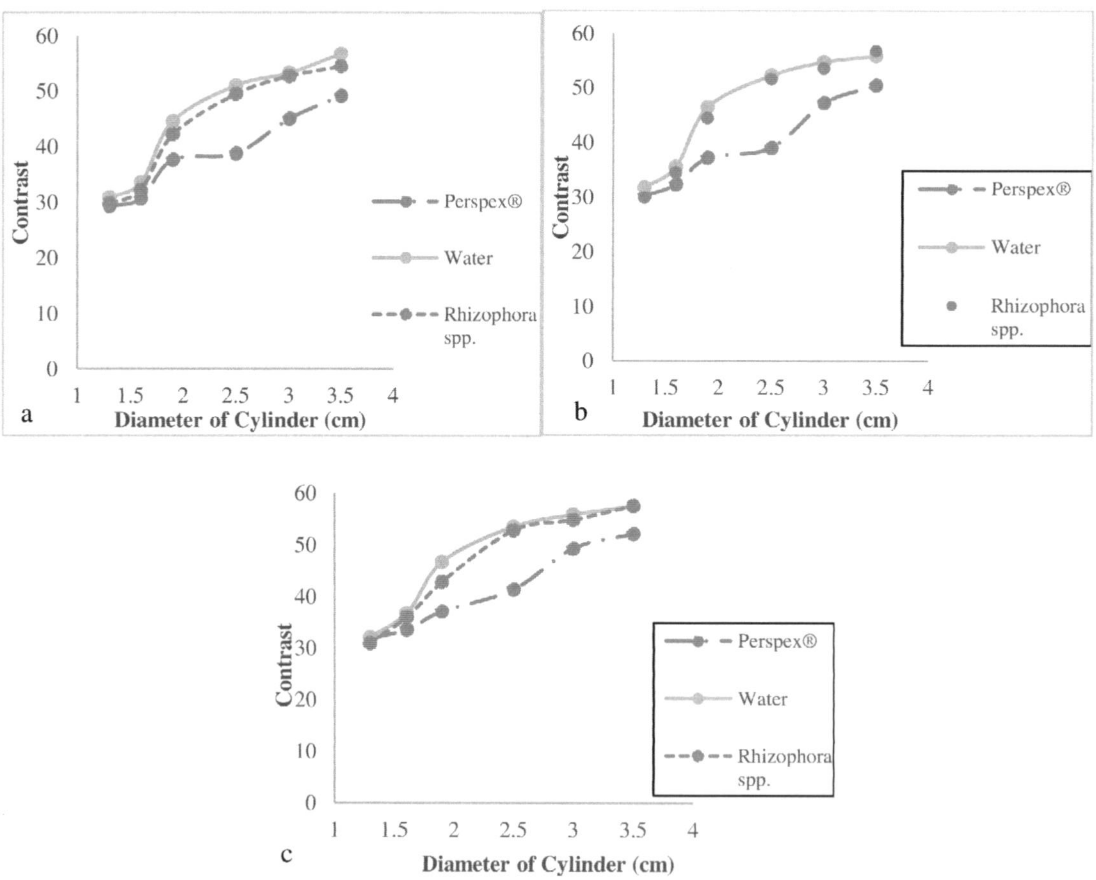

Figure 3. The contrast values at different vial volumes on *Rhizophora* spp. particleboards, Perspex ® and water at (a) 10 sec, (b) 20 sec and (c) 30 sec times of projection.

A paired sample *t*-test was performed to compare the contrast between water-*Rhizophora* spp. particleboards and water-Perspex® and presented in Table 4. The results showed that there were no significant different of contrast values between images of water and *Rhizophora* spp. particleboard phantoms when 20 sec and 30 sec of times of projection were used shown by the *p*-values higher than 0.05. This had indicated the similarity of attenuation and construction of SPECT/CT images between the particleboards and water. The overall results had indicated the suitability of corn starch bonded *Rhizophora* spp. particleboards to be fabricated and constructed as phantom for SPECT/CT imaging.

Table 4. The paired sample *t*-test of average contrast values of water-*Rhizophora* spp. particleboards and water-Perspex at different time of projections.

Time per projection (sec)	*p*-value of paired sample *t*-test (2 tailed)	
	Water - *Rhizophora* spp. particleboards	**Water - Perspex®**
10	0.003	0.009
20	0.079	0.016
30	0.070	0.017

4. Conclusions

The corn starch bonded *Rhizophora* spp. particleboards showed close value of effective atomic number to water. The constructed SPECT/CT phantom made of *Rhizophora* spp. showed remarkable attenuation properties and contrast resolution in comparison to water and Perspex® at all sizes and volumes of radionuclide using ^{99m}Tc. The contrast of SPECT/CT images in *Rhizophora* spp. particleboards also showed an excellent agreement to water and Perspex® at all times of projection set ups. The results had indicated the potential use of *Rhizophora* spp. particleboards as phantom for SPECT/CT imaging.

5. References

[1] Banjade D P, Tajuddin A A and Shukri A 2001 A study of *Rhizophora* spp. wood phantom for dosimetric purposes using high energy photon and electron beams *Appl. Radiat. Isotopes* **55 (3)** 297-302

[2] Bradley D A, Tajuddin A A, Che Wan Sudin C W A and Bauk S 1991 Photon attenuation studies on tropical hardwoods *Appl. Radiat. Isotopes* **42** 771-773

[3] Shakhreet B Z, Bauk S, Tajuddin A A and Shukri A 2009 Mass attenuation coefficients of natural *Rhizophora* spp. wood for X-rays in the 15.77 to 25.27 keV range *Radiat. Prot. Dosim.* **135** 47-53

[4] Marashdeh M W, Bauk S, Tajuddin A A and Hashim R 2012 Measurement of mass attenuation coefficients of *Rhizophora* spp. binderless particleboards in the 16.59–25.26 keV photon energy range and their density profile using X-ray computed tomography *Appl. Radiat. Isotopes* **70** 656-62

[5] Ababneh B, Tajuddin A A, Hashim R and Shuaib I L 2016 Investigation of mass attenuation coefficient of almond gum bonded *Rhizophora* spp. particleboards equivalent human tissue using XRF technique in the 16.6 – 25.3 keV photon energy *Australas. Phys. Eng. Sci. Med.* **39(4)** 871-76

[6] Abuarra A, Hashim R, Bauk S, Kandaiya S and Tousi E T 2014 Fabrication and characterization of gum Arabic bonded *Rhizophora* spp. particleboards *Mater. Design* **60** 108-15

[7] Tousi E T, Hashim R, Bauk S, Jaafar M S, Abuarra A and Ababneh B 2014 Some properties of particleboards produced from *Rhizophora* spp. as a tissue-equivalent phantom material bonded with *Eremurus* spp. *Measurement* **54** 14-21

[8] Mohd Yusof M F, Abd Hamid P N K, Tajuddin A A, Hashim R, Bauk S and Sulaiman O 2017 Characterization of tannin-added *Rhizophora* spp. particleboards as phantom materials for

photons beams *Ind. Crop Prod.* **95** 467-74
[9] Artlett M L, Bacharach S L, Voipio-Pulkki L M and Dilsizain V 1995 Artifactual in-homogeneties in myocardial PET and SPECT scans in normal subjects *J. Nucl. Med.* **36** 188-95
[10] Scott C W 2001 General nuclear medicine *J Nucl. Med.* **42** 1499-1507
[11] Duvauchelle P, Peix G and Babot D 1999 Effective atomic number in the Rayleigh to Compton scattering ratio *Nucl. Instrum. Meth. B* **155** 221-28
[12] Abd Hamid P N K, Mohd Yusof M F, Tajuddin A A and Hashim R 2017 Measurement of mass attenuation coefficients for corn starch-bonded *Rhizophora* spp. particleboards at 16.59 – 25.26 keV photons using X-ray fluorescence configuration *Jurnal Sains Nuklear Malaysia* **29(2)** 47-54
[13] AAPM-21 1983 Protocol for the determination of absorbed dose from high energy photon and electron beams, Task Group 21 *Med. Phys.* **10** 741-71

Acknowledgements
The authors would like to acknowledge the financial aid of Research University grant no. PSPTEKIND1000000004178 by Universiti Sains Malaysia. The authors also are grateful to the Advanced Medical and Dental Institute, Universiti Sains Malaysia for the use of CT scanner.

A preliminary study to metaheuristic approach in multilayer radiation shielding optimization

Muhammad Arif Sazali[1], Nahrul Khair Alang Md Rashid[1], and Khaidzir Hamzah[1]

[1]Faculty of Chemical and Energy Engineering, Universiti Teknologi Malaysia, 81310 Johor Bahru, Johor

marif69@live.utm.my

Abstract. Metaheuristics are high-level algorithmic concepts that can be used to develop heuristic optimization algorithms. One of their applications is to find optimal or near optimal solutions to combinatorial optimization problems (COPs) such as scheduling, vehicle routing, and timetabling. Combinatorial optimization deals with finding optimal combinations or permutations in a given set of problem components when exhaustive search is not feasible. A radiation shield made of several layers of different materials can be regarded as a COP. The time taken to optimize the shield may be too high when several parameters are involved such as the number of materials, the thickness of layers, and the arrangement of materials. Metaheuristics can be applied to reduce the optimization time, trading guaranteed optimal solutions for near-optimal solutions in comparably short amount of time. The application of metaheuristics for radiation shield optimization is lacking. In this paper, we present a review on the suitability of using metaheuristics in multilayer shielding design, specifically the genetic algorithm and ant colony optimization algorithm (ACO). We would also like to propose an optimization model based on the ACO method.

1. Introduction

Technology involving ionizing radiation is widely used in many fields such as medical application, manufacturing industries, and research. However, ionizing radiation is harmful to humans as it can cause biological changes in the body. The International Commission of Radiation Protection (ICRP) recommended three basic principles to control radiation exposure to the public which are justification, optimization, and dose limitation [1]. The dose can be limited using any or combination of three of the following methods of protection: minimizing the time of handling the radiation source, maximizing the distance from the source, and putting a shield between people and the source.

A shield is defined as a physical entity placed between an ionizing radiation source and an object to be protected with the aim of reducing the radiation level at the object's position [2]. It can be made of composite material, multilayer of various materials, or combinations of both. The advantage of composites is that the shielding capabilities of different materials can be merged in a single material, reducing the weight of the shield as seen in polymer composites [3]. However, they may lack uniformity, resulting in inconsistent shielding performances [4][5][6]. Non-uniform composites may also result in pin holes, pure polymer areas which radiation can penetrate through [7]. On the other hand, multilayer configuration can be advantageous in the case of mixed radiations, for example a californium-252 emitting both neutrons and gamma rays. Each of the layer can attenuate the types of radiation which it has high macroscopic cross section or attenuation coefficient. The alternate layers of shielding can also suppress secondary radiation [8].

With the advances of technology, the requirements for the shield also increases. Different applications have different design objectives. For stationary systems, the primary consideration is cost, while for mobile applications, the total weight of the shield is more concerned. Limitations in resources would also mean that the shield design needs to be optimized and tailored to its application. A shield can be regarded as a combinatorial optimization problem (COP) when it is made up of various materials. For example, there should be an optimum weightage of fillers in a composite shield, or an optimum arrangement of layers in a multilayer shield. Metaheuristics were developed to tackle such COPs in many different fields such as economy, engineering, and industry [9]. Examples of metaheuristic algorithms are simulated annealing, tabu search, genetic algorithm (GA), and ant colony optimization (ACO). This paper focuses on GA and ACO since only GA had been implemented in optimizing composite shield. Meanwhile, we intent to propose on the optimization of a multilayer shield using ACO.

2. Combinatorial optimization problem

A combinatorial optimization deals with searching optimal combinations or permutations of available problem components. A COP is defined as P=(S, Ω, f) in which S is a search space defined over a finite set of discrete decision variables. Ω is a set of constraints among the variables and f is the objective function that assigns a positive cost value to each of the solutions [10].

An example of a COP is the travelling salesman problem (TSP) which was formulated in 1930 as a mathematical problem. It is represented as an undirected weighted graph, where vertices represent cities and edges represent the path connecting the cities. The goal of the optimization problem is to find the shortest total distance travelled with a condition that the salesman visits each city only once. The problem is known to be difficult computationally, but easily solvable with the use of heuristic (search) algorithms even with thousands of cities [11].

3. Metaheuristics

Optimization algorithms are classified as either complete or approximate algorithms. Optimal solutions are guaranteed when using complete algorithms, but for certain COPs that are considered NP-hard, they make take an exponential computation time. Approximate algorithms or heuristics were developed where the guarantee is sacrificed for good near-optimal solutions in a relatively reasonable computation cost. The word metaheuristics originates from meta (Greek: 'beyond, in an upper level') and heuristic (Greek:'to find'). They were also called modern heuristics. The difference between a heuristic and a metaheuristic is that the latter is problem-independent. It can be designed to solve approximately a wide range of COPs without the need of adapting deeply to each problem [9][12][13].

Genetic algorithm was popularized by John Holland in the 1970s. It was based on natural evolution and the concept of survival of the fittest. The algorithm selects the solution randomly, picks the most fit solutions, and uses them to create a new generation which is better than the previous generation. The basic GA is very general hence it can be modified to adapt to the problem at hand through implementations such as representation of solution (chromosomes), types of crossover (the recombination operator), selection strategies, and mutation operators. The pseudocode of GA is shown in figure 1(a). After recombination, individuals go through mutation where randomness was introduced into the search to prevent the optimization from getting trapped in local optima. Then, based on their fitness value, they are used for successive generations, assuring the survival of the fittest individuals [9][11].

Ant colony optimization is a technique introduced by Dorigo Marco in 1992 who took inspiration from the foraging behavior of ants. They always seek the shortest path between their colony and food sources. The unique feature in ACO is the implementation of pheromone parameter. It is associated to every solution explored by the artificial ants. Updating the pheromone value after each iteration helps in separating promising solutions from the bad ones. The pheromone also "evaporates", meaning that it can decrease over time. The goal is to make good solutions more attractive for the ants in the next iteration. The pseudocode of ACO is given in figure 1(b) [9][14].

```
Initialize the population with random individuals
Evaluate each individual
repeat
    Select parents
    Recombine pairs of parents
    Mutate the resulting offspring
    Evaluate new individuals
    Select individuals for the next generation
until a termination condition is satisfied
```

(a)

```
Initialize pheromone values
while termination condition not met do
    Construct Ants Solutions
    Update Pheromones
    Daemon Actions
end
```

(b)

Figure 1. Pseudocodes of selected metaheuristics. (a) GA; (b) ACO.

3.1. *Comparisons of GA and ACO*

There are many comparisons that have been done between the two metaheuristics for many problems in various fields. Gill *et al.* [15] have applied both methods for very-large-scale integration (VLSI) circuit partitioning where GA produces consistently better average result than ACO over all instances [15]. Putha *et al.* [16] test them for solving traffic signal coordination and they conclude that ACO is better for solving very complicated networks [16]. ACO is also observed to perform better than GA in solving energy efficient coverage in wireless sensor networks [17]. Jiang *et al.* [18] use the metaheuristics to optimize automatic seeding transplanters. The performance of ACO is found to be more efficient than GA, but it is more time-consuming [18]. Katona *et al.* [19] work on the parallel implementations of both GA and ACO for multimodal transport systems optimization. They have determined that ACO is suitable when the network structure is unknown, while GA is robust against failures and network errors [19]. In the case of TSP, there are two recent studies. Alhanjouri and Alfarra [20] deduce that GA is better than ACO. However, K. Arora and M. Arora [11] results suggest otherwise. GA is found to be better in lower numbers of cities which agreed the previous study, but it is bested by ACO at higher complexity. In terms of execution times, ACO is always better [11].

3.2. *Suitability of ACO for shielding optimization*

The ACO was developed to solve COPs which have practical implementation in the real world. One of the original COPs was the TSP mentioned in Section 2. In recent studies, it has been applied in many engineering problems. They include job scheduling, structure, digital image processing, power engineering, clustering, and routing [21]. Maier *et al.* [22] showed that ACO was an attractive alternative to genetic algorithm (GA) for water distribution systems [22]. Their work was expanded to include pumping station, storage, multiple extended period loading conditions [23]. In structural design, ACO had been applied to optimize the weight of transmission tower [24], space trusses [25], steel frames [26][27], truss topology [28], and cost of tall bridge piers [29]. Aydoğdu & Saka [30] extended the optimization of steel frames by including the effect of warping in irregular frames which was shown to possibly yield an unsafe design if it is not taken into consideration. Multilayer shield problem, like other engineering problems, can also be considered as a COP as the designer need to choose between combinations of materials that can fulfill the requirements of the shield.

Most of the above examples were modified from the original TSP optimization. While the TSP concerns the total distance travelled by the salesman, the structural optimizations were focused on achieving the least weight possible which is directly related to the cost of the structure. For example, as shown in the work of [26], they use ACO to determine the best combinations of W-shape members that result in minimum weight while satisfying standard specifications. This can be applied to multilayer shield optimization where the ACO can be used to identify the most optimum combination of materials, arrangements, and layer thickness that result in the best shielding performance.

4. Proposed Work with ACO

Based on the previous examples, it can be said that GA and ACO have their own advantages for different applications. However, in terms of radiation shield optimization, it is yet to be determined how ACO would perform. The only work that had been done using GA was the study by Hu et al. in [31]. They designed a multilayer composite shield for mixed neutron and gamma rays using GA combined with the Monte Carlo N-Particle (MCNP) code which is an internationally recognized code for analyzing the transport of neutrons and gamma rays. It also deals with coupled transport which is the transport of secondary gamma rays resulting from neutron interactions. The flow chart of the algorithm and MCNP simulation are shown in figure 2(a) and 1(b) respectively. The algorithm was written to produce combinations of materials contents of composites that can minimize the amount of radiation dose. However, the authors did not include the details of the GA itself. Pseudocode of a typical GA is provided in figure 1(a).

The algorithm managed to generate five composite materials. Each of them had different percentages of elements and densities. These new materials showed good shielding performance when compared to lead oxide. They had also designed multilayer shields named "cakes" where GA was used to determine the optimum thickness ratios of individual layers. Four of the composites were manufactured and experimentally evaluated. The MCNP simulation data showed good agreement with experimental data for thickness less than 10 cm before dose buildup effect occurred. The authors concluded that the GA and MCNP code can be applied to optimize lightweight, compact, and heat resistant material for mixed radiation shields.

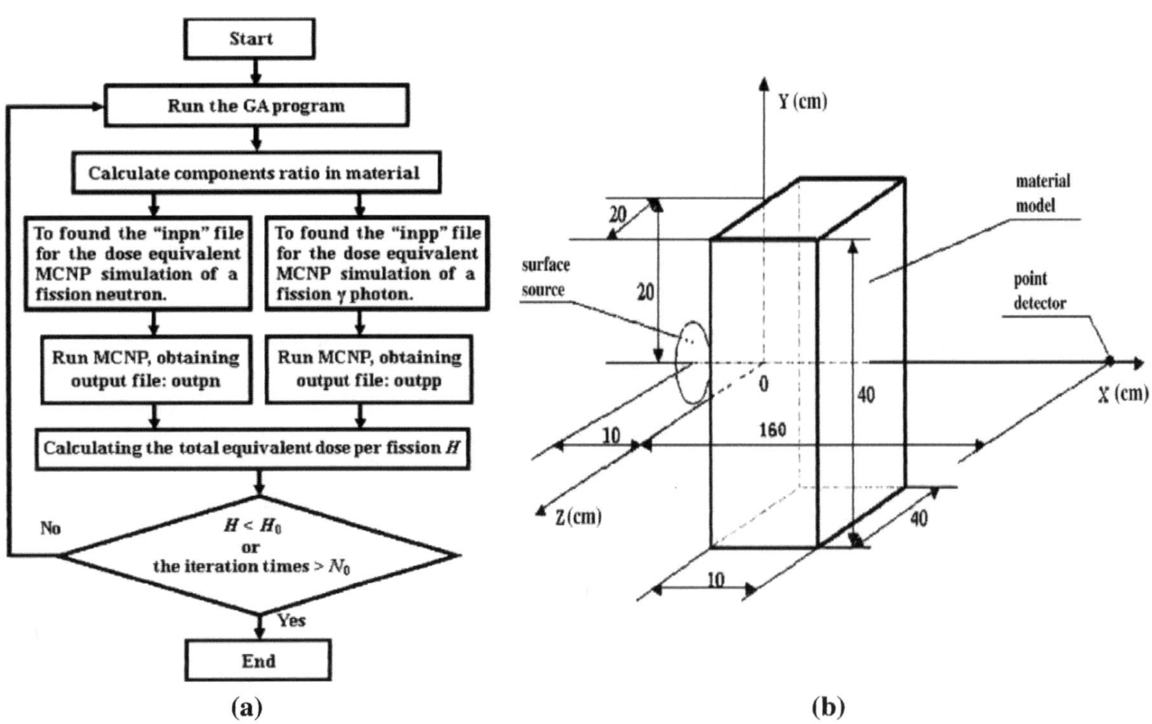

Figure 2. The details of the algorithm [21]. (a) Simulation flow chart; (b) MCNP simulation model.

IOP Conf. Series: Materials Science and Engineering **298** (2018) 012042 doi:10.1088/1757-899X/298/1/012042

Based on the previous work, we would like to outline the approach of developing ACO for multilayer shield design. The flow chart of the algorithm is shown in figure 3, while the MCNP model is shown in figure 4. The objective function is shown below.

$$\text{minimize } H(A,X)=H_n(A,X)+H_g(A,X) \tag{1}$$

where,

- $H(A,X)$: Total equivalent dose from neutrons and gamma-rays penetrating the shield of arrangement A and layer thickness X, Sv
- $H_n(A,X)$: Equivalent dose from neutrons, Sv
- $H_g(A,X)$: Equivalent dose from gamma-rays, Sv
- A: Arrangement of materials
- X: Set of thicknesses of individual layers in A

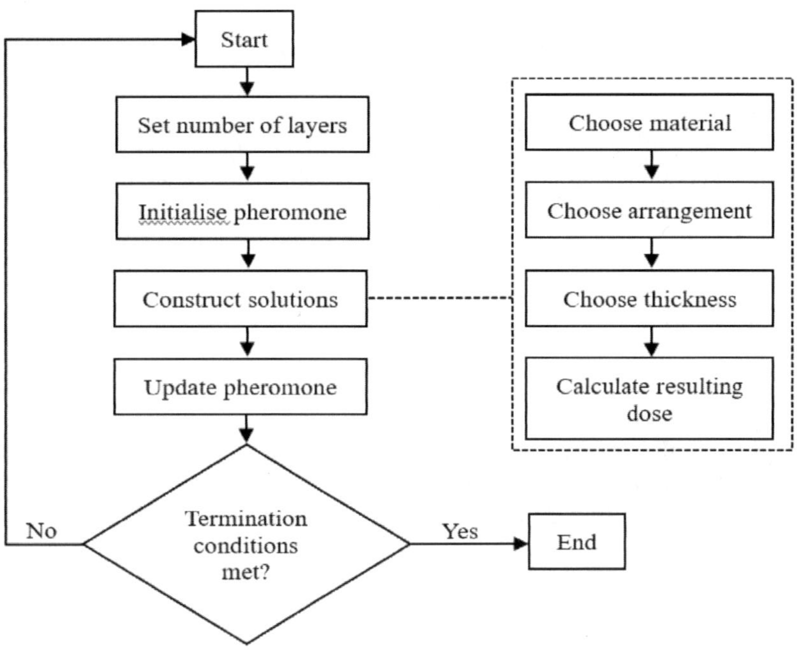

Figure 3. ACO algorithm flow chart for a multilayer shield

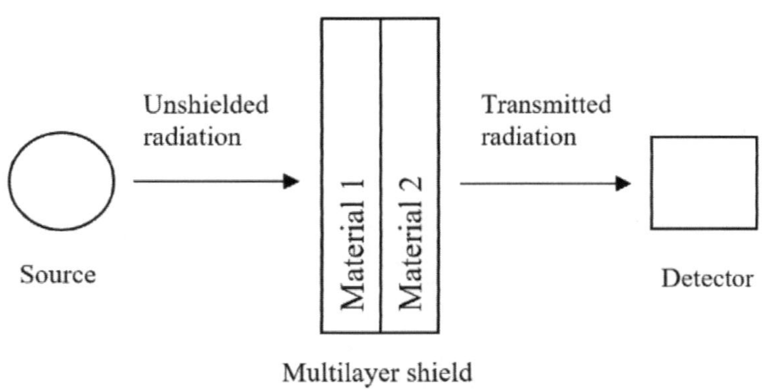

Figure 4. Illustration of MCNP simulation

IOP Conf. Series: Materials Science and Engineering **298** (2018) 012042 doi:10.1088/1757-899X/298/1/012042

Based on the pseudocode in figure 1(b), the algorithm will start by initializing the pheromone value for each of the possible solutions which are usually set to a value greater than zero [32]. In the first iteration, all of the possible solutions will have equal probability to be chosen by the artificial ants. A large number of ants is recommended to reduce the random fluctuations and to reduce the number of iterations for convergence [13]. Then, each of the ants will randomly choose one of the shield design solutions. In this case, a candidate solution is generated when an ant has picked the candidate materials, their arrangement, and the thickness of each layer. This three information will be exported to MCNP to calculate the equivalent dose on the outer layer of the shield as shown in figure 4.

The pheromone update will be carried out based on the dose of each of the solutions chosen by the ants. The lower the dose of a solution, the higher the pheromone will be added to the solution, following the rule of equation 2. Furthermore, the pheromone value for each solution will also undergo evaporation at the end of each iteration at the rate based on Equation 3. This step is crucial and must be carefully set, otherwise the algorithm may not converge to the most optimum solution [11][14].

$$\tau(A_i, X_i) \leftarrow \tau(A_i, X_i) + \frac{Q}{H(A_i, X_i)} \tag{2}$$

$$\tau(A_i, X_i) \leftarrow \tau(A_i, X_i)(1 - \rho) \tag{3}$$

where

$\tau(A_i, X_i)$	Pheromone value associated with solution of arrangement A_i and thickness X_i
Q	Constant value
$H(A_i, X_i)$	Total equivalent dose, Sv
ρ	Evaporation rate of pheromone values

After the pheromone update, the algorithm will check for termination condition. In this case, the algorithm may be terminated after a certain number of iterations. If the condition is not met, a new iteration will begin with the previous update pheromone values. This is one of the main characteristics of ACO where a history of the previous generation is preserved for future improvements of solutions as opposed to GA where the history of the parent generation will be eliminated when the daughter generation is created. After the algorithm terminates, the solution that has the highest amount of pheromone value should be the best design for the radiation shield for the given set of materials.

5. Conclusion

The review of metaheuristics GA and ACO were provided. The application of metaheuristics for shielding optimization is found to be lacking as only GA had been implemented for such COP. ACO is chosen because it has been compared to GA in many optimization problems. A proposal for the development of an optimization algorithm based on ACO for multilayer shield against mixed neutron and gamma radiations is also given. The versatility and problem-independence of metaheuristics mean that they can be modified to optimize shields for other applications. The algorithm should be helpful to researchers in creating new shield materials, and engineers in designing a shield with any material available to them.

6. References

[1] Ghilea S and Vasilescu F 1990 Recommendations of the International Commission on Radiological Protection *ICRP publication* **60**

[2] Chilton A B, Shultis J K, and Faw R E 1984 *Principles of radiation shielding* (New Jersey: Prentice Hall Inc)

[3] Nambiar S and Yeow J T 2012 Polymer-composite materials for radiation protection *ACS applied materials and interfaces* **4** 5717-26

[4] Osman A M, El-Sarraf M A, Abdel-Monem AM , and Abdo A E 2015 Studying the shielding properties of lead glass composites using neutrons and gamma rays *Annals of Nuclear Energy* **78** 146-51

[5] Waly E S, and Bourham M A 2015 Comparative study of different concrete composition as gamma-ray shielding materials *Annals of Nuclear Energy* **85** 306-10

[6] Wang P, Tang X, Chai H, Chen D, and Qiu Y 2015 Design, fabrication, and properties of a continuous carbon-fiber reinforced Sm_2O_3/polyimide gamma ray/neutron shielding material *Fusion Engineering and Design* **101** 218-25

[7] Kim Y, Park S, and Seo Y 2015 Enhanced X-ray Shielding Ability of Polymer–Nonleaded Metal Composites by Multilayer Structuring *Industrial & Engineering Chemistry Research* **54** 5968-73

[8] Gaber FA, El-Sarraf MA, and Kansouh WA 2013 Utilization of boron oxide glass and epoxy/ilmenite assembly as two layer shield *Annals of Nuclear Energy* **57** 106-10

[9] Boussaïd I, Lepagnot J, and Siarry P 2013 A survey on optimization metaheuristics *Information Sciences* **237** 82-117

[10] Papadimitriou CH, and Steiglitz K 1998 *Combinatorial optimization: algorithms and complexity* (US: Courier Corporation)

[11] Arora K, Arora M 2016 Better result for solving TSP: GA versus ACO *Int. J. Adv. Res. Comput. Sci. Softw. Eng.* **6** 219-24

[12] Cordón García O, Herrera Triguero F, and Stützle T 2002 A review on the ant colony optimization metaheuristic: Basis, models and new trends *Mathware & soft computing* **9**

[13] Blum C 2005 Ant colony optimization: Introduction and recent trends *Physics of Life reviews* **2** 353-73

[14] Socha K, and Dorigo M 2008 Ant colony optimization for continuous domains *European journal of operational research* **185** 1155-73

[15] Gill S S, Chandel R, and Chandel A 2009 Comparative study of ant colony and genetic algorithms for VLSI circuit partitioning *Int. J. of Electrical and Computer Engineering* **4**

[16] Putha R, Quadrifoglio L, and Zechman E 2012 Comparing ant colony optimization and genetic algorithm approaches for solving traffic signal coordination under oversaturation conditions *Computer-Aided Civil and Infrastructure Engineering* **27** 14-28

[17] Anjali S, and Singh G 2013 Comparing Ant Colony Optimization & Genetic Algorithm for Solving Energy Efficient Coverage in WSNs *Int. Journal for Advance Research in Engineering and Technology* **1** 64-71

[18] Jiang Z, Zhou M, Tong J, Jiang H, Yang Y, Wang A, and You Z 2015 Comparing an ant colony algorithm with a genetic algorithm for replugging tour planning of seedling transplanter *Computers and Electronics in Agriculture* **113** 225-33

[19] Katona G, Juhasz J, and Lenart B 2015 Compare Ant-colony and Genetic algorithm for shortest path problem and introduce their parallel implementations *Models and Technologies for Intelligent Transportation Systems (MT-ITS) 2015 Int. Conf.* (IEEE) 312-19

[20] Alhanjouri M, and Alfarra B 2013 Ant colony versus genetic algorithm based on travelling salesman problem *Int. J. Comput. Tech. Appl* **2** 570-8

[21] Mohan B C, Baskaran R 2012 A survey: Ant Colony Optimization based recent research and implementation on several engineering domain *Expert Systems with Applications* **39** 4618-27

[22] Maier H R, Simpson A R, Zecchin A C, Foong W K, Phang K Y, Seah H Y, Tan C L 2003 Ant

colony optimization for design of water distribution systems *J. of water resources planning and management* **129** 200-9
[23] Ostfeld A, Tubaltzev A 2008 Ant colony optimization for least-cost design and operation of pumping water distribution systems *J.l of Water Resources Planning and Management* **134** 107-18
[24] Bland J A 2001 Optimal structural design by ant colony optimization *Eng. Opt.* **33** 425-43
[25] Camp C V, Bichon B J 2004 Design of space trusses using ant colony optimization *J. of Struc. Eng.* **130** 741-51
[26] Camp C V, Bichon BJ , Stovall SP 2005 Design of steel frames using ant colony optimization *J. of Struc. Eng.* **131** 369-79
[27] Kaveh A, Talatahari S 2010 An improved ant colony optimization for the design of planar steel frames *Eng. Struc.* **32** 864-73
[28] Christodoulou S 2005 Optimal truss design using ant colony optimization *5th GRACM Int. Congress on Comp. Mechanics*
[29] Martínez F J, González-Vidosa F, Hospitaler A, Alcalá J 2011 Design of tall bridge piers by ant colony optimization *Eng. Struc.* **33** 2320-9
[30] Aydoğdu İ, Saka M P 2012 Ant colony optimization of irregular steel frames including elemental warping effect *Adv. in Eng. Soft.* **44** 150-69
[31] Hu H, Wang Q, Qin J, Wu Y, Zhang T, Xie Z, Jiang X, Zhang G, Xu H, Zheng X, and Zhang J 2008 Study on Composite Material for Shielding Mixed Neutron and γ-Rays *IEEE Transactions on Nuc. Sci.* **55** 2376-84

Acknowledgment
We would like to thank Universiti Teknologi Malaysia for providing the necessary facilities for this review.

IOP Conf. Series: Materials Science and Engineering **298** (2018) 012043 doi:10.1088/1757-899X/298/1/012043

Measurement of backscatter factor for kilovoltage x-ray beam using ionization chamber and Gafchromic XR-QA2 film

N S Ab Shukor[1], N H Mohd Asri[1], S M F Syed Jaafar[1], M S Rosnan[2] and S A Abdul Aziz[1]

[1]School of Health Sciences, Health Campus USM, Universiti Sains Malaysia, Kelantan, Malaysia

[2]Radiology Department, Hospital Universiti Sains Malaysia, Kelantan, Malaysia.

shazzleen@usm.my

Abstract. Backscatter factor (BSF) is an important parameter in the determination of surface dose for kilovoltage X-ray beam. The purpose of this study was to measure the BSF for kilovoltage diagnostic X-ray beam, and compare the measured BSF between Gafchromic XR-QA2 film and shadow free (SFD) ionization chamber (IC). The parameters that may affect the BSF, such as tube voltage (kVp) and field size, were also studied. The results were in good agreement with the TRS 457, with deviation of less than 12 %. Based on the film study, the BSF obtained from the film measurement were found to be lower than that of the IC, with average difference of 26 %. It was also found that smaller field size resulted in lower effective energy, and the amount of photons which scattered back onto the surface were also smaller. This study demonstrated that the Gafchromic XR-QA2 film was not suitable for the application of small field size.

1. Introduction

Backscatter factor (BSF) is defined as the ratio of the collision kerma of a phantom material; at the surface of a full scatter phantom located at a point in the beam axis, to the collision kerma of the same material; at the same point in the primary beam, with no phantom present [1][7]. Alternatively, BSF is defined as the ratio of a tissue [e.g. water, ICRU tissue or poly (methyl methacrylate) (PMMA)] kerma on the surface of a phantom consisting of the same material, to the tissue kerma at the same point in space, in the absence of the phantom [1][2][3][7]. The numerical values of BSF vary depending on both the definition used, and the chosen reference phantom material. Values of BSF were provided by ICRU-74 and IAEA TRS 457, for 21 diagnostic beam qualities with kilovoltages of between 50.0 and 150.0 kV, and for different types of filtration (2.5 mm aluminum (Al); 3.0 mm Al; 3.0 mm Al + 0.1 mm copper (Cu)). For each kV and filtration, the BSF were given in three field sizes and phantoms (i.e. 15.0 cm thick water, ICRU tissue and PMMA), respectively. The data obtained in these reports were taken from published studies [2][3]. The BSF is an important parameter in the determination of dose for kilovoltage X-ray beam. Equation (1) represents the definition of the BSF in radiodiagnostic dosimetry:

$$BSFair(Q)=\frac{(Kair,Q)w}{(Kair,Q)air} \quad (1)$$

where *(Kair, Q)air* is the incident air kerma free-in-air for the primary incident beam of quality, *Q*, at the point of measurement. This is achieved by multiplying the free-in-air ionization reading in the *Q* beam, *Mair,Q* (corrected for the influenced quantities; pressure, temperature and recombination), by the corresponding calibration coefficient *NK,Q*. On the other hand, *(Kair,Q)w* is the air kerma at the surface of the water phantom [3].

Unfortunately, BSF data for beams in the diagnostic range were not as widely available, compared with the beams in the therapeutic range. The importance of BSF in determining the absorbed dose at the phantom's surface was clear, but achieving an accurate measurement has not been simple [4]. BSF measurement using an ionization chamber (IC) has always been a challenge, due to: (i) the effect of the displaced material inside the chamber volume, which results in the lack of scatter radiation being produced within the chamber; (ii) more scattering event occurring at the posterior of the chamber, as a result of lower attenuation; (iii) extra scatter radiation produced within the wall, due to the physical size of the apparatus. Gafchromic XR-QA2 films possess several appealing features, which might be suitable for the measurement of BSF. The film was designed specifically as a quality assurance (QA) tool for radiological applications. It is suitable for application in the energy range of between 20.0 and 200.0 kVp, with sensitivity to the dose range of between 0.1 and 20.0 cGy. As opposed to its predecessor, i.e. XR-QA film with Caesium Bromide (CsBr) sensitive layer, the sensitive layer of the XR-QA2 film contains Bismuth(III) Oxide (Bi2O3), which is intended to boost the photoelectric effect at lower energies. A previous study found that the sensitivity of the XR-QA2 film increased at the energy range of 18.0 to 39.0 keV, with maximum variation of about 170 %, and decreased at the energy range of 38.0 to 46.5 keV [5]. The film was well suited for the kilovoltage X-ray beam study, with the effective energy range from 33.7 to 39.0 keV. Currently, there is no general consensus amongst published literature pertaining to the accuracy of this type of film for BSF measurement.

2. Materials and methods

This study was carried out at the Satellite X-ray Unit in the Radiology Department, Hospital Universiti Sains Malaysia using a static X-ray machine (Optimus 80, Philips Healthcare, MA, USA) with three-phase power supply

2.1 Determination of effective energy

The effective energy of the kilovoltage X-ray machine was measured by determining the half-value layer (HVL) of the nominal X-ray energies, i.e. 70, 81, 90, 102 and 117 kVp. A shadow free (SFD) IC (PTW, Freiburg, Germany) was connected to an electrometer (Unidose E, PTW, Freiburg, Germany), and placed on the surface of a solid water phantom. The distance of measurement between the X-ray tube and the IC was 100 cm, with 10 × 10 cm2 field size. Al sheets were placed under the collimator and broad focus setting was used. The measurements were made in an open field, and the Al sheets were added until the IC reading was reduced to half of its initial value. The HVL value was obtained from the graph generated from the relationship of the electrometer reading versus the Al thickness used [6].The value of linear attenuation coefficient for the energy can be calculated using equation (2) as follows:

$$\mu=\frac{0.693}{HVL} \quad (2)$$

where, μ/ρ was calculated and converted to the corresponding effective energy, by using the X-ray mass attenuation coefficient table for Al (National Institute of Standard and Technology, 2000).

2.2 Measurement of backscatter factor

The SFD IC with a sensitive volume of 6 cm^3 was used for the BSF measurement. In-air measurements for no-scatter condition were carried out, by placing the IC at 85 cm distance from the surface of the light beam diaphragm of the X-ray machine. This IC was placed on top of a polystyrene slab which was positioned hanging at the height of 15 cm (supported by a pair of retort stands), from the top of the X-ray table. This resulted in a source to image-receptor distance (SID) of 100 cm, as recommended by the TRS 457. The field sizes chosen were 10 × 10, 15 × 15 and 20 × 20 cm^2. The IC and Gafchromic XR-QA2 films (Lot # 02201501, Ashland Inc., KY, USA) with field size of 5 × 5 cm^2, were placed at the centre of the polystyrene, and perpendicular to the beam direction (figure 1). For the phantom measurement setup, solid water phantoms with 15 cm thickness (constructed by stacking several solid water slabs) were placed on top of the X-ray table. Next, the IC and film were each placed on the surface of the solid water phantom (figure 2), followed by the placement of a bolus on top of the detectors. Similar SID and field sizes were used, with all the measurements taken at 70, 81, 90, 102 and 117 kVp, and 20 mAs.

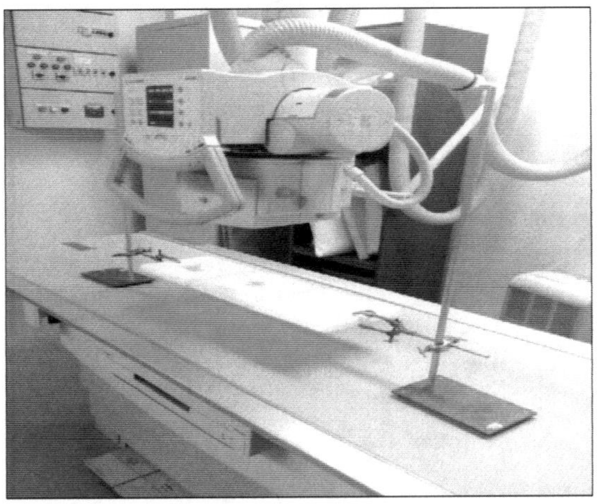

Figure 1. In air measurement setup.

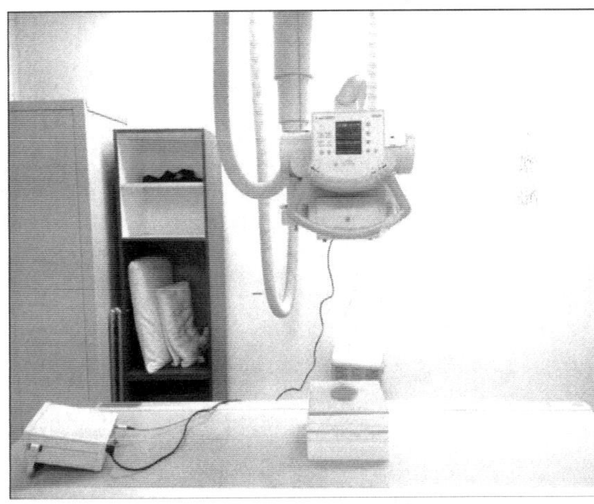

Figure 2. Phantom measurement setup.

The exposed films were read using a flatbed film scanner (Epson Expression 10000XL, Seiko Epson Corporation, Owa, Japan) with Verisoft software (PTW, Freiburg, Germany) at 24 hours post-irradiation, to ensure that the film chemicals have been stabilised. The images were scanned using a scanning resolution of 300 dots per inch (dpi), and saved as 16 bit grayscale in tagged image file (TIF) format. Each film's response in pixel value was taken using the 'Film Calibration' option, as provided by the Verisoft software. The BSF for these conditions were then calculated using equation (1).

3. Results and discussions

Table 1 shows the measured effective energies (keV) for the X-ray machine, in comparison with the energies obtained from the study by Petoussi-Henss *et. al.* [2]. Based on the results, the difference between the effective energies obtained from both studies were about 20.91 %. The effective energies of the current study were found to be lower, due to the difference in the energy spectra used [4]. As a consequent, the effective energy affects the BSF in these studies. Table 2.0 shows that the BSF increase linearly with increased kVp. The BSF also increase with increased field size, since bigger field size results in the increase number of photons that scattered back from the phantom. The BSF obtained from the current study were found to be lower by an average of ± 12 %, as compared with the previous study [2]. Since the effective energy of the X-ray spectrum, i.e. monoenergetic photon beams, were lower, the tendency for the photons to be scattered in the forward direction was also lower, thus decreasing the backscatter. The scattered photons were only able to reach the detector in a limited distance (due to less penetrating power), which would then decrease the backscatter.

IOP Conf. Series: Materials Science and Engineering **298** (2018) 012043 doi:10.1088/1757-899X/298/1/012043

Table 1. The effective energy (keV) obtained in this current study, compared with the study by Petoussi–Henss et al. [2] as published in the TRS 457.

Tube Voltage (kVp)	Effective Energy (keV)		
	Current study	TRS 457	Difference (%)
70	33.7	39.3	14.24
81	35.5	42.9	17.24
90	35.8	46.3	22.60
102	37.9	48.1	21.20
117	39.0	55.4	29.60
	Average difference (%)		20.91

Table 2: The BSF (measured using the IC) obtained in this current study, compared with the study by Petoussi-Henss et al. [2] as published in the TRS 457.

Tube Voltage (kVp)	Backscatter Factor			
	10×10 cm^2		20×20 cm^2	
	Current study	TRS 457	Current study	TRS 457
70	1.059	1.300	1.216	1.340
81	1.089	1.320	1.265	1.370
90	1.127	1.340	1.300	1.400
102	1.145	1.340	1.316	1.400
117	1.147	1.370	1.354	1.460

Table 3 shows that the BSF increase with increased field size, for both the IC and films. The BSF increased by about 11.50 and 15.82 %, for 15×15 and 20×20 cm^2 field sizes, respectively. This was due to more X-ray photons detected as the field size increase, reflecting the increased number of photons that scattered back from the phantom. From this study, the BSF measured using the Gafchromic XR-QA2 films were found to be lower than the BSF measured using the IC, by an average of ± 26.17 %, for all field sizes. At 10×10 cm^2 field size, the BSF were observed to be fluctuated, with the average difference of 46.7 %. As the field size gets bigger, the BSF become closer to the BSF obtained using the IC, by an average difference of 17.4 and 14.42 %, for 15×15 and 20×20 cm^2 field sizes, respectively. The results showed that the Gafchromic XR-QA2 film was not a suitable dosimeter for the measurement of BSF in small field size, since lesser effective energy passes through the collimator, and the probability of photons to scatter back to the surface will be lower than that of a bigger field size. From this study, we also found that the Gafchromic XR-QA2 film was energy dependent, since even a small increment in the beam qualities may lead to the variation in the film response by ± 12 %.

Table 3.The comparison between the BSF obtained using the IC and Gafchromic XR-QA2 films, at various field sizes.

Tube Voltage (kVp)	Backscatter Factor (BSF)					
	$10 \times 10\ cm^2$		$15 \times 15\ cm^2$		$20 \times 20\ cm^2$	
	IC	Film	IC	Film	IC	Film
70	1.059	0.800	1.167	0.840	1.216	1.120
81	1.089	0.472	1.229	0.984	1.265	1.161
90	1.127	0.594	1.237	1.028	1.300	1.389
102	1.145	1.120	1.270	1.132	1.316	1.156
117	1.147	0.611	12.87	1.163	1.354	1.389

4. Conclusion

As a conclusion, the BSF for kilovoltage X-ray beam increased linearly with the kVp and field size. In comparison with the TRS 457, the BSF obtained from this study were found to be lower by an average of ± 12 %. Based on the film study, the BSF measured by the Gafchromic XR-QA2 films were found to be lower than that of the IC by an average of ± 26.17 %, for all field sizes. The Gafchromic XR-QA2 film was found to be energy dependent and thus, not suitable to be used as a dosimeter for the measurement of BSF for small field size. This was due to lower effective energy passes through the collimator, and the probability of photons to scatter back onto the surface was lower than that of a bigger field size.

5. References

[1] Ma C M, Li X A, and Seuntjens J P 1998 Study of dosimetry consistency for kilovoltage x-ray beams *Med. Phys.* **25** 12 2376-84

[2] Petoussi-Henss N, Zankl M, Drexler G , Panzer W and Regulla D 1998 Calculation of backscatter factors for diagnostic radiology using Monte Carlo methods *Phys.Med Biol.* **43** 2237–50

[3] Hamza B, Hugo B, Annette F and Pedro A 2011 Backscatter factors and mass energy-absorption coefficient ratio for diagnostic radiology dosimetry *Phys. Med. Biol.,* **56** 7179-7204

[4] Chris J M, Howard R E and Michael A S L 2012 Measurement of low-energy backscatter factors using GAFCHROMIC film and OSLs J. *Appl. Clin. Med. Phys.* **13** 6 126-132

[5] Di Lillo F, Mettivier G, Sarno A, Tromba G ,Tomic N , Devic S and Russo P 2016 Energy dependent calibration of XR-QA2 radiochromic film with monochromatic and polychromatic x-ray beams *Med. Phys.* **43** 583-8

[6] Mohd Fahmi M Y, Mohd Hadzmi Y, Muhammad Syazwan R, Reduan A, and Ahmad Bazlie A K 2017 Dose measurement using Al_2O_3 dosimeter in comparison to LiF:Mg,Ti dosimeter and ionization chamber at low and high energy x-ray *AIP Conf. Proc.* **1799** 040007-1-7

[7] Robin H, Brendan H, Lois H, Zdenka K, David T and Clive B 2014 Advances in kilovoltage x-ray beam dosimetry *Phys. Med. Biol.* **59** R183-231

[8] IAEA 2007 Dosimetry in diagnostic radiology: an international code of practise *Technical Reports Series No.457* International Atomic Energy Agency

[9] Smith L, Hill R, Nakano M and Kim J 2011 The measurement of backscatter factors of kilovoltage x-ray beams using Gafchromic EBT2 film *Australas. Phys. Eng. Sci. Med.*, **34** 261-6

[10] Nada T, Chrystian Q, Bruce R W, Saad A, Hamed B, LiHeng L, Francois D, Jan S and Slobodan D 2014 Characterization of calibration curves and energy dependence Gafchromic XR-QA2 model based radiochromic film dosimetry system *Med. Phys.* **41** 6 062105-1-9

[11] Shimizu K, Koshida K and Miyati T 2001 Monte Carlo simulation analysis of backscatter factor for low energy x-ray *Proceedings of the Ninth EGS4 User' Meeting in Japan* **22** 115-18

[12] Guerda M, Sou Tung C T, Ivan D M and Maria F C 2012 energy dependence of the new Gafchromic EBT3 film:dose response curves for 50kV,6 and 15MV x-ray beams *Int.J. Med. Phys. Clin. Eng. Radiant Oncol.* **1** 60-5

[13] Carl A C 1993 Differences in reported backscatter factors for low energy x-rays: a literature study *Phys. Med. Biol.* **38** 521-31

[14] Veneziani G R , Correa E L , Potiens M P A, Campos L L 2015 Attenuation coefficient determination of printed ABS and PLA samples in diagnostic radiology standard beams *J. Phys. Conf. Ser.* **733** 012088 1-5

[15] Ahmed M, David R D, Donald M and Hans-Micheal K 2010 Dosimetry in diagnostic radiology *Eur. J. Radiol.* 76 11-4

Acknowledgments

This work was financially supported by Universiti Sains Malaysia Short Term Research Grant No. 304/PPSK/61313177. The authors would like to acknowledge the Radiology Department, Hospital Universiti Sains Malaysia, for the X-ray machine and equipment used in this study.

A study of PC-based ultrasonic goniometer system of surface properties and characterization of materials

S Sani[1,2], M H Md Saad[2], N Jamaludin[2], M P Ismail[1], S Mohd[1], I Mustapha[1,2], N A Masenwat[1], T S Tengku Amran[1] and M H A Megat Ahmad[1]

[1] Industrial Technology Division, Malaysian Nuclear Agency, Bangi 43000 Kajang, Selangor, Malaysia.
[2] Department of Mechanical & Materials Engineering, Faculty of Engineering and Built Environment, National University of Malaysia, Bangi 43000 Kajang, Selangor, Malaysia.

suhairy@nm.gov.my

Abstract. This paper discussed the design and development of a portable PC-based ultrasonic goniometer system that can be used to study material properties using ultrasonic wave. The system utilizes an ultrasonic pulse-receiver card model attached to computer notebook for signal display. A new specific software package (GoNIO) was developed to control the operation of the scanner, displaying the data and analyze characteristics of materials. System testing was carried out using samples with cubic dimension of about 10 mm x 20 mm x 30 mm. This size allows the sample to be fitted into the goniometer specimen holder and immersed in a liquid during measurement. The sample was rotated from incident angle of 0° to 90° during measurement and the amplitude reflected signals were recorded at every one degree of rotation. Immersion transducers were used to generate and receive the ultrasounds that pass through the samples. Longitudinal, shear and Rayleigh wave measurements were performed on the samples to determine the Dynamic Young's Modulus. Results of measurements are explained and discussed.

1. Introduction

An ultrasonic goniometer is based on the principle of reflection and transmission of ultrasound at oblique incidence at plane boundary [1]. It can be used to measure the speeds of compression, transverse and Rayleigh waves irrespective of the shape of the test object provided that the object has at least one plane surface. The method can be used to its best advantage with pulse-echo technique using two transducers, one transmitting and the other receiving and each of which are mounted on rotating arms, the angular positions of which can be accurately determined using vernier on a circular scale. The complete arrangement is immersed in water or liquid tank. The test object is positioned in such a way that the axels of the probes intersect at its surface, which is perpendicular to the plane containing these axes. These arrangements allow the determination of Dynamic Young's modulus through calculation.

In principle, the ultrasonic characterization of materials can be performed in two ways, *i.e.* by measuring sound velocity and sound attenuation. In homogeneous, isotropic and elastic media, the wave velocity, V is given by:

IOP Conf. Series: Materials Science and Engineering **298** (2018) 012044 doi:10.1088/1757-899X/298/1/012044

$$V = \sqrt{\frac{E}{\rho} \frac{(1-\upsilon)}{(1+\upsilon)(1-\upsilon)}} \tag{1}$$

where E is the Young modulus, ρ the material density and υ the Poisson ratio.

Ultrasonic waves are attenuated as the waves pass through a material. This attenuation can be caused by beam divergence (distance effect), absorption (heat dissipation) and scattering. Only scattering is affected by the characteristics of the material, mostly because the degree of inhomogeneity in the material and the frequency of the transducer. The attenuation caused by scattering was given as [2],

$$\alpha_s = \begin{cases} 1/D & \text{for diffusion range} \quad \lambda \leq D \\ Df^2 & \text{for stochastic range} \quad \lambda \approx D \\ D^3 f^4 & \text{for Rayleigh range} \quad \lambda >> D \end{cases} \tag{2}$$

where, f is the wave frequency, λ the wavelength and D the average inhomogeneity.

The principle of operation is illustrated in Figure 1 [3].

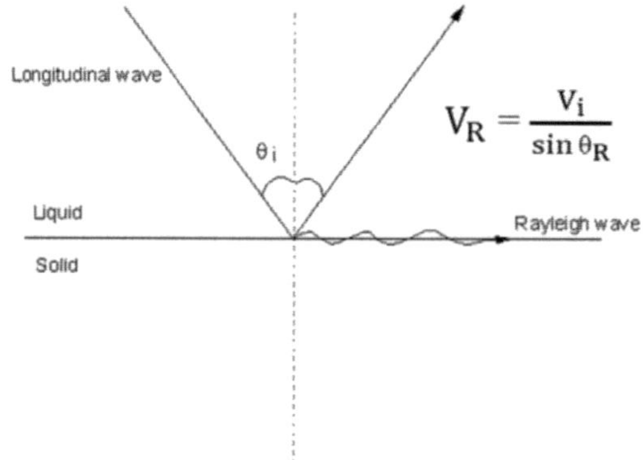

Figure 1. An ultrasonic beam is incident at a critical angle θ_i and Rayleigh wave propagated on the solid surface.

The ultrasonic goniometer was developed for applications that need a variable incidence and detection angle as well as an adjustable distance of incidence and detection point at the region near the transducer also known as Fresnel field [4]. It is very difficult to accurately investigate defects in materials. This is because it can produce multiple indications and the signal amplitude of the reflected signal from the flaw may vary considerably if the flaw located within effective distance from the transducers varies.

Manual ultrasonic measurement of sample using ultrasonic goniometer technique is physically intensive and time consuming. For these reasons, we developed a PC-based ultrasonic goniometer system and used it on a few samples to evaluate its performance. This paper discussed the system developed and its measurement results.

2. Material and Method

The longitudinal wave velocity was measured by conventional ultrasonic flaw detector using longitudinal probe whereas Rayleigh wave velocity was measured using ultrasonic goniometer system developed. Figure 2 shows the system, the samples as well as the control software (GoNIO). The shear wave velocity was then calculated according to the formula [2];

$$q^6 - 8q^4 + 8q^2(3 - 2u^2) - 16(1-u^2) = 0 \tag{3}$$

where $q = V_R/ V_S$, $u = V_S/V_L$, V_L = longitudinal wave velocity, V_S = shear wave velocity and V_R = Rayleigh wave velocity.

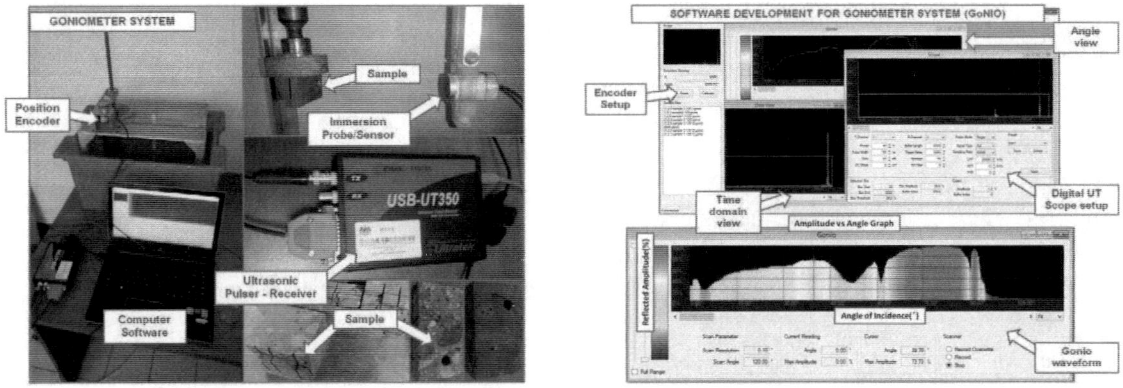

Figure 2. The ultrasonic goniometer system.

The developed ultrasonic goniometer system utilized an ultrasonic pulse-receiver card model attached to computer notebook for signal display showing maximum amplitude against degree of rotation. The immersion transducer frequency of 2.25MHz with 20mm diameter size was used to generate and receive Rayleigh wave ultrasound through the samples. The sample attached to goniometer was rotated from incident angle of 0° to 90° and the amplitudes of reflected signal from the sample were recorded at every one degree of rotation. The distance between transducer and sample is set to 152 mm to obtain maximum amplitude echo. This was based on the near field length calculation according to equation 4:

$$\text{Near Field} = \frac{D^2}{4\lambda} = \frac{D^2 f}{4V} \tag{4}$$

where, D = diameter of the probe, f = frequency of the probe and V = velocity of water ($1485 ms^{-1}$)

Figure 3 shows the prepared samples used in the experiments. These are steel, hematite concrete, aluminum and lead glass. The samples were cut to a cubic dimension of about 10 mm x 20 mm x 30 mm so that can be fitted into the goniometer specimen holder. The densities of samples were determined by weighing the samples and measuring the volume water displaced.

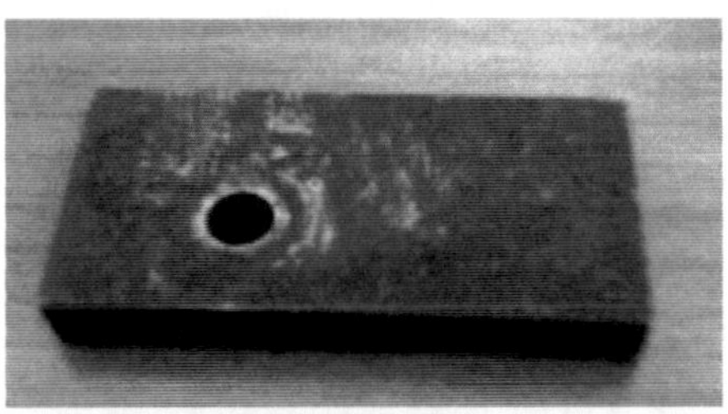

(a)

(b)

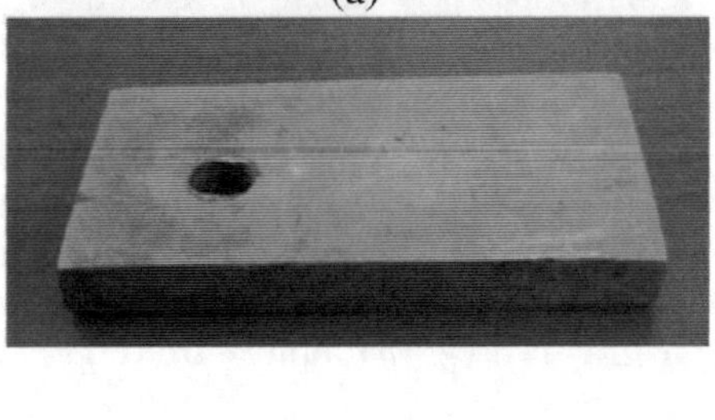

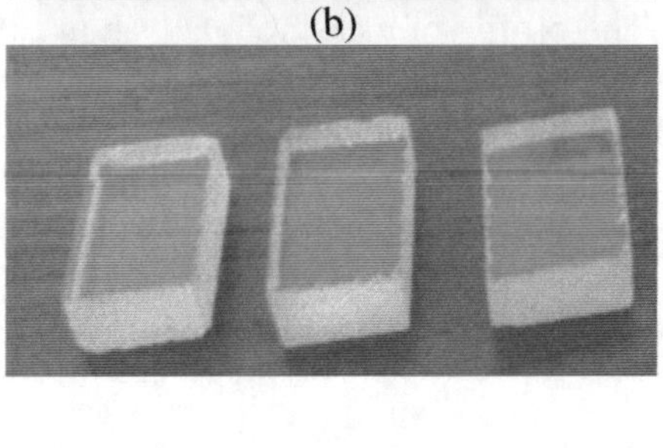

(c) (d)

Figure 3. The material of (a) steel, (b) hematite concrete, (c) aluminum and (d) lead glass are used in this experiment.

3. Results and Discussion

A typical ultrasonic goniometer result displayed by the system is as shown in Figure 4. It shows the Rayleigh wave critical angle for steel (the vertical green line). Using Snell's law, the Rayleigh wave velocities of the samples were calculated. Figure 5 shows the ultrasonic goniometer data for all four samples. The vertical dashed line indicates the Rayleigh wave critical angle which is caused by the increase of attenuation (Rayleigh scattering) for each sample respectively.

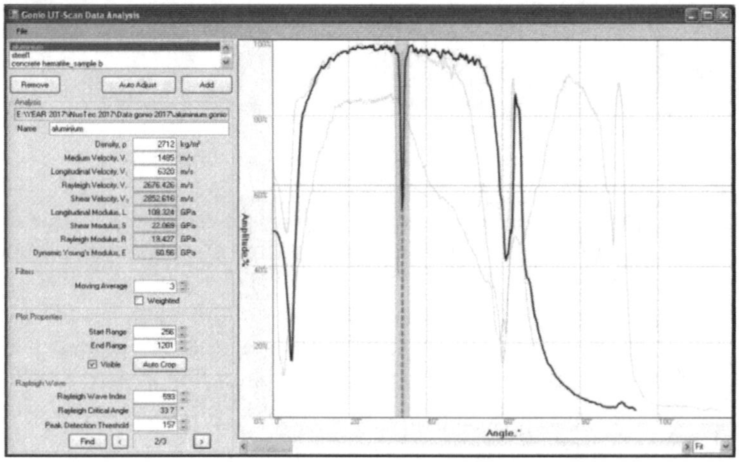

Figure 4. The (GoNIO) software display of goniometer curve characteristic of steel.

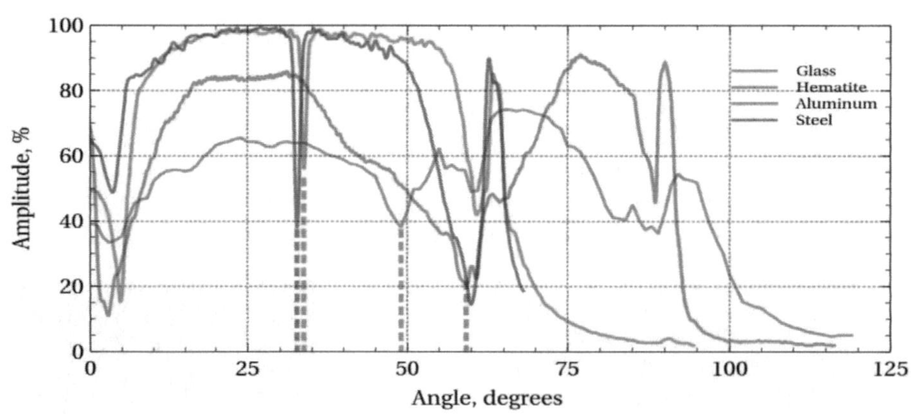

Figure 5 shows the goniometer curve characteristics for all the samples.

The profile of measured data shows the reduction in amplitude at low angle of reflection. This is caused by the geometrical setting and sound divergence causing only partial reflection of ultrasonic wave back to the transducer. Further increase of the reflected angle resulted in better reflection of ultrasonic wave thus increasing the amplitude. The first attenuation of the amplitude is caused by the occurrence of Rayleigh scattering on sample surface that is characteristics of the sample properties.

The ultrasonic goniometer system provides the calculated properties of all the samples based on the amplitude-angle profile data. These values are shown in Table 1 with accompanying referenced values for comparison for Tensile Modulus and Dynamic Young Modulus from goniometer measurement (Thbe Tensile Modulus for these materials were determined using tensile test, a destructive testing method).

Table 1. Comparison Tensile Modulus and Dynamic Young Modulus from goniometer measurement.

Tensile Modulus (Young's Modulus, Modulus of Elasticity) [5]		Dynamic Young Modulus (Goniometer measurement)	
Material	*(10^9 N/m^2, GPa)*	Material	*(10^9 N/m^2, GPa)*
Steel, stainless AISI 302	180	Steel	174.15
Steel, Structural ASTM-A36	200		
Concrete	17	Hematite concrete	20.76
Concrete, High Strength (compression)	30		
Aluminum	69	Aluminum	60.56
Glass	50 - 90	Lead glass	49.73

Table 1 show that the tensile and calculated Young's modulus for steel is very similar, even though the compared steels are of different types. All the other materials show similar traits with the calculated Young's modulus within the range of tensile values. This indicates the ultrasonic measurement can provide reliable values of Young's modulus.

4. Conclusion

A portable PC-based ultrasonic goniometer system was developed to study material properties using ultrasonic wave. The system was tested using samples that can be compared tensile Young's modulus. The ultrasonic goniometer measured the surface Rayleigh wave velocity of the samples and with the combination of conventional ultrasonic pulse echo method gave the complete velocities values that were needed to determined Dynamic Young's modulus. These values were found to be comparably similar, indicating the effectiveness of the ultrasonic method to determine Young's modulus of material.

5. References

[1] Mayer, W. G. 1965 *Energy partition of ultrasonic waves at flat boundaries*, Ultrasonic Vol. 3, pp 62-68
[2] Szilard, J. 1982 *Ultrasonic Testing* (USA: John Wiley & Sons.)
[3] Ismail, M P, Bloser, D.S., 1983 *Ultrasonic Goniometer Measurement of the Properties of Maraging Steel Using Rayleigh Waves* (Jurnal Sains Nuklear, Vol. 1, No. 2.)
[4] Blitz, J., Simpson, G., 1996, *Ultrasonic Methods of Non-destructive Testing. the Ultrasonic Goniometer* (London: Chapman and Hall) p 103
[5] *Elastic Properties and Young Modulus for some Materials*. The Engineering ToolBox. Retrieved from https://www.engineeringtoolbox.com/young-modulus-d_417.html

Acknowledgement
The authors would like to express his appreciation to Ministry of Science, Technology and Innovation also to Malaysian Nuclear Agency for their support to this project.

IOP Conf. Series: Materials Science and Engineering **298** (2018) 012045 doi:10.1088/1757-899X/298/1/012045

Study on the window cooling system of 300 keV Electron Accelerator

K.W. Leo, R.M. Chulan, S.A. Hashim, M. Azhar, Y. Dalim, M. Mokhtar, A. H. Baijan, R. M. Sabri, M. Faiz, A. Azaman, R.C. Rosli

Malaysian Nuclear Agency, Bangi, 43000 Kajang. Selangor. Malaysia.

leo@nm.gov.my

Abstract. This paper describes the method used to identify the parameters required for the window cooling system. This locally designed low energy electron accelerator with the present energy of 140 keV will be upgraded to 300 keV. The heat will be increased due to the increment of the beam energy; therefore an appropriate cooling system is required to prevent the breaking of the few µmm titanium window. The broken window will disrupt the beam transportation inside the vacuum environment of the accelerating tube and also the high voltage will be breakdown if the vacuum environment couldn't be sustained. Therefore, a desired air cooling system to cool down the 9 kwatt (30 mA, 300 keV) beam power has been designed. As the result, the window cooling system with the required pressure based on the beam powers have been calculated and identified.

1. Introduction

Accelerator is a promising technology widely used in various applications such as flue gases treatment and cable cross linking [1], medical applications such as cancer therapy and isotope production [2][3], as a collider for high energy physic experiment and study [4][5] and agriculture for mutation breeding [6]. In Nuclear Malaysia, a localized own developed accelerator technology [7] has been started with the initial energy of 140 keV as shown in figure 1, the so-called the low energy electron accelerator. In the present status, this accelerator is under-going upgrade works from the energy 140 keV to 300 keV with the replacement of the high voltage power supply, including the development or modification of the individual supporting systems such as scanning system, gas insulating system, window cooling system and etc. In this paper, the principle of the window cooling system will be described and discussed.

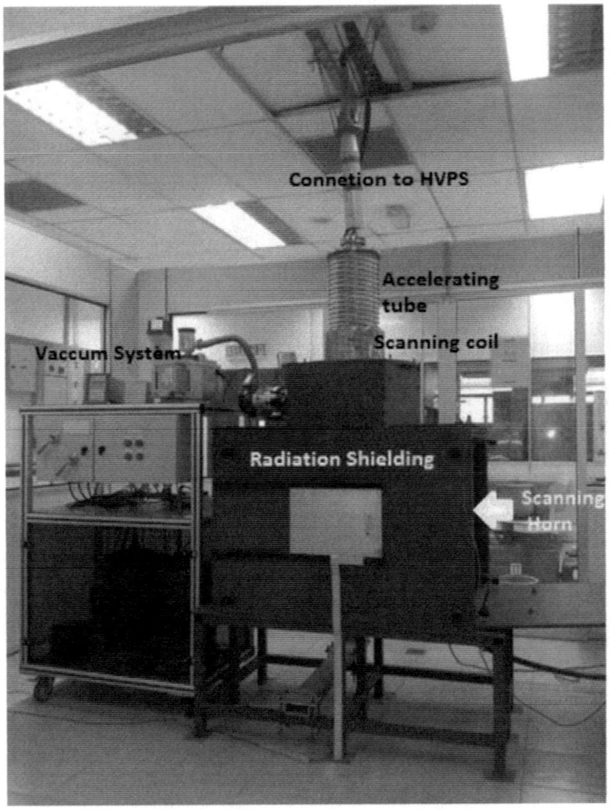

Figure 1. 140 keV low energy electron accelerator

2. Principle of Electron Accelerator

Basically, an electron accelerator is a tool to accelerate the electron beam with the required energy and beam current which are mainly determined by the applications. The industrial electrostatic electron accelerator (ELV-4) [8] comprises of a source (cathode) to produce an electron beam by filament heating system, an accelerating element to accelerate the electron beam by the accelerating tube supplying with the high voltage supply and shielded by the insulation gas system, guiding element to guide an electron beam (beam scanning) and auxiliary systems such as vacuum system to provide a vacuum environment, a window cooling system (air cooling) to avoid the breakdown of the window due to the beam heating and several power supplies to heat up the filament and scan the beam horizontally and vertically. The combination of the all individual system is required to deliver the energetic electron beam to the user. In this paper, we're focused on the window cooling system and the detail will be discussed in the next section.

3. Design and requirement of the window cooling system

Basically, the window cooling system of the low energy electron accelerator consists of the air blower and coupling with the air ducting as shown in figure 2. It is required to cool down the titanium window with the thickness of 25 μm. The titanium window must be cooled down below the titanium foil breaking point temperature of 400 ^{0}C [8], it is essential to avoid the vacuum breakdown which could disturb the electron beam transportation and moreover the breakdown of the vacuum environment will cause the corona discharge inside the accelerating tube [8]. Therefore, the design of the window cooling system must be considered those parameters such as thickness of the titanium window, heating temperature is depending on the beam energy and beam power; the required air velocity for window cooling must be identified according to those parameters.

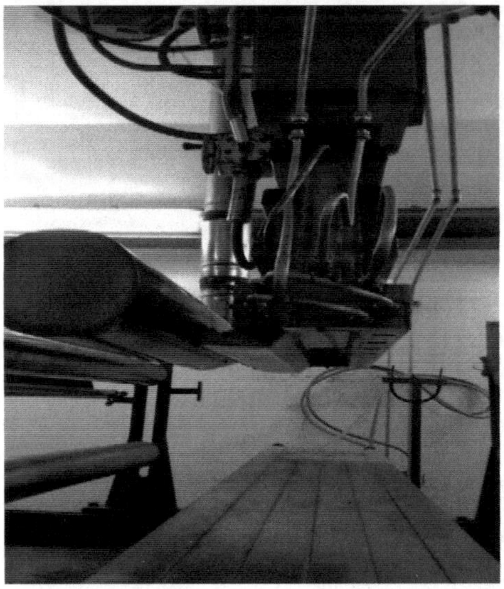

Figure 2. Window Cooling System with air ducting of ELV-4

Basically, the required air velocity of the window cooling depends on the heat loss at the titanium window. The increment of the titanium foil thickness will increase the energy loss which will be transferred to the heat energy. Subsequently, it will increase the temperature of the titanium window. The details expression and calculation are given as below:

Energy loss E_{Ti} in Titanium foil of 50 μm thickness (d_{Ti}) is [8]:

$$dE_{Ti,50} = 31\ keV \tag{1}$$

If $dE/d(d_{Ti}) \approx$ constant and, with the thickness of titanium foil ~25 μm, then:

$$dE_{Ti,25} = E_f = 15.5\ keV \tag{2}$$

Therefore, power loss P_{Ti} (watt) into the titanium foil can be estimated as:

$$P_{Ti} = \frac{E_f I_B}{e} \tag{3}$$

Moreover, maximum temperature of the titanium foil can be determined as:

$$T_{Ti} = \frac{P_{Ti}}{\alpha A} \quad (4)$$

and therefore,

$$\alpha = \frac{P_{Ti}}{T_{Ti} A} \quad (5)$$

where,

E= Electron beam energy (MeV)
d_{Ti} = Titanium foil thickness (cm)
ρ_{Ti} = Titanium foil density , 4.4 (gcm^{-3})
α= Heat transmission coefficient ($Wm^{-2}{}^{\circ}C^{-1}$)
A=Area of scanning surface (In our system, A=9.7 cm × 19.25 cm)

From equation (1) to (5), the heat transmission coefficient can be estimated. The heat transmission coefficient dependence on the cooling air velocity is shown in table 1.

Table 1. Foil heat transmission coefficient dependence on cooling air velocity [8]

V [ms^{-1}]	**0**	**10**	**30**	**100**
α[$Wm^{-2}{}^{\circ}C^{-1}$]	**0**	**40**	**110**	**250**

From the table 1, a polynomial fitting can be applied to express the relationship velocity as, cooling air velocity, v:

$$v = 0.0007\alpha^2 + 0.2327\alpha - 0.1283 \quad (6)$$

and, relative pressure, P [Pa] can be determined as:

$$P = \frac{1}{2}\rho v^2 \quad (7)$$

To sustain the titanium foil without being broken, a temperature must be maintained below 400°C [8]. Using equation (1) to (6), the required cooling air velocity and pressure are estimated with three beam currents (10 mA, 20 mA and 30 mA) and with titanium window temperature below than 400°C as shown in figure 3.

From results as shown in Figure 3, the required v_{min} for foil temperature < 300°C and 10 mA beam current is~ 7.1 ms^{-1}; v_{min} for foil temperature < 300 °C and 20 mA beam current is ~ 15.4 ms^{-1}. v_{min} for foil temperature < 300 °C and 30 mA beam current is ~ 25.0 ms^{-1} . An air blower coupling with the inappropriate air ducting and air nozzle has been fabricated. In next section, the capability of the existing window cooling system will be discussed with the experimental setup and result.

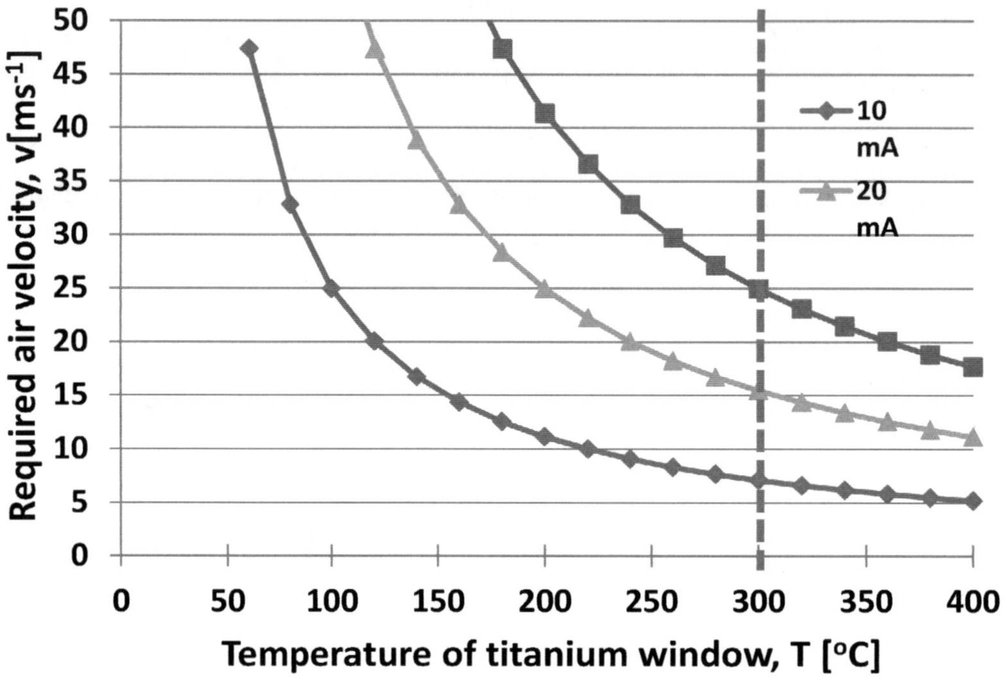

Figure 3. Required cooling air velocity and pressure of 10 mA, 20 mA and 30 mA beam current.

By taking into account the breaking point of titanium foil at 400 ^{0}C and the maximum beam parameter of the system with 30 mA, the required air velocity in our system has been set at the reference point of 300^{0}C as indicated by the vertical red broken line. Therefore, we must ensure that the window cooling system must be able to provide the air velocity of 25 ms^{-1} at the maximum beam current of 30 mA.

4. Experimental results of the window cooling system

The requirement of air velocity for the window cooling has been identified as shown in figure 3. Therefore, we must ensure that the present system must be fulfilled the identified magnitude. For this purpose, two experiments have been carried out as i) air velocity measurement with the different position of air ducting , ii) air velocity measurement within the window surface . The experimental setups are shown in figure 4 and figure 6. Meanwhile the experimental results are shown in the figure 5 and figure 7.

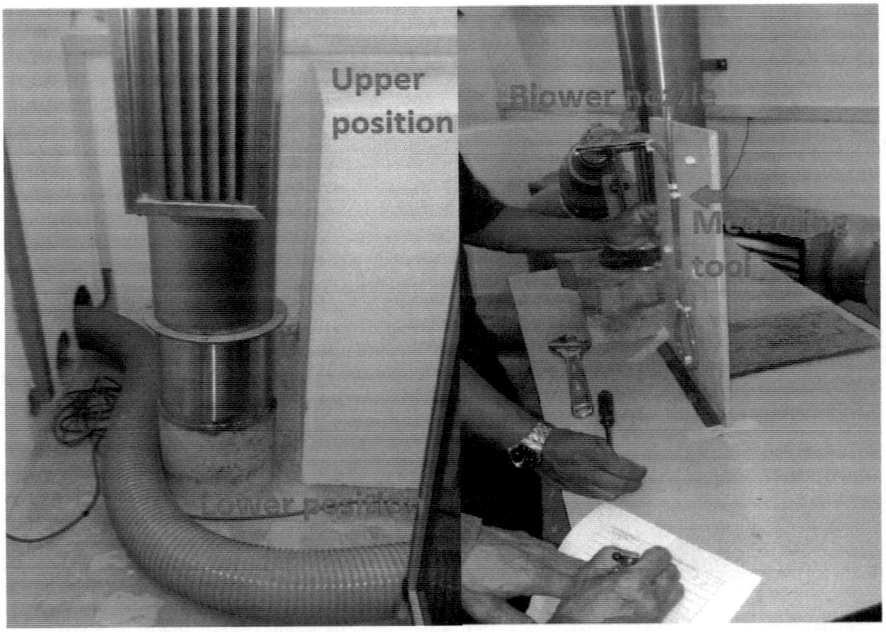

Figure 4. Experimental setup air velocity measurement with the different position of air ducting.

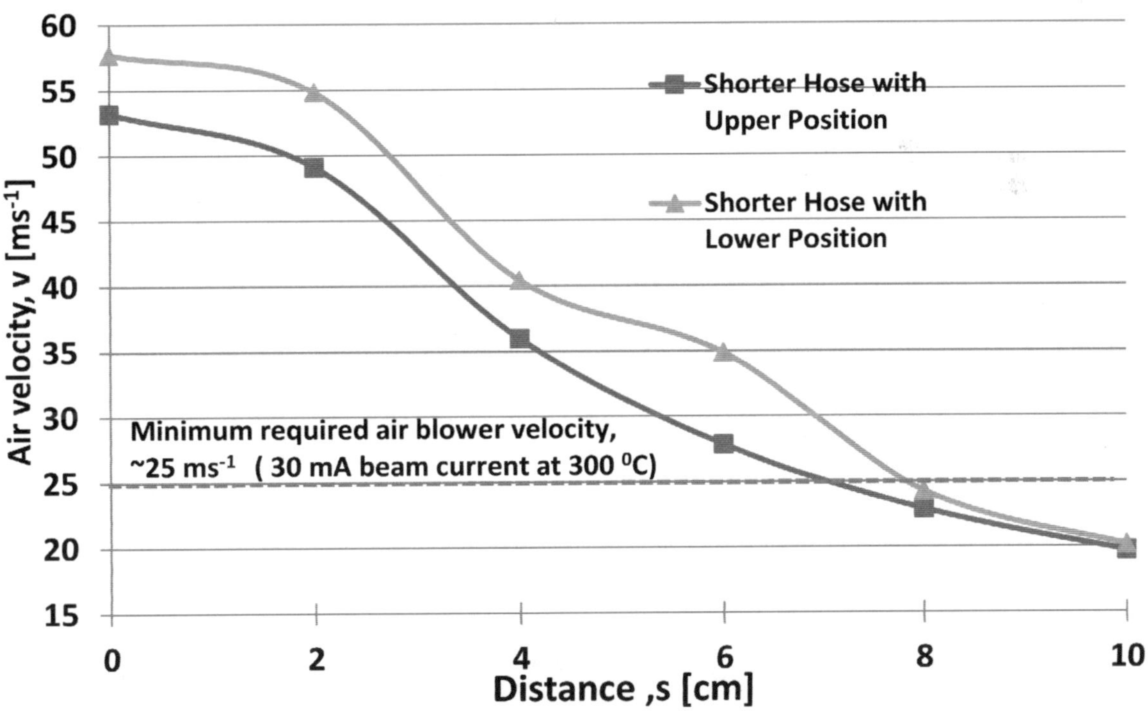

Figure 5. Experimental result air velocity measurement with the different position of air ducting.

IOP Conf. Series: Materials Science and Engineering **298** (2018) 012045 doi:10.1088/1757-899X/298/1/012045

Figure 6. Experimental setup air velocity measurement within the window surface.

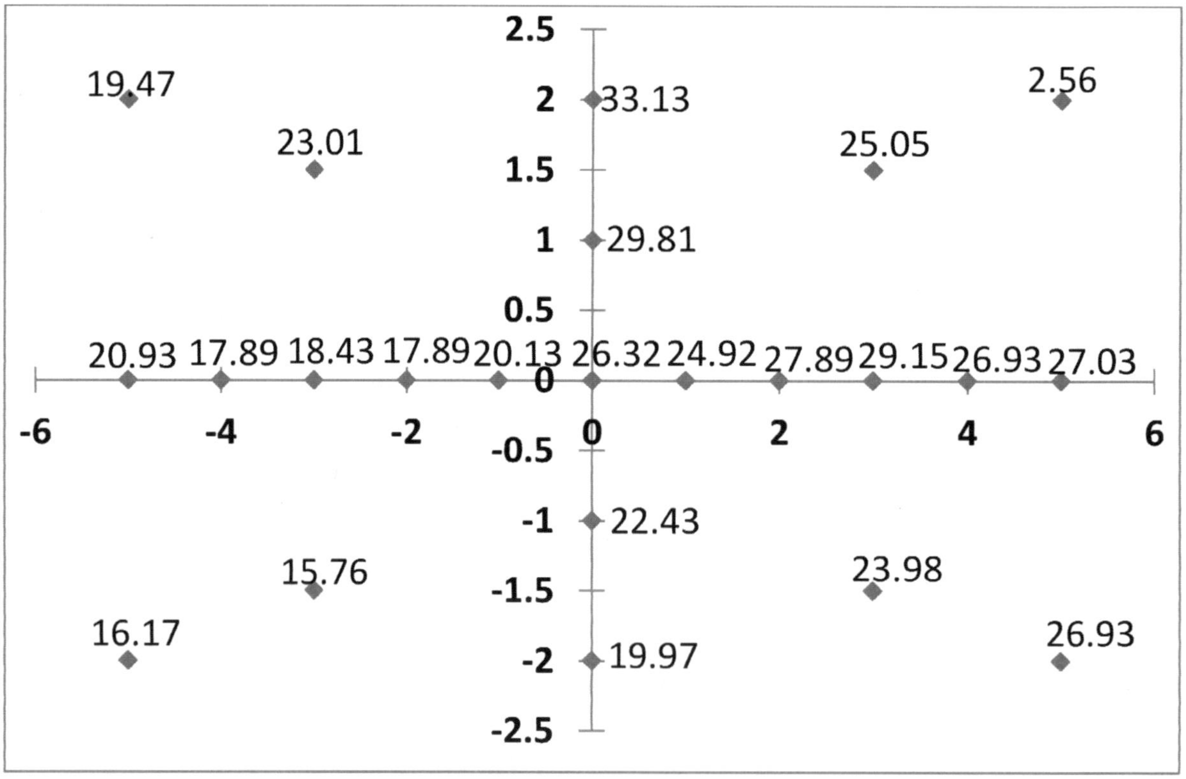

Figure 7. Experimental result air velocity measurement within the window surface.

From the experimental result, with the different length of air ducting, it is clearly shown that the position of the ducting is one of the essential parameter to increase the air velocity within the titanium window.

From the experimental result, within the window surface, we could state that the maximum air velocity is 33 ms^{-1} and minimum air velocity is 2.56 ms^{-1} respectively. Therefore, we could know clearly that the magnitude differ is caused by the inappropriate air nozzle without the uneven and unfixed dimension of the air nozzle. In addition, the air reluctant could be reduced with the modification of the air ducting.

5. Conclusions

In this paper, parameters of the heat generated by the beam current of 10 mA, 20 mA and 30 mA of the electron beam accelerator have been identified. Furthermore the air velocity generated by the existing window cooling system has been measured and identified experimentally. In conclusion, the existing window cooling system must be improved and modified with the additional air nozzle to provide the required air velocity equally within the titanium window.

6. References

[1] Machi S 2011 Trends for electron beam accelerator applications in industry *Rev. of Accelerator Science and Technology,* **4** 1–10, World Scientific

[2] Dorman E and Emery R 2013The University of Washington clinical cyclotron: a summary of current particles and energies used in therapy isotope production, and clinical research, *Proc. of 20th Int. Conf. on Cyclotrons and Their Applications* Vancouver, BC, Canada.

[3] Tsutsui H 2014 KEK-Nuclear Malaysia-Sokendai Accelerator School in Nuclear Malaysia, 1st-4th December 2014

[4] Edwards D A and Edwards H T 2008 Particle colliders for high energy physics *Rev. of Accelerator Science and Technology,* **1** 99–120 World Scientific

[5] Wilson P B 2008 Electron linacs for high energy physics *Rev. of Accelerator Science and Technology,* **1** 7–41, World Scientific

[6] Hase Y KEK-Nuclear Malaysia-Sokendai Accelerator School in Nuclear Malaysia, 1st-4th December, 2014

[7] Ghazali A B, Mohd Rizal M, Rosli D, Mohd Rizal M C, Lee C H, Ghazali A B, Leo K W and Zahidee M T 2005 Development of electron beam machine at Nuclear Malaysia, *Technology and Application Accelerator Meeting VIII, P3TM-BATAN,* Yogyakarta, 21 – 22 Nov. 2005.

[8] Fadeev S N 2013 Workshop on electron accelerator technology in Nuclear Malaysia, 10th October -11st November, 2013.

Acknowledgments

We would like to express our gratitude to Dr. Fadeev, from the Budker Institute of Nuclear Physics, Russia, for his guidance and willingness to join us at the Workshop on Electron Accelerator Technology during which he had shared his experience and knowledge with us theoretically and practically.

IOP Conf. Series: Materials Science and Engineering **298** (2018) 012046 doi:10.1088/1757-899X/298/1/012046

Evaluation of radiation dose in pediatric head CT examination: a phantom study

Nik Norhasrina Nik Din[1, a)], Rafidah Zainon[2, b)] and Ahmad Taufek Abdul Rahman[1, 3, c)]

[1]Department of Physics and Material Studies, Faculty of Applied Sciences, Universiti Teknologi MARA, 40450 Shah Alam, Selangor, Malaysia
[2]Oncological and Radiological Sciences Cluster, Advanced Medical and Dental Institute, Universiti Sains Malaysia, Bertam, 13200 Kepala Batas, Pulau Pinang, Malaysia
[3]Institute of Science, Universiti Teknologi MARA, 40450 Shah Alam, Selangor, Malaysia

[a)]nikhasrina@gmail.com,
[b)]rafidahzainon@usm.my,
[c)]ahmadtaufek@salam.uitm.edu.my

Abstract. The aim of this study was to evaluate the radiation dose in pediatric head Computed Tomography examination. It was reported that decreasing tube voltage in CT examination can reduce the dose to patients significantly. A head phantom was scanned with dual-energy CT at 80 kV and 120 kV. The tube current was set using automatic exposure control mode and manual setting. The pitch was adjusted to 1.4, 1.45 and 1.5 while the slice thickness was set at 5 mm. The dose was measured based on CT Dose Index (CTDI). Results from this study have shown that the image noise increases substantially with low tube voltage. The average dose was 2.60 mGy at CT imaging parameters of 80 kV and 10 – 30 mAs. The dose increases up to 17.19 mGy when the CT tube voltage increases to 120 kV. With the reduction of tube voltage from 120 kV to 80 kV, the radiation dose can be reduced by 12.1% to 15.1% without degradation of contrast-to-noise ratio.

1. Introduction

The Computed Tomography (CT) technology, performance and clinical application undergoes tremendous evaluation. Numerous procedures have been done to reduce radiation dose to children during CT including modulation of tube current and lowering the tube voltage [13]. There is a need to investigate the CT imaging protocol to evaluate the radiation dose received by pediatric patients. CT is widely used in medical imaging to treat adult and pediatric. The use of low tube voltage in CT imaging is recommended to increase the image contrast enhancement. However, the radiation dose varies according to tube voltage and patient size. Theoretically, the tube current increases when lower tube voltage is used in the study. This is to maintain image quality obtains from the CT imaging.

IOP Conf. Series: Materials Science and Engineering **298** (2018) 012046 doi:10.1088/1757-899X/298/1/012046

2. Materials and methods

2.1 CT phantom imaging

A CT Dose Index phantom (figure 1) with 10 cm diameter was scanned with Siemens Somatom Definition AS+128 Slice CT scanner. This phantom was used to mimic the pediatric head in this study. The *CTDI* phantom has four holes at periphery and a hole in the center. The phantom was scanned at 80 kV and 120 kV while the tube current was set manually and using AEC mode. The pitch was adjusted to 1.4, 1.45 and 1.5 while the slice thickness was set at 5 mm. The radiation dose was measured with ion chamber detector that was inserted in each hole during the experimental work. The *CTDI* weighted values was computed based on equation 1.

$$CTDI_W = \frac{1}{3} CTDI_{100CENTER} + \frac{2}{3} CTDI_{100PERIPHERY} \quad (1)$$

Equation (1) shows the calculation for *CTDI* weighted values. Each hole of *CTDI* phantom was scanned to obtain the value of *CTDI* weighted.

3. Results

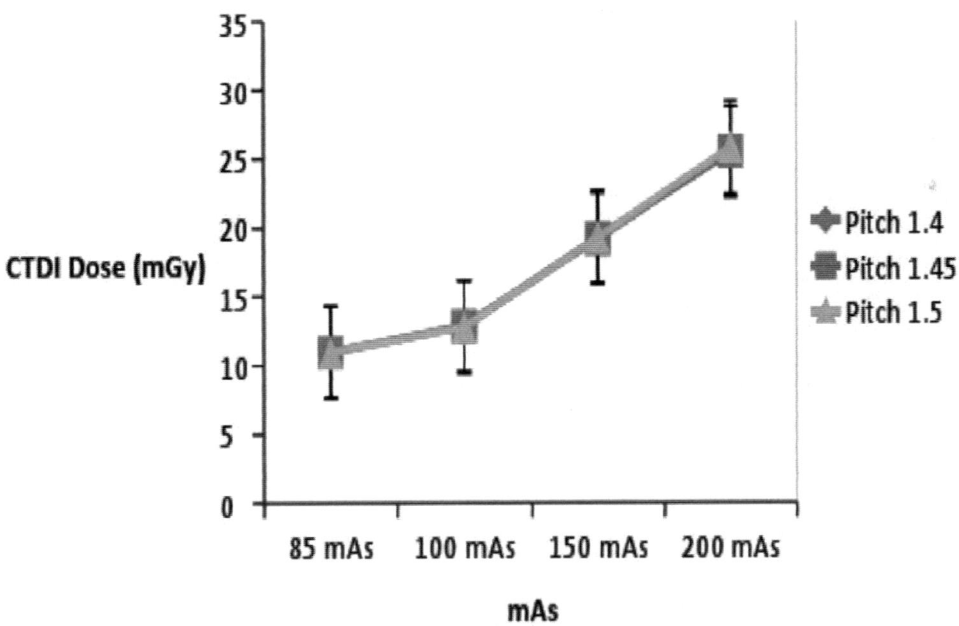

Figure 1. *CTDI* dose against mAs at 120 kV for care dose setting for head phantom

3.1 Care Dose

CTDI dose against mAs (in figure 2) shows that not too much different, even changing the pitch. Abdomen caredose for pediatric patients by using Care Dose (used in hospitals) was not too much different. Scan time 2.49 by the rotation time 0.5 and slice thickness 5mm.

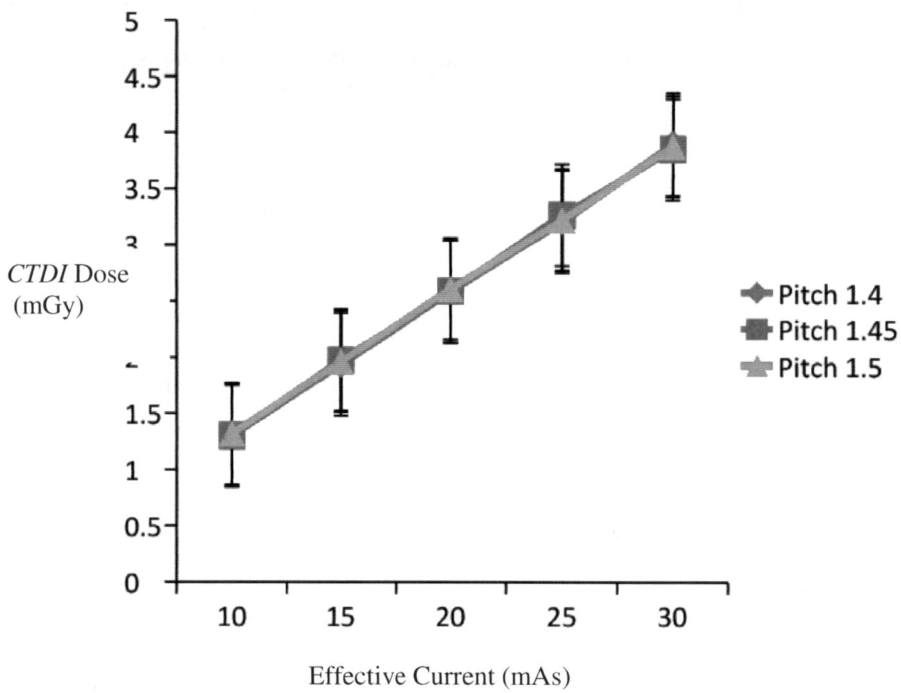

Figure 2. *CTDI* dose against mAs in 120 kV for manual setting.

A Mann Whitney U Test revealed no significant difference between Caredose 80kV (Md= 16.06,n=4) and manual setting at 80 kV (Md=0.6437, n=4), U=8 z= 0 p=1, r=0 (figure 3). This would be considered as large effect. This because the tube current used for caredose were 85 to 200, while tube current used for manual setting were 10-30 mAs. This shows that it was not suitable to choose lowest manual setting because the dose will absorbed into the pediatric patient's body. Whereas for 120 kV Caredose shows (Md= 16.06,n=4) and manual setting for 120 kV shows (Md=2.261, n=4), U=8 z= -2.309, p=0.21, r=-1.15, this would considered as small effect where the value of r shows less than 0.1 from Cohen (1988) criteria, while in between care dose energy 120 kV and 80 kV both have the same value of median while in between 120 kV and 80 kV both have different values of median 2.2610 and 0.6437. It has 1.61735 differences.

For a second finding, a Mann Whitney U test shows that pitch values of caredose 120 kV were; pitch 1.4 (Md= 16.06, n=4) , 1.45 (Md =16.03) and 1.5 Pitch (Md=16.01). This shows that using pitch 1.4, 1.45 or 1.50 still considered for 120 kV. Besides that, by using manual setting where the energy chooses 80 kV; pitch 1.4 (Md=2.261) 1.45 (Md=2.282) and 1.5 (Md=2.2865r). These shows that pitch 1.4 used to have a large difference compared to pitch 1.45 and 1.5.

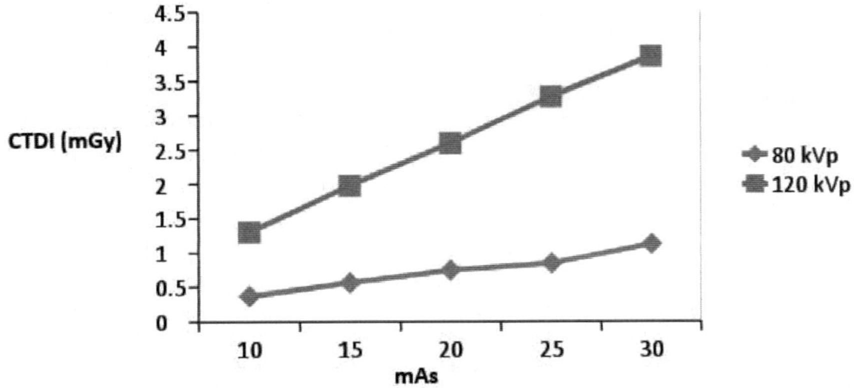

Figure 3. Pitch 1.45 used in different mAs for ionization dosimeter (mGy) at 80 kV and 120 kV.

4. Discussion

Choosing pitch value > 1 decreased patient dose and image quality. However, pitch value < 1 results in better image quality, but a higher patient dose. Using 1.5 pitch still considered either in manual setting or caredose compared to choosing 1.4 and 1.45 pitch values since the median values were small difference. However, choosing 80 kV with pitch 1.5 is better compared to choosing 120 kV for 1.5 pitch value for the pediatric patient. These because the values choice of pitch affect both image quality and patient dose. [12][11]. From this study it is shown that the radiation dose in pediatric head CT examination varies with tube voltage and other CT imaging parameters. The tube current plays an important role in determining the optimal radiation dose in pediatric head phantom.

5. Conclusion

This study shows that the radiation dose is reduced significantly with the use of low tube voltage in pediatric head CT imaging. The use of AEC mode offers the best service with lowest radiation exposure to pediatric patients.

6. References

[1] Leidholdt J, and Bone J 2002 Computed Tomography *The Essentials Physics of Medical Imaging* (2nd ed. Philadelphia,USA: Lippincott Williams & Wilkins).

[2] Bushberg J T, Seibert. J A, Leidholdt E M 2012 X-ray Dosimetry in Projection Imaging and Computed Tomography *The Essentials Physics of Medical Imaging* (3rd ed. Philadelphia,USA: Lippincott Williams&Wilkins).

[3] Funama Y, Awai K, Nakayama Y, Kakei K, Nagasue N, Shimamura M, Yamashita Y 2005 Radiation dose reduction without degradation of low-contrast detectability at abdominal multisection CT with a low-tube voltage technique: phantom study *Radiology* 237

[4] Goodenough D 2012 Catphan 500 and 600 manual, ed. Huda W, Scalzetti E M and Levin G (2000 Technique Factors and Image Quality Greenwich, NY: The Phantom Laboratory).

[5] Goldman L W 2007 Principles of CT: radiation dose and image quality *J. of Nuclear Medicine Technology* **35(4)** 213–225; quiz 226–228.

[6] Holmberg O, Malone J, Rehani M, McLean, D and Czarwinski R 2010 Current issues and actions in radiation protection of patients *European J. of Radiology* **76(1)** 15–19.

[7] Jost G, Lengsfeld P, Voth M, Schmid E and Pietsch H 2010 The influence of tube voltage and phantom size in computed tomography on the dose-response relationship of dicentrics in human blood samples *Phys. in Medicine and Biology* **55(11)** 3237–3248.
[8] Martin C J, Sutton D G and Sharp P F 1999 Balancing patient dose and image quality *Appl. Radiation and Isotopes* **50(1)** 1–19.
[9] Schindera S T, Nelson R C, Yoshizumi T, Toncheva G, Nguyen G, DeLong D M and Szucs-Farkas Z 2009 Effect of Automatic Tube Current Modulation on Radiation Dose and Image Quality for Low Tube Voltage Multidetector Row CT Angiography *Phantom Study* **16(8)** 997–1002 (. Academic Radiology).
[10] Tang K, Wang L, Li R, Lin J, Zheng X and Cao G 2012 Effect of low tube voltage on image quality, radiation dose, and low-contrast detectability at abdominal multidetector CT: Phantom study *J. of Biomedicine and Biotechnology*.
[11] Kalender W A 2014 Dose in x-ray computed tomography.
[12] Silverman P M, Kalender W A, Hazle J D 2001 Common terminology for single and multisliceheical CT *AJR Am J Roentgenol* **176 (5)** 1135-6 (Pubmed citation).
[13] Singh S., Kalra M K, Matthew M D, Gilman D, Jiang Hsieh M D, Pien H H, Digumarthy S R, Shepard J A O 2011 Adaptive statistical iterative reconstruction technique for radiation dose reduction in chest CT: A Pilot Study *Radiology* **259(2)** (radiology.rsna.org).

Acknowledgments
We would like to express our special thanks to radiographers and science officer in Imaging Unit at Advanced Medical and Dental Institute, USM for the technical assistance in this study. Thanks to UiTM for the opportunity and facilities provided for MSc study (Nik Norhasrina Nik Din). We also would like to acknowledge BESTARI research grant (600-IRMI/MyRA 5/3/BESTARI (030/2017)) that supports this study.

IOP Conf. Series: Materials Science and Engineering **298** (2018) 012047 doi:10.1088/1757-899X/298/1/012047

Spatial interpolation of gamma dose in radioactive waste storage facility

Nazran Harun[1], Muhammad Fathi Sujan[1] and Mohd Zaidi Ibrahim[1]

[1]Waste Technology Development Centre (WasTeC), Waste and Environmental Technology Division, Malaysian Nuclear Agency, Bangi, 43000 Kajang, Malaysia

nazranharun@nm.gov.my

Abstract. External radiation measurement for a radioactive waste storage facility in Malaysian Nuclear Agency is a part of Class G License requirement under Atomic Licensing Energy Board (AELB). The objectives of this paper are to obtain the distribution of radiation dose, create dose database and generate dose map in the storage facility. The radiation dose measurement is important to fulfil the radiation protection requirement to ensure the safety of the workers. There are 118 sampling points that had been recorded in the storage facility. The highest and lowest reading for external radiation recorded is 651 microSv/hr and 0.648 microSv/hour respectively. The calculated annual dose shows the highest and lowest reading is 1302 mSv/year and 1.3 mSv/year while the highest and lowest effective dose reading is 260.4 mSv/year and 0.26 mSv/year. The result shows that the ALARA concept along time, distance and shield principles shall be adopted to ensure the dose for the workers is kept below the dose limit regulated by AELB which is 20 mSv/year for radiation workers. This study is important for the improvement of planning and the development of shielding design for the facility.

1. Introduction The first section in your paper

Radioactive waste storage facility is a controlled working area and the only radioactive waste storage facility in Malaysia under management of Waste Technology Development Centre Nuclear Malaysia. Low-level waste comes from reactor operations, medical, academic, industrial, and other commercial uses of radioactive materials [1]. The storage of radioactive waste must is provided in order to ensure that both human health and the environment will be protected, both now and in the future, without imposing undue burdens on future generations. Radiation monitoring should be conducted routinely to determine the external radiation levels and surface contamination levels inside the waste storage facility, along the boundaries of the storage facility and on the surface of waste packages [2]. It is compulsory to comply and fulfil the dose rate at the storage facility in order to apply Class G license (Storing waste, dispose of, dissolve milling installations) (LEM-SS-14 Class G) for new, amendment and renew license applications [3]. Rigorous and continuous radiation monitoring is important to ensure that environmental radiation remains within internationally agreed and legal safety limits. This monitoring will ensured that the organisation involved quickly aware of any change in environmental radiation and able to provide any health warnings and protection necessary advice to the public [4]. Dose limit criteria are divided into 2 categories which dose limit for public is 1mSv/year and radiation workers is 20mSv/year , where the maximum effective dose on the worker averaged over a period of five consecutive years shall not exceed 20 mSv [5].

Interpolation is a statistical technique potentially capable of generating the intermediate unknown points of independent variables for spatial data.[6] Inverse distance weighting estimates the variable of interest by assigning more weight to closer points. It is a simple technique that does not require prior information to be applied to spatial prediction. [7] IDW's general idea is based on the assumption that the attribute value of an unsampled point is the weighted average of known values within the neighborhood, and the weights are inversely related to the distances between the prediction location and the sampled locations.[8]

This study aims to measure external radiation doses received by the workers in radioactive waste storage facility. Spatial documentation of the dose rate is essential as a baseline. It is also important for radiation safety and environmental radiological monitoring programme. The information is important to assure workers wear proper personnel protective equipment in the facility. This study was conducted to measure external radiation data in the radioactive waste storage facility, establish local dosimetry database and to generate local dosimetry map. In addition, the result of the study can be used in planning and developing shielding and as additional supporting reference for storage facility licensing.

2. Material and method

The gamma radiation dose was measured using Ludlum Model 2241-3 and a probe 44-2 which is a 2.5 cm x 2.5 cm scintillation crystal. The scale of survey meter is 0.001 microSievert per hour to 9999 Sv per hour where it is suitable to measure low, intermediate and high radiation [9]. The survey meter has been calibrated at the Secondary Standard Dosimeter Laboratory at Nuclear Malaysia. Measurements were conducted at 1 m above the flat land surface [10, 11, 12].

Annual dose data obtained when every dose data were calculated using the following equation:

$$\text{Annual dose (mSv/a)} = \text{Dose } (\mu\text{Sv/hr}) \text{ x working hours/year} \tag{1}$$

where the annual doses for a working 2,000 hours in a year in such areas [13]. Effective dose values at indoor were determined by using annual dose times the indoor occupancy fraction where indoor occupancy fraction is 0.8[13]. The value of effective dose mSv/year were calculated using the following equation [13][14]:

$$\text{Effective dose (mSv/year)} = \text{Annual dose (mSv/year)} \times 0.80 \tag{2}$$

There were 118 sampling points that had been recorded in storage facility. The survey points were transferred to a basemap to create geodatabase. Site measurement provided an hourly dose rate value only at the sampling points. The map was plotted according to the following steps;

- The spatial information was created in the digital maps and the dose rate, annual rate and effective rates were recorded in the attributes.
- The IDW method, with dose rate data as the variable, was applied to the digitized map for projection of the contour based on interpolation.
- The value range were reclassified according to the following intervals;
  a. Hourly dose rate: below than 1, 5, 10, 15, 20, 50 and more than 100 microSv/hr.
  b. Effective dose and annual dose rate: below than 2, 5, 10, 15, 20, 50, 100 and more than 200 miliSv/year

IOP Conf. Series: Materials Science and Engineering **298** (2018) 012047 doi:10.1088/1757-899X/298/1/012047

3. Result and discussion

This study covered an area approximately 720 m^2 in the storage facility. Each sampling point reading showed different dose rate. Several part of the facility cannot be access due to the radioactive drum and device. Based on figure 1, the arrangement of the Disused Sealed Radioactive Sources (DSRS) and solid waste showed comprehensible visual dose on the map. Sealed radioactive sources (SRSs) were used extensively in agriculture, industry, medicine and various research areas. SRS that is no longer in used is termed as DSRS. DSRS located at the centre to the south of the building. The highest dose rate was located at south of the block which contained radium needle in the concrete shielded drum while the lowest dose rate was located to north of the block which contained radioactive solid waste. Mostly, solid wastes were below than 6 μSv/hr.

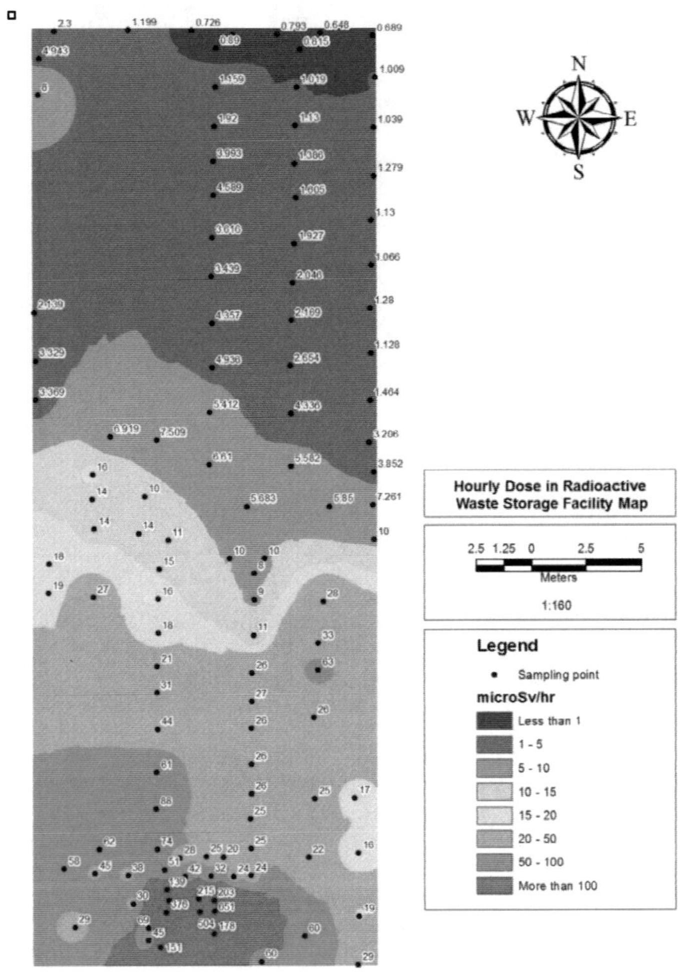

Figure 1. Hourly dose in radioactive waste storage facility.

Figure 2 and figure. 3 show the high dose value at the DSRS storage area; the workers shall received high dose if they work in long time at the specific area. Effective dose (mSv/year) is a calculation that considers the occupancy fraction which is 0.8. Effective dose showed wider area for dose ranging from15 to 50 mSv/year. Therefore,it is important to make the workers aware of the importnace to use approriate personnel protective equipment and personnel monitoring in the facility. The workers shall schedule and limit working time in the facility to reduce exposure to the external radiation dose.

The outcome of this study shows that distribution of high and low dose rate of radioactive waste in Block 33. The map is important as a visual guide for planning and development programme for personnel working in the controlled area. Planning using ALARA concept and safety principle is crucial in an area with radiation sources. Due to high dose in the working area, every personnel who has been given access to the controlled area shall comply with prevailing instructions applicable to the area issued. All workers must be provided with suitable and adequate personal protective equipment and personnel monitoring where the doses received during normal, operation, accidental exposure and emergency exposure recorded [7].

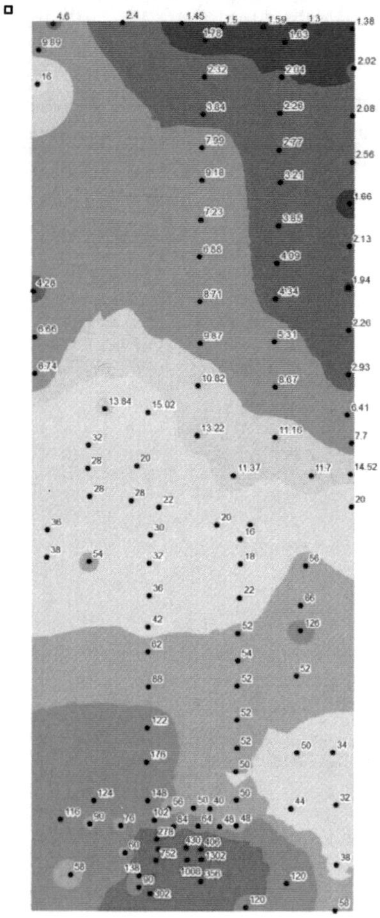

Figure 2. Annual dose in radioactive waste storage facility.

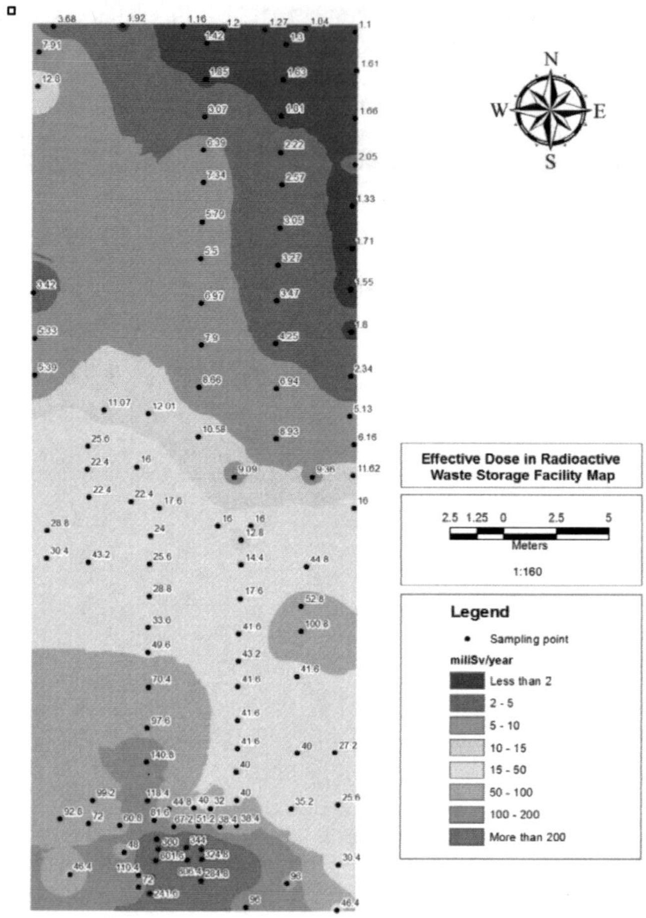

Figure 3. Effective dose in radioactive waste storage facility.

4. Conclusion

This study shows level and distribution of external radiation for public and workers concerned, in controlled working area such as the radioactive waste storage facility in Nuclear Malaysia. The annual and effective dose shows that workers shall adopt ALARA along time, distance and shield principles to that ensure the dose received by the workers is kept below dose limit regulated by AELB, which is 20 mSv/year for radiation workers. Ludlum survey meter can be used to measure external radiation level for monitoring workplace and environmental radiation. This study is important as a baseline for external radiation data, process and method to monitor environmental radiological effect in a controlled area in Nuclear Malaysia. It is also fulfils and comply the environmental radiological monitoring program for licensing purposes.

5. References

[1] US NRC, 2015 Backgrounder on radioactive waste (https://www.nrc.gov/reading-rm/doc-collections/fact-sheets/radwaste.html).

[2] IAEA 2006 Storage of radioactive wast, *IAEA Safety Standards Series No. WS-G-6.1* (IAEA, Austria) pp 19.

[3] AELB 2016 Senarai semak - kelas G: Menstor sisa, melupus, membubar pepasangan pengilangan LEM-SS-14 kelas G (http://portal.aelb.gov.my/sites/aelb/SiteAssets/elesen/Senarai%20Semak%20-%20Kelas%20G.pdf 07 June 2017).

[4] Environmental Protection Agency (EPA) 2017 (http://www.epa.ie/radiation/monassess/ radmonitor/ 07 June 2017).

[5] Atomic Energy Licensing Act 1984 (Atomic Energy Licensing: Basic Safety Radiation Protection) Regulations 2010).

[6] Ashraf I, Hur S and Park Y 2017 An Investigation of interpolation techniques to generate 2D intensity image from LIDAR data *Open Access J.* (IEEE Access; http://ieeexplore.ieee.org/stamp/stamp.jsp?arnumber=7915679 18 July 2017).

[7] Babak O and Deutsch C V 2008 Statistical approach to inverse distance interpolation (http://www.ccgalberta.com/ccgresources/report10/2008-131_statistical _inverse_distance.pdf 14 Feb 2017).

[8] Lu G Y and David W W 2008 An adaptive inverse-distance weighting spatial interpolation technique *Earth Systems and GeoInformation Sciences* **34(9)** 1044-1055.

[9] Ludlum 2017 General purpose ratemeter (http://ludlums.com/component/virtuemart/area-monitoring-5/handheld-instruments-29/general-purpose-meters-30/ratemeter-scaler-digital-34/general-purpose-ratemeter-scaler-84-detail?keyword=2241-3&Itemid=2657 07 June 2017).

[10] UNSCEAR 2000 Sources and effects of ionizing radiation (United Nations Publication, New York).

[11] Sanusi M S M, Ramli A T, Gabdo H T, Garba NN, Heryanshah A, Wagiran H, Said M N 2014 Isodose mapping of terrestrial gamma radiation dose rate of Selangor state, Kuala Lumpur and Putrajaya, Malaysia *J. of Environmental Radioactivity* **135** 67-74.

[12] IAEA 2003Guidelines for radioelement mapping using gamma ray spectrometry data *IAEA-Techdoc-1363* (IAEA, Austria)

[13] UNSCEAR 2010 Sources and effects of ionizing radiation Annex B: Exposures of the public and workers from various sources of radiation (United Nation, New York) **I** 226.

[14] Zahidi N K, Jalal N A, Kadir A B A, Abdul Razalim F A and Ibrahim S 2016 Environmental Dose Mapping at Raymintex Plant Area Using 3 Types of OSL Dosimeter *Nuclear Malaysia R&D Seminar*. (Malaysian Nuclear Agency)

IOP Conf. Series: Materials Science and Engineering **298** (2018) 012048 doi:10.1088/1757-899X/298/1/012048

Development of the automated bunker door by using a microcontroller-system

M A Ahmad[1,a), K W Leo[1], G H P Mohamad[2], A Ahmad[1], S A Hashim[1], R M Chulan[1] and A H Baijan[1]

[1]Accelerator Development Centre, Technical Support Division, Malaysian Nuclear Agency, Bangi, 43000 Kajang, Selangor, Malaysia.
[2]Instrumentation and Automation Center, Technical Support Division, Malaysian Nuclear Agency, Bangi, 43000 Kajang, Selangor, Malaysia.

[a)]mohdazhar@nm.gov.my

Abstract. The new low energy electron beam accelerator bunker was designed and built locally to allocate a 500 keV electron beam accelerator at Block 43T in Malaysian Nuclear Agency. This bunker is equipped with a locally made radiation shielding door of 10 tons. Originally, this door is moving manually by a wheel and fitted with a gear system. However, it is still heavy and need longer time to operate it manually. To overcome those issues, a new automated control system has been designed and developed. In this paper, the complete steps and design of automated control system based on the microcontroller (PIC16F84A) is described.

1. Introduction

A bunker was completely constructed in 43T building in Malaysian Nuclear Agency. The bunker was built for the use of electron beam (EB) accelerator which has a capacity of 500 keV. This bunker is equipped with plug-in type door that acts as a shielding and it has an estimated weight of 10 tons.

The weakness of the existing system of the bunker door is the operations for opening and closing the door which are fully manually operated. Consequently, more time is needed to open and close the door which is not effective given the limited manpower available. For that reason, the bunker door operating system is upgraded so that it is more effective and utilises minimal manpower. To overcome and solve these problems, a new system has been developed to upgrade from manually operated to automated motor controller system for the bunker door. The system is built based on using microcontroller PIC16F84A.

2. Bunker design

The building that placed the EB accelerator and its auxiliaries can be divided into three areas. The areas are working area, ventilation room and irradiation room (bunker). The working area includes control panel monitoring and the open spaces inside the building. These working area is classified as clean area in terms of radiation safety. The ventilation room is where the blower and the exhaust motor are placed.

The bunker was designed to prevent radiation from entering working areas. It housed EB accelerator, high voltage transformers and also ventilation system. This bunker was constructed using high density concrete, with thickness of 85 cm and 500 cm high. During operation, the bunker door must be secured and interlocked with the EB accelerator. Figure 1 shows the layout of EB Bunker.

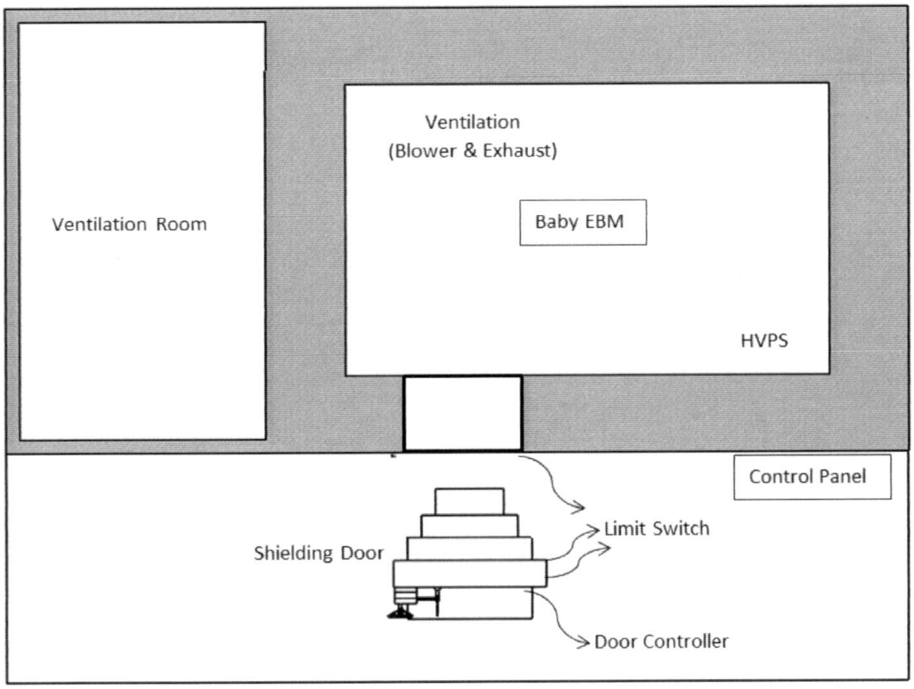

Figure 1. Layout of EB Bunker.

3. Microcontroller PIC16F84A

Microcontroller (μc) refers to computer chips that have been created to control electronic devices. It is one of the branch micro-conscious processing operations that is easy and with low cost-effective. The microcontroller generally contains all the memory and interface I/O for simple application.

The flash type memory of PIC16F84 program can theoretically be removed and re-programmed to up to 10 thousand times. This makes PIC16F84A suitable for program development, learning and application in life. In comparison, programmable logic controller (PLC) is usually used for wide range of tasks and more suited for industrial applications due to its various number of I/O. In addition, PLC is more expensive because it uses many chips. In this simple controlling project, it is sufficient to use microcontroller PIC16F84A.

Mapping pins for PIC microcontroller PIC16F84A is shown in figure 2. Further information regarding PIC16F84A may be referred to the data sheet MicrochipPIC16f84A. For the programming of this controller system, MicroBasic Pro v1.45 compiler software is used.

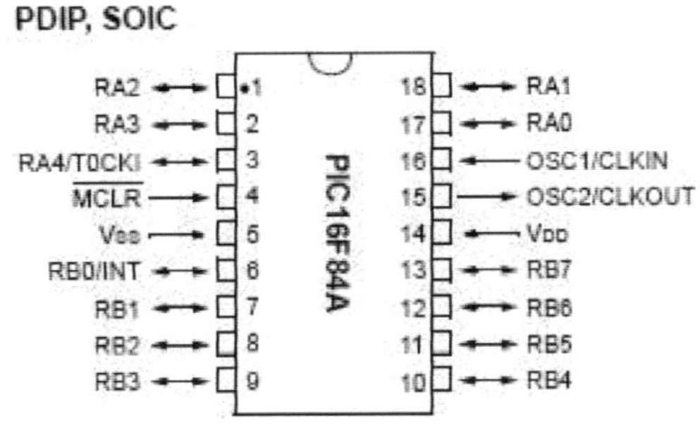

Figure 2. Mapping pins PIC16F84A.

4. Design Concept (Logic Flowchart)

Figure 3 represents the flowchart or process of the motorised bunker system, which shows the idea on how to operate the bunker door. This diagram represents overall movement, decision and selection of this automated door operation from start until end. Therefore, the programming development on this controller system should follow the logic of this flowchart.

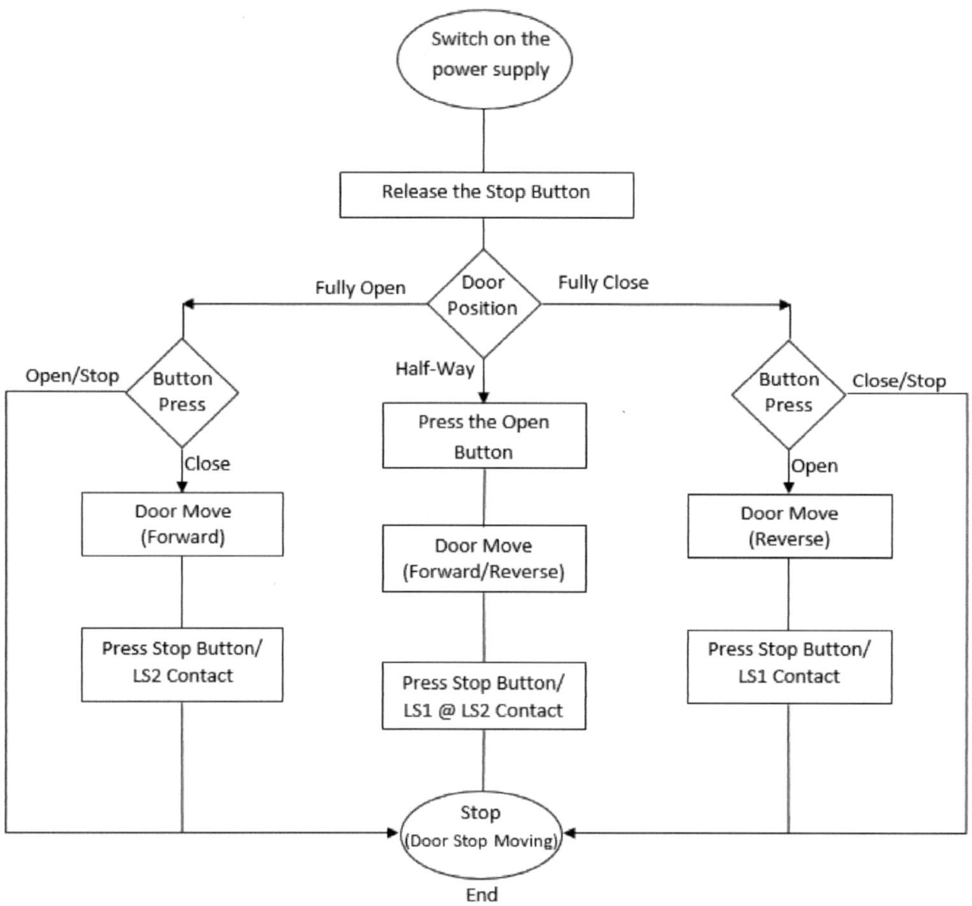

Figure 3. Flowchart of the motorized bunker system.

5. Mechanism (Block diagram)

The block diagram of automated bunker door control system is shown in figure 4. This control system is programmed into a PIC micro-controller with five input data: forward button, reverse button, stop button, upper limit and lower limit instruction.

This input data is connected to the PIC Microcontrolle,r programmed and link with two data output for the movement of reverse or forward of the bunker door. This information is communicated with the high torque reversible motor and it is connected to the pulley system. The pulley is attached to the existing steering wheel to move the door of the bunker. The type of reversible motor which will be used in this system is induction motor with capacity of 1 HP, 5.7 Ampere, 4 poles and 1430 r/min. The completed control board is shown in figure 5 and, the completed motorized automated bunker door is shown in figure 6.

IOP Conf. Series: Materials Science and Engineering **298** (2018) 012048 doi:10.1088/1757-899X/298/1/012048

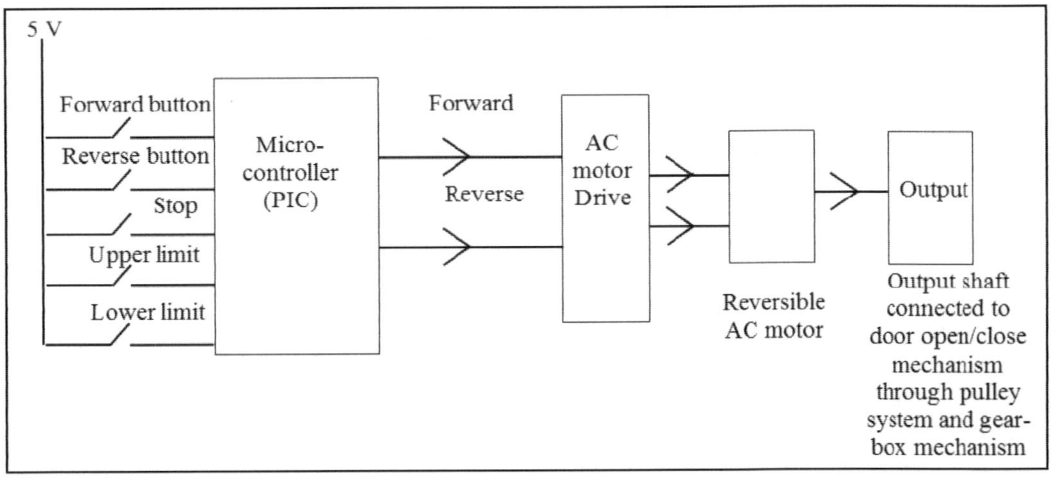

Figure 4. Block diagram.

Figure 5. Controller Board.

Figure 6. Automated bunker door.

6. Summary

The system has been developed, implemented and tested for its functionality. The result showed that the automated bunker door operated as required together with it safety implementation systems. In term of time and performance, by using this new automatic system, it only took less than one minute to open or close the door, as compared with previous manual system which took more than five minutes. This project also could be considered as cost effective because the whole controller system was locally designed and developed, including the programming and installation. The importance of the development of this automated bunker door is to increase effectiveness of time and work job so that utilization of manpower could be minimized.

7. References

[1] Ghazali A B, Mamat M R, Darmawan R, Chulan R M, Lee C. H, Leo K W and Taat M Z 2005 Development of electron beam machine at Nuclear Malaysia (Tech. and Appl. Accelerator Meeting VIII P3TM-BATAN Yogyakarta)

[2] Ahmad M A, Hashim S A, Ahmad A, Leo K W, Chulan R M, Dalim Y, Baijan A H, Zain M F and Ros R C 2017 The implementation of physical safety system in bunker of the electron beam accelerator *AIP Conf. Proc.* **1799** pp 030012

[3] Hashim S A, Jahar S, Muhamad A and Idris S 2011 Safety aspects of EPS-3000 electron beam machine (Nuclear Technical Convention Malaysia)

Acknowledgments

The authors would like to gratefully acknowledge to Dr. Glam Hadzir from Instrumentation and Automation Centre (PIA) and all staff in Accelerator Development Centre (ADC) of Nuclear Malaysia who give guidance and knowledge that contributed for completing this project.

IOP Conf. Series: Materials Science and Engineering **298** (2018) 012049 doi:10.1088/1757-899X/298/1/012049

Public acceptance of nuclear power among Malaysian students

Anas Muhamad Pauzi[1,a)], Juniza Md Saad[1], Asyraf Arif Abu Bakar[1], Abdul Hannan Damahuri[1], Nur Syamim Mohd Syukri[1]

Nuclear Engineering & Energy Group, Institute of Sustainable Energy, Universiti Tenaga Nasional, Jalan IKRAM-UNITEN, 43000 Kajang, Selangor, Malaysia

[a]anas@uniten.edu.my

Abstract. Malaysian government's aim to include nuclear energy for electricity generation has triggered various reactions from all especially the public. The objective of this study is to have a better understanding on the knowledge, sources of information of nuclear power and sources of energy chosen by Malaysian in 20 years' time. Besides that, we want to examine the level of acceptance and perception of Malaysian towards nuclear energy and we want to identify the correlation between public perceptions with the acceptance towards nuclear power in Malaysia, and also to study the differences between perception and acceptance of nuclear power with gender and educational level. For this research methodology, the research questions are given orally or through paper-pencil and also social networking site such as Facebook or through electronic media application such as WhatsApp and Google docs. The data were analysed using a SPSS version 22.0 (Statistical Package for the Social Sciences). Results showed that more than 50% of the respondents have the knowledge of nuclear energy. A part of from that, only 39 % are confident government can afford to build NPP in Malaysia and 41 % disagree nuclear energy is the best option for future energy. From analysis using SPSS 22 we estimate negative perception will give a negative acceptance in term of support towards the use of nuclear energy in power generation in Malaysia. There are also slight correlation that the higher the level of education of Malaysian, the more negative the perception of Malaysian in accepting nuclear energy as source of power in Malaysia. Therefore in shaping a positive acceptance of NPP in Malaysia, the authorities need to educate the people with the knowledge of nuclear in order to overcome the negative perception towards nuclear power.

1. Introduction

Nuclear energy have become more popular for certain country in order to diverse its energy production. Besides, nuclear energy can be used to overcome the rising pollutions that causes by releasing carbon dioxide from the burning fossil fuel to generate electricity. Unfortunately, apart from being safe, clean and efficient source of energy, nuclear energy also been bedeviled as an extremely dangerous and highly polluted energy source among general public [1].

H. M. Visschers and L. Wallquist [2] and also the OECD [3] state that public attitudes had constantly molded the arrangement of nuclear power since the late 1960s. However, after a few nuclear accidents and growing environmental concerns regarding the safe storage of radioactive waste, these events had gradually reduced acceptance levels of nuclear power in the western world up to the turn of the century. In the beginning of 21st century climate change had entered nuclear debate and gave an environmental advantage for nuclear technology as the acceptance level in the US and Europe showed a rise in opinion polls. Unfortunately after nuclear accident in Fukushima, general public began to be hesitant of the nuclear power. The same thing happens in Switzerland as the public are showing fewer acceptances directly after Fukushima accident.

C. Sun and X. Zhu [4] discussed that even though it is clear that nuclear power offered more benefits, but due to uncertainty of nuclear energy and "not in my backyard"(NIMBY) perception, local citizen in China resist the nuclear power program in their neighborhood which is being indicated by anti- nuclear movement in Jiangmen in 2013. As public attitudes towards nuclear energy keeps changing over time, Y. Kim et al. [5] found that limitation of previous studies in taking perspective on people attitudes also contributes into this consideration. Previous studies limited their consideration into accept or reject, without taking reluctantly accept, which means acceptance of the use of nuclear energy without a friendly attitude towards it because of a high level of dependence on nuclear energy, and a lack of alternative energy sources within that country as part of the consideration even though the proportion of those who "reluctantly accept" is fairly large.

G. He et al. [5] in public participation and trust in nuclear power development in China observed that Chinese citizen still have high expectant in governmental authorities in term of nuclear information assistance, quick response to nuclear crisis and country's commitment towards nuclear future even after several nuclear accidents such as Three Miles Island (1979), Chernobyl (1986) and recently Fukushima (2013) accident in Japan.

The purpose of public acceptance research is to alleviate the potential conflict between technological development and social development by investigating characteristic of public awareness. At present, the methodology of spreading assumption has been altogether different from that past in China. People can get to the assortment of against nuclear energy through the Internet, TV show and other media. Execution of an arrangement of the framework, for example, hearings and open enactment makes people in general take part in social issues habitually, and the administration will broadly counselled and receive the general supposition amid the choice making methodology [7].

Public acceptance give a great impact on the nuclear development policy, technology, economy and other issues related to nuclear energy. Government have to consider this issue in decision making as if there is a controversy even a conflict on nuclear power among the people, it will create the tremendous social cost. This can be seen in Germany when the government announced "abandon nuclear program" after Fukushima accident, this has caused a huge economic losses to a country [7,8].

1.1. Nuclear power knowledge

Basically, knowledge of nuclear power suggest on how much people actually know about nuclear power, nuclear technologies, and its operation or inspection of nuclear facilities. With high level of knowledge about nuclear power, people are enabled to value the advantages and consequences of nuclear energy instead of misinterpreting the risks and being anxiety of nuclear power generation. Besides, individual willingness to gain knowledge and a certain educational level as well as government policy factor in such that there should be more opportunities for accessing knowledge and learning about nuclear power and technology must be taken into consideration in increasing public knowledge about nuclear power [6].

Even though it is clear that knowledge of nuclear power is necessary in preparing a country in implementing nuclear power plant, G. He et al. [7] examined that generally Chinese citizen is ineffectively educated about nuclear security and risk management of nuclear power since the first nuclear power plant was connected to the Chinese power grid in 1991 and governmental authorities seldom educate and exposed any nuclear incidents or accidents to the publics.

Therefore in the process of developing nuclear power, knowledge of nuclear power and technologies should be continuously publicize to the public and also should be able to accomplish people requirement in nuclear information. As a result, it could enhance public confidence about nuclear energy and technology, thus enlarge the public acceptance of nuclear power in some extent by improving the safety of nuclear power plant and establishing the quantitative safety goals suitably [7]

1.2. Nuclear power development status in Malaysia

Since 2009, a Nuclear Power Development Steering Committee driven by *Kementerian Tenaga, Teknologi Hijau dan Air* (KeTTHA) has been conducting various studies towards preparing a Nuclear Power Infrastructure Development Plan (NPIDP) [8]. During that particular period, Malaysia was determined to be very enthusiastic towards developing nuclear power generation as deliberated in the Economic Transformation Programme (ETP), Entry Point Project (EPP) 11: Deploying Nuclear Energy for Power Generation. Hence, Malaysia Nuclear Power Cooperation (MNPC) was established in January 2011 as Nuclear Energy Programme Implementing Organization (NEPIO) in Malaysia, as recommended by International Atomic Energy Agency (IAEA). Huge mission was assigned to the MNPC to establish comprehensive groundworks for a successful, sustainable, safe, secure and peaceful national nuclear power programme within time, on budget and in a transparent manner. Figure 1 shows the summary of Government Decisions at 2012.

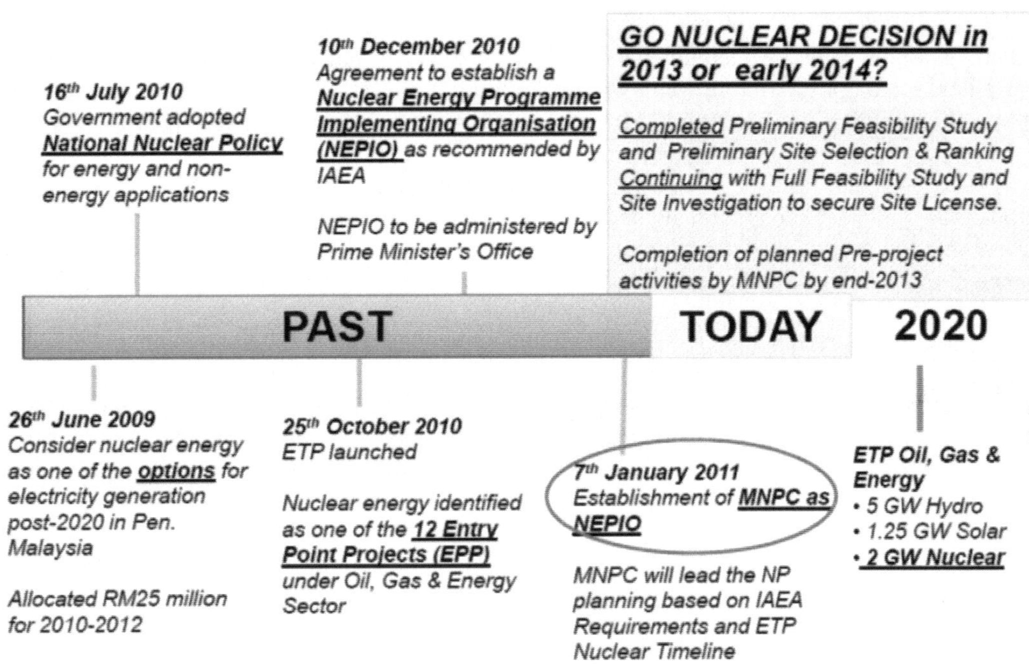

Figure 1. Summary of Government Decisions [9].

However, in March 2017, the Minister in the Prime Minister's Department, Dato Sri Nancy Shukri said that Malaysia is ready for nuclear power and it could have its own nuclear power plant by 2030 by focusing the development in the Peninsula Malaysia due to its high power consumption. Despite of numerous difficulties faced by the government as well as MNPC to make a decision on the timing to embark on the nuclear power project, the primary challenge is still to convince the public on safety, viability and cost effective of nuclear power for the country. Public perceptions towards necessity of having nuclear power is highly necessary in order for government's political decision making and to identify the basic and ground issues that affect the community or public [10].

Recently, the said government endeavor had been critically debated among Malaysian people as there are pros and cons with regards to this development. Each one of them has concrete proof for each of statement and everyone contributes their thoughts and finding for the better understanding.

Therefore, this study was conducted to evaluate on the reaction and the readiness of Malaysian students should a nuclear power plant is being considered in the near future.

2. Methodology

The structure of questionnaire is divided into six categories based on the current influential areas related to nuclear energy raised by Malaysian. The categories were as follows: 1. Demographic information; 2. Knowledge of nuclear power; 3. Sources of energy that would be used in Malaysia in 20 years' time; 4. Nuclear power information sources; 5. Perception of nuclear power; 6. Acceptance of nuclear power. The targeted respondents were chosen based on their education level. The questionnaires were distributed to four (4) different groups; secondary school students, foundation/diploma students, degree students and postgraduate students. Data were collected from Sekolah Menengah Kubang Bemban in Kelantan, Universiti Kebangsaan Malaysia (UKM) and Universiti Tenaga Nasional (UNITEN), both in Selangor. A total of 280 random respondents participated in this study. Social networking media were also being used in distributing the questionnaires to the respondents. The questionnaires were uploaded in Google docs form and the link was circulated in Facebook and WhatsApp.

The data collected was simulated using SPSS Statistic version 22.0 software. The data were then analysed in three (3) different methods based on the objective of this study. Pearson Correlation test is used to analyze the correlation between public perception and acceptance towards nuclear power in Malaysia. This method is suitable to be used for examining the relationship between two variables in linear fashion, which positive value indicates the constructive relationship and vice versa. In addition, one-way Analysis of Variance (ANOVA) test and Turkey's Post Hoc analysis were utilized to determine the difference between perception and acceptance of nuclear power with regards to education level proportions. For the final objective, the three test (T-test) method is used to study the difference between perception and acceptance of nuclear power with regards to gender proportions.

3. Results and discussion

Figure 2 shows the demographic breakdown according to gender, age group and education level for the 280 respondents participated in the study. Data were collected at a school in Kelantan which is Sekolah Menengah Kubang Bemban and two educational institutions in Selangor, Universiti Kebangsaan Malaysia (UKM) and Universiti Tenaga Nasional (UNITEN). The data were well distributed and would be used to determine the polarity of perceptions.

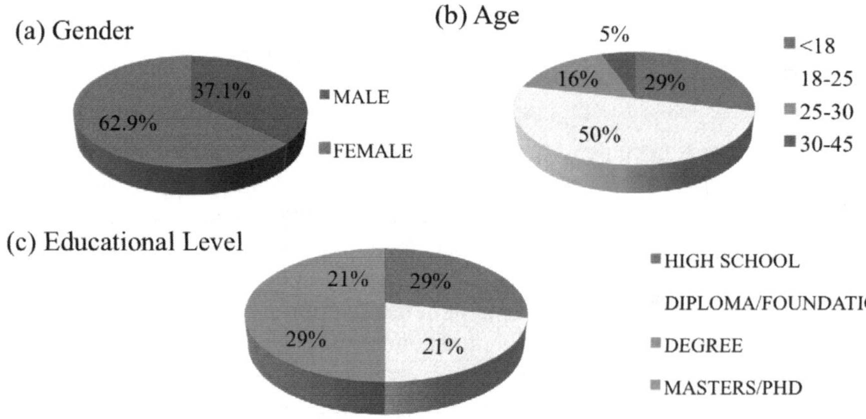

Figure 2. Demographic information (a) Gender, (b) Age, and (c) Education level.

IOP Conf. Series: Materials Science and Engineering **298** (2018) 012049 doi:10.1088/1757-899X/298/1/012049

Figure 3 provides understanding on the knowledge of the respondent about nuclear power. Basically, the majority of the respondents have minimal knowledge about nuclear energy. The 5% respondents, primarily derived from the high school respondents, are not even aware about the trefoil warning symbol that symbolized ionizing radiation which demonstrates a very low level of knowledge in relation to nuclear.

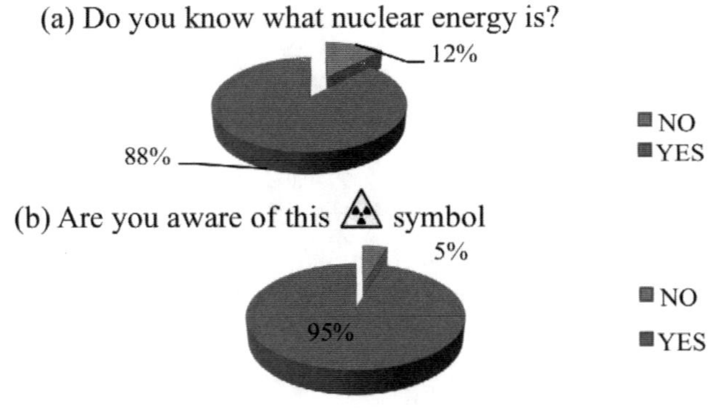

Figure 3. Knowledge on NPP

Figure 4 shows the result of the statistical test using SPSS regarding perceptions toward nuclear power from survey questions regarding their perceptions towards nuclear energy by referring to past nuclear accidents. The result shows that majority of the respondents 42.9% and 50.7% have medium high and high negative perception towards Malaysia building its first nuclear power plant and only 1.1% shows positive perception. Meanwhile, Figure 5 shows slightly different level of acceptance where 45.7% of respondents have medium low acceptance, while only 8.6% support Malaysia to build its first nuclear power plant (NPP).

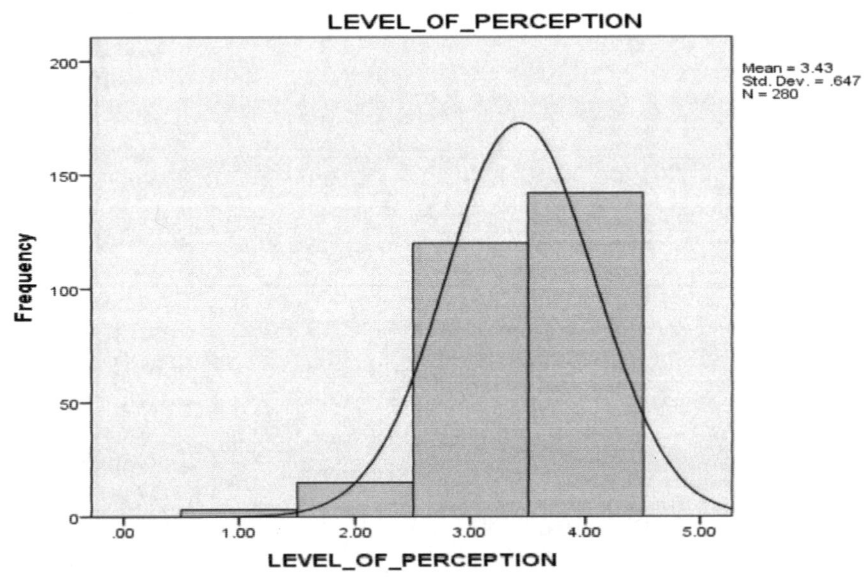

Figure 4. SPSS Statistical test on Negative perception toward nuclear power plant (NPP)

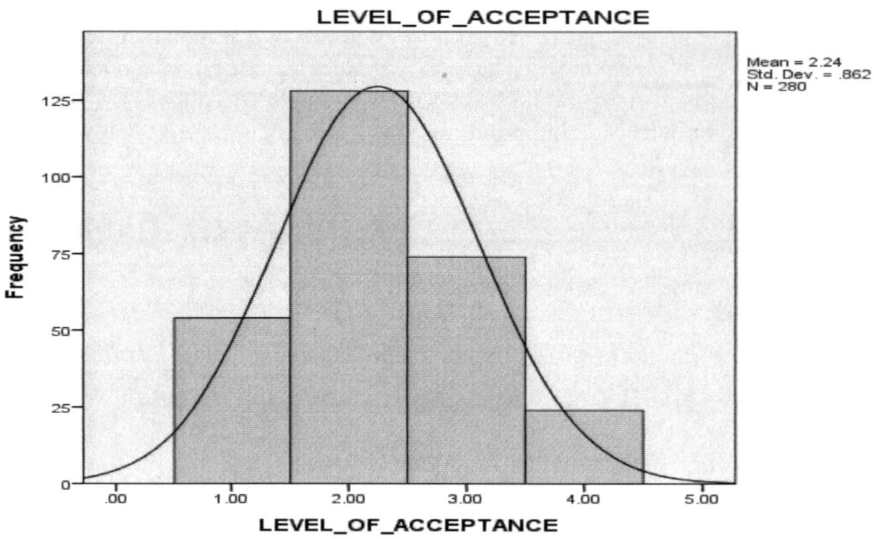

Figure 5. SPSS Statistical test on Acceptance toward nuclear power plant (NPP)

Comparing to other conducted surveys reported by H. Hassan [10] as depicted in Table 1 and Figure 6 shows that the nuclear accident in Fukushima in 2011 had significantly influenced public opinions for Malaysia to build its first nuclear power plant (NPP). The acceptance percentage was dropped sharply by 68 % in 2011 to 17%. It can be suggested that lack of information about nuclear power and its impact towards the country influence public opinion about the use of nuclear power plant in Malaysia.

Table 1. Comparison of Acceptance towards NPP in Malaysia between 2009-2012 [10] and 2017.

Title	Year	Institution	# Respondents	Result
The Level of Public Acceptance of Nuclear Power as a Source of Electric Power Generation in Malaysia	2008-2009	UKM	504	i. Agree: 74% ii. Disagree: 22% iii. Neutral: 4%
Siri Jelajah Malaysia – Program Kesedaran dan Penerimaan "I Love Nuclear"	2010	NUKLEAR MALAYSIA	4000	i. Support: 85% ii. Reject: 7% iii. Neutral: 8%
The SERIKANDI Plan	2011	UNITEN	42	i. A good alternative: 17% ii. Will consider: 64% iii. Bad idea: 19%
MANUPA	2012	UKM	565	i. Support: 47% ii. Not sure: 7% iii. Not ready: 17% iv. Does not support: 29%
Public Perception of Nuclear Power among Malaysian Student	2017	UNITEN	280	i. High acceptance: 8.6% ii. Medium acceptance: 72.1% iv. Low acceptance: 19.3%

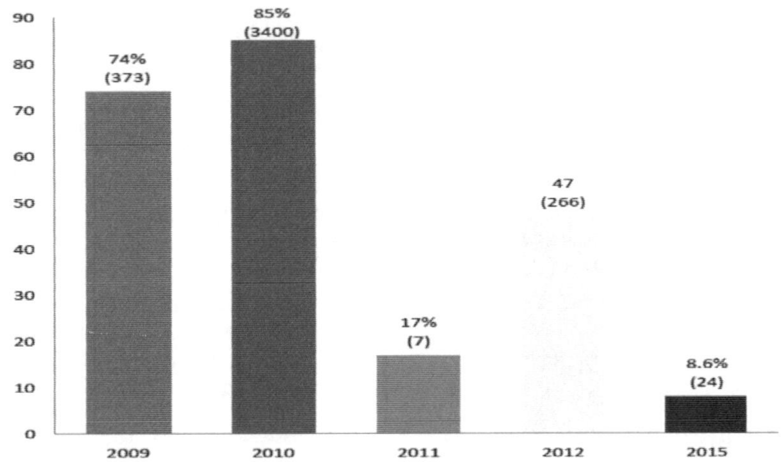

Figure 6. Public acceptance towards NPP between 2009-2012 [11] and 2017.

T-test was conducted to study correlation between gender and NPP acceptance. The result shows no significant difference between perception and gender of respondent. Meanwhile, study regarding correlation between education level and NPP acceptance using Analysis of Variance (ANOVA) test shows that there are slight differences between perception and educational level [$F(3, 276) = 3.329$, $p < 0.05$] but no significant difference between acceptance and educational level as [$F(3, 276) = 1.680$, $p= 0.171$, $p > 0.05$]. Next, a Turkey's Post Hoc test revealed that significant differences could only be seen in the perception of high school and degree students, while postgraduate shows no correlation.

4. Conclusion

Based on this study, only two groups which are high school and degree students make a difference in public perception of nuclear power. Degree students have more negative perception towards nuclear power compared to high school students. The Analysis of Variance (ANOVA) test conducted shows slight trend where the higher educational level, the more likely they have negative perception towards the idea of implementation of nuclear power plant.

The study has open-up an opportunity to those who concerned to know the current status and also serves as an update of the level of public perception and acceptance of nuclear technology in Malaysia.

5. References

[1] Tick H O, Shen Y P, Shing C C 2010 Energy policy and alternative energy in Malaysia: Issues and challenges for sustainable growth *Renewable and Sustainable Energy Reviews* **14** pp. 1241–1252.

[2] Visschers H M, Wallquist L 2013 Nuclear power before and after Fukushima: The relations between acceptance, ambivalence and knowledge *J. of Environmental Psychology* **36** pp. 77-86.

[3] OECD 2013 Public Attitutes Towards Nuclear Power (NEA No. 6859, Accessed 19th Dec 2017, https://www.oecd-nea.org/ndd/pubs/2010/6859-public-attitudes.pdf).

[4] Sun C, Zhu X 2014 Evaluating the public perceptions of nuclear power in China: Evidence from a contingent valuation survey *EnergyPolicy* **69** pp. 397–405.

[5] Kim Y, Kim W, Kim M 2014 An international comparative analysis of public acceptance of nuclear energy*Energy Policy* **66** pp. 475–483.

[6] He G, Mol A P J, Zhang L, Lo Y Public participation and trust in nuclear power development in China *Renewable and Sustainable Energy Reviews* **23** pp. 1–11, 2013.

[7] Li C J, Zhang C M, Chan Y, Zuo J, Chen J 2013 The Study on Safety Goals and Public

Acceptance of Nuclear Power *Energy Procedia* **39**, pp. 415 – 422,
[8] Kidd S 2013Nuclear power Economics and public acceptance *Energy Strategy Reviews* **1** pp. 277-281
[9] Malaysia Nuclear Power Corporation 2014 Nuclear Infrastructure and Instituitional Arrangements for the Development and Deployment of Sustainable Nuclear Energy Systems – Experience of MNPC, as a Dedicated NEPIO (INPRO Dialogue Forum on Int. Collaboration on Innovations to Support Globally Sustainable Nuclear Energy Systems, ed. Jaafar M Z, IAEA Vienna) pp. 26.
[10] Fauzan Amin Misnon, Yeoh Siong Hu, Irman Abd. Rahman and Samudi Yasir 2017 Malaysia Public Perception Towards Nuclear Power Energy-Related Issues (American Institute of Physics)
[11] Hasfazilah Binti Hassan 2012 Penerimaan Orang Awam Terhadap Pembangunan Loji Kuasa Nuklear Di Malaysia (MANUPA) Menerusi Laman Sosial Facebook (thesis BSc. Universiti Kebangsaan Malaysia).

IOP Conf. Series: Materials Science and Engineering **298** (2018) 012050 doi:10.1088/1757-899X/298/1/012050

Limitations and challenges towards an effective business continuity management in Nuklear Malaysia

A.H.A Hamid[1]

[1]Malaysian Nuclear Agency (Nuklear Malaysia), Ministry of Science, Technology and Innovation (MOSTI), Bangi, 43000 Kajang, Selangor, Malaysia

amyhamijah@nm.gov.my

Abstract. One of Nuklear Malaysia's top concerns is radiological and nuclear safety as well as security preparedness of its operational facility management, which was bonded by Act 304, Directive 20 and International Atomic Energy Agency (IAEA) guidelines. In 2012, the Malaysian government initialised the Business Continuity Management System under the supervision of Malaysian Administrative Modernization and Management Planning Unit (MAMPU), referring to MAMPU.BPICT.700-4/2/11 (3), ISO 22301:2012 and Business Continuity Good Practice Guidelines 2013 documentation. These standards are integral to the implementation of a resilient management program that indicates an organisation's capability to prevent any accident from occurring and spreading its impact, which includes sufficient recovery action to post-accident situation towards a normal operational and managerial state. Unfortunately, there is a lack of certified Business Continuity Management standard among the public sector agencies compared to local private sectors. Subsequently, Nuklear Malaysia has been selected by MAMPU and CyberSecurity Malaysia as one of the pioneering agencies to be certified accordingly. This paper significantly recognized Nuklear Malaysia's effort to plan, analyse, design, implement, review and validate the establishment of this standard currently. The project was implemented using a case study approach to complete the required certification activities. As a result, this paper proposed benchmarking the selected literature reviews against the Nuklear Malaysia experience to determine best practices in implementing and managing Business Continuity effectively. It concluded that a resilient Business Continuity Management program needs to be incorporated into Nuclear Malaysia's capabilities in ensuring its mitigation capacities to survive any unexpected event and subsequently overcome future challenges.

1. Introduction

Business Continuity Institute (BCI) recently investigated adverse organisational threats associated with cyber-attacks and data breaches, which includes unplanned Information Technology (IT) and telecom outages [1]. These threats caused by inappropriate data protection in managing and accessing current technology. Unfortunately, these threats are real and adhering to these four significant disruptions, namely the unplanned IT and telecom outages (72 percent), adverse weather (43 percent), and security incidents (24 percent) such as utility supply interruption (40 percent) and cyber-attacks (35 percent). Unplanned IT and telecom outages are the most common disruption in Asia (55 percent) that affects the Public Administration and Defense Sector (78 percent).

Following that, other disruptions in Asia and the Public Administration and Defense Sector accordingly consist of adverse weather (36 and 45 percent), and security incidents (28 and 52 percent). Most security incidents in Asia are made up of the utility supply interruption (28 percent), and cyber-attacks (56 percent); affecting the Public Administration and Defense Sector at 52 and 49 percent respectively. The current study indicated that 726 responding organisations worldwide intended to increase their budgets for business continuity (21 percent) by taking up ISO 22301 as their guiding business continuity program (63 percent). Asian organisations indicated 56 percent budget contributions towards business continuity and 59 percent had applied the ISO 22301 guidance; and likewise the Public Administration and Defense Sector granted 67 and 66 percent cordially.

MyCert reported an increment of adverse threats in Malaysia, namely content-related (52 percent), cyber harassment (20 percent), Denial-of-Service (DoS) attacks (74 percent), fraud (20 percent), intrusions (45 percent) and vulnerability reports (59 percent) [2]. Content-related threats encompassed intellectual property and trademark infringement, false notifications and hoax warning emails on official information. Cyber harassment incidents involved cyber bullying, intrusion and modification of misleading personal profiles and religious information that includes cyber stalking. DoS incidents involved Distributed-Denial-of-Service (DDoS) amplification attacks and log triggers originated from DDoS mitigation services using Malaysian IP addresses. Malicious codes incidents are related to botnets, bots and malware threats through web hosting of ransomware reported through Cyber999. Cyber999 is the Malaysian Computer Emergency Response and Help Centre services provided by CyberSecurity Malaysia. There is no denial that cyber-attacks have outgrown the sophistication and method of currently available defence mechanisms. Therefore, organisational enterprises should improve their defence.

In 2016, Malaysian Nuclear Agency (Nuklear Malaysia) was chosen to be the pioneer public sector agency to adopt Business Continuity Management (BCM) that refers to ISO 22301 as guidance. This project spearheaded by Malaysian Administrative Modernization and Management Planning Unit (MAMPU) and CyberSecurity Malaysia. This project implemented organisational resilience focused on the organisational ability to anticipate, prepare for, respond and adapt to changes and sudden disruptions to survive and sustain organisational services and operations [3]. This action is integral to protect, access and dispose of information-based assets directly related to organisational resilience or ability to safeguard sensitive information [1].

The BCM adaptation based on the holistic management process and program involving training, exercises and reviews to ensure business continuity stays relevant and up-to-date. BCM adaptation began with the identification of possible threatening events in organisational activities which transformed into a workable framework [3-10]. This framework provides resilient, effective response and recovery. The framework safeguards the interests of key stakeholders, reputation, brand and value-creating activities. The framework also improves organisation's ability to successfully and appropriately react to disruptive events. An efficient and resilient BCM program incorporates organisational capabilities to prevent accidents from occurring and spreading its impacts, and to recover to normal operational state, after an accident has happened. This program requires active efforts in business continuity and disaster recovery activities to fulfill the proper resilience level of critical operations and IT infrastructure throughout the organisation [11].

This paper discusses the current trending of BCM development to preserve organisational resilience. Literature reviews suggested the conceptual Business Continuity (BC) Planning Workflow and Organisational Chart for further investigation. Further to these conceptual recommendations, this paper includes discussions on the limitations, challenges and resolutions which ended with concluding remarks for future improvements.

2. The development of BCM

This section describes the trending of BCM development since the 1970s. BCM has evolved from disaster recovery planning and risk management into planning and managing of Business Continuity (BC). This section also compares integral elements of Good Practice Guidelines.

As Information Technology (IT) is a subset of BCM, therefore, this paper discusses the relationship between BCM and IT support to maintain organisational resilience. The current outcome of BCM development has produced an integrated BCM framework that can be applied worldwide. According to figure 1, BCM was initiated as the IT disaster recovery in the 1960s, then progressively developed into emerging legislation and standards from 1970s until 2001 [4][10][12]. During the development stage, BCM framework integrated the practices of the Disaster Recovery Planning (DRP) and Crisis Management (CM). The past practices addressed organisational mission and management objectives to recover core systems, data and communication technologies that support the business, as the value-added organisational representation to their customers and interested parties which included increasing the enterprise's ability toward risk preparedness [4, 10]. BCM key strategies and solutions based on the risk management activities and assessment centred on data processing or information technology department (e.g. hardware, software, data up and running after a disruption) [12][13].

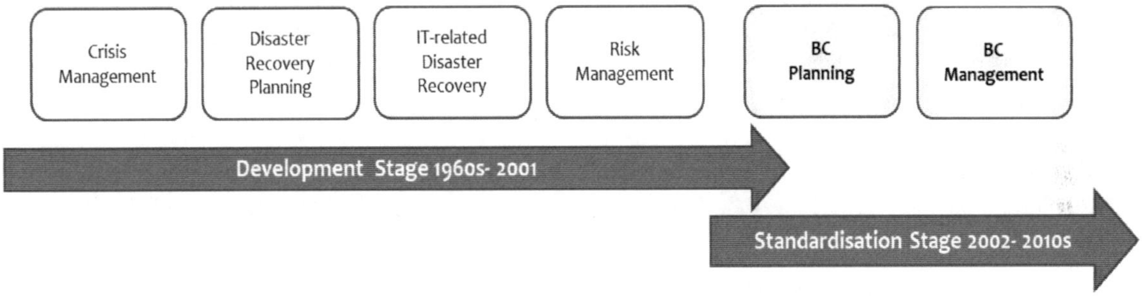

Figure 1. Historical Development of BCM [4][10][12].

Figure 1 also explains the current standardization stage of competing standards and breakouts from 2002 onwards. The ideal BCM was initiated as an approach to keep running business services and operations during occurrences of disruptive events that are IT related, business-related, as well as natural disasters [10][11]. The current stage expanded DRP into Business Continuity Planning (BCP) that emphasized beyond IT-related continuity efforts throughout the organisation, recovery planning of core processes and business functions consisting of emergency response, business continuity, disaster recovery and crisis management [4][12]. BCM and BCP are subsets of enterprise risk management meanwhile DRP is a subset of business continuity [4][12]. Table 1 shows a good practice guideline that integrates BC functional activities required for the success and effectiveness of a BCM resilience program as discussed earlier.

Table 2 displays related BCM functional roles associated with IT support that refer to multiple types of loss events [8][11]. This table explains the rapid societal vulnerabilities related to organisational awareness to adopt a resilient BCM program accordingly. The integration of BC and IT continuity management applied during incidents and emergency events consist of the integration between the enterprise, national and international parties internally and externally. Incidental disruptive event occurs by chance or due to a combination of unforeseen circumstances that can be escalated into an emergency, disaster or crisis if not adequately managed, whereas an emergency is a sudden, unexpected event that requires immediate action to restrain potential threats to health and safety of the environment or properties. Figure 2 suggests the integrated BCM framework that provides the IT aspect as a part of BCM, risk management as a precondition for useful BCM which include prioritized core documents such as BC Plan, DRP, and Business Impact Analysis (BIA) [8][12]. These key components indicate efficiently and resilient BCM in an organisation.

Table 1. Good Practice Guidelines between Disaster Recovery Plan, Risk Management, including BC Planning and Management [4][10][12].

Activities	Good Practice Guidelines		
	Key Method	Key Parameters	Type of Incident
Disaster Recovery Plan	Technical components of BCP	Addresses the recovery of core systems, data and communication technologies that support the business	Disaster recovery that is subset of business continuity
Risk Management	Risk analysis	Impact and probability	All types of events, usually segmented
BC Planning	Recovery planning of processes and business functions	Covering emergency response, business continuity, disaster recovery and also managing crises	Events causing significant business interruption
BC Management	Business impact analysis	Impact and time	Events causing significant business interruption

Table 2. Relationships between BCM and IT Support [8][11].

Types of Loss Events	Level of Awareness	IT Support	
		Methodological	Technical
Incident	**Enterprise (internal)**	Business and IT Continuity Management	• Traditional file-based backup and restore • Image-based backup and restore • Bare metal backup and restore • Point in time snapshots • Data replication • Continuous Data Protection (CDP) • Local high availability (HA) configurations • Remote business continuity (hot standby) • Hosted or cloud-based backup • Recovery to hosted/cloud-based services • Managed DR services • Virtualisation enabled recovery
Emergency	**Enterprise (internal) integrated with national and/or international (external)**		
Disaster		• Integrated protection system • Crisis management • Early warning systems • Legal support	• GIS and remote sensing o Drought o Earthquake o Flood o Landslides o Search and rescue • Internet • Forecasting systems (flood, cyclone, stock exchange)
Crisis	**National (internal) or International (external)**		

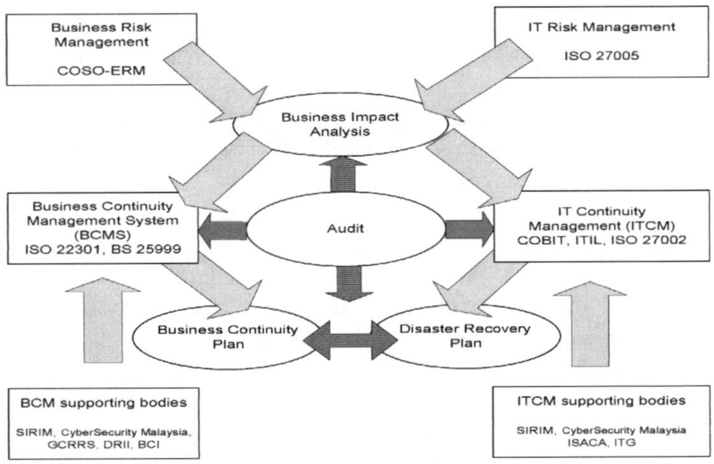

Figure 2. Integrated BCM Framework [8]

3. Conceptual framework

This section proposes the conceptual framework of an integrated BC planning workflow and organisational chart as follows. There are four main steps in developing the BC research methodology. These steps are the identification of the problem, defining the problem, designing and developing possible solutions and also, presentation of the proposed solutions. Figure 3 suggests research methodology of the BCM lifecycle. The lifecycle is divided into five phases consist of the Policy and Program Management and Embedding BC in the organisation, Initial Analysis, Core Processes and Activities Analysis, Design and Implementation and Validation [3][4][6][12].

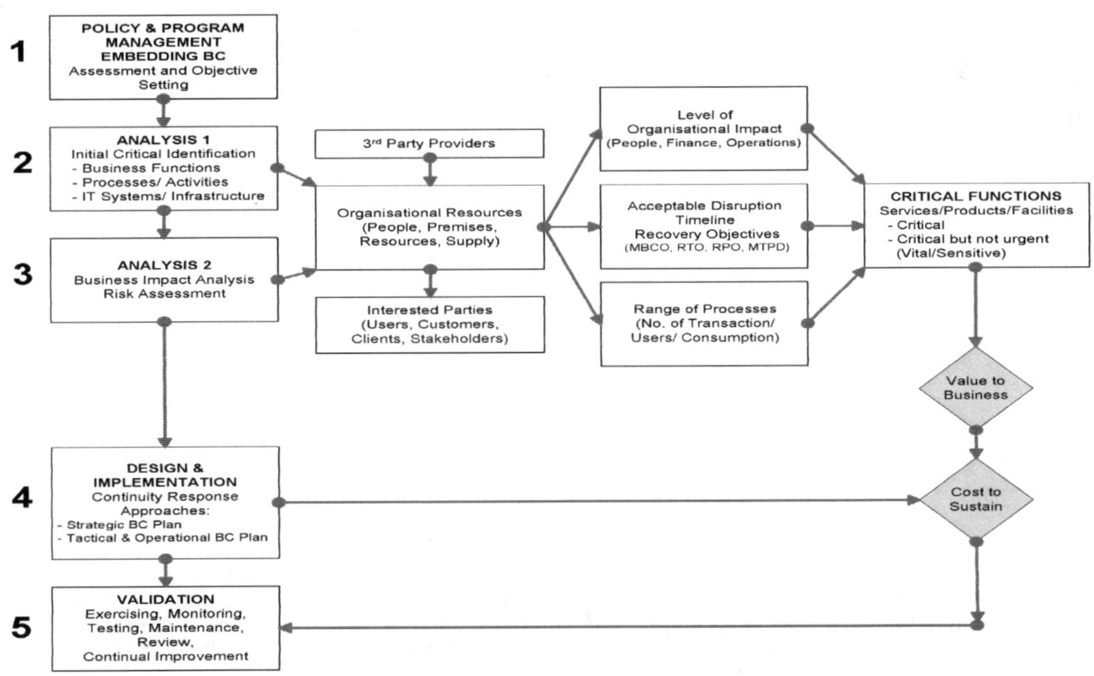

Figure 3. Integrated BC planning workflow.

IOP Conf. Series: Materials Science and Engineering **298** (2018) 012050 doi:10.1088/1757-899X/298/1/012050

First and foremost, the first phase explains the BC planning starts with executive-level assessments of organisational continuity objectives embedded in the organisational Policy and Program Management.

Next, the second phase is the initial analysis conducted to identify prioritized organisational business functions, processes and IT systems (e.g. applications and infrastructure).

The third phase involves further analysis of multiple departments and facilities associated with competent departmental managers according to PPRS evaluation process of the indicated items as follows, People, Premises, Resources and Utility Supply. They have to evaluate appropriate capabilities and capacities of their people (e.g. skill and knowledge), dedicated premises (buildings and facilities), resources (IT, data, information, equipment and material) and suppliers (utilities, product and services provided by third parties). The capability and capacity evaluation coordinate critical functions of organisational resources through mutual optimization and collaboration of the itemized PPRS. The critical functions comprise of data and documentation which retain legal bonded process and regulations in the applied key business processes or during the restoration of minimum acceptable work levels during disruptive events. The analysis also considers how unexpected business interruption might affect suppliers, customers and other interested parties. Consequently, the core processes and activities analysis based on the critical functions of provided PPRS organisational resources which indicate the level of organisational impact, recovery objective and range of processes.

The fourth phase indicated the design and implementation of the proposed continuity response approach regarding strategic, tactical and operational BC plans alongside providing manual, procedures, working instructions, records and forms.

The fifth and final phase of the lifecycle emphasizes on the documented validation and continual improvement of the critical functions in the BC Plans which includes the manual, procedures, working instructions, records and forms.

4. Proposed Organisational Chart

Figure 4 proposes an ideal Business Continuity Team Organisational Chart, comprising of the Continuity Management Team (CMT), Coordinator Team (CT), Crises Command Team (CCT), Business Recovery Team (BRT), IT Recovery Team (IRT), Systems/Applications Recovery Team (SRT) and the Administrative Support Team (AST). The roles and responsibilities of each team are explained below [3][4][6].

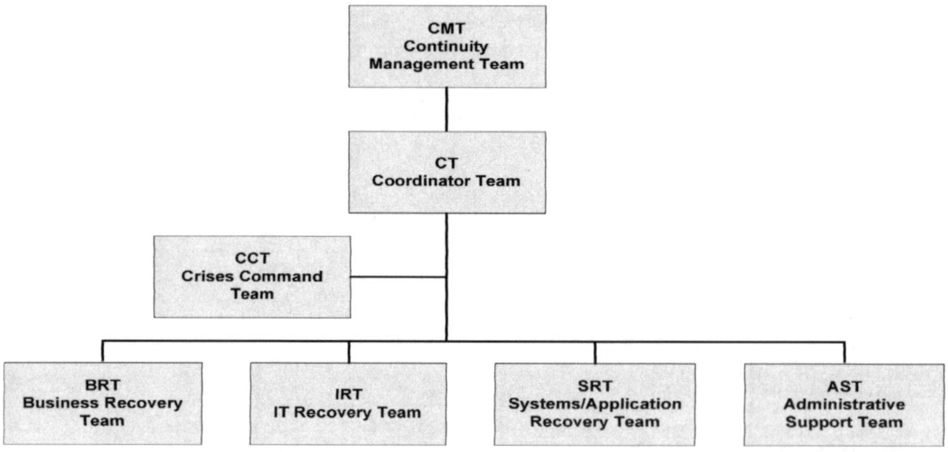

Figure 4. Proposed Business Continuity Team Organisational Chart.

IOP Conf. Series: Materials Science and Engineering **298** (2018) 012050 doi:10.1088/1757-899X/298/1/012050

4.1. Continuity management team (CMT)

- Consists of senior managers and top-ranked officers
- Includes all managerial board elements
- Responsible for decision-making and supporting business continuity program implementations
- Appointment confirmation of each BC Team members
- Monitor overall management in each phase of the recovery processes
- Responsible for the headquarters recovery to the rebuild alternative operational site at another location

4.2. Coordinator team (CT)

- Organize and integrate the business continuity program
- Report on recovery processes and activities

4.3. Crises command team (CCT)

- CMT representatives during the activation and operation of BCP
- Additional to CMT members
- Authoritative source of information during disaster or disruptive situations and for recovery processes
- Responsible for coordinating public authorities
- Control all internal and external communication according to CMT Leader's approval

4.4. Business recovery team (BRT)

- Responsible for operationalizing recovery processes and activities that refer to process recovery strategies

4.5. IT recovery team (IRT)

- Ensure recovery of the IT infrastructure necessary to support the recovery processes and activities related to core businesses
- Responsible for integrating voice and data communications with other parties

4.6. Systems/applications recovery team (SRT)

- Responsible for systems and applications recovery to support necessary business processes and activities

4.7. Administrative support team (AST)

- Provide logistical and managerial support to the recovery processes and activities of core business activities

5. Limitations, challenges and resolutions

This section discusses the BCM limitations and challenges, and suggests resolutions that contribute as key success factors in pursuing essential BCM resilience in an organisation [1][4][10][14][15].

IOP Conf. Series: Materials Science and Engineering **298** (2018) 012050 doi:10.1088/1757-899X/298/1/012050

5.1. Key Issues

The explanation of each BCM key issues covers the commitment and involvement issues, misunderstanding of data recovery by BCM practitioners, the inappropriate approach in executing BCM processes, and also, incorrect and inappropriate assumptions in formulating business continuity and disaster recovery plans.

5.1.1. Commitment and involvement issues

- Time-consuming implementations that require significant financial resources and specialists engagements
- External events that underscore the interconnected nature of risks and demonstrate the need for practitioners' policy planning such as unexpected increment of adverse threats that out-grown the sophistication, method and defence mechanism
- The disruptive effects of adverse weather as a growing concern for many organizations
- Lack of resources and inconsistencies of the BCM adaptation
- Inadequate incremental changes due to more focusing on compliance and regulatory requirements rather than the workable recovery objectives, levels and timeline during disruptive events
- Employees and experts differ in their perceptions of critical employee designation due to overestimation of their abilities in offering workplace recovery solutions
- A gap in attitude between experts and users regarding working from home as a workplace recovery arrangement

5.1.2. Misunderstanding of the data recovery by BCM practitioners.

- Cyber issues continue to figure significantly as one of the practitioners' main concerns
- Some organisations exposed to significant operational risk due to lack of workplace recovery arrangements
- Noncompliant data dynamics and dependencies
- Too much focus on hot/alternate site rather than to move operations back to a restored primary location
- Data recovery and fault-tolerant highly depends on the IT level of capability within the organisation to deliver uninterrupted services and processes in restoring data losses during hardware and software failure or logic errors

5.1.3. Inappropriate approach in executing BCM processes.

- Organisations which have workplace recovery arrangements in place face risks and uncertainties when it comes to actual recovery plan implementation, i.e. one of five experts and 17 percent of employees feel uncomfortable to execute their recovery solution plan for area-wide event services
- Organisations adopt different workplace recovery arrangements, relying on their own or third-party locations, i.e. 45 percent of experts and 20 percent end users expect to recover at their alternative locations owned by their organisation, meanwhile 21 percent of experts and 13 percent end users plan to recover at their alternative locations provided by third party service providers
- Business Impact Analysis (BIA) sessions conducted in silos while analysing the level of organisational impact, acceptable disruptive timeline, acceptable recovery objectives and also, range of processes by functional area

IOP Conf. Series: Materials Science and Engineering **298** (2018) 012050 doi:10.1088/1757-899X/298/1/012050

- A wide range of risk assessment refers to multiple functional areas engaging processes and activities involving management, infrastructure and technical support services, operational, production, third-party services and also, research and development activities
- Out of context of disaster impacts and durations using different BCM tools

5.1.4. Incorrect and inappropriate assumptions in formulating business continuity and disaster recovery plans.

- Insufficient consideration of employee preferences on workplace recovery may contribute to implementation risk in case of disruptive event
- Differences in priorities between employees and business continuity practitioners in the event of a disaster may lead to challenges in executing recovery plans
- Shortage of resources brings to irrelevant and unstated risk assumptions and limiting factors
- Impaired recovery efforts to operate an alternate site with the affected local area or regional disasters

5.2. Key Success Factors

The explanation of each BCM key success factors covers the understanding of roles and responsibilities of the top management, applied single framework across the organisation, applied approaches for unique skills and experience, applied tools to automate and support BCM processes and emerging organisational "good practices".

5.2.1. Understanding of the roles and responsibilities of the senior directors and operations management.

- Effective business continuity and resilience heavily rely on human factors
- Periodical meetings of the cross-functional project steering committee
- Address sustainable awareness campaign
- The company image and reputation positively represented
- The need to engage key stakeholders in implementing BC activities on a yearly basis
- Investments in resilience should be sustained for organizations to build and maintain adaptive capacity

5.2.2. Applied single framework across the organisation.

- Organisations should focus on the objective of appraisal threats and its particular impacts
- Strategic against Tactical and Operational Business Impact Analysis sessions
- Business and IT connected BC planning
- Identify risks and single point of failure
- Identify risk mitigation measures
- Incorporate BCM processes and primary operational tasks
- Adopt international BCM standard and framework

5.2.3. Applied approaches to unique skills and experience.

- Working from home remains a common tactic in delivering workplace recovery
- Establish measurable degree of organisational preparedness to recover from disasters regarding recoverable processes and operations
- Arrange new ways of representing the business processes by analysing the critical dependency factors, according to the PPRS analysis.

- Establish appropriate risk acceptance criteria
- Understanding the complete data recovery that is incorporated into end-user computing resources as the backup cycle to ensure reliable data retrieval
- Incorporate end-user computing systems and enterprise applications

5.2.4. Applied tools to automate and support BCM processes.

- Obtain business continuity-oriented quality metrics from the procedures provided by the prioritized critical processes and activities
- Adopt service and product based risk assessment
- Identify the prioritized critical processes and activities that support organisational service and product delivery
- Adequate knowledge of the BCM tools and workflows
- Business Impact Analysis based on site-wide disasters

5.2.5. Emerging organisational "good practices".

- The safety of employees remains the key priority for both workplace recovery experts and users
- Adverse weather is a key driver for invoking workplace recovery arrangements
- Validation is the key success of workplace recovery or any strategy deployed as part of the business continuity plan
- Get the techniques or incentives to engage participants and executives properly
- Address organisational expectations and requirements and permission that it will engage and attract the employees
- Recognise and document relevant risk assumptions and their recovery restrictions
- Long-term continuity includes on-going recovery activities
- Maintain and retain essential BC resources
- Identified organisational buffer against the circumstances of drastic and radical changes as BC opportunities to protect and steer the corporate projects and regulatory processes
- BC as part of organisational culture
- Assurance of recovery and return to normal operations

6. Conclusions

This paper concludes that the main principles in achieving a resilient BCM program requires a thorough understanding of the defined business requirements to survive any unexpected events. Organisational stakeholders, employees and other interested parties should know how to plan solutions for unexpected events and sudden challenges at any time. The implementation of these strategic plans will ensure the achievement of principle awareness of a resilient BCM program regarding disaster recovery. The disaster recovery plan should emphasize on the mitigation and monitoring of the indicated business continuity performance as well as sufficient preparations in anticipating expected future threats.

7. References

[1] Alcantara P and Riglietti G 2017 *Horizon Scan Report 2017* (Business Continuity Institute (BCI): United Kingdom).
[2] Syarifah Roziah Mohd Kassim and Kilausuria Abdullah 2017 Incident trend analysis for 2016 (eSecurity 2017 CyberSecurity Malaysia: Seri Kembangan).
[3] Bird L and Higgins D 2013 *Good practice guidelines 2013: Global* (United Kingdom: The Business Continuity Institute (BCI)).
[4] Bras J and Ribeiro R 2016 Business continuity and disaster recovery: an overview, trends and

challenges (13th Int. Conf. of Information Systems and Tech. Management CONTECSI 2016 Brazil).
[5] Kozlowski M 2016 The concept of method for determining the minimum level of airport business continuity J. of KONBiN **1(37)** 2083-4608.
[6] Malaysian Administrative Modernization Planning Unit (MAMPU) 2009 *Pengurusan Kesinambungan Perkhidmatan Agensi Sektor Awam* (MAMPU.BPICT.700-4/2/11(3) Kuala Lumpur: MAMPU).
[7] Momani N M 2010 Business continuity planning: are we prepared for future disasters. American J. of Economics and Business Administ. **2(3)** 272-279.
[8] Svata V 2013 System view of business continuity management *J. of Sys. Integ.* **4(2)** 19-35.
[9] The International Organization for Standardization (ISO) 2012 *ISO 22301 International Standard- Societal Security- Business Continuity Management Systems- Requirements* (ISO Switzerland).
[10] Venclova K, Urbancova H, and Vydrova H V 2013 Advantages and disadvantages of business continuity management *Int. J. of Social, Behavioral, Educational, Business and Industrial Eng.* **7(4)** 895-899.
[11] Giacchero A, Giordano F and Schiraldi M 2013 From business continuity to design of critical infrastructures: ensuring the proper resilience level to datacentres *Int. J. of Eng. and Tech. (IJET)* **5(4)** 3544-3553.
[12] Krell E 2006 *Management Accounting Guideline: Business Continuity Management*, (S.o.M.A.o.C. ed. (CMA-Canada) Canada: The Society of Management Accountants of Canada and The American Institute of Certified Public Accountants.
[13] Kaddarova J and Bajus R 2009 Business continuity management *Intercathedra Ann. Sci. Bulletin of Plant- Economic Dep. of the European Wood Technology University Studies*, (Nature University of Poznan: Poland) **25**.
[14] Alcantara P and Riglietti 2016G *Workplace Recovery Report* (Business Continuity Institute (BCI): United Kingdom).
[15] Lingeswara R 2012 Key Issues, Challenges and Resolutions in Implementing Business Continuity Projects. ISACA J. **1**.

Acknowledgments
This work would not have been possible without the official financial support from MAMPU, CyberSecurity Malaysia and Nuklear Malaysia. The author is especially indebted to the leaders and members of Nuklear Malaysia PKP-BCMS Project Adaptation from ISO 22301 who provided extensive guidance more than she could ever give credit for here.

IOP Conf. Series: Materials Science and Engineering **298** (2018) 012051 doi:10.1088/1757-899X/298/1/012051

The role of women in nuclear - attracting public participation in regulatory decision-making process

Azlina Mohamad Jais and Najwa Hassan

Nuclear Power Regulatory Coordination Unit, Malaysia Nuclear Power Corporation (MNPC), Cyberjaya, Selangor, Malaysia.

azlina@mnpc.org.my

Abstract. Public participation is vital in demonstrating transparency and enhancing effectiveness of a nuclear regulatory process. As such, it is necessary for nuclear practitioners to involve the public in key nuclear delivery milestones. This paper specifically discusses challenges faced in attracting public participation throughout the nuclear regulatory decision-making process, and highlights the roles of women in nuclear (WiN) in initiating the said public discourse.

1. Introduction

The fundamental objective of a nuclear regulatory authority is to ensure an adequate level of protection of the public and the environment against any possible harmful effects of ionising radiation that could result from the use of nuclear energy and radiation sources. The nuclear regulatory authority strives to ensure that all its regulatory decisions are technically sound, consistent and timely. The nuclear regulator is also committed to ensure its regulatory decisions are transparent, have a clear basis in laws and regulations, and are fair to all parties. The nuclear regulator must consider itself as being accountable to the public, especially when it involves regulatory decisions, and be open to public scrutiny. Therefore, the public has the rights to be informed on the procedures and the results of the nuclear regulatory activities, and to participate in nuclear regulatory decision-making processes undertaken by the nuclear regulatory authority.

It was actually observed during the Fukushima nuclear incident that the Japanese public interest in nuclear piqued post-accident, especially when it involved the dissemination of information during regulatory decision-making processes. In fact, the Japanese public at the time thought that the dissemination of information was slow and untimely [1]. This adversely contributed to the public distrust and the loss of confidence in Japanese nuclear regulatory authority.

Nuclear regulators must therefore consider public participation as one of the key regulatory tools to improve transparency, efficiency and effectiveness of its regulatory process. Public participation must be pursued with the aim of improving the quality of regulatory decisions. This is especially important now since information, regardless of whether it is accurate or not, is truly at the tip of our fingers; public participation during the regulatory decision-making process is thereby particularly important to assure public confidence and understanding in the regulatory decision itself.

This paper attempts to discuss the challenges of attracting public participation during the regulatory decision-making processes and explore potential roles that women in nuclear can possibly play to help initiate the nuclear public discourse.

2. The needs for public participation in nuclear milestones

One of the key lessons learned from the Fukushima accident is that the public always has the right to be informed and consulted by the nuclear regulatory authority. This comes from one of the action plans highlighted by IAEA post-Fukushima, which focuses on communication and information dissemination. It was clearly stated that Member States should enhance transparency and effectiveness of communication and improve dissemination of information among operators, regulators and various international organizations. This should involve possibly the freest possible flow and wide dissemination of safety-related technical and technological information to enhance the nuclear safety [2].

Therefore, the regulator should always be open and transparent about how the decisions are obtained; in fact, the public can freely ask about anything so that the public are fairly confident that the regulatory authority is not hiding anything from them. This is actually in line with the basic principles of the regulatory authority relations with the public, in which a practical mean to provide the public an access to the information on its regulatory activities and procedures must be established. By having public that is fully aware of its decision-making process, the regulatory authority can thus be self-assured of its transparency, can have a clear basis in its laws and regulations, and can be deemed fair to all parties.

Nuclear safety is the primary concern for the regulator who needs to ensure that the licensee or industry is fulfilling their obligations. Eventhough the public does not have the technical background to understand regulatory requirements for nuclear safety, it is still an obligation for the regulator to inform and explain to the public what the requirements are for [3]. By having this platform of communication, the regulator now has the tool to approach the public directly which possibly establishes trust and accountability between the public and the regulator.

Figure 1. A simple model of a robust national nuclear system [4]. (Note: 'Regulation' includes all regulatory activities and control, but a prime method of interaction and feedback is regulatory inspection activities.)

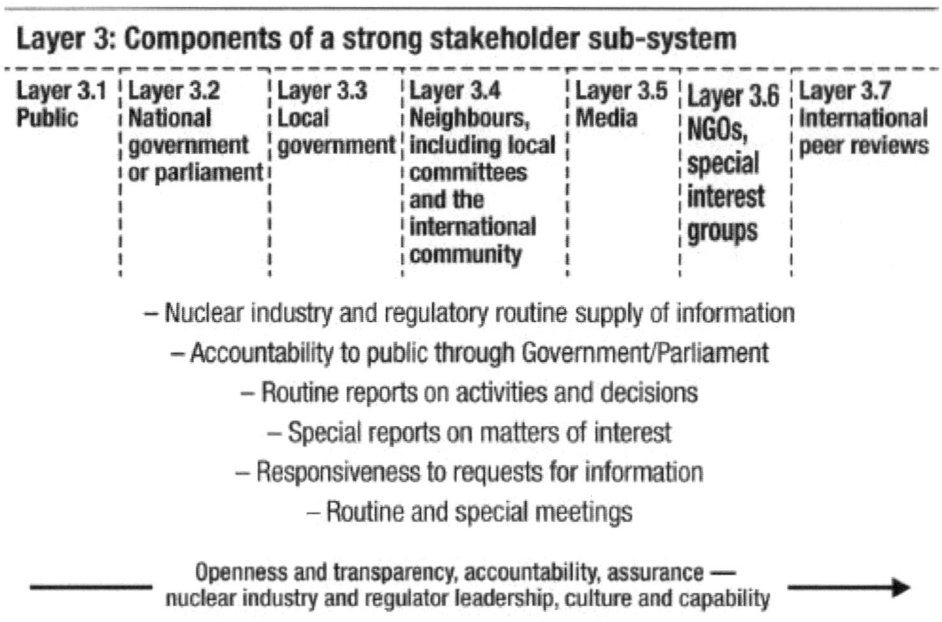

Figure 2. Components of a strong stakeholder sub-system.

Figure 1 shows the interconnection between the licensee or industry, regulator and stakeholders. None of the entities can stand on their own, as a successful nuclear power programme requires close interaction between the different nuclear players [4]. Meanwhile, Figure 2 illustrates a possible holistic mechanism of public participation in regulatory decision-making process which involves multi layers of nuclear stakeholders [5].

Unfortunately, addressing the public during any regulatory decision-making processes is not a common practice in Malaysia. For example, in the Department of Environment (DOE) Environmental Impact Assessment (EIA) manual, regarding public participation, the public can review and provide comments for the EIA report at relevant DOE state offices, public libraries and local authority offices [6]; it is thus a one-way public participation route. The authors thereby suggest a two-way communication practise for the nuclear regulatory authority in Malaysia so that the authority can directly hear from the public and thus readily learn their worries and concerns. At the same time, the regulatory authority should have a ready-made platform to address any of the public concerns and possibly misinformation. This is especially important to gain public trust and understanding in the credibility and decision-making of the nuclear regulator. It should however be clear here that the motivation to involve public participation during the regulatory decision-making process is mainly to inform them of the correct information; public support is actually an indirect result of the successful public participation and should be the objective of this activity.

3. Roles of women in nuclear

Those who have been involved in the nuclear industry may easily notice that the industry is dominated by men. In fact, even though women make up half of the human population and there are significant numbers of women scientists and engineers around the globe, only 22.4% of the nuclear workforce is comprised of women [7]. This raises up few questions: does this imply that women are less interested in nuclear, and does the nuclear industry actually scare women off?

It is upon this observation that the three following objectives of Women in Nuclear (WiN) were crafted: (1) to attract more women to choose a career in nuclear, (2) to support women's career progression in nuclear related fields, and (3) to promote dialogue within the nuclear industry that is in

IOP Conf. Series: Materials Science and Engineering **298** (2018) 012051 doi:10.1088/1757-899X/298/1/012051

line with WiN [8]. These objectives are necessary and timely to assure women can still play significant roles in the fight for clean energy against global warming. The role of women is especially important in attracting public participation during key nuclear delivery milestones, for example during regulatory decision-making process, so that the nuclear industry does not only appear inclusive but is also truly inclusive.

3.1. Women and Nuclear Public Participation

Nuclear is still a very sensitive subject. More often than not, the word 'nuclear' itself incites negative sentiments among the public because to them 'nuclear' is either a bomb or related to the Fukushima accident; the benefits of nuclear energy are thus lost in bad impressions. As such, feeding the public with the right information about nuclear is very important especially for a nuclear newcomer state like Malaysia.

It is however admittedly not easy to disseminate correct information to the public. This is unavoidable due to negative impressions of 'nuclear'. The nuclear industry must be strategic and selectively diplomatic in disseminating the correct information to the right audience. In fact, the authors are of the opinion that the right spokesperson for the nuclear regulatory authority should be a woman. This is because, from what we learned from Fukushima, women are in general more in tune with the perceived emotion of the public and thus are more relatable. A dialogue between women of common understanding can then be more productive especially when the discourse is centred around the safety of family and children.

Table 1 succinctly illustrates characteristic differences between men and women in communication [9]. It can be inferred from the table that women tend to look for commonalities in order to create connection with the audience while men tend to share experiences as a way of providing information rather than asking questions. In addition, men are generally selective listeners who focus mostly on the main points while women listen to each and every word and are more attentive to the both verbal and non-verbal cues during the communication. Thereby, women may possibly make a more relatable nuclear spokesperson than men. Furthermore, a female "spokeperson" can be more persuasive to mothers who have a big influence in the welfare of their children, whom may one day be future decision makers in the country.

It is upon the above observation that the authors are of the opinion that public participation during any nuclear regulatory decision-making processes can be more effective and more successful when the communication team involves women than an "all men" team. Men can nonetheless still be of a great assistance; men can help explain the technical aspects of the decisions while women can translate the points into laymen terms as women are naturally more nurturing and caring compared to men.

IOP Conf. Series: Materials Science and Engineering **298** (2018) 012051 doi:10.1088/1757-899X/298/1/012051

Table 1. Levels of communication between men and women.

No.	Men	Women
1	Men keep their problems to themselves and don't see the point in sharing personal issues	Women are more likely to talk to other women when they have a problem or need to make a decision
2	Men tend to relate to other men on one-up and one-down basis. Status and dominance is important.	They are more relationship oriented, and look for commonalities and ways to connect with other women.
3	Men focus on talking and providing information rather than asking questions. They share experiences as a way of being one-up.	They focus on building rapport, by sharing experiences and asking questions.
4	Men can have a disagreement, move on to another subject and go get a drink together.	If women have a disagreement with each other it affects all aspects of their relationship.
5	Men build relationships while they are working on tasks with each other.	Get things done at work by building relationships.
6	Men move to solutions and problem solving right away.	Women want to talk about the problems and solve them collaboratively.
7	For men asking for help reflects instability to achieve on one's merit.	Offering help and advice is a sign of care.
8	Men listen to the main points. They are selective listeners.	They listen to each and every word: they show attentiveness through verbal and non-verbal cues.

4. Conclusions

The Fukushima accident had somewhat negatively blemished public trust in nuclear regulators around the globe and this had detrimentally affected the nuclear regulator's credibility to regulate effectively. As Timothy Sloan, President and CEO of the Wells Fargo, remarked, "to regain the trust we have lost, we must continue to be transparent with all our stakeholders and go beyond what has been asked of us by our regulators by reviewing all of our operations — leaving no stone unturned — so we can be confident we have done all that we can do to build a better, stronger Wells Fargo".

The Malaysian nuclear industry is at a similar predicament; we should not leave any stone unturned in our mission to win public acceptance post-Fukushima for the successful delivery of nuclear power in Peninsular Malaysia. One of the key lessons learned from the accident is to consistently and openly consult the public in major nuclear milestones, especially in nuclear regulatory decision-making processes. The authors are of the opinion that in order to properly attract public participation in nuclear projects, women should be given a significant role, if not a leading role, in the nuclear communication team. This is because women, by nature, are more effective and relatable communicators than men. A more relatable and effective nuclear communication team can help to ensure the success of Malaysia's nuclear power programme.

5. References

[1] Shigemura J, Tanichi M, Yoshino A, Nagamine M and Shimizu K 2017 The Fukushima Disaster: Lessons Learned from GEJE Catastrophe *The Journal of Employee Assistance* **47** pp 16-19

[2] International Atomic Energy Agency (n.d) *IAEA Action Plan on Nuclear Safety* Retrieved from https://www.iaea.org/sites/default/files/actionplanns.pdf

[3] International Nuclear Safety Group 2017 Ensuring Robust National Nuclear Safety Systems - Institutional Strength In Depth *International Atomic Energy Agency* p 6

[4] International Nuclear Safety Group 2017 Ensuring Robust National Nuclear Safety Systems - Institutional Strength In Depth *International Atomic Energy Agency* p 7

[5] International Nuclear Safety Group 2017 Ensuring Robust National Nuclear Safety Systems - Institutional Strength In Depth *International Atomic Energy Agency* p 14

[6] Department of Environment 2016 Environmental Impact Assessment Guideline In Malaysia p 56

[7] Cardenas-Mazurkiewic, David 2010 *Women Changing the Face of Nuclear Industry* Retrieved from https://www.iaea.org/newscenter/news/women-changing-face-nuclear-industry

[8] Women in Nuclear, Global. (n.d) *WIN Mission* Retrieved from http://www.win-global.org/about/what

[9] Mohindra, PV 2012 Gender Communication: A Comparative Analysis of Communicational Approaches of Men and Women at Workplaces *IOSR Journal of Humanities and Social Science* **2(1)** pp 18-27 doi:10.9790/0837-021182

IOP Conf. Series: Materials Science and Engineering **298** (2018) 012052 doi:10.1088/1757-899X/298/1/012052

Study on relationship of performance shaping factor in human error probability with prevalent stress of PUSPATI TRIGA reactor operators

Ahmad Nabil Bin Ab Rahim, Faizal Mohamed, Mohd Fairus Abdul Farid, Mohd Fazli Zakaria, Alfred Sangau Ligam, Nurhayati Binti Ramli

Reactor Technology Centre, Technical Support Division, Malaysian Nuclear Agency, Bangi, 43000 Kajang, Selangor, Malaysia

ahmadnabil@nuclearmalaysia.gov.my

Abstract. Human factor can be affected by prevalence stress measured using Depression, Anxiety and Stress Scale (DASS). From the respondents feedback can be summarized that the main factor causes the highest prevalence stress is due to the working conditions that require operators to handle critical situation and make a prompt critical decisions. The relationship between the prevalence stress and performance shaping factors found that $PSF_{Fitness}$ and $PSF_{Work\ Process}$ showed positive Pearson's Correlation with the score of .763 and .826 while the level of significance, p = .028 and p = .012. These positive correlations with good significant values between prevalence stress and human performance shaping factor (PSF) related to fitness, work processes and procedures. The higher the stress level of the respondents, the higher the score of selected for the PSFs. This is due to the higher levels of stress lead to deteriorating physical health and cognitive also worsened. In addition, the lack of understanding in the work procedures can also be a factor that causes a growing stress. The higher these values will lead to the higher the probabilities of human error occur. Thus, monitoring the level of stress among operators RTP is important to ensure the safety of RTP.

1. Introduction

Based on past nuclear accident cases, human error is said to be the main cause for accidents in nuclear power plants. However, it is also associated with other factors such as less effective training, complicated procedures and also natural disasters. These factors have resulted additional pressure or stress on operators in nuclear power plants. If nuclear plant operators have a high and complete engineering knowledge to operate technology systems, it will definitely help to reduce the risk of undesirable accident.

The subject of human factor is very important to reduce the risk of nuclear accidents. One of the methods to study human factor is through a Performance Shaping Factors (PSF) based on the standardized Plant Analysis Risk-Human (SPAR-H) method. Lesson leant form nuclear power plant can also be applied to nuclear research reactor. This study was conducted on the factors of human performance shaping factor, accident sequence analysis, and probability of human error. Human performance shaping factors are factors that affect the probability of human error as considered in the

IOP Conf. Series: Materials Science and Engineering **298** (2018) 012052 doi:10.1088/1757-899X/298/1/012052

analysis of human reliability such as the level of training, quality or the existence of procedural guidelines, the time available to perform the action and others.

According to the Department of Occupational Safety and Health, JKKP Negeri Johor (2011), there are four major elements related to industrial accidents in Malaysia namely employers, laws, machinery machines and workplace environmental factors. However, most employers often overlook the environmental factors in the workplace as it is considered a trivial problem. In these workplace environmental factors, stress or stress problems, uncomfortable working conditions, ventilation systems, inadequate lighting and ergonomics are hidden problems that will lead to chronic health problems and consequently will result in accidents that not only endanger them but also affecting other employees, the public and the environment.

The Quality of Worklife (QWL) module was developed by the National Institute of Occupational Safety and Health (NIOSH) or better known as the Occupational Health and Safety Institute to measure the symptoms and causes of stress in the workforce industry. Whereas DASS is a set of three self-report scales designed to measure the negative emotional state for depression, anxiety, and stress used in this study to quantify the prevalence stress. Since each operator reacts differently to the same case given, the prevalence stress is chosen as individual factor that contribute to individual HEP. The main scope of this study is to find out any significant correlations between prevalence stress and PSFs

2. Method

The design of the study is a survey method that uses quantitative approaches to gather information to achieve the objectives of the study. The survey method can be defined as the acquisition of information directly from a group of individuals by submitting a question via a questionnaire.

In this study, the SPAR-H worksheet was used to assess the performance shaping factor (PSF) for each respondent. Then, using the Depression, Anxiety and Stress Scale (DASS) form and the Quality of Work Life Module (QWL Module) were used to evaluate prevalence stress level, identify and analyse the causes and factors that cause worker's prevalence stress.

Respondents were eight RTP operators who also involved in RTP maintenance work. Respondents usually involve in the daily inspection of RTP known as early morning checklist activity and after-hours checklist activities.

All data from the SPAR-H worksheet will be analysed to obtain the value of the performance shaping factor (PSF) for each operator. Subsequently, human reliability can be seen from the probability of human error (HEP). In the survey form, all data and information will be analysed using the Social Science Statistics Package (SPSS) software. This analysis is used to identify factors that influence the prevalence stress among RTP controllers and identify the correlation relation between prevalence stress and performance shaping factors (PSFs).

3. Results and Discussions

SPAR-H worksheet enables the HEP_{Action} to be predicted between respondents based on the condition imposed on the case study. Table 1 shows the value of HEP_{Action} for this case study.

Table 1. Human Error Probability.

Respondent	HEP (Action)
1	1
2	0.0013
3	0.0005
4	0.0040
5	0.0200
6	0.0020
7	0.0100
8	0.0050

It was found only respondent 1 would definitely fail to perform correction task to the given for this case. This is due to the fact that the respondent 1 has assessed $PSF_{Procedures}$, $PSF_{Fitness}$, and $PSF_{Working\ Process}$ with higher score than most of other operators. One of the most stand out factor for this operator was due to the lesser amount of service at RTP compared to other operators. Thus the respondent 1 has lesser time and exposure to perform such Initiating Event selected for this study. This also help to explains why the respondent 1 select a higher score for $PSF_{Working\ Process}$ because the respondent still could not completely comprehend all the RTP working process. Other respondent's probability of error was found with different values. This variety is due to different values were chosen for the all 8 PSFs based on the respondents choices. The different background the respondents may effect different values for PSFs chosen that results such HEP_{Action}.

Out of many background factors, Prevalence Stress was chosen to identify whether there are any correlation with the eight of the Performance Shaping Factor (PSF) that contributed to HEP_{Action}. The relationship of PSF for HEP_{Action} with Prevalence Stress is shown in the Table 2 below.

Table 2. Correlation of Performance Shaping Factors and Prevalence Stress Measured from (DASS).

PSF		Values
Time	Pearson Correlation	.022
	Sig. (2-tailed)	.958
Stress	Pearson Correlation	-.455
	Sig. (2-tailed)	.257
Complexity	Pearson Correlation	-.482
	Sig. (2-tailed)	.227
Experience	Pearson Correlation	.285
	Sig. (2-tailed)	.494
Procedure	Pearson Correlation	.740*
	Sig. (2-tailed)	.036
Ergonomics	Pearson Correlation	.520
	Sig. (2-tailed)	.186
Fitness	Pearson Correlation	.763*
	Sig. (2-tailed)	.028
Working Process	Pearson Correlation	.826*
	Sig. (2-tailed)	.012

Only $PSF_{Fitness}$, $PSF_{Procedure}$, $PSF_{Working\ Process}$ were found to have good significant values with positive correlation that are .740, .763 and .826. This value means that the higher states of Prevalence Stress may cause higher score selection of the identified PSFs among respondents. The higher selections on PSF composites would results in higher value of Human Error Probability.

4. Conclusion

The positive correlation between DASS and PSF the procedure for HEP measures showed that the higher the stress the higher would be the respondent's negative perception of the procedure. Individuals who experienced this high stress might feel that the present procedure was incomplete as compared to those with low stress that gave positive perceptions when the respondent assumed that the procedure was normal and oriented. A person with high stress may experience cognitive impairment. When the cognitive function was disrupted, the respondent could not properly evaluate the existing procedure. When we did not understand something, this would cause the occurrence of stress. One of the probable causes of experiencing high stress was many emergency procedures that were not well understood by the respondents. This situation is very parallel with one's service period. For new operators in the service, this is one of the main reasons they are not very fluent about certain procedure, especially the rarely implemented procedures. Therefore it is important to propose adequate training activity with related with abnormal situation to the RTP operators.

5. References

[1] Blackman H ., Gertma, D I and Boring R L 2008 Human Error Quantification Using Performance Shaping Factors in the SPAR-H Method *Proc.of the Human Factors and Ergonomics Society Annual Meeting 2008 (*Idaho Falls, USA, 1 September 2008) **52(21)** 1733-1736.

[2] Gertman D, Blackman H, Marble J, Byers J and Smith C 2005 *The SPAR-H Human Reliability Analysis Method* (U.S. Nuclear Regulatory Commission. Washington, DC) NUREG/CR 6883.

[3] Gravetter F J and Forzano L B 2003 Research Methods for Behavioral Science (USA: Wadsworth/Thompson Learning).

[4] Hirschberg S 1990 Dependencies, human interactions and uncertainties in Probabilistic Safety Analysis (Final report NKA project RAS-470 Nordic Liaison Committee for Atomic Energy).

[5] Bell J and Holroyd J 2009 Review of Human Reliability Assessment Methods (Health and Safety Executive)

[6] Lovibond SH; Lovibond P F 1995 Manual for the Depression Anxiety Stress Scales (2nd ed. Sydney: Psychology Foundation).

[7] La Montagne N 2006 Measuring occupational stress: development of pressure management indicator *J. of Occupational Health Psychology* **3(3)** 61-65.

[8] NRC 1983 PRA Procedures Guide: A Guide to the Performance of Probabilistic Risk Assessments for Nuclear Power Plants (NUREG/CR-2300 Washington, D.C.: U.S. Nuclear Regulatory Commission).

IOP Conf. Series: Materials Science and Engineering **298** (2018) 012053 doi:10.1088/1757-899X/298/1/012053

Study on detection geometry and detector shielding for portable PGNAA system using PHITS

H Ithnin[1, 4], L N S Dahing[1], N M Lip[2], I Q Abd Rashid[3], E J Mohamad[4]

[1]Industrial Technology Division, Malaysia Nuclear Agency, Malaysia
[2]Faculty of Computer and Mathematical Science, Universiti Teknologi Mara
[3]Department of Nuclear Science, Universiti Kebangsaan Malaysia
[4]Faculty of Electrical and Electronic Engineering, Universiti Tun Hussein Onn, Malaysia

hanafi_i@nm.gov.my

Abstract. Prompt gamma-ray neutron activation analysis (PGNAA) measurements require efficient detectors for gamma-ray detection. Apart from experimental studies, the Monte Carlo (MC) method has become one of the most popular tools in detector studies. The absolute efficiency for a 2 × 2 inch cylindrical Sodium Iodide (NaI) detector has been modelled using the PHITS software and compared with previous studies in literature. In the present work, PHITS code is used for optimization of portable PGNAA system using the validated NaI detector. The detection geometry is optimized by moving the detector along the sample to find the highest intensity of the prompt gamma generated from the sample. Shielding material for the validated NaI detector is also studied to find the best option for the PGNAA system setup. The result shows the optimum distance for detector is on the surface of the sample and around 15 cm from the source. The results specify that this process can be followed to determine the best setup for PGNAA system for a different sample size and detector type. It can be concluded that data from PHITS code is a strong tool not only for efficiency studies but also for optimization of PGNAA system.

1. Introduction

Neutron-capture gamma ray has been introduced as a method of elemental analysis years ago. This method is particularly useful for determining nondestructively elements which absorb neutrons but do not produce radioactive products [1]. The design of prompt gamma neutron activation analysis (PGNAA) has been optimized multiple times for high gamma detection efficiency and low background. Scintillation-based Sodium Iodide, NaI detectors have been used widely for radiation spectroscopy and radioisotope-based applications including medical and industrial areas [2]. These types of detectors have shown reliable results for low level radioactive source measurements due to its higher detection efficiency and suitable features for operating detection at room temperature. However, the accuracy of measurement depends strongly on some detection properties of scintillation detectors such as detection geometry and detection shielding. Experimenting with the geometry is an effective method to evaluate the efficiency of NaI detectors for optimizing portable prompt gamma neutron activation analysis.

For the past years, the Monte Carlo (MC) method has been widely utilized for efficiency determination studies [3-6]. Absolute efficiency is defined as the ratio of the number of gamma rays emitted by the source (in all directions). Some experimental and calculation studies have also reported on detection efficiency determination [1]. This study is performed by calculating the efficiency of 2x2 inch NaI scintillation detector and using the same detector model for geometry and shielding study for portable PGNAA system. Optimisation of photon detection for portable PGNAA system is important as the neutron source use such as Carlifornium-252 emitting fast neutron thus the intensity of the prompt gamma is very low comparing to reactor based PGNAA. Current systems were using high-purity Germanium detector HPGe and optimisation were ran by experimentally, using this simulation study limit the exposure time for the workers in order to setup the best geometry for the portable PGNAA system measurement. The aim of this study is to simulate the portable PGNAA system and to find the best detection geometry with the best shielding configuration for the system.

2. Methodology

The PGNAA system presented in this paper was simulated and optimized using Particle and Heavy Ion Transport code System (PHITS) software [7]. In this study, PHITS has been used for detector efficiency calculation, geometry design and optimization. First the absolute efficiency of the modeled detector is calculated. Absolute efficiency must be known especially in radioactivity measurements. The definition of the absolute efficiency is shown in;

$$\varepsilon_{abs} = \frac{N_c}{N_g} \tag{1}$$

where, N_c is defined as the number of counts recorded by detector and N_g is the number of radiations emitted by the source d (all directions such as isotropic source) [1]. Efficiency of the NaI detector is calculated using model shown in figure 1. The efficiency of the detector is measured for different energy of the sealed source such as Na-22, Mn-54, Cs-137 and Co-60.

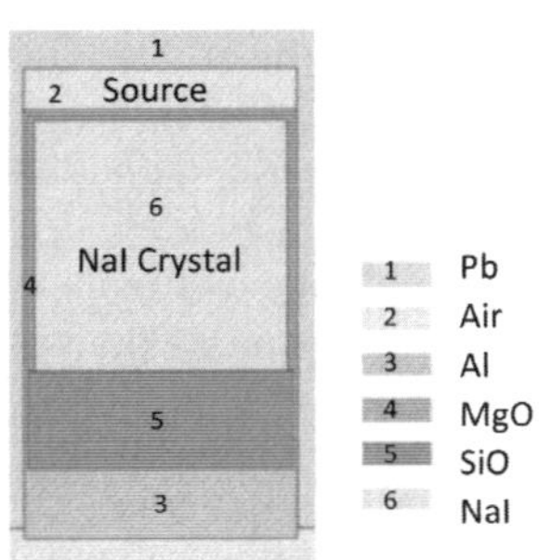

Figure 1. 2x2" NaI detector model.

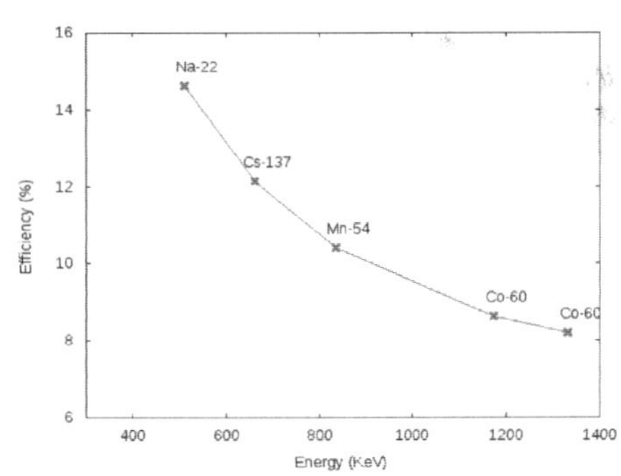

Figure 2. Energy vs Efficiency graph.

Figure 2 shows the graph of the efficiency for NaI detector simulated using PHITS. The trend for the efficiency graph from this detector model is in agreement with previous study [8][9] where at the higher energy of photon the detection efficiency is reduced. This is due to higher energy gamma-ray photon has high penetrating power and less photon imparted its energy towards the detector.

Using the modeled NaI detector, the geometry of the PGNAA system is modeled according to experimental setup (figure 3). The neuron source is placed inside neutron howitzer so that the neutron is shielded and collimated to only one direction. Energy distribution for neutron source is model according to Watt fission spectrum. The sample is assumed to be silicon, Si cuboid and the space between the neutron source and sample is the polyethylene moderator. The detector crystals arranged 90 degree from the neutron source so that no direct hit by passing neutron. The detector crystal is

IOP Conf. Series: Materials Science and Engineering **298** (2018) 012053 doi:10.1088/1757-899X/298/1/012053

design to be shielded by lead so that only gamma rays from sample are detected. The result that we obtained from this simulation is reflecting the capability of the detector chosen to produce the intensity of prompt gamma counted and hence producing the full spectrum of intensity of gamma ray versus the energy of gamma ray.

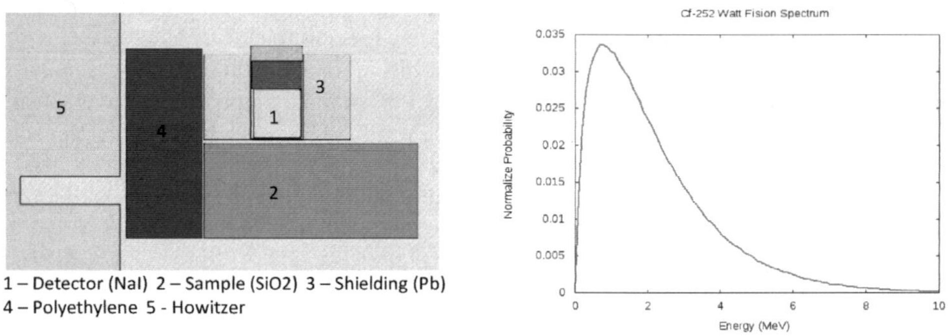

Figure 3. (a) Portable PGNAA system setup and (b) spectrun of Cf-252 source used

In details of this work, the dimensions used for the moderator is 10 cm in length and has a thick of 8 cm. The target or known as sample is a Silicon (Si) cuboid with a dimension of 10 cm × 10 cm × 30 cm is bombarded with neutron beam with an energy distribution shown in figure 3(b) at 180 degree angle from the neutron source. The axis of cylindrical-shaped detector, NaI is located perpendicular or 90 degree from the position of the neutron generator. For the first objective, position of the detector is varied along the sample surface to find maximum prompt gamma intensity detected. Afterwards, using optimized distance of detector placement, shielding effect on the detector is studied. Two geometries setup is present in this study to find the effect towards the detection of the prompt gamma that is, with lead shield and without lead shielding referring to figure 3(a).

3. Result and Discussion

PGNAA technique has been simulated using PHITS software which applied the Monte-Carlo code to find the probability of neutron activation of photon flux produced to identify element exist in a sample. Thermal neutron flux in prompt-gamma neutron activation analysis setting is the key factor to determine a portable PGNAA performance. In this study 8 cm polyethylene block is simulated as neutron moderator. The bigger the size of the moderator between the source and the sample gives higher thermal neutron produce hence higher gamma ray intensities. This is due to the increasing of thickness space of the moderator in between source and sample. A moderator in the system carried a function to reduce the initial kinetic energy of the free neutron from the source. In this process, the energy released is conserved due to the neutron kinetic energy passing through the material of the moderator. The speed of the neutron released undergoes reduction, as a product of decreasing kinetic energy.

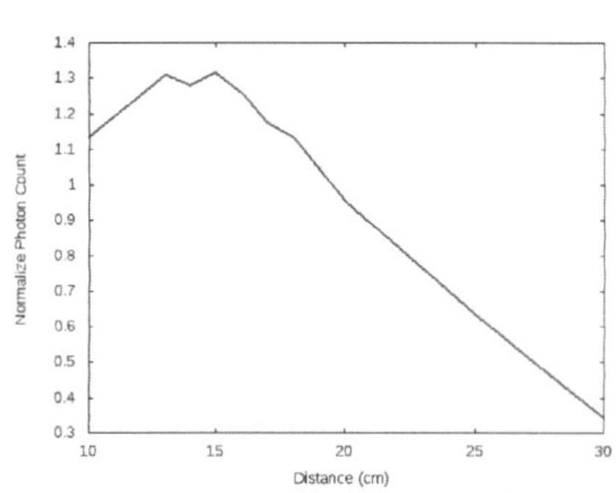

Figure 4. Normalize photon count with varying detector distance from source.

The first objective is to study the effect of distance of detector from the source to observe the intensity of gamma ray produced. The higher yield of gamma ray, the higher the sensitivity of the system in which PGNAA system can optimized to construct a better non-destructive technique for a bigger sample size. Graph in figure 4 shows the highest intensity of photon counting is around 15 cm thus suggesting it is the optimum distance for detector placement. This can be explained as near the source the neutron had minimum interaction with the sample. The interaction of thermal neutron is at maximum until the neutrons penetrate around 15 cm. After 15 cm the graph shows decreasing of photon counting as most of the thermal neutron had imparted their energy to the sample.

The other objective is to study the effect of detector shielding using lead cylinder for gamma-ray shielding and polyethylene block for neutron shielding. Figure 5 shows the result from the simulation with two different geometries. The red spectrum on the graph represents setup with no lead shielding. While the blue spectrum which is plotted on top of red spectrum represent setup with 5 cm lead shielding. Table on figure 5 shows photon flux at the detector for five different peaks for the two geometry setup. Without the present of lead shielding overall photon counting were higher than with the present of lead shielding. This is due to the isotropic nature of detection at the detector crystal. When lead shielding is present, overall photon counting is lower which is not good for prompt gamma detection.

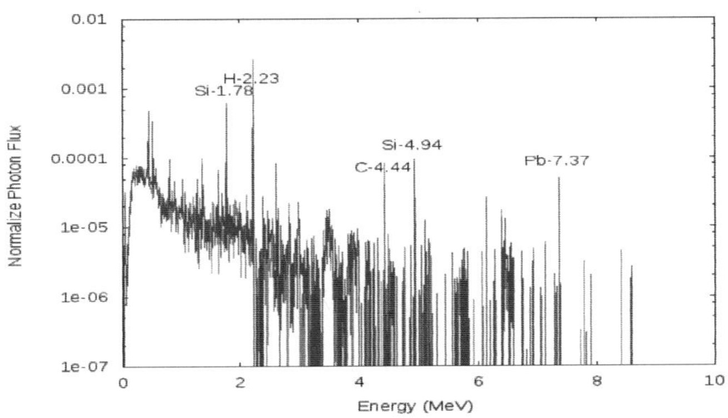

Element - Energy, MeV	Peak Flux Geometry 1	Peak Flux Geometry 2
Si-1.78	6.1448E-04	5.2934E-04
H-2.23	2.5786E-03	7.8644E-04
C-4.44	8.3970E-05	4.0198E-05
Si-4.94	9.3370E-05	8.6551E-05
Pb-7.37	-	4.9375E-05

Figure 5. Prompt gamma spectrum from SiO_2 sample with two different arrangements; Geometry 1: No-lead shielding (red spectrum); Geometry 2: 5cm lead shielding (blue spectrum).

Analysing the result for peak to peak counting we found that lead shielding functioning as shielding for unwanted prompt gamma peak from Hydrogen at 2.23 MeV and Carbon at 4.44 MeV which activated from the moderator. Another observation made is the present of Lead peak at 7.37 MeV when using the lead as detector shielding. This cannot be dismissed as it is the prompt gamma which is directly activated from the shielding material itself. For the interested peak which is Si peak, there is not much difference as the small difference is only due to the collimated nature of the detection when the shielding is present.

4. Conclusion

A portable PGNAA system has been modelled and simulated using PHITS software. The simulated system used Cf-252 neutron source with silicon, Si sample and NaI detector. The prompt gamma spectrum for the sample was successfully detected. The effect of moderator towards the quality of the spectrum detected in the simulation was studied. The result showed that in the present of the moderator, more prompt gamma was detected. This is due to the presence of more thermal neutron in the sample. On the other hand the noise or background signal was higher with the use of moderator. For this, shielding effect was studied and it is found that lead shielding functioning as shielding for unwanted prompt gamma peak from moderator material. On the other hand the present of shielding also activated the prompt gamma from lead. The results also shows the detector needed to arrange close to the sample so the detection of prompt gamma rays is increased. For an optimum system, the best distance is recorded to be at 15 cm. The simulation have confirm the two parameters studies in this work and it can be used for the setup of real experimental of portable prompt gamma activation analysis (PGNAA) system in future.

5. References

[1] Richard M L 1993Prompt-Gamma Activation Analysis *J. of Research of the National Institute of Standards and Technology* **98(1)** 127-133
[2] Yi C Y and Hah S H 2012 Monte Carlo calculation of response functions to gamma-ray point sources for a spherical NaI(Tl) detector *Appl. Radiation and Isotopes* **70(9)** 2133–2136
[3] Baccouche S, Al-Azmi D, Karunakara N, and Trabelsi A 2012 Application of the Monte Carlo method for the efficiency calibration of CsI and NaI detectors for gamma-ray measurements from terrestrial samples *Applied Radiation and Isotopes*, **70(1)** 227–232
[4] Moreira M C F, Conti C C and Schirru R 2010 A new NaI(Tl) four-detector layout for field contamination assessment using artificial neural networks and the Monte Carlo method for system calibration *Nuc. Instr. and Meth. in Phys. Research Section: A* **621(1–3)** 302–309
[5] Vegors Jr. S H, Marsden L L and Heath R L 1958 (U.S. Atomic Energy Commission (USAEC) Report 17 IDO-16370).
[6] Miller W F and Snow 1961 W J NaI and CSI efficiencies and photo fractions for gamma-ray detection *Nucleonics* **19(11)** 174
[7] Sato T, Niita K, Matsuda N, Hashimoto S, Iwamoto Y, Noda S, Ogawa T, Iwase H, Nakashima H, Fukahori T, Okumura K, Kai T, Chiba S, Furuta T and Sihver L 2013 Particle and heavy ion transport code system PHITS Version 2.52 *J. Nucl. Sci. Technol.* **50(9)** 913-923
[8] Akkurt I, Tekin H O, Mesbahi A 2015 Calculation of detection efficiency for gamma detector using MCNPX *Acta Physica Polonica A*, **128(28)**
[9] Akkurt I, Gunoglu K, Arda S S 2014 Detection efficiency of NaI(TI) detector in 511-1332keV Energy Range *Sci. and Tech. of Nuclear Installation*, **5** 186798

Acknowledgements
The authors would like to extend their appreciation to all personnel of the Plant Assessment Technology group, Industrial Technology Division, Malaysian Nuclear Agency. The authors also would like to thanks all personnel that involve directly or indirectly in completing this study.

IOP Conf. Series: Materials Science and Engineering **298** (2018) 012054 doi:10.1088/1757-899X/298/1/012054

Stakeholder engagement for promoting the Comprehensive Nuclear-Test-Ban Treaty (CTBT): Malaysia's experience

F I A Rashid[1, a), M Z Zolkaffly [1, b) , N Jamal [1,c)

[1] Planning and International Relations Division, Malaysian Nuclear Agency, MOSTI, Bangi, 43000 Kajang, Malaysia

[a) faisal_izwan@nm.gov.my
[b) zulfakar@nm.gov.my
[c) noriahj@nm.gov.my

Abstract. In order to keep abreast on issues related to CTBT in Malaysia, Malaysian Nuclear Agency (Nuklear Malaysia), as the CTBT National Authority in Malaysia, has collaborated with local partners to implement various stakeholder engagement programme. This paper aims at highlighting Malaysia's approach in promoting CTBT through stakeholder engagement programme targeted at multilevel stakeholders, both national and international. Such programmes includes participation in the international forums, inter-agency meetings, awareness seminars, training courses, technical visits to IMS station, promoting civil and scientific application of International Monitoring System (IMS) data and International Data Centre (IDC) products using Virtual Data Exploitation Center (vDEC), inviting youth groups to participate in the CTBTO Youth Group ,and publications of CTBT-related topics. This approach has successfully fortify Malaysia's commitments at the international level, enhanced national awareness of global multilateral framework, increased stakeholders awareness and their roles related to CTBT, as well as building domestic capacity on CTBT matters. In conclusion, stakeholder engagement is crucial in promoting and enhancing stakeholders understanding on CTBT. Continuous engagement with relevant stakeholders will enable effective dissemination and smooth implementation of CTBT related matters that will eventually support global universalization of CTBT.

1. Introduction

Malaysia has always been a strong advocate for disarmament and world free of nuclear weapon. This commitment is reflected through Malaysia signature, ratification and accession to a number of key international legal instruments related to non-proliferation, disarmament and peaceful uses of nuclear energy, including the Non-Proliferation Treaty (NPT), Southeast Asian Nuclear-Weapon-Free Zone Treaty (SEANWFZ) and Comprehensive Nuclear-Test-Ban-Treaty (CTBT). In the context of nuclear non-proliferation and disarmament, CTBT was recognised as one of the thirteen practical steps for the systematic and progressive efforts to implement Article VI of the NPT.

CTBT is a multilateral treaty banning all nuclear explosions – everywhere and by anyone. Presently, the CTBT has not entered into force due to pending ratifications from remaining eighth countries under the Annex 2 of the CTBT namely, China, Egypt, India, Iran, Israel, North Korea, Pakistan, and United States [1]. Malaysia signed the CTBT on 23 July 1998 and ratified it on 17 January 2008. Subsequently, Malaysian Nuclear Agency (Nuklear Malaysia) was entrusted by the

IOP Conf. Series: Materials Science and Engineering **298** (2018) 012054 doi:10.1088/1757-899X/298/1/012054

Government of Malaysia to act as the National Authority for the implementation of the CTBT in the country. In Malaysia, the implementation of CTBT consist two aspects; 1) policy and legal 2) technology as shown in Figure 1. The policy and legal aspect covers the formulation of policy and legal framework on CTBT. Meanwhile, the technology aspect covers the Radionuclide Monitoring Station (RN42) and CTBT National Data Centre (MY-NDC).

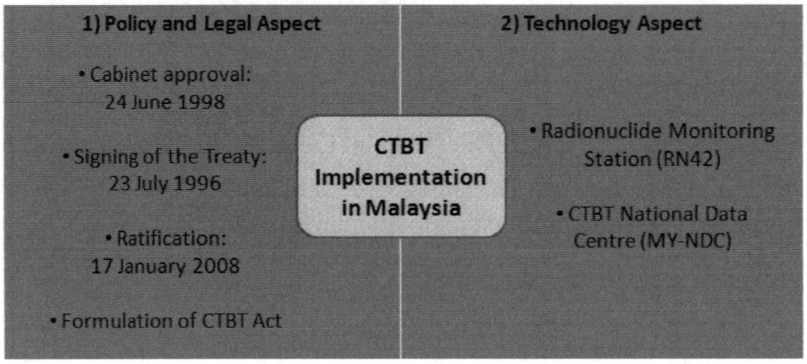

Figure 1. Two aspects of CTBT implementation in Malaysia.

As the CTBT National Authority, Nuklear Malaysia has implemented various programmes together with the engagement from relevant stakeholders in Malaysia. These programmes aimed at promoting and increasing awareness among relevant national stakeholders on CTBT. Every CTBT programmes is carefully plan and design to achieve its intended objectives and allow effective stakeholder engagement. In engaging stakeholder, there no single 'magic bullet' solution exists because each of them has unique purpose and is influenced by several factors [2]. Therefore, it requires thorough design and planning specifically tailored to the objectives sought for each programmes. On this note, the engagement with relevant stakeholders in CTBT programmes is critical in ensuring the stakeholders are able to grasp the issues surrounding CTBT as well as enabling them to communicate their feedbacks to improve the implementation of CTBT programmes in the future.

This paper aims at highlighting Malaysia's approach in promoting CTBT through stakeholder engagement programmes targeted at multilevel stakeholders, both local and international.

2. Framework for stakeholder engagement

Figure 2 illustrates the framework for stakeholder engagement adopted for the implementation of CTBT related programmes in Malaysia. It consists of three stages namely, 1) the identification of stakeholder and when to engage; 2) implementation of stakeholder engagement programmes; and 3) stakeholder feedback. From the implementation stage and onward, the review and improvement process will take place to assess the effectiveness of the implemented programmes and to acquire feedbacks from stakeholders. The feedbacks will be used to improve the implementation of future programmes. As a result, this framework ensures the CTBT programmes are subject to continuous improvement and stakeholder oriented.

IOP Conf. Series: Materials Science and Engineering **298** (2018) 012054 doi:10.1088/1757-899X/298/1/012054

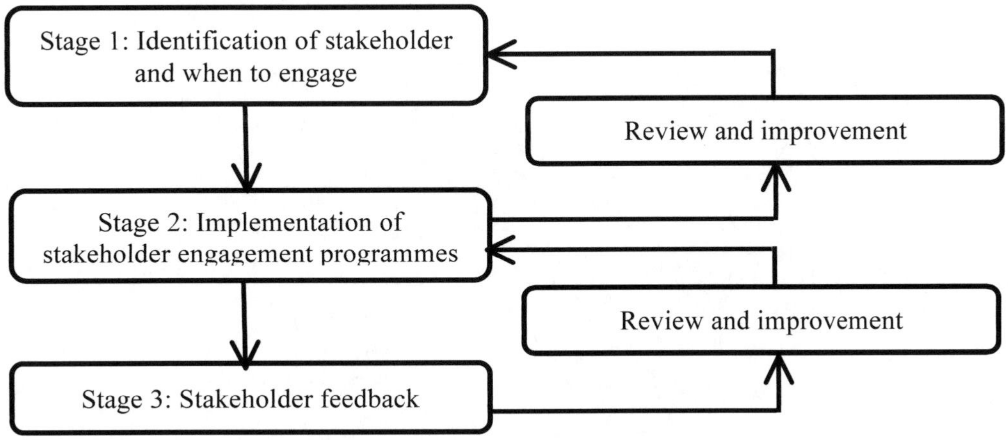

Figure 2. Framework for stakeholder engagement.

3. Stakeholder categories and level of engagement

3.1. Stakeholder Categories

In general, there are many types of stakeholder. The attributes and interests of each stakeholder vary from one another. For example, the national stakeholder's interests are mainly on legal and implementation aspects of the CTBT that shall be implemented in accordance to the national framework. Meanwhile, for international stakeholders, they are more interested on Malaysia's commitment on CTBT and the means of implementing CTBT programmes. Taking into consideration the diversity of stakeholder's interests, the CTBT programmes are carefully designed to suit the stakeholders so that it can stimulate their participation hence creating a 'buy-in' to the programme and thus to its outcomes.

In Malaysia, the stakeholders are categorised into two groups - international and national. Figure 3 shows the graphical representation of these stakeholder groups and its entities. At the international level, the stakeholders include the international organisations such as the United Nations and the Preparatory Commission of the Comprehensive Nuclear-Test-Ban Treaty Organisation, regional organisations such as ASEAN, individual country, and international community. Meanwhile, at the national level, the stakeholders are government agencies, local authorities, universities and the public including the youth groups. The communication is always a two way process between Nuklear Malaysia and the stakeholders. It allows Nuklear Malaysia to deliver the programmes and at the same time gather feedbacks from the stakeholders.

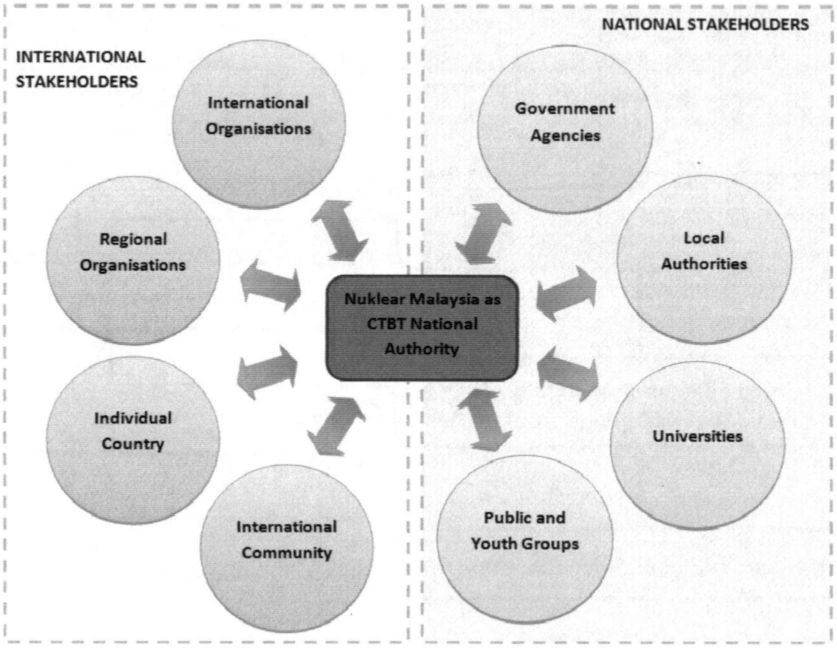

Figure 3. Stakeholders categories.

3.2. Level of Engagement

According to the International Association for Public Participation, public participation consists of five elements with increasing level of engagement as shown in Table 1 [3]. Those elements are inform, consult, involve, collaborate and empower. The further down the table, the more influence the stakeholders have on CTBT programmes. In Table 1, the stakeholders are mapped into various levels of engagement. In addition, a variety of mechanisms were employed by Nuklear Malaysia to engage with different stakeholders, depending on level of engagement. Using this hierarchy of engagement mechanisms, the stakeholders were presented with numerous avenues to engage and deliver their feedbacks and inputs.

4. Implementation of stakeholder engagement programmes

In promoting the CTBT, Nuklear Malaysia has implemented various CTBT programmes to engage relevant stakeholders, both national and international. The programmes allow two-way communications between Nuklear Malaysia and stakeholders. The programmes are made up of awareness seminars, meetings and workshops, conferences, webnews and social media, virtual data exploitation centre (VDEC), CTBTO Youth Group and publications.

4.1. Meetings, Workshops and Conferences

Meeting, workshops and conferences offer the platforms for stakeholders to communicate their inputs and feedbacks from their perspective. Nuklear Malaysia frequently conducted coordination meetings with relevant stakeholders in order to ensure their inputs were taken into consideration in the CTBT programmes. Since the implementation of CTBT covers legal and technical matters, views from stakeholders, particularly the government agencies and local authorities are pertinent to ensure that CTBT is implemented in accordance with national framework and regulations. Meanwhile, at the international level, meetings, workshops and conferences are usually organised by the international organisations to review information, define issues, brainstorm ideas and define solutions on CTBT related matters. Malaysia usually participated in the international meetings, workshops and conferences to convey its views as well as engaging with international stakeholders on CTBT matters.

IOP Conf. Series: Materials Science and Engineering **298** (2018) 012054 doi:10.1088/1757-899X/298/1/012054

The international meeting, workshops and conferences also allowed Malaysia to broaden its network on CTBT and engage in universal issue such as nuclear disarmament and non-proliferation.

Table 1. Level of stakeholder engagement.

Level of Engagement	Stakeholders Involved	Engagement Mechanisms
Inform To provide the stakeholders with balanced and objective information to assist them in understanding the issues	All international stakeholders	International conferences and workshops
	All national stakeholders	National seminars, web news and article in website and social media, publications
Consult To obtain feedback or decision stakeholders	International stakeholders United Nations, CTBTO Preparatory Commission	International workshops, meetings, conferences
	National stakeholders Government Agencies, Local Authorities	Meetings
Involve To work directly with the stakeholders to ensure their views and concerns are consistently understood and considered	National stakeholders Government Agencies, Local Authorities	Meetings
	Universities	Virtual Data Exploitation Centre (VDEC)
	Youth Groups	CTBTO Youth Group
Collaborate To partner with the stakeholders in implementation	International stakeholders CTBTO Preparatory Commission	International training courses and workshops
	National stakeholders Government Agencies, Local Authorities	Operation of RN42, training courses, CTBT authorised users, VDEC
Empower To equip the stakeholders with appropriate capacity and capability to enable them to play effective role	National stakeholders CTBT authorised users	Training courses

4.2. Awareness Seminars

Awareness seminars are the platform to disseminate information on CTBT. At the national level, Nuklear Malaysia often organised awareness seminar on CTBT. The seminars are targeted at all types of national stakeholders. In order to build awareness and instill interest among stakeholders, the materials presented in the seminar are designed to be general and broad. The seminars also allow the audience to express their queries and views.

4.3. Training Courses
Training courses were conducted to develop stakeholder's capacity and capability in subjects related of CTBT. At the international level, training courses are organised by the international organsations, particularly the Preparatory Commission of the Comprehensive Nuclear-Test-Ban Treaty Organisation. The international training courses are mainly focusing on the operation of International Monitoring System (IMS) stations and national data centre. At the national level, Nuklear Malaysia conducted the training course for CTBT authorised users in Malaysia. The training course aimed at building the awareness on CTBT as well as developing the CTBT authorised user's capability to access and use the International Monitoring System (IMS) data and International Data Centre (IDC) products. The training course used e-learning materials developed by the Preparatory Commission of the Comprehensive Nuclear-Test-Ban Treaty Organisation. Through the training courses, the CTBT authorised users in Malaysia are able to perform their roles more effectively and keep abreast with CTBT issues.

4.4. Operation of RN42
Under the IMS network of CTBT, Malaysia hosted RN42 at Cameron Highland, Pahang. Since the beginning of its operation in November 2008, Nuklear Malaysia has formed a close collaboration with the Malaysian Meteorological Department (MMD) in operating the RN42. The activities of RN42 have engaged the MMD as an important partner on CTBT issues. In addition, MMD staffs have the opportunity to undergo technical trainings on the operation of RN42 organised at the international level.

4.5. CTBT Authorised Users
Presently, Malaysia has twenty-eight CTBT authorised users. These users were nominated by Malaysia to the Preparatory Commission of the Comprehensive Nuclear-Test-Ban Treaty Organisation to enable them to have access to IMS data and IDC products. Since IMS data and IDC products are confidential, access are only limited to these users. These users are coming from Nuklear Malaysia, Atomic Energy Licensing Board, Malaysian Meteorological Department and National Security Council. By engaging users outside Nuklear Malaysia, they are able to keep abreast with CTBT matters and develop their capability to utilise data and products of CTBT for verification of nuclear explosion as well as for civil and scientific applications.

4.6.VDEC
VDEC is a dedicated platform developed by the Preparatory Commission of the Comprehensive Nuclear-Test-Ban Treaty Organisation to provide scientists and researchers from many different disciplines around the globe with access to data to conduct research and to publish new findings [4]. In order to help promote the application of IMS data for civil and scientific application, Nuklear Malaysia has extended the information on VDEC to several universities in Malaysia. This effort aimed to include the academic community, particularly the university in CTBT programme and to make them aware and use the IMS data for conducting scientific research. The VDEC also enable them to exchange knowledge and information at a range of forums, including the CTBTO's Science and Technology Conference series, workshops and technical meetings.

4.7 CTBTO Youth Group
Recognising the importance of youth involvement as key to advancing the Treaty's entry into force, the CTBTO Youth Group was established in 2016. The group is open to all students and young graduates who are interested in directing their careers towards global peace and security and who wish

IOP Conf. Series: Materials Science and Engineering **298** (2018) 012054 doi:10.1088/1757-899X/298/1/012054

to actively engage in promoting the CTBT and its verification regime [5]. Sharing the same spirit, Nuklear Malaysia extended the information on CTBTO Youth Group to youth groups at universities. The universities include the Universiti Teknologi Mara, Universiti Putra Malaysia, Universiti Malaya, This effort aimed at engaging Malaysian youth as agent for promoting CTBT in Malaysia.

4.8. Publications

Stakeholders in Malaysia are also engaged via publications on CTBT. Nuklear Malaysia has published a few publications on CTBT and is available at public domains. Through publications, it could provide information related to CTBT to the stakeholders. As the publications are available at public domain, a large crowd of stakeholders can be engaged. Despite the engagement through publications is a one way communication, it allows early engagement to broader stakeholders and may lift their interest to be more engaged on CTBT issues.

4.9. Articles in Web and Social Media

As we are living in a digital world, we could not rule out the influence of website and social media on stakeholders. In this relation, Nuklear Malaysia disseminates information on CTBT via articles in web and social media. Every CTBT programme and activity conducted by Nuklear Malaysia was posted on the website and social media. As a result, more stakeholders can be reached and it increased Nuklear Malaysia's visibility as the CTBT National Authority in Malaysia.

5. Outcome and benefits

The outcomes of engaging stakeholders in CTBT programmes are as follow:

- fortify Malaysia's commitments at the international level
- enhance national awareness of global multilateral framework including CTBT
- increase stakeholders awareness and their roles related to CTBT
- build domestic capacity on CTBT matters
- increase Nuklear Malaysia's visibility as CTBT National Authority in Malaysia

The benefits of engaging stakeholders in CTBT programmes are listed in Table 2.

Table 2. Benefits to Nuklear Malaysia and stakeholders.

Benefits to Nuklear Malaysia	Benefits to Stakeholders
• Improved communication between Nuklear Malaysia as CTBT National Authority and stakeholders • Greater cooperation, collaboration and mutual support with stakeholders • Accommodate stakeholder's inputs for improved CTBT programme • Ensure the programmes are implemented in collaboration and support with stakeholders • Address stakeholder's concerns and needs • Improved relationship and a wider network of support • Enhanced implementation of CTBT programme • Increased stakeholder awareness	• Greater opportunities for stakeholders to contribute to CTBT programmes and related matters • Provide a platform to have their inputs heard and addressed • Opportunity to participate in decision making process • More open and transparent line of communication • Early identification of connections between stakeholder's interests and CTBT programmes • Increase stakeholders awareness and understanding

6. Conclusion

Engaging relevant stakeholder in CTBT programmes is pertinent to ensure the implemented programmes are well-suited to the targeted stakeholders and creating an environment that encourage their engagement in the programmes. Malaysia through Nuklear Malaysia as the National Authority for the implementation of the CTBT in Malaysia has implemented various programmes with engagement from relevant stakeholders. Through those programmes, it has enabled wider promotion of CTBT among stakeholders, both national and international. In addition, stakeholders are able to engage and share their inputs. It also widen Nuklear Malaysia's network of support as well as helping Nuklear Malaysia's capacity to implement and sustain the CTBT programmes. Consequently, it will not only benefit Nuklear Malaysia alone, but Malaysia as a whole.

7. References

[1] M. B. D. Nikitin 1 September 2016 Comprehensive Nuclear-Test-Ban Treaty: Background and Current Developments Congressional Research Service RL3354

[2] S. Yee May 2010 Stakeholder engagement and public participation in environmental flows and river health assessment A Report from the Australia-China Environment Development Partnership

[3] International Association for Public Participation 14 February 2017 What is the spectrum of public participation? available url: https://sustainingcommunity.wordpress.com/2017/02/14/spectrum-of-public-participation/

[4] CTBTO 20 September 2017 Virtual Data Exploitation Centre (VDEC) available url: https://www.ctbto.org/specials/vdec/

[5] CTBTO 20 September 2017, "CTBTO Youth Group" available url: https://youthgroup.ctbto.org/youth-group-homepage/

IOP Conf. Series: Materials Science and Engineering **298** (2018) 012055 doi:10.1088/1757-899X/298/1/012055

Effectiveness of mathematics education in secondary schools to meet the local universities missions in producing quality engineering and science undergraduates

Abu Bakar Hasan[1, a)], Abdul Fatah Abdul[1] and Zalilah Selamat[1]

[1]Faculty of Engineering and Built Environment, Universiti Sains Islam Malaysia, Bandar Baru Nilai, Negeri Sembilan, Malaysia

abakarh@usim.edu.my

Abstract. Critical claims by certain quarters that our local undergraduates are not performing well in Mathematics, Statistics and Numerical Methods needs a serious thinking and actions. Yearly examinations results from the Sijil Pelajaran Malaysia (SPM equivalent to A-Level) and Sijil Tinggi Pelajaran Malaysia (STPM equivalent to O-Level) levels have been splendid whereby it is either increasing or decreasing in a very tight range. A good foundation in mathematics and additional mathematics will tremendously benefit these students when they enter their university education especially in engineering and science courses. This paper uses SPM results as the primary data, questionnaires as secondary, and apply the Fish Bones technique for analysis. The outcome shows that there is a clear correlation between the causes and effect.

1. Introduction

Mathematics education, especially in secondary schools in Malaysia, are taught in the form of the basic syllabi for Form One to Three, and the advanced syllabi for Form Four to Five. While in the upper secondary schools, that is, Form 6, mathematics education is basically at par with the one in the first year at university. Additional mathematics that a student took the secondary school is useful for those students taking engineering and science bachelor degrees. Undergraduates who took engineering, physics and mathematics as their field of studies may find that the knowledge and understanding in additional mathematics are very helpful.

Mathematical-based subjects such as Mathematics, Numerical Methods and Statistics are the three subjects usually offered to those students majoring in engineering, physics and mathematics at university. These subjects usually involve theorems, formulae and laws, and need a clear understanding on the subject matters which requires derivations, proofing and applications. Mathematics subject alone covers algebra, calculus, integration and derivations [1]. Numerical Methods is basically another ways for solving mathematical problems [2], while Statistics covers probability [3].

2. Issues and problem statements

The SPM results from the state of Melaka (table 1) for Additional Mathematics were very encouraging with more than 70% passes has been achieved yearly but at the same time the total number of candidates taking the subject were declining and, this is not a good sign.

IOP Conf. Series: Materials Science and Engineering **298** (2018) 012055 doi:10.1088/1757-899X/298/1/012055

Table 1. Yearly SPM Results for Additional Mathematics from Melaka.

Year	% Passes	Grade A+	Grade A	Grade A-	Grade G(fail)	Total No. of Candidates
2012	76.89	1013	51	485	1515	6532
2013	79.0	876	222	419	1386	6602
2014	75.5	362	417	389	1581	6463
2015	72.4	191	435	290	1637	5936
2016	70.6	242	373	259	1536	5220

Malaysia Government is introducing STEM (Science, Technology, Engineering and Mathematics) at the school level. The decreasing numbers of students taking additional mathematics at the SPM level means the chances of the local university in Malaysia getting the right quota for STEM courses are at risk. More students are moving to courses that are non-technical or requires less or non-mathematical type of subjects such as information technology, computer science or hotel management.

Undergraduates may find that mathematics, numerical methods and statistics are subjects that requires them to be good in problem solving, derivation and analysis in order to excel in those subjects. Thus, weak students in mathematics will find mathematics at the university level a subject difficult to understand. Therefore, the local universities are producing indirectly graduates who are afraid of numbers.

The main objective of this paper is to show that the above claim or argument has bases. A case study at Universiti Sains Islam Malaysia (USIM), Bandar Baru Nilai, Negeri Sembilan was selected where the primary data was taken from the Faculty of Engineering and Built Environment (FKAB).

3. Analytical and critical analysis

The paper is proposing the Fish Bone technique [4] for analysis purpose that comes together with a hypothesis as follows: Smart students in Mathematics at secondary school will become smart undergraduates in Mathematics at university.

For the primary data collection, data are being collected from USIM undergraduates. From table 1, smart students are defines as those students who obtained grade A+, A and A- in their respective subjects at SPM level. Figure 1 shows the smart students in Additional Mathematics made up 15% to 24% in the state of Melaka, which is quite high in terms of percentage. These students were assumed to continue their studies after finishing their SPM even though this work has no proof to back up the statement. Anyway, in order to support the mentioned hypothesis, data are collected from the smart students (who not necessary came from Melaka) that managed to get a place to study Electronic Engineering degree at USIM.

Figure 2 shows the smart students' results for Mathematics I (KEH1132) are nothing to shout loud where nobody obtained Grade A in their examination. The two intakes were taught by two different Mathematics lecturers in English.

Secondary data collections are conducted by giving three different sets of questionnaires to three groups, mainly (i) Bachelor of Electronics Engineering undergraduates at FKAB (USIM), (ii) Additional Mathematics secondary school teachers from Melaka and Selangor, and lastly (iii) Mathematic Lecturers from USIM.

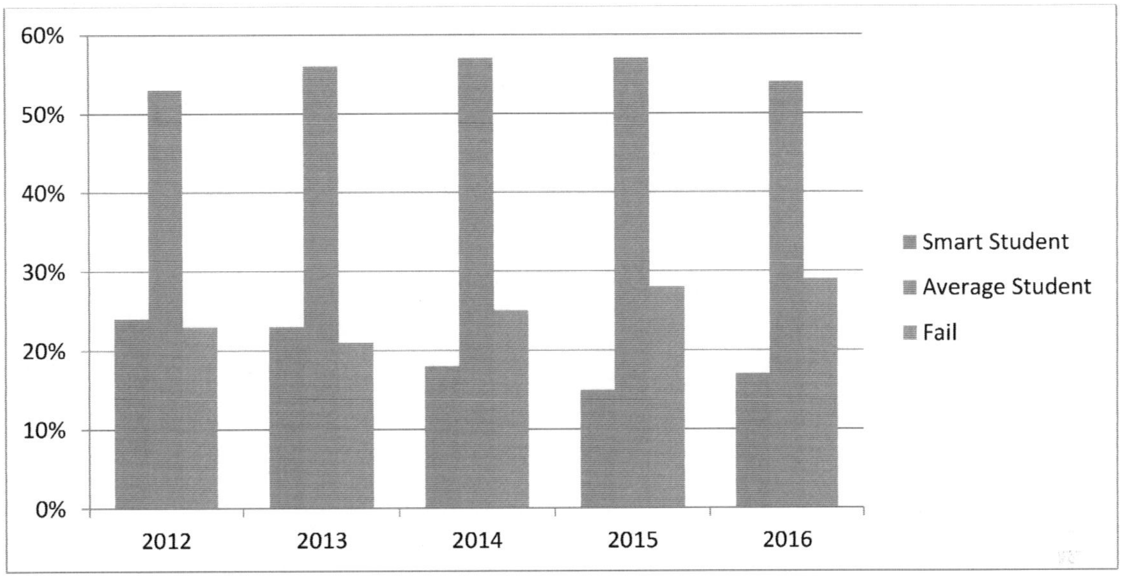

Figure 1. Yearly smart students and failures SPM results for Additional Mathematics.

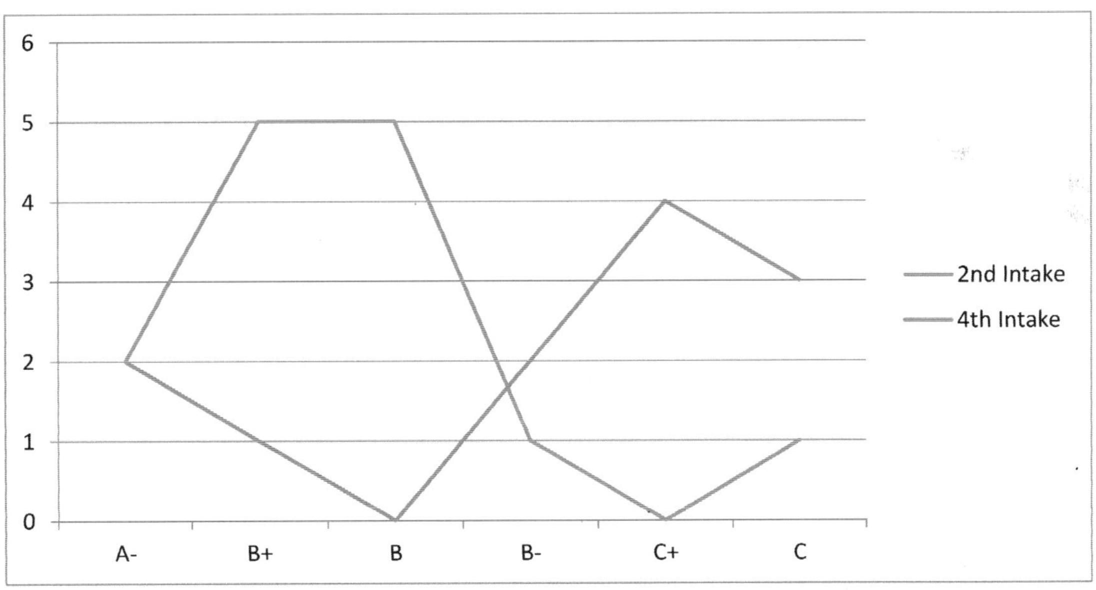

Figure 2. Smart Students at FKAB, USIM for Mathematics I (KEH1132).

3.1. Bachelor of Electronics Engineering Undergraduates

From table 2 below, 25 out of 29 undergraduates agreed to Question no. 2 which states that "Do you find any difficulties in excelling Year 1 Mathematics subjects?". The other four undergraduates who have no difficulties in taking KEH1132 were assumed to be the smart students, and have no means to counter check on this since the questionnaires did not carry any names. Again, all of the undergraduates said "yes" when it came to Question no. 5 which stated that "Do you think that your secondary school Mathematic teachers are good enough?". Lastly, the undergraduates were about even to Question no. 3: "Do you agreed that the enhancement exercises in Additional Mathematics' secondary school classes will develop the students to understand less?".

Table 2. FKAB Undergraduates.

Set A	TT	T	ST	R	Y	T	S	TS
Q1	3	25	0	1				
Q2					25	4		
Q3							14	15
Q4							8	6
Q5					29	0		

3.2. Additional Mathematic Secondary School Teachers

From table 3 below, around 83% of the secondary school teachers did not agree Q1 which ask "Do you think that the enhancement exercises in Additional Mathematics classes will help the student to memorise the question and answer patterns?". For Q2, all teachers agreed to a more creative technique to be used in class. The teachers were on even when comes to Q3: "Do you agreed that your secondary school teachers should come from first graders". All the teachers were saying "not agree" to Q4: "Do you think that your secondary school teachers are not keen in teaching Additional Mathematics?" None has answered Q5.

Table 3. Additional Mathematic Teacherswe

Set B	S	TS
Q1	1	5
Q2	6	0
Q3	3	3
Q4	0	6
Q5	No answer	

3.3. University Mathematic Lecturers

From table 4 below all lecturers tended to say "Moderate (S)" rather than "Not Difficult (TS)" when answering Q2: "What is your comments on the level of understanding amongst your Mathematic Year 1 undergraduates?" and Q3: "What is your comment on your undergraduate willingness to master Mathematic Year 1 subject?". To Q5, most of the lecturers said that the undergraduates needed a better understanding in order to excel in the subject.

Table 4. University Mathematics Lecturers

Set C	Y	T	R	S	T	S	TS
Q1	3	1					
Q2			2	2	0		
Q3			1	3	0		
Q4						2	2
Q5			Not good in understanding the subject				

4. Suggested strategies, plan of action and solution

For the short term strategy, it is better to look at the pattern in setting up the SPM examination questions. The present pattern of questions can be predicted rightly by any candidates, that is, by looking through the past year question papers. The pattern has been static for years. This phenomenon is strongly supported by the enhancement exercise carried out by the teachers at schools. The proposed medium term strategy is to look at the criteria in taking in teachers especially at the secondary school level. The criteria should be in the common belief that good students should be creamed by a group of good teachers. Our neighbouring country, Singapore, has put a condition into her teaching profession criteria, that is, only excellent graduates are being considered. Today, Singapore education system is one of the best in the world. Courage implementers at the ministry level must be prepare to work without being influenced by politics is one of the long term strategy that needs consideration.

The Ministry of Education should come out with a plan of action to improve the present scenario of education. The scenario of having more than one type of secondary schools, that is, national and non-national type, has indirectly caused the smart students to register with a particular school type. It is well known that in the Chinese school certain subjects are being taught in Chinese and English at two different times. Thus, Chinese school students are better prepared when they entered the examination hall compared to the national type of schools. Again, looking at the develop countries in the world such as Japan, etc. whereby the Japanese school education system is well developed and teachings are in Japanese. The government must believe in herself that one type of schools has been proved successful in other countries such as Germany, USA, Sweden, China and Korea.

Solutions to the problems must be found as soon as possible. Local universities are finding it harder to fill up the intake quotas for engineering courses which has forced some of them to streamline or cancel certain undergraduate courses. Table 2 shows the failures rate in additional mathematics is increasing yearly and this is very serious indicator. This means less students are eligible to enrol in engineering and science courses at the tertiary level.

5. Recommendations for actions

The paper is proposing the following main causes as the inputs to the Fish Bone Diagram (figure 3) as stated below, and these causes came from the three questionnaires' findings and Melaka SPM results:

F1 - The additional mathematics teachers at the secondary schools are capable to deliver the teaching materials.
F2 - In Melaka, roughly 75% passes in additional mathematics at SPM level yearly.
F3 - Around 86% of the engineering undergraduates claimed that Year 1 mathematics subject are difficult.
F4 - None of the so-called smart students obtained an 'A' in their first year for Mathematic I.
F5 - A creative technique should be used in the school classes was 100% supported by all the teachers.
F6 - Subject interest and understanding amongst the undergraduates are questionable.

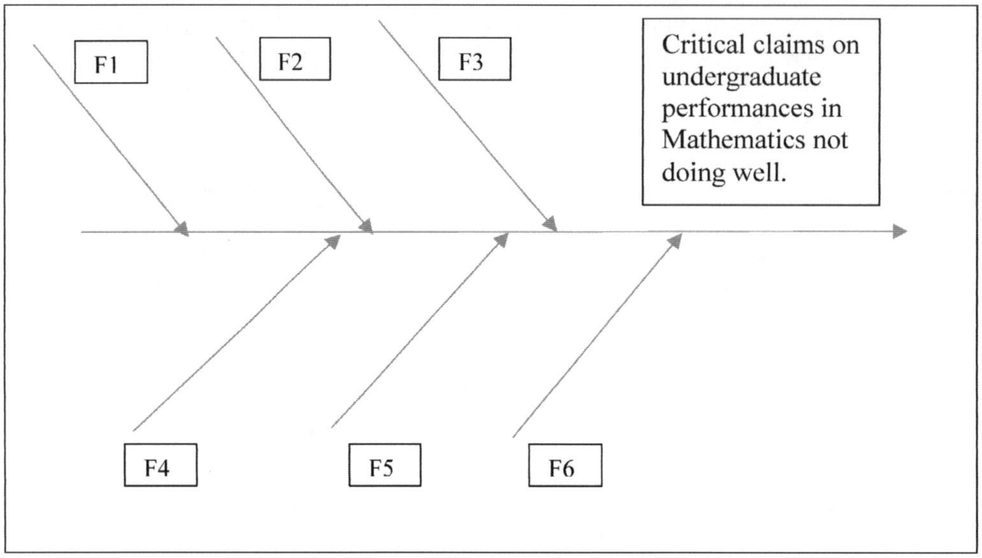

Figure 3. Fish Bone Diagram.

The outcome or effect of figure 3 is that there was a basis on the claim that mathematics at the university level needs a better understanding in order to master it; whereby the enhancement exercise is no more applicable at the university education level.

6. Conclusion

From this study, there is no truth in the proposed hypothesis, that is, any smart students at SPM level do not necessarily turned smart when they further their studies into the local university when comes to Mathematics. Our country is producing less and less mathematicians, engineers or scientists who are capable in challenging Albert Einstein's caliber and ingenuity. Vision 2020 and TN2050 are the two milestones that Malaysia needs to achieve with the supports of local universities in producing quality STEM graduates.

7. References

[1] Sullivan M 2015 *Pre-calculus* 10th Ed. Prentice Hall
[2] Rao K S 2007 *Numerical Methods for Scientists and Engineers* 3rd Ed. Prentice-Hall India
[3] Navidi W 2014 *Statistics for Scientists and Engineers* 4th Ed. Mc-Graw Hill Higher Education,.
[4] Fons T and Piet H C 2016 100+ Management Models. MPH Bookstores

IOP Conf. Series: Materials Science and Engineering **298** (2018) 012056 doi:10.1088/1757-899X/298/1/012056

Adsorption of thorium onto soil around coal-fired power plant

Dgku Kamilah Pg. Ismail , Nik Azlin Nik Ariffin , Nurzulaifa Shaheera Erne Md Yasim

Faculty of Applied Sciences, Universiti Teknologi MARA, 40450 Shah Alam, Selangor, Malaysia

deekay1705@yahoo.com

Abstract. A study has been performed to determine the adsorption of Thorium (Th-232) onto three soil sample around Jimah Energy Venture coal-fired power plant located in Jimah, Port Dickson. Parameters that were set constant include soil dosage at 2.5g each; 50 ml of Th-232 solution. Parameters tested for optimum adsorption was based on initial concentration and pH of Th-232, and contact time. Highest percentage removal of Th-232 obtained for sample 1 and 3 at initial concentration of 25 ppm, pH 5 for 140 minutes are 99.69% and 99.98% respectively, while for sample 2, the highest percentage removal obtained at initial concentration of 25 ppm, pH 7 for 120 minutes are 99.92%. Freundlich and Langmuir isotherms were used to determine the adsorption isotherm for Th-232 adsorption. Based on this study, the result can be used as future reference regarding soil capability of adsorbing radionuclide. It is concluded that all soil sample has considerable ability of adsorbing Th-232 from entering the groundwater.

1. Introduction

Coal is one of the largest contributors for power generation in Malaysia. Ramasamy (2012) stated that the estimation for coal demand in Malaysia can be up to 10280MW in 2020. The coal-fired power plant produces huge amounts of fly ash and bottom ash containing natural radionuclides (Pandit *et al,* 2011). When coal is burned into fly ash, uranium and thorium are concentrated at up to 10 times from their original levels (Mara, 2007). The main radiation hazard from solid fallout will result in elevated concentration of natural radionuclide in surface of soils around the power plant (Zakaria et al., 2010). The two principal processes that can contaminate soil from coal industry is through precipitation of airborne dusts and land disposal of uranium or thorium containing wastes (Abdul Rashid et al., 2016). Thorium accumulation in soil can likely cause phytotoxicity and poisoning the food chain (Guo et al., 2008).

This study investigated the adsorption of the Th-232 onto the soil around Jimah coal-fired power plant. Jimah East Power Sdn. Bhd (JEP) is a special-purpose vehicle set up to develop the 2,000MW coal-fired power plant in Jimah, Negeri Sembilan, known as Project 3B (Yee, 2016). The coal-fired power plant is located far from residential area but the surrounding is packed with other industrial activities. According to U.S. Agency for Toxic Substances and Disease Registry, workers are exposed to higher levels of thorium and other radionuclides in certain thorium related industries, as indicated by the measurement of Th-232 exhaled breath of some Th-232 plant workers is indirect evidence of

IOP Conf. Series: Materials Science and Engineering **298** (2018) 012056 doi:10.1088/1757-899X/298/1/012056

higher thorium intakes. Thorium can increase probability of lung cancer, pancreas, and blood in workers exposed to high levels of it in the air (ASTDR, 2014).

In this study, soil act as the adsorbent of the Th-232. The surface of the soil has net negative charge and radionuclides are attracted from the solution to balance the surface charge by adsorption (Soderlund et al., 2011). Thorium was most soluble in acidic soils and the highest concentrations of dissolved Th and U were found close to the soil surface (Ahmed et al., 2012). Radionuclides mobility in the soil and their availability to be uptake by plants depend on the mineralogical composition and soil texture, pH, organic matter content, cation composition of soil solution and Ca and K concentrations. Soil adsorption capability of Th and U depended on the soil composition and pH of the soil solution (Vandenhove et al., 2007). For example, Vandenhove et al., (2007) observed that soil-liquid distribution coefficients were significantly related to organic matter content and amorphous Fe. Other studies also found that uranium sorption in soils were highly influenced by natural organic matter content, followed by Fe, Mn, and Al oxides content in soil (Bednar et al., 2007). This shows that in suitable condition, soil can become potential adsorbent of Th-232. The result from this study can become future reference in selecting new coal-fired power plants location which can adsorb Th-232 from leaching into the groundwater. Previous studies focuses more on thorium adsorption in wastewater using activated carbon, mesoporous molecular sieves and oxidized carbon nanotubes (Cheng et al., 2014). The lack of study on capability of soil on removing Th-232 around coal-fired power plants is the reason why this study is conducted.

2. Methodology

2.1 Sampling Site

Sample soils were collected at 3 different points around Jimah coal-fired power plant that was located at Port Dickson, Negeri Sembilan. The sampling point is located 2 km from the power plant and the distance between each sampling point was 500 meter. The coordinate of the sampling area is given in Table 1.

Table 1. Coordinate of sampling location.

Sampling point	Latitude	Longitude
Sample 1	2.597517°	101.723235°
Sample 2	2.602260°	101.723496°
Sample 3	2.608739°	101.729737°

3. Sample preparation and analysis

Collected soil samples were air-dried for two to three days, sieved to less than 2 mm size and oven dried overnight at 110 °C. Next, the soil samples were ground and sieved further for XRD and FTIR analysis to determine the functional groups in soil samples and also the chemical compounds present in the sample. For physical and chemical characteristic of soil samples, analysis for organic content, pH, cation exchange capacity (CEC) and particle size distribution clay were carried out.

3.1 Batch analysis

The adsorption study method was based on US-EPA (1992, 1999) using 50 mL of respective Th-232 concentrations each at 5, 10, 15, 20, and 25 and 40mg/L were mixed with 2.5 g soil in conical flask and the mixture were put on shaker for 24 hours. Next the mixture was filtered from the solid phase and the concentration of Th-232 remaining in the solutions was measured. After optimum initial concentration obtained, the experiment proceed with optimum pH determination at range from 2-10 for 24 hours. Then, at the optimum pH and initial concentration, the adsorption process conducted with different contact time period between 60-140 min.

IOP Conf. Series: Materials Science and Engineering **298** (2018) 012056 doi:10.1088/1757-899X/298/1/012056

The concentration of Th-232 in solution determined using Inductively Coupled Plasma Optical Emission Spectrometer (ICP-OES). Percentage of Th-232 removal was calculated using the calculation i) below whereby C_i is the initial concentration and C_f is the final concentration of Th-232.

$$(\%)removal = \frac{C_i - C_f}{C_i} \times 100 \tag{1}$$

3.2 Adsorption isotherm

Analysis of adsorption was studied using Langmuir and Freundlich equations. Langmuir isotherm assumes that the binding sites is in infinite numbers having the same affinity for adsorption and the adsorption is mono layer and have no interaction between the adsorbed molecules. Langmuir isotherm linear equation to determine adsorption parameters described as below:

$$\frac{1}{q_e} = \frac{1}{q_m} + (\frac{1}{q_m} KL)\frac{1}{Q} \tag{2}$$

where q_e (mg/g) and q_m (mg/g) are the amount of adsorbed Th-232 per unit mass of adsorbent and adsorption capacity, respectively. *KL* was the Langmuir constant, related to the bonding energy between the adsorb ion and adsorbent (L/mg) and C_c is a unadsorbed Th-232 concentration in solution at equilibrium (mg/L).

The Freundlich isotherm used to describe heterogeneous systems and reversible adsorption and is not necessarily to form monolayers. The linear form equation was defined as follows:

$$\log q_e = \log K_f + \frac{1}{n}\log C_e \tag{3}$$

where q_e was the amount of adsorb Th-232 per unit weight of soil at equilibrium (mg/g), C_e was the equilibrium concentration of Th-232 (mg/L), K_f was Freundlich constant (mg/g), and $1/n$ was adsorption intensity respectively. n is a measurement of the deviation from linearity of adsorption.

4. Result and discussion

4.1 Soil physical and chemical properties

The results of physical and chemical characterization of soils were tabulated in table 2. The organic matter percentage in sample 1 is the highest compared to sample 2 and 3 while the CEC is averagely same in all soil samples. Organic matter occurred by decomposition process of dead organisms from physical and biochemical complex organic matter into simple organic matter as they reverted back to soil. CEC is the capacity of soil surface in holding nutrient composition. High CEC is equal to high organic matter percentage and this is tally with the clay percentage is the highest for sample 1. The average pH of sample soil are between pH 4 and pH 5 which is suitable for Th-232 adsorption onto the soil because Th-232 is most stable in acidic soil condition [4]. The collected soil was categorized as sandy clay loam soil as there were highest percentage of sandy soil compare to silt and clay soil texture (table 2). From XRD result it was found that silica dioxide, SiO_2 percentages are high in all soil samples which are 87%, 85% and 80% respectively and this is concurrent with percentage of clay in each sample. Other mineral present in sample soil are magnesite, hematite, pyrolusite and spinel in less than 12% of each mineral.

Table 2. Soil texture and other physical-chemical properties from three different locations.

Soil Sample	*Organic matter (%)*	*pH*	*Cation exchange capacity (cmol/kg)*	*Sand (%)*	*Silt (%)*	*Clay (%)*
Sample 1	9.82	4.88	0.004	45.99	12.50	28.49
Sample 2	1.30	5.75	0.004	47.20	14.50	27.26
Sample 3	2.85	5.57	0.003	43.46	13.42	21.00

Three different absorption peaks were detected from the FTIR spectrum. The peak at 3651.19 cm^{-1} shows the OH functional group. The adsorption band at 1507.86 cm^{-1} occurred for primary amine N-H functional group while the carboxylic acid (-COOH) group were detected at 1751.01 cm^{-1} wavelength. From the previous study had stated that hydroxyl and carboxylic acid groups played a major role in the adsorption process [9]'.

According to Wu *et. al.* [6], the hydroxyl group is highly effective towards adsorption rate and efficient through the exchange of radionuclide ions. Besides that, the carboxylic acid group adsorption rate is relatively high due to the presence of more negatively charged groups on its surface [19].

4.2 Effect of initial concentration

The effect of initial concentration on Th-232 removal was shown in figure 1. The percentage removal of Th-232 is directly proportional with the increasing of initial concentration from 5 ppm to 25 ppm for all collected samples. The increase in Th-232 adsorption with the concentration can be attributed to the availability of greater surface area and larger number of adsorption site. The maximum percentage removal of Th-232 for sample 1, sample 2 and sample 3 are 99.92%, 99.88%, and 99.95% respectively at initial concentration of 25 ppm.

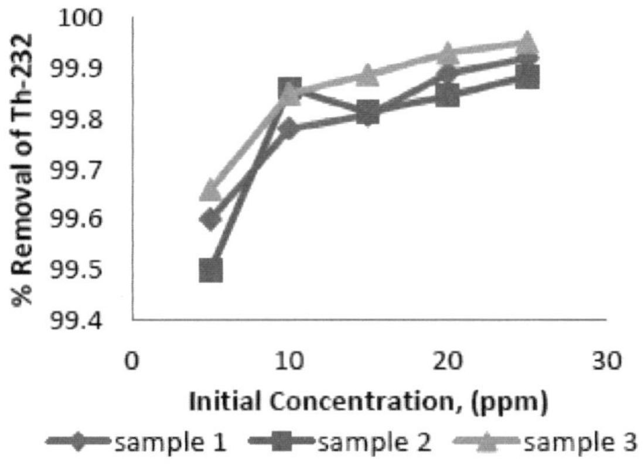

Figure 1. Graph of initial concentration versus percentage removal of Th-232.

4.3 Effect of pH

Figure 2 shows the percentage removal of Th-232 increase from pH 2 to pH 5 before fluctuated at pH above 5. This is due to resistant minerals were decomposed under stronger acidic conditions, and the presence of acid radical ions in soil enhanced the solubility of ThO_2 by forming thorium-sulfate, fluoride and phosphate complexes below pH 7 [16]. The optimum pH 5 was determined for sample 1 and sample 3, but for sample 2 the optimum pH is 7. At acidic condition, the proton of the functional group such as carboxyl, hydroxyl, and amine group in the adsorbent does not dissociate easily and thereby decreases the interaction of Th-232 with the adsorbent. However, when pH increases to 5, due to de-protonation of these functional groups, the increase in negative charge on the adsorbent surface offers more Th-232 adsorption sites and thus increases Th-232 adsorption . The maximum percentage removal of Th-232 is 99.70%, 99.72%, and 99.77% respectively. The maximum adsorption rate due to development negative charged on the surface of adsorbent.

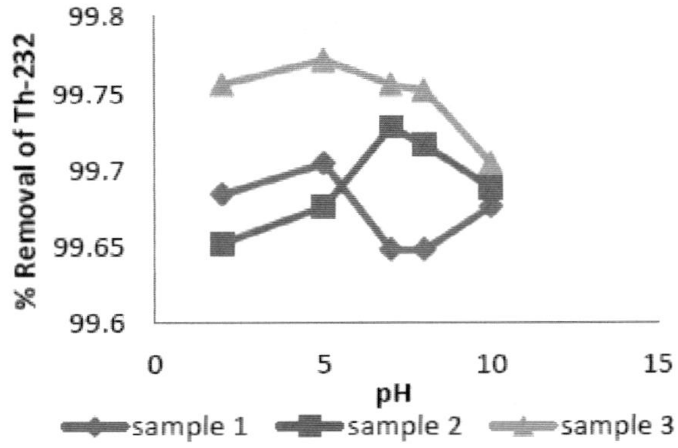

Figure 2. Graph of pH versus percentage removal of Th-232.

4.4 Effect of Contact Time

Figure 3 shows the percentage removal of Th-232 influence by the contact time between adsorbent and Th-232 solution. It can be seen that the percentage removal of Th-232 solutions is increasing in period of 60 to 140 minutes for sample 1 and 3. This happened because there were abundant active sites on the soil surface in early period of adsorption [11]. The equilibrium times of Th-232 by sample 1 and sample 3 occurred at 140 minutes, as the maximum percentage removal of Th-232 are 99.96% and 99.98%. Whereby for sample 2, the optimum contact time is at 120 minutes and it start declining approaching 140 minutes implying to saturation of adsorbent site causing inefficiency in adsorption process as the time extended

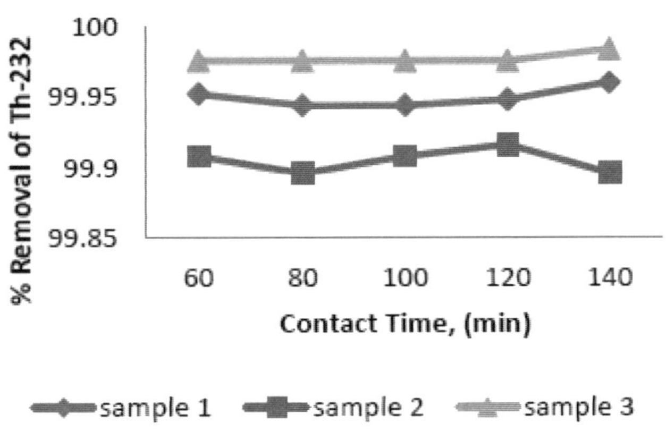

Figure 3. Graph of contact time versus percentage removal of Th-232.

4.5 Isotherm Studies

Based on table 3, it was evident that sample 1 follows Langmuir isotherm as the value of R^2 is closest to 1 compare to the Freundlich isotherm model. Therefore it is suggested that sample 1 adsorption process is homogenous and occurs on one layer only. In this experiment, sample 1 obtained the higher value of K_L indicating high affinity towards adsorbent. Sample 2 and 3 follows Freundlich isotherm as the value of R^2 is closest to 1. Freundlich isotherm suggests that the adsorption process is favourable when the n value is greater than 1 and sample 3 shows a value of n greater than 1. As for sample 2, the n value is less than 1 indicating that the adsorption process is chemical.

Table 3. Langmuir and Freundlich parameters and correlation coefficient for Th-232

Th-232		Langmuir Isotherm			Freundlich Isotherm	
	q_m	Kl	R^2	n	Kf	R^2
Sample 1	19.10	0.35	0.1042	0.38	4.25	0.0310
Sample 2	83.02	0.13	0.0968	0.29	36.29	0.2548
Sample 3	86.71	0.05	0.3877	2.19	$1.04 \times 10'^{6}$	0.4709

5. Conclusion

This study concluded that all soil samples collected have the potential capability of adsorbing Th-232. Several parameters displayed huge part in removal of Th-232. As pH increased the ability to adsorb decrease and so does the extension of contact time as it reached saturation point. Highest percentage removal of Th-232 for sample 1 and 3 obtained at initial concentration of 25 ppm, pH 5 for 140 minutes which are99.69‘% and 99.98% respectively, while for sample 2, the highest percentage removal of Th-232 at initial concentration of 25 ppm, pH 7 for 120 minutes are 99.92%. In addition, all the soil samples have acidic pH which is one of the suitable characteristic of soil on Th-232 adsorption. All collected soil sample have high SiO_2 percentage which means slightly clayey and have the potential in prevention of Th-232 from entering the groundwater. The capability of soil as potential adsorbent for Th-232 is proven based on FTIR result by the presence of three functional groups namely hydroxyl groups, carboxylic groups and amine groups which played a major role in adsorption. From isotherm studies, sample 1 follows Langmuir isotherm while sample 2 and 3 follows Freundlich isotherm. Based on these experiment, all collected soil sample around Jimah coal-fired power plant has considerable ability to adsorb Th-232.

6. References

[1] Bednar J, Medina V F, Ulmer-Scholle D S, Frey B A, Johnson B .L, Brostoff W N, Larson S L 2007 Effects of organic matter on the distribution of uranium in soil and plant matrices *Chemosphere* **70(2)** 237-247.

[2] Agency for Toxic Substances and Disease Registry (ATSDR) 2014 Addendum to the Toxicological Profile for Thorium (U.S. Department of Health and Human Services, Atlanta, GA).

[3] Pandit G G, Sahu S K, Puranik V D 2011Natural radionuclides from coal fired thermal power plants - estimation of atmospheric release and inhalation risk *Radioprotection* **46** 173 - 179 ().

[4] Ahmed H, Young S, Shaw G 2012 Solubility and mobility of thorium and uranium in soils: the effect of soil properties on Th and U concentrations in soils solution *Geophysical Research,* **14** 2012-2994.

[5] Mara H 2007 Coal Ash Is More Radioactive than Nuclear Waste *Scientific American.* **99** 122-126.

[6] Wu H X, Wang T J, Chen L, and Jin Y 2009 The roles of the surface charge and hydroxyl group on a Fe-Al- Ce adsorbent in fluoride adsorption *Industrial Engineering Chemistry Research* **48** 4530 – 4534.

[7] Cheng L, Zhai L, Liao W, Huang X, Niu B and Yu S 2014 An investigation on the behaviors of thorium(IV) adsorption onto chrysotile nanotubes in *J. of Environmental Chemical Engineering 2,* 1236-1242

[8] Yee L H 2016 Jimah power plant valued at RM2.4bil The Star 23 January 2016 pp8

[9] Hossain M A, Ngo H H, Guo W S, and Nguyen T V 2012 Biosorption of Cu (II) from water by banana peel based biosorbent: experiments and models of adsorption and desorption *J. of Water Sustainability* **2(1)** 87 – 104.

[10] Soderlund M, Lusa M, Lehto J, Hakanen M, Vaarama K, Lahdenpera 2011 A Sorption of Iodine, Chlorine, Technetium and Cesium in Soil

[11] Haro N K, Vecchio P D, Marcilio N R, Feris L A 2017Removal of atenolol by adsorption - Study of kinetics and equilibrium" in *J.of Cleaner Production* **154** 214 - 219

[12] N. Mitchell D, Perez-Sanchez M C Thorne 2013 A Review of The Behaviour of U-238 Series Radionuclides in Soils and Plants *J. of Radiological Protection* **33** 17- 48

[13] Abdul Rashid N S et. al. 2016Solubility assessment of 232Th from various types of soil in Malaysia using USP and DIN In Vitro digestion method **1784**.

[14] N. Zakaria R. Ba'an, Kathiravale S 2010 Radiological Impact from Airbone Routine Discharge of Coal-Fired Poweer Plant (Bahagian Teknologi Alam Sekitar, Agensi Nuklear Malaysi*a).*

[15] Melson N H 2011 Sorption of Thorium Onto Subsurface Geomedia (Auburn University).

[16] Guo P, Duana T, Song X, Xua J, Chena H 2008 Effects of soil pH and organic matter on distribution of thorium fractions in soil contaminated by rare-earth industries *Talanta* **77** 624-627.

[17] Ramasamy S 2012. Coal Policy in Malaysia (Retrieved from http:// www.egcfe.ewg.apec.org/publications/ proceedings/CFE/Austrailia_2012/4C-1_Ramasamy.pdf)

[18] Vandenhove H, VanHees M, Wouters K, Wannijn J 2007 Can we predict uranium bioavailability based on soil parameters Part 1: effect of soil parameters on soil solution uranium concentration *Environ. Pollut.* **145(2)** 587-595.

[19] Chang WC, Deka J R, Wu H Y, Shieh F K, Huang S Y and Kao H K 2013 Synthesis and characterization of large pore cubic mesoporous silicas functionalized with high contents of carboxylic acid groups and their use as adsorbents *Appl. Catalysis B: Environmental* **142 - 143** 817 – 827.

Acknowledgment

The authors would like to express gratitude to Ministry of Higher Education for allocating Research Acculturation Grant Scheme (RAGS) 600-IRMI/RAGS5/3(59(2015)) and thank the Universiti Teknologi MARA (UiTM) Shah Alam especially to Faculty of Applied Science, for giving the opportunity and providing facilities to made this study successful.

Institute of Physics
Dirac House, Temple Back
Bristol BS1 6BE UK

ISSN: 1757-8981
ISBN 978-1-5108-5837-4

CURRAN ASSOCIATES INC.
proceedings.com